DIFFERENTIAL TOPOLOGY AND GENERAL EQUILIBRIUM WITH COMPLETE AND INCOMPLETE MARKETS

Differential Topology and General Equilibrium with Complete and Incomplete Markets

by

Antonio Villanacci
Università degli Studi di Firenze, Italy

Laura Carosi
Università degli Studi di Pisa, Italy

Pierluigi Benevieri
Università degli Studi di Firenze, Italy

and

Andrea Battinelli
Università degli Studi di Siena, Italy

KLUWER ACADEMIC PUBLISHERS
BOSTON / DORDRECHT / LONDON

A C.I.P. Catalogue record for this book is available from the Library of Congress.

ISBN 1-4020-7201-5

Published by Kluwer Academic Publishers,
P.O. Box 17, 3300 AA Dordrecht, The Netherlands.

Sold and distributed in North, Central and South America
by Kluwer Academic Publishers,
101 Philip Drive, Norwell, MA 02061, U.S.A.

In all other countries, sold and distributed
by Kluwer Academic Publishers,
P.O. Box 322, 3300 AH Dordrecht, The Netherlands.

Printed on acid-free paper

Printed in the Netherlands.

Contents

List of Figures

This book is dedicated to
David Cass

Acknowledgments

Antonio Villanacci is greatly indebted to David Cass who taught him not only what general equilibrium and incomplete markets are, but also how to make research in economic theory a passionate part of his everyday life. Without David Cass, this book could not have been written. He also wants to give a special thank you to Alessandro Citanna, coauthor of his in the GEI papers which use the approach adopted in the book; that longstanding cooperation has been very fruitful for many parts of the book.

It is a great pleasure for Laura Carosi to acknowledge her debt to Alberto Cambini and Laura Martein. They have always been very generous in providing advice, encouragement and a stimulating atmosphere to her research activity.

Pierluigi Benevieri owes special thanks to Massimo Furi, from whom he learned differential topology and who introduced him to the analysis of the topological degree. He has constantly helped the authors in solving several mathematical problems along the writing.

The authors thank the following friends for comments and cooperation: Yves Balasko, Riccardo Cambini, Ilaria Cingottini, Hervé Crès, Julio Davila, Elena Del Mercato, Maria Patrizia Pera, Tito Pietra, Francesco Ruscitti, Paolo Siconolfi, Mauro Sodini, Marco Spadini, Ünal Zenginobuz, Pierluigi Zezza.

Our work greatly benefited from some unpublished lecture notes on differential topology by Michael Magill and Martine Quinzii.

Andrea Battinelli is not coauthor of the second part of the book.

Introduction

General equilibrium

In this book we try to cope with the challenging task of reviewing the so called general equilibrium model and of discussing one specific aspect of the approach underlying it, namely, market completeness.

With the denomination "general equilibrium" (from now on in short GE) we shall mainly refer to two different things. On one hand, in particular when using the expression "GE approach", we shall refer to a long established methodological tradition in building and developing economic models, which includes, as of today, an enormous amount of contributions, ranging in number by several thousands[1]. On the other hand, in particular when using the expression "standard differentiable GE model", we refer to a very specific version of economic model of exchange and production, to be presented in Chapters 8 and 9, and to be modified in Chapters 10 to 15. Such a version is certainly formulated within the GE approach, but it is generated by making several quite restrictive assumptions[2].

Even to list and review very shortly all the collective work which can be ascribed to the GE approach would be a formidable task for several coauthors in a lifetime perspective. The book instead intends to address just a single issue. Before providing an illustration of its main topic, we feel the obligation to say a word on the controversial character of GE.

First of all, we should say that we identify the GE approach as being based on three principles[3].

[1]There are of course some contributions which are generally considered hallmarks in the history of the discipline. Not to mention classical sources of the second-last century, we indicate here [5] and [88].

[2]Notice the qualification "differentiable". By the expression "standard continuous GE model", or "standard Arrow-Debreu-McKenzie model", it is usually intended a model which is more general than the one discussed in this book. See the references in the previous footnote. We shall come back to this later on.

[3]Although our definition does not coincide with his, we have benefited by the reading of [53].

1) Methodological individualism. Any theory of the economic process has to be founded on a description of the behavior of individual agents; terms and concepts describing collective entities should be related to, and ultimately explained by, terms and concepts describing individuals.

2) Maximization principle. Individuals act consciously and deliberately. Deliberation of individual agents can be described as achieved through maximization of some objective function which characterizes their preferences, inclinations and desires. Agents' decisions take into account the existence of natural, institutional, and technological constraints; they result in the formulation of well defined plans of action.

3) Consistency principle. Any observable state exhibits consistency of the individual plans of action. The aggregate effects of the individual plans of actions allow each individual plan to be executed as intended, thereby validating the deliberation leading to that plan. This is generally achieved by the presence of variables representing such aggregate effects, which appear as exogenously given in the decision problems of the individuals, but which are endogenously determined in the overall model.

More specifically, the GE approach we refer to corresponds to the particular instance of application of the above three principles, in which the variables describing aggregate effects of the decisions of all individuals reduce to price signals, the individual plans of action consist in transaction offers concerning goods, services and productive engagements, and consistency reduces to market clearing[4].

It is certainly true that several GE inspired contributions exist, which have addressed specific issues in a non standard way, trying to amend the basic GE formulation in order to cope with received criticisms. In a sense this book could be looked at as trying to do the same thing. Yet it is a fact (or at least that's what we believe) that in rather general terms the GE approach exhibits many highly controversial aspects. In this introduction we are definitely not even taking up the much more limited (but still rather demanding) topic consisting in the critical assessment of the GE approach; nevertheless, we would like here to mention in very sketchy form a list, by no means an exhaustive one, of controversial aspects which we consider important, and providing a serious challenge to the use of GE altogether.

It is essential that we clearly distinguish between controversial aspects which we regard as relevant to the GE approach as a whole, and controversial aspects

[4] It could be said that what we are in fact dealing with is *walrasian* GE; in broad sense, many other theoretical constructions may be considered part of the GE approach, such as, e.g., temporary economic equilibrium with rationing (with price and quantity constraints as representative variables), or bankruptcy models (with the same role played by anonymous aggregate rates of default).

which can reasonably be ascribed to the standard differential GE model only. In the former class we mention just three.

1 A description of social and economic affairs, even if centered on a specific concept of equilibrium, needs to include the possibility that non equilibrium states occur. The importance of an equilibrium concept has to do with its being supported by convincing arguments describing equilibrium states (better still, some of them) as attracting motion through non equilibrium states[5]. This is, or seems to be, in opposition to the consistency principle; it is certainly not by chance that most GE models disregard it entirely (and we aren't going to do any better).

2 The assumption of unbounded rationality of economic agents is certainly appealing for the model builder in that, once rationality is appropriately defined, agents' behavior is right away amenable to analysis by logical induction and mathematical characterization. This allows to prove theorems, but it does not show how fruitful such theorems are for understanding economic processes. Many scholars feel that to drop the unbounded rationality assumption may open the way to an abyss of indetermination, leaving the model builder with almost nothing in hand. According to such a feeling, mistake and foolishness may take any form whatever... or should they not? We are quite sympathetic with that feeling, but we believe that systematic deviations from perfect rationality cannot fail to be at least part of adequate models of human behavior.

3 Time and uncertainty can be analyzed in the GE approach. Then, agents choices of an action today require an effort to forecast the future. Expectation formation becomes a central issue and it is generally modelled as follows. Agents know the characteristics of future states of the world in terms of some relevant exogenous, private variables, like their own endowments and tastes. They are also assumed to form expectations about some economy-wide endogenous variables, like future good prices. In fact, in the GE framework, market clearing prices which arise from decisions based on the above described information do confirm individual predictions: in accordance with the consistency principle, agents' expectations are self-fulfilling. Whereas optimizing agents should not make systematic and persistent mistakes in expectations, simply because they are costly, it is anyway clear

[5]To take a very simple and classical example from physics, consider the rigid pendulum. Classical mechanics tries to explain first of all how it moves, and then to identify its rest positions. Of the latter (there are two of them, rod up and rod down) one is definitely more meaningful than the other; this has nothing to do with properties of the rest positions *per se*; it is explained by describing *motion*, and by the *dynamic* notion of stability.

that they do often fail in their predictions: a perfect foresight or "rational expectations" assumption is quite unrealistic.

Concerning the standard differential GE model, our list is longer.

1 Exchange and production take place in at times highly, and at times less (but not so much less), institutionalized environments, where individuals face restrictions on their overall access to trade, resources, and technology, on the form and timing of their transaction and productive engagement offers, and on the way in which their offers are transformed into actual trades and productive activities[6]. There is reason to believe that the institutional shape of trading has some influence on trade outcomes[7], similarly for productive activities, and even conversely.

2 There is a contradiction in assuming perfectly flexible prices in an universe of price takers. If everybody takes prices as given, why on earth do they come to change[8]? Widespread reference to entirely fictitious figures, such as the celebrated auctioneer, testifies the seriousness of the problem much more than giving an answer to it.

3 The assumptions on households preferences and endowments have raised objections ever since. To mention a few:

   (a) (transitivity) empirical observations apparently contradicting transitivity of individual preferences have been extensively reported[9];

   (b) (completeness) evaluation of courses of action involving departure from everyday experience is often impossible, or at least temporarily suspended; investment in information gathering in order to clarify one's own perception of the available alternatives is common practice;

   (c) (monotonicity) strict preference for commodity bundles offering slightly more of just one commodity and nothing less, i.e., desirability of every commodity, characterizes models which are hardly usable for the analysis of waste, hence for the discussion of ecological issues[10];

   (d) (strictly positive endowments) to envisage an economic environment where each individual is initially endowed of positive quantities of all

[6]Resort to intermediaries or representative bodies is at times mandatory. Bids are often collected according to standardized rules, for which there exist deadlines. Simultaneous trade of different commodities is hardly practically possible even if not explicitly banned. Terms of trade, property rights, and access to natural resources are very often established by means of war.

[7]On this respect see [124], pp. 45-46 and the whole Chapter 3.

[8]A paradox already pointed out in [6], p. 322.

[9]This goes back at least to [116].

[10]It must be stressed that monotonicity is not an element characterizing the standard continuous GE model.

conceivable goods sounds even derisive of several millions of human beings currently spending their whole life in starvation[11].

4 Price taking behavior of agents is in sharp contrast with the general class of phenomena which go under the name of "strategic interaction", i.e., a situation in which agents' decisions influence and are influenced by the decisions of other agents, and everyone is aware of that influence. It can be somehow argued that the absence of strategic interaction is consistent with an economy with many and small agents. On the other hand, we do observe "big" agents, and small agents can form big enough coalitions which can behave strategically. It is then clear that strategic behavior is hard to be assumed away from the model. We present below some other important instances of the pervasiveness of strategic behavior, namely, asymmetric information and externalities and public goods.

(a) The way exchanges are described in a GE model requires both buyers and sellers to be perfectly informed about the quality of the goods being sold in the market. That assumption is quite far to be satisfied for a wide class of goods, ranging from labor to technologically sophisticated products or even chemically or genetically modified food items. While the literature on those topics modelled in a partial equilibrium framework is part of undergraduate economics education (see, for example, [118]), attempts to fully incorporate asymmetric information in a GE framework have been made only recently[12].

(b) Agents are assumed to take decisions on the basis of their tastes, endowments and technological capabilities without worrying about what other agents consume or produce. Consumption or production externalities are on the other hand present in many everyday aspect of life: from negative or positive effects coming from neighbors' consumption, to congestion effects, or pollution problems, or the presence of public goods[13].

5 The assumption that there is a unique price for each commodity rules out first of all the possibility that exchange ratios depend on the quantities of the commodities involved, which is common commercial practice. Second, in extensively decentralized economic environments, exchange ratios vary not only with time but also (even when parallel transactions occur at equal

[11]Versions of the standard continuous GE model with nonnegative (but nonnull) initial endowments are widespread; it remains to be seen whether these versions can actually accommodate the poverty issue.

[12]See a forthcoming special issue on the Journal of Mathematical Economics.

[13]Public goods are characterized by the fact that consumption of a unit of the good by one individual does not preclude its consumption by another.

time instants) with the specific agents involved in the transaction. Indeed, economic competition is greatly enhanced by the possibility to stipulate transactions at freely chosen exchange ratios, and part of the economic activity of individual agents consists in acquiring information concerning the price distribution of specific commodities[14]. This, too, is completely ruled out.

6 Since individual agents determine their transaction offers for all commodities simultaneously, it is even questionable to talk about markets for individual commodities as conceptually separable from one another. The notion building an intuitive counterpart to the model is that of a big, unique clearing house, rather than a constellation of trading conveniences.

Even if we do consider the above listed criticisms well posed, we have not discussed them in the least at length. This certainly has to do with our present incapability to do so in a decent and satisfactory way; nevertheless, we believe that we have a different and important reason for that.

Simply, we want to concentrate our attention on what we consider one of the most important and, probably, the most studied market imperfection: financial market incompleteness. In fact, since general equilibrium models with incomplete markets (or GEI models) share with GE models the fundamental structure, basically all the above listed criticisms to models in the latter class apply to models in the former as well[15].

To say why we have chosen to analyze incomplete markets "only" is simple. We do believe that the general methodology used there can be applied to the standard GE model with other imperfections. On the other hand, substantial path-breaking work needs to be done to cope with the other fundamentals problems mentioned in the above list.

Having so strictly narrowed our topics of interest, in the remainder of this introduction we discuss the basic aspects of GEI models.

Our goal is to try to explain what are incomplete markets, why they do arise in a modern economic system and why their analysis is important. We then describe the meaning and the relevance of a generalization of the incomplete market model: the case of restricted participation.

What are incomplete markets?

[14]The issue must be kept distinct from the price taking issue, even if the two are connected: there exist models where simultaneous price setting behavior determines unique prices for all commodities, and models where partially informed agents observe constant valued samples of dispersed price variables.

[15]It should be clear at this point which criticisms refer to standard differentiable GEI (and GE) models only, and which ones to standard continuous GEI (and GE) models as well.

The aim of the theory of general equilibrium with incomplete asset markets is to study the simultaneous pricing of commodities and financial securities, emphasizing the role of interactions between asset markets and commodity markets in determining consumption and financial investment.[16]

Let's compare and contrast the GEI model with the standard GE model.

In the latter model it is assumed that there are markets for all commodities[17]. Commodities are very strictly defined and distinguished from one another, in order to accommodate for explicit consideration of future delivery times and of conditional commitments (the latter making execution of stipulated transactions dependent on the realization of uncertain future events)[18]. Individual agents face a single decision problem, even if with a large number of variables. A solution to that problem, to be determined in current time, specifies the complete profile of purchases and sales for all future times in any possible event. Admissible solutions have to satisfy a single, overall budget constraint. To each admissible solution it is associated a specific pattern of wealth transfers across time and across events. Similarly, markets for the same commodity at different times, or at the same time but in presence of different events, are completely different markets, which have to clear separately in equilibrium. An equilibrium determines a complete and consistent temporal sequence of executions of delivery commitments, whatever event happens to realize. There is no room, nor need, to envisage the possibility of recontracting as time goes by.

On the other hand, if we feel the urgency to describe a world in which the role of time and uncertainty is definitely more essential, we may try to do it by dropping the assumption that a complete set of markets exists; in other words, we may describe individual agents as entitled to stipulate transactions only for current delivery ("spot" markets), or for future delivery contingent on uncertain events *subject to* some limitations. The nature of such limitations precisely affects the extent by which individuals may achieve transfer of their own purchasing power across different states and different dates. In other words, it defines the form of imperfection of the existing system of financial markets. If we do so, one way or another, we have started to develop a model of general equilibrium with incomplete asset markets[19].

Once we admit that transactions for delivery of some commodity at some future date and contingent on some event are not currently possible, but we retain the assumption that such a commodity is desirable at that date in that event, there

[16]A very close definition has been put forth by [58], p. 1.

[17]We use the word "markets" with the proviso outlined at point 6 above.

[18]"A contract for the transfer of a commodity now specifies, in addition to its physical properties, its location and its date, an event on the transfer of which the transfer is conditional. This new definition of commodity allows one to obtain a theory of uncertainty free from any probability concept and formally identical with the theory of certainty developed in the preceding chapters", [40], p. 98

[19]See [83], p. 1524.

is a simple consequence: we are bound to accept that if and when that time comes and that event realizes, some transactions concerning that commodity shall have to take place, which have not been possible before. That is, generalizing for all commodities, in each future state and in each possible event at that state, all markets for current delivery, or spot markets[20], may have to be open whether or not an equilibrium has obtained in the previous times and in the states realized at those times.

As a consequence, even if individual agents were able to transfer wealth in some way to any future date in any event at that date, they would still need to consider what their budget constraint at that date and in that event will be, in order to be able to stipulate all the spot transactions they expect to desire. That is, at each date and in each event individual agents face a different decision problem, still with a large number of variables, and with a full sequence of future and contingent budget constraints. Thus current transaction offers for current and future goods and assets must be affected from expectations about future states and future prices.

Different expectations about what state is more likely to occur will be reflected by different properties of the preferences of distinct individual agents; different expectations about what prices are more likely to prevail in each state will affect agents plans to demand or supply current and future goods and assets. An equilibrium concept may only require market clearing for currently open markets, possibly sustained by conflicting or inconsistent expectations concerning prices to prevail in markets for future delivery which are not yet open. However, the equilibrium concept adopted here is in the spirit of the consistency principle; namely, perfect foresight.

> "An equilibrium will therefore consist of a consumption plan for each agent, which is optimal in his/her budget set, spot prices for current and future consumption, and asset prices such that all markets clear and agents' price expectations are borne out in equilibrium."[21]

A formal description of the GEI model is presented and discussed in Chapter 10.

Why is the incomplete market model important?

As discussed in the previous subsection, there are several reasons to support the idea that incompleteness of asset markets adds some form of realism to the model. There are other reasons as well. Some new phenomena that cannot be described in the standard GE model arise in the GEI framework. Before

[20] Any couple (t, s) where t is a time date and s is one of the events possibly occurring at time t is usually referred to as a "spot".

[21] [101], p. 3.

briefly introducing the ones which will be analyzed in detail in Chapters 10-15, we want at least to report the following others from a paper ([58], pp. 1-2) by Geanakoplos:

> "...the limitations (for understanding local uniqueness and efficiency) of representative agents and single commodity models, the non-neutrality (and occasional neutrality) of money, the importance of diversification, the dependence of asset prices on covariances and not variances, the 'random walk' or martingale property of asset prices, the relevance (or possible irrelevance) of corporate financial policy, the rationale for lenient bankruptcy and default penalties, the speculative, precautionary, and transactions demands for money, the ambiguity of firm objectives[22], ...".

By analyzing the validity of each of the main results of the standard GE model (i.e., existence, Pareto optimality and generic local uniqueness of equilibria[23]) in the GEI framework, the importance of the latter can be greatly appreciated.

GEI models studied in the literature differ from one another for the types of returns promised by available assets. *Nominal* assets deliver units of account or "money"; *numeraire* assets deliver units of a pre-specified good; *real* assets deliver bundles of goods. Each of the three main results of the Arrow-Debreu model fails to hold in one or more of the above defined type of incomplete market frameworks, as roughly[24] summarized in the following table

	existence	generic local uniqueness	Pareto optimality
numéraire assets	Yes	Yes*	No*
nominal assets	Yes	No*	No*
real assets	Yes*	Yes*	No*

(I.1)

where "Yes" means "it is always ensured", "Yes*" means "it is generically[25], but not always ensured", and "No*" means "it is generically not ensured".

In fact, if either all types of assets are available, or existing assets have mixed returns - i.e., they promise to deliver both units of account and goods - then all main results of the complete market model fail to hold - see for example, [126], [74] and [94].

One of the main goals of this book is to carefully prove each result in the table; this is done in Chapters 10 to 13. For the time being, let's comment on the relevance of each "negative" result.

Existence. The proof of existence is an important check for the consistency of the model. As shown in table (I.1), that result is ensured for models with nominal

[22]On the last topic see Remark 23 in Chapter 10.

[23]The first two in both the continuous and the differential case, the last one in the second case only.

[24]The results in the real asset model with respect to generic local uniqueness and Pareto optimality refer to a subset of the equilibrium set - see Chapter 13.

[25]Generically means in an open and full measure subset of the endowment space for the case of nominal and numeraire assets, and in an open and full measure subset of the space of endowments and *asset structures* in the case of real assets.

and numeraire assets, and for models with real assets outside a negligible set. The proof of generic existence in the last case has forced economists to use new methodology involving homotopy and degree theory.

Local uniqueness. As described at some length in Chapter 12, the generic lack of local uniqueness simply means that the set of equilibrium allocations associated with almost all economies contains a continuum of equilibria. That result is important for at least the following three reasons.

Indeterminacy is a logical hurdle for the rational expectation hypothesis.

> "...is it plausible to maintain that households are capable of concentrating their beliefs (correctly) on one among a surfeit of possible market outcomes?"[26]

In the second place, it is sometimes argued that a model with a continuum of equilibria is limited in its predictive and explanatory power[27]. If one takes that view, and we certainly agree on the first limitation, she has to recognize that the indeterminacy result suggests either the need for adding some missing element in the model, or that there is room for some institution to "guide" the economy in the choice of a particular equilibrium with desirable properties.

Finally, the presence of a continuum of equilibria shows that there is a non-trivial relationship between nominal and real sectors of the economy: rescaling of all prices in some states may change equilibrium allocations.

Pareto Optimality. A reason to study the GEI model which is good enough by itself, is to make a point against the received view that competitive equilibrium and Pareto efficiency are systematically related concepts. Such a relation is a distinguished feature of the standard GE model, which is stated and proved in the so called welfare theorems (see Section 5 in Chapter 8). The circumstance that in the GEI model equilibria are typically inefficient is important to contrast arguments sometimes put forth in new classical macroeconomics and finance theory, according to which market efficiency is a general property of the GE approach.

Generic Pareto non optimality is not the end of the story. As a consequence of the above observations, it is quite natural to investigate if equilibria are at least some sort of second best or, as many authors say, if they are at least *constrained* optimal. In other words, it is natural to try to compare equilibria with allocations which can be implemented by a planner who is *constrained* to use only existing assets. As we are going to illustrate in Chapter 15, typically (in the space of economies) equilibria are not even constrained Pareto optimal. A powerful and informed planner can do better than the market. Arguments for criticizing planner intervention must then be based either on informational limitations on the part of the planner, or on her being affected by the benefits of the intervention.

[26][29], p. 264.

[27]Magill and Quinzii (1996), page 438.

We conclude this subsection with the description of a fruitful generalization of the concept of incomplete markets: the so called restricted participation model, to be analyzed in Chapter 14.

> "While there might be some disagreement over whether, in a modern developed economy, financial markets are actually incomplete, there can hardly be any disagreement over whether at least some economic agents are variously constrained in transacting on those financial markets."[28]

In fact, it is possible to extend the model with incomplete markets to incorporate the presence of different restrictions to the access to financial markets across households, and to obtain the results described in table (I.1) in this more general framework.

Our mathematical approach to GE

The standard GE model originates from an effort to find very general conditions guaranteeing existence and Pareto optimality of equilibria. In the standard continuous GE model, those conditions amount to continuity, local nonsatiation and quasiconcavity of each agent's utility function, and other assumptions on consumption sets and endowment vectors - see Remark 50 in Chapter 8.

The seminal papers by Debreu ([41], [42]) and the related work by E. Dierker and H. Dierker ([52]), ([51]), Smale ([108]) and Balasko ([8], [9]), tried to go beyond the two mentioned classical results in order to study uniqueness of equilibria. Stronger assumptions were then imposed to be able to use a differential topology approach; they consist in assuming that utility functions are sufficiently differentiable and satisfy properties which, together with some related requirements on consumption sets and endowments, allow to get differentiable demand functions - see Chapter 8.

Beside allowing to formulate existence proofs in a different (and probably easier) way, and to show (generic local) uniqueness of equilibria, these assumptions and techniques are also used to study generic inefficiency and some relevant local properties of the GEI model.

Since equilibria of the model are solutions to parametrized systems of non linear equations, we use here two fundamental differential topology techniques.

The first one consists in a class of theorems which, using the concept of homotopy, enable us to state conditions under which for all values of the parameter, a solution to the system exists - see Theorems 57 and 59 in Chapter 7.

The second tool, a "parametric" transversality theorem (Theorem 26 in Chapter 6), is a consequence of Sard's theorem and gives conditions under which a

[28] [29], p. 274.

system has typically no solutions in the space of the parameters. It turns out that the transversality theorem can be used to obtain both negative results - see the theorems in Section 2 of Chapter 15 - and positive ones - see Theorem 25 in Chapter 12, or Section 5.2 in Chapter 15.

The peculiarity of our approach is based on the fact that in almost all standard GE models and GEI models we know[29], it is possible to describe equilibria in terms of the first order conditions for the agents' maximization problems and of the market clearing conditions. That approach was used by Smale ([108]) and, more recently, in some papers written by a group of general equilibrium theorists associated with David Cass, at the University of Pennsylvania. The so called extended approach is different from the usual one, which describes equilibria in terms of aggregate excess demand functions; the relationship between the two approaches is discussed in Section 8 of Chapter 8.

The mathematical tools introduced in the first part of the book are useful for both approaches.

More precisely, the mathematics of the book was mainly thought as a set of tools to allow the above described thorough analysis of a parametric system of equations. In the attempt to making this part consistent and, as much as possible, self-contained, it finally developed as an introduction to elementary calculus on manifolds in euclidean spaces and to global analysis of manifolds in the form given by Sard's theorem and the transversality theorem (Theorem 26).

An introduction to differential topology

The first part of the book introduces the reader to the knowledge of a body of mathematical instruments which are necessary in the second part, and which belong to the discipline of differential topology. It is definitely not a treatise in differential topology, not even a very concise one, for two reasons: it is thought for graduate students in an economics program, not in a mathematics one; it remains on a very basic ground, and stops before really entering the core of the discipline. On the other hand, it is written with the firm intention to put the reader in the position to master the basic elements and techniques introduced, in order to handle their applications to specific modelling problems at benefit of her own research inquiries. Therefore, the concepts introduced are thoroughly discussed, and many proofs are carried through in a good bit of detail.

What is the object of differential topology?

It may be answered that the discipline arises with the concept of manifold. Very often in calculus we encounter objects having a structure which is locally

[29] The only exception seems to be the model with incomplete financial markets and real assets studied in Chapter 13.

amenable to the structure of some euclidean space, without being itself a linear structure, and with different global topological properties. These objects are often obtained, for example, as solution sets to nonlinear systems of equations satisfying appropriate regularity conditions. Classical differential topology deals with the appropriate definition and with the classification of such objects, to be called manifolds; it studies their global properties, and extends the concept in order to encompass in the definition more abstract objects, which are not included in euclidean spaces, but are still locally homeomorphic to them. As a matter of fact, it provides some additional structure to the objects of the latter type, in order to define in a meaningful way the concept of diffeomorphism.

While Chapter 1 is devoted to some mathematical prerequisites, Chapters 2 and 3 are concerned with the extension of the fundamental ideas of calculus to functions of manifolds into manifolds. They contain the definition of function of class C^r on an arbitrary subset of euclidean spaces, the definition of manifold, and the definition of differential and derivative for functions between manifolds. The last item requires the concept of tangent space to a manifold at a given point, and the global concept of tangent bundle, as environments which are appropriate to play the role of domain and value set for differentials and derivatives (respectively). Tangent spaces can be conveniently looked at as local first order approximations of the manifolds themselves. Differentials work then as in ordinary calculus, generally providing a meaningful characterization of the local behavior of functions.

The concept of regular value is introduced and discussed in the Chapter 4. This leads to the regular value theorem, which may be seen as justifying by itself the study of differentiable manifolds, in that it shows how they naturally enter in the solution to standard calculus problems. The study of regular values and more generally of the concept of transversality (which belongs to Chapter 5) give existence results for manifolds, and can be seen in this quite narrow sense as providing an answer to some very specific global question. Differential boundaries of manifolds, a concept to be kept distinct from that of topological boundary, and manifolds with boundary are also introduced in Chapter 5, and the developed machinery is extended in order to work for this case as well.

Chapter 6 deals with the concept of Lebesgue measure zero for subsets of euclidean spaces and for subsets of manifolds and then with the celebrated Sard's theorem. Since the theory of Lebesgue measure clearly requires a particularly careful approach and, in addition, it is out of the scope of the book, we did not insert it in the chapter. This has been possible because the concept of measure zero can be defined with no connections to the more general notion of measure. Moreover, part of the not easy proof of Sard's theorem is given.

The theory of the topological degree, an already classical instrument, is presented in detail in Chapter 7. The version presented here is degree mod 2, with no attention to the concept of orientation. The emphasis on this topic is justified by its central role in the existence proofs of the second part.

The reader with some background in differential topology will not fail to notice that some hard choices have been definitely made in deciding the material to be presented, thereby simply omitting a vast majority of topics. This was intended in order to keep the style of the presentation sufficiently elementary, taking into account the average level of mathematical background of our prospective readers. We would like to discuss at this point some of our choices explicitly.

1 We have restrained ourselves to treat only manifolds of euclidean spaces, avoiding as much as possible to deal with abstract manifolds. The latter could not be completely left out of the picture, for the presence of a few circumscribed (but essential) issues arising in the chapter on real assets[30]. Thus the presentation of a few absolutely necessary definitions and the statement of a very limited number of results has been confined to three sketchy appendices. In particular, the definition of functions of class C^r has been formulated with no explicit reference to the concept of tangent space for abstract manifolds, which would have otherwise entailed a fairly more advanced level of sophistication. Yet we found it mandatory to give at least the statement of Whitney's theorem, both for purely informative purposes and because it allows us to keep using the approach inside euclidean spaces to prove some facts concerning the mentioned issues.

2 We have almost exclusively concentrated our attention in the use of calculus techniques for functions of manifolds into manifolds, thereby mostly addressing purely local issues. This is certainly contrary to the true spirit of differential topology, as far as we understand it. We would be certainly satisfied if our presentation had resulted more in a enticing introduction than in a pure abridgement, and we hope we have succeeded in this endeavor.

3 The treatment of the theory of the topological degree is at the same time narrow in scope but extended in accuracy, particularly in relation to other parallel topics which are hinted at or even overlooked. The chosen topic has been the simplest, namely, degree mod 2. We have not taken up the problem of orientation. This choice was sufficient for the purposes of our applications, which were limited to existence results for solutions to nonlinear systems of equations. At the same time, in comparison with other treatments, our definition of degree allows us to deal with systems for which the solution set only, not the domain of the function involved, is known to be

[30] A formulation of the constraints on transactions in future and contingent markets faced by individual agents in terms of grassmannian manifolds seems to be particularly appropriate there.

compact[31]. This slightly generalizes other constructions, where the degree is defined for functions having domain which is either a compact manifold or the closure of an open subset of some euclidean space. This generalization is well known among the scholars of degree theory, but it is not so easily found in the literature. In addition, we deal here with functions which are only of class C^r and not of class C^∞.

4 From the point of view of the pedagogical methodology, we have chosen to keep a substantially detailed level of accuracy in the presentation of demonstrative arguments. This may have led us into running two risks. On one hand, some mathematical tools properly belonging to other disciplinary areas have been conveyed and discussed together with the material strictly belonging to the main subject. For example, we give some information concerning the concept of properness but we take for granted several other topics in general topology, we discuss some properties of zero measure sets without entering the area of Lebesgue measure, and we even end up repeating here and there some arguments which are typical of linear algebra. On the other hand, some slightly more trained readers may find themselves deprived of the right stimulus to provide their own argumentations in the course of reading; and we are well aware of the importance of such stimulus for a productive learning. Yet we feel that by running those risks we have taken a chance to approach a somewhat larger audience, which is in need for its own research activity to be gradually introduced to the field, and to be reassured in the attempt to come nearer to it.

[31] This does not imply that such set is nonempty, of course, which is precisely the point.

I

Chapter 1

PREREQUISITES

The chapter is devoted to the presentation of the mathematical prerequisites which are necessary for a productive reading of the book. The topics, which the prospective reader is assumed to be familiar with, are arranged under three general headings: calculus, linear algebra and general topology; for each heading we propose a set of chapters from one (or two) widely used textbooks. Even if our appreciation for the proposed textbooks is truly deep, it goes without saying that adequate knowledge of the listed topics may have been acquired by our readers anywhere else. On the other hand, throughout the book we shall be constantly in need to refer to specific results (theorems, propositions, lemmas, and so on) from these topics, and we shall do so by giving each time a detailed reference to some specific item contained in one of the four books.

Calculus:

[3] Apostol T. M., *Mathematical Analysis*, second edition, Addison-Wesley Publishing Company, Reading, MA, 1974: Chapters 12 and 13.

[77] Lang, S., *A First Course in Calculus*, Springer-Verlag, New York, NY, 1986.

Linear algebra:

[78] Lang, S., *Linear Algebra*, Addison-Wesley, Reading, MA, 1970: Chapters 1 to 7 and Appendix 1.

General topology:

[90] Munkres, J. M., *General Topology, A First Course*, Prentice-Hall, Englewood Cliffs, NJ, 1975: Chapters 1, 2, 3, Sections 4.1 and 4.2 in Chapter 4, Section 6.4 in Chapter 6, Chapter 7.

Section 1 contains a list of symbols used throughout the book. The reader will first find a set of symbols used in calculus, linear algebra and general topology, and then the set of symbols introduced in the first part of the book. Every item in the second list is followed by a page number indicating the item's first occurrence.

As a general principle we have tried to follow as much as possible the notation used in the four textbooks. Our attempt has not been entirely successful, for the simple reason that the notation used by the different authors is at times conflicting. In such cases we have been forced to make our choice. The chosen items are the only ones appearing in the list. In addition, in a few cases we have been in need to establish our own notation, because we could not find it in our sources.

Section 2 contains brief reviews of additional material which belongs to the three above listed headings, but which is not contained in the main reference sources.

In the second part of the book, we present several general equilibrium models. All of them share the same basic structural elements - households, commodity and asset markets - and differ for specific time and financial characteristics. To use different notation in each chapter would have been impossible. We have therefore chosen to give different meanings to the same symbol in the various chapters. In fact, to denote terms which are formally different across chapters, but economically similar, we have used the same symbol in the following main cases: utility functions (u), consumption vector (x), endowments (e), good prices (p), asset prices (q), asset yields (y), asset demand (b), extended equilibrium function (F).

1. Notation and symbols

$\equiv$	definition symbol	$x \equiv y$, x equal to y by definition
$\neg P$	negation of proposition P	true (only) if P false
$X \subseteq Y$	set X is included in set Y	$\forall x, x \in X \Rightarrow x \in Y$
$\#X$	cardinality of set X	
$X \subset Y$	set X is strictly included in set Y	$X \subseteq Y$ and $X \neq Y$
$X \backslash Y$	difference of sets X and Y	$\{x : x \in X \text{ and } x \notin Y\}$
$\times_{\lambda \in \Lambda} X_\lambda$	cartesian product of family of sets	$\{(x_\lambda)_{\lambda \in \Lambda} : x_\lambda \in X_\lambda,\ \lambda \in \Lambda\}$
X_t	t-section of set $X \equiv T \times S$	$X_t \equiv \{s \in S : (t, s) \in X\}$
$X/\sim$	quotient set of set X w.r. to equiv. relation $\sim$	
$[z]$	equivalence class of z	$\{x \in X : x \sim z\}$
$\text{Int}\, A$	topological interior of A	

δA	topological boundary of A	
$f : X \to Y,$ $x \mapsto f(x)$	function with domain X and value set Y	the value of f at x is $f(x)$
$\operatorname{Im} f$	image of function f	$\{y \in Y : \exists x \in X,\ y = f(x)\}$
$f_{\mid M,N} : M \to N$	restriction of f to M as domain and N as value set	
id_X	identity function of set X	$id_X : X \to X,\ x \mapsto x$
$i_{X,Y}$	canonical insertion of $X \subseteq Y$ in Y	$i_{X,Y} : X \to Y,\ x \mapsto x(x) = x$
$\mathbb{N}$	set of positive integers	
$\mathbb{N}_0$	set of nonnegative integers	
$\mathbb{N}^*, \mathbb{N}_0^*$	extended sets of positive, nonnegative integers	$\mathbb{N} \cup \{\infty\}, \mathbb{N}_0 \cup \{\infty\}$
$\mathbb{Z}$	set of integers	
$\mathbb{Q}$	set of rational numbers	
$\mathbb{R}$	set of real numbers	
$\mathbb{C}$	set of complex numbers	
$\mathbb{R}^n$	vector space of real n-tuples	
$\mathbb{R}^m \times \mathbb{R}^n = \mathbb{R}^{m+n}$	canonical identification of $\mathbb{R}^m \times \mathbb{R}^n$ with $\mathbb{R}^{m+n}$	
$L^\perp$	orthogonal space to the subspace L of $\mathbb{R}^m$	
$Aff(B)$	affine hull of $B \subseteq \mathbb{R}^m$	intersection of all affine subspaces of $\mathbb{R}^m$ containing B
$ci_{m,q}$	canonical immersion of $\mathbb{R}^m$ into $\mathbb{R}^{m+q}$	$ci_{m,q} : \mathbb{R}^m \to \mathbb{R}^{m+q}$, $x \mapsto (x, 0)$
$\mathbf{e}^i$ (bold face)	i-th vector of the canonical basis of $\mathbb{R}^n$	$(0, \ldots, 0, 1, 0, \ldots, 0)$
$\mathbf{1}$ (bold face)	vector of ones in $\mathbb{R}^n$	$(1, ..., 1, ..., 1)$
$x \geq y$	$x \in \mathbb{R}^n$ weakly dominates $y \in \mathbb{R}^n$	$\forall i \in \{1, \ldots, n\}$, $x_i \geq y_i$
$x > y$	$x \in \mathbb{R}^n$ semistrictly dominates $y \in \mathbb{R}^n$	$x \geq y$ and $x \neq y$
$x \gg y$	$x \in \mathbb{R}^n$ strictly dominates $y \in \mathbb{R}^n$	$\forall i \in \{1, \ldots, n\}$, $x_i > y_i$
$\mathbb{R}^n_+$	nonnegative orthant of $\mathbb{R}^n$	$\{x \in \mathbb{R}^n : x \geq 0\}$
$\mathbb{R}^n_{++}$	positive orthant of $\mathbb{R}^n$	$\{x \in \mathbb{R}^n : x > 0\}$
$Df(x), D^2f(x)$	jacobian and hessian matrix of f at x	
$D^3f(x)$	array of third order partial derivatives of f at x	

$f \times g$	product function of functions $f : X \to Y$ and $g : W \to Z$	$X \times W \to Y \times Z$, $(x, y) \mapsto (f(x), g(y))$
(f, g)	vector function with components $f : X \to Y$ and $g : X \to Z$	$X \to Y \times Z$, $x \mapsto (f(x), g(x))$
π_i	i-th projection of cartesian product $\times_{i=1}^{n} X_i$	$\times_{i=1}^{n} X_i \to X_i$, $(x_1, \ldots, x_i, \ldots, x_n) \mapsto x_i$
f_x	partial function at x of $f : X \times Y \to Z$	$Y \to Z,\ y \mapsto f(x, y)$
f_y	partial function at y of $f : X \times Y \to Z$	$X \to Z,\ x \mapsto f(x, y)$
$\mathcal{M}_{m,n}$	space of $m \times n$ matrices	
$\mathcal{M}_{m,n}^{*}$	space of $m \times n$ matrices of full rank	
$A_{m \times n}$	$m \times n$ matrix	
a^i	i-th row of a matrix A	
${}^{j}a$	j-th column of a matrix A	
A^T	transpose of the matrix A	
$diag\,(v)$	diagonal matrix A with $v \in \mathbb{R}^n$ on the diagonal	
rank A	rank of the matrix A	
ker A	kernel of the matrix A	the kernel of the linear function represented by A
rank L	rank of the linear operator L	
$\begin{bmatrix} A & E \end{bmatrix}$ or $\begin{bmatrix} A \mid & E \end{bmatrix}$	block row matrices	
$vec\,A$	vector of ordered columns of A	$A \in \mathcal{M}_{m,n}$, $vec\,A \equiv (a_{11}, ..., a_{m1}, a_{21}, ..., a_{mn}) \in \mathbb{R}^{mn}$
$B_r^n(x)$	n-dimensional open ball of center x and radius r	
S^n	n-dimensional unit sphere in $\mathbb{R}^{n+1}$	
T^2	torus in $\mathbb{R}^4$	p. 43
$\mathcal{T}^2$	torus in $\mathbb{R}^3$	p. 43
H^m	closed euclidean halfspace	p. 107
(ϕ, U, V), (χ, S, T)	local parametrizations of manifolds	p. 35
(ψ, V, U), (ξ, T, S)	charts of manifolds	p. 35
dim	dimension of a manifold	p. 35

codim	codimension of a manifold	p. 36
T_xM	tangent space of M at x	p. 51
N_xM	normal space of M at x	p. 57
df_x	differential of f at x	p. 58
df	derivative of f	p. 60
TM	tangent bundle of M	p. 55
NM	normal bundle of M	p. 122
$M \pitchfork N$	M is transverse to N	p. 57
CP_f	set of critical points of f	p. 80
C_f	set of critical values of f	p. 80
RP_f	set of regular points of f	p. 80
R_f	set of regular values of f	p. 80
∂M	differential boundary of M	p. 109
$I^n(a,b)$	closed n-dimensional interval	p. 134
$J^n(a,b)$	open n-dimensional interval	p. 134
$\mu(I)$	Lebesgue measure of I	p. 134
deg	degree	p. 166
$(\mathcal{N},\rho)$	tubular neighborhood	p. 189
$G_{I,S}$	Grassmannian manifold of I-dimensional subspaces of $\mathbb{R}^S$	p. 387

Sequences

Given a set X, a sequence in X is a function $f : \mathbb{N} \to X$. Occasionally, given $m \in \mathbb{N}$, we call sequence any function having $\mathbb{N}\setminus\{1,...,m\}$ or $\mathbb{N}_0$ as its domain.

When reference to a specific sequence f is clear, the symbol x_n is mostly used instead of $f(n)$. In such a case, the sequence itself is denoted $(x_n)_{n\in\mathbb{N}}$ and the image of the sequence is denoted $\{x_n : n \in \mathbb{N}\}$. The former is a function and the latter is a set. Sometimes, a sequence is denoted by (x_n), to simplify the notation.

Jacobian matrices

For $j \in \{1,...,n\}$ and $i \in \{1,...,m\}$, consider an open subset U_j of $\mathbb{R}^{l_j}$ and a differentiable function

$$f_i : \times_{j=1}^n U_j \to \mathbb{R}^{k_i}, \quad x \mapsto f_i(x)$$

where $x \equiv (x_j)_{j=1}^n$ and for each j, $x_j \in U_j$. Define

$$f : \times_{j=1}^n U_j \to \times_{i=1}^m \mathbb{R}^{k_i}, \quad x \mapsto (f_i(x))_{i=1}^m$$

To compute the jacobian matrix of f it is convenient to construct a table as follows - see Table (1.1) below.

1 In the first two columns, write the component functions $f_1, ..., f_i, ..., f_m$ of f and the dimensions $k_1, ..., k_i, ..., k_m$ of the value set of each one.

2 In the first two rows, write the subvectors $x_1, ..., x_j, ..., x_n$ of x or the dimensions $l_1, ..., l_j, ..., l_n$ of the euclidean spaces they belong to.

3 For each i and j, write the partial jacobian matrix $D_{x_j} f_i(x)$ in the entry at the intersection of the i-th superrow and the j-th supercolumn.[1]

We then obtain the following table

$$\begin{array}{ll} & \\ k_1 & f_1 \\ ... & \\ k_i & f_i \\ ... & \\ k_m & f_m \end{array} \begin{array}{c} \begin{array}{ccccc} l_1 \text{ (or } x_1) & ... & l_j \text{ (or } x_j) & ... & l_n \text{ (or } x_n) \end{array} \\ \begin{pmatrix} D_{x_1} f_1(x) & ... & D_{x_j} f_1(x) & ... & D_{x_n} f_1(x) \\ & & & & \\ D_{x_1} f_i(x) & & D_{x_j} f_i(x) & & D_{x_n} f_i(x) \\ & & & & \\ D_{x_1} f_m(x) & & D_{x_j} f_m(x) & & D_{x_n} f_m(x) \end{pmatrix} \end{array} \tag{1.1}$$

where the jacobian matrix is the part of the table between parentheses [2]. When we follow the above procedures, we use a sentence of the form: "The computation of the jacobian matrix of f is *described* below." Observe that each element of type $\left[D_{x_j} f_i(x)\right]$ is a $k_i \times l_j$ submatrix of $Df(x)$. We sometimes omit to specify the dimensions of the value sets and of the euclidean spaces.

In a matrix as that in formula (1.1), if some entries are zero, we will sometimes omit to write them.

Our conventions on vectors

We denote the scalar product between two vectors $x, y \in \mathbb{R}^n$ by xy.

Given a matrix $A \in \mathcal{M}_{m,n}$, premultiplication and postmultiplication by a vector are denoted by

$$Ax \quad \text{and} \quad yA$$

respectively, where it is then implied that $x \in \mathbb{R}^n$ and $y \in \mathbb{R}^m$. In other words, in operations involving vectors and matrices we avoid using the symbol denoting transposed vectors.

A vector $x \in \mathbb{R}^n$ which is part of a partitioned matrix is treated as a $1 \times n$ matrix, i.e., a row vector, as shown below where $A \in \mathcal{M}_{n,n}$, α is a scalar and T is the transposition symbol.

$$\begin{bmatrix} A & x^T \\ x & \alpha \end{bmatrix}$$

[1] By superrow (supercolumn) of a matrix we mean a set of adjacent rows (columns) of that matrix.
[2] In general we use both brackets and parentheses to denote matrices.

2. Miscellaneous results

In this section we collect some results which are not explicitly mentioned in the textbooks we have cited above.

2.1 Functions and sets

Proposition 1 *Consider two sets X and Y, and a function $f : X \to Y$. Let $A \subseteq X$ and $C \subseteq Y$. Then*

$$f\left[A \cap f^{-1}(C)\right] = f(A) \cap f\left[f^{-1}(C)\right] = f(A) \cap C$$

Proof. The reader can easily verify the inclusions $f\left[A \cap f^{-1}(C)\right] \subseteq f(A) \cap f\left[f^{-1}(C)\right] \subseteq f(A) \cap C$. Let $c \in f(A) \cap C$; since $c \in f(A)$, there exists $a \in A$ such that $f(a) = c$; since $c \in C$, $a \in f^{-1}(C)$, and hence $c \in f\left[A \cap f^{-1}(C)\right]$. Thus $f(A) \cap C \subseteq f\left[A \cap f^{-1}(C)\right]$. ■

2.2 Calculus

Proposition 2 *Let $f : X \to W$ and $g : Y \to Z$ be functions. Then $f \times g : X \times Y \to W \times Z$ is of class C^r if and only if f and g are of class C^r. Moreover,*

$$d(f \times g)_{(x,y)} = df_x \times dg_y$$

Let $f : X \to W$ and $g : X \to Z$ be functions. Then $(f, g) : X \to W \times Z$ is of class C^r if and only if f and g are of class C^r. Moreover,

$$d(f, g)_x = (df_x, dg_x)$$

The proof of this result is left as an exercise.

2.3 Linear algebra

The word rank is often used with three different meanings. If A is an $m \times n$ matrix, the column (row) rank of A is the maximum number of linearly independent columns (rows) of A. If V and W are finite dimensional vector spaces and $F : V \to W$ is a linear function, the rank of F is the dimension of Im F, a subspace of W ([78], p. 98). These meanings in an appropriate sense coincide, as it can be seen in the following proposition. The first equality in its statement is the content of Theorem 10, [78], p. 147, allowing one to talk unambiguously about the rank of a matrix. The second equality is a consequence of the fact that the choice of a basis in W establishes a linear isomorphism between W and $\mathbb{R}^{\dim W}$ ([78], Theorem 4, pp. 48 and 52), and hence between Im F and the subspace of $\mathbb{R}^{\dim W}$ generated by the columns of the matrix A representing F.

Proposition 3 *Let V and W be finite dimensional vector spaces, let $\mathcal{B} \equiv \{v_1, \ldots, v_n\}$ be a basis of V and let $\mathcal{C} \equiv \{w_1, \ldots, w_m\}$ be a basis of W. If $F : V \to W$ is a linear function, and A is the matrix representing F with respect to $\mathcal{B}$ and $\mathcal{C}$ ([78], pp. 119-120), then*

$$\text{row rank } A = \text{column rank } A = \text{rank } F$$

The following lemma is a direct consequence of the definition of rank.

Lemma 4 *If $F : V \to W$ is a linear function,* $\text{rank } F \leq \min \{\dim V, \dim W\}$.

Definition 5 If $\text{rank } F = \min \{\dim V, \dim W\}$, F is said of **full** rank.

The rank of the composite of two linear functions is related to the ranks of the component functions, as stated by the following proposition. This in turn translates into a relation between the rank of a product matrix and the ranks of the factor matrices.

Proposition 6 *Consider the two linear functions $G : V \to W$ and $F : W \to Z$. Then*

$$\text{rank}\,(F \circ G) \leq \min \{\text{rank } F, \text{rank } G\}$$

In particular, $\text{rank}\,(F \circ G) = \text{rank } F$ *if* $\text{rank } G = \dim W$, *and* $\text{rank}\,(F \circ G) = \text{rank } G$ *if* $\ker F = \{0\}$.

The proposition is understood if it is kept in mind on one hand that $\text{Im}\,(F \circ G) = F(G(V))$ is contained in $\text{Im } F$, and on the other that $\text{Im}\,(F \circ G) = \text{Im}\,\left(F_{|\,\text{Im } G}\right)$ (see Lemma 4).

Corollary 7 *Let A and B be two matrices such that the product AB is well defined. Then*

$$\text{rank } AB \leq \min \{\text{rank } A, \text{rank } B\}$$

In particular, $\text{rank } AB = \text{rank } A$ *if B is square and nonsingular, and* $\text{rank } AB = \text{rank } B$ *if A is square and nonsingular.*

The computation of the rank of a matrix is greatly simplified by performing elementary operations which give it a block triangular form.

Lemma 8 (Gaussian elimination in block form) *Consider the following real matrix in partitioned form:*

$$M \equiv \begin{pmatrix} A & B \\ C & D \end{pmatrix}$$

where the matrices A, B, C, D are of order $n \times n$, $n \times m$, $l \times n$, and $l \times m$, respectively. If A is nonsingular, then

$$\begin{aligned}\text{rank } M &= \text{rank}\begin{pmatrix} A & B \\ 0 & D - CA^{-1}B \end{pmatrix} = \text{rank}\begin{pmatrix} A & 0 \\ C & D - CA^{-1}B \end{pmatrix} \\ &= \text{rank } A + \text{rank } \left(D - CA^{-1}B\right)\end{aligned}$$

It must be kept in mind that the nonsingularity assumption on A is crucial even if M is already in block triangular form, as shown by the matrix $\begin{pmatrix} 0 & B \\ 0 & 0 \end{pmatrix}$. The result is understood by observing that

$$\begin{aligned}\text{rank}\begin{pmatrix} A & B \\ 0 & D - CA^{-1}B \end{pmatrix} &= \text{rank}\begin{pmatrix} A & B \\ C & D \end{pmatrix}\begin{pmatrix} I_n & -A^{-1}B \\ 0 & I_l \end{pmatrix} \\ &= \text{rank}\begin{pmatrix} I_n & 0 \\ -CA^{-1} & I_l \end{pmatrix}\begin{pmatrix} A & B \\ C & D \end{pmatrix}\end{aligned}$$

is equal to rank M by the last corollary. Moreover, in the matrix

$$\begin{pmatrix} A & B \\ 0 & D - CA^{-1}B \end{pmatrix}$$

the first n rows are independent since the rows of A are so; hence none of the last l rows, having the first n components equal to zero, can be a linear combination of the first n rows, unless it is the trivial combination, i.e., the null vector.

2.4 General topology

Proposition 9 *Let X and Y be two topological spaces and $f : X \to Y$ be a closed map. Given any element $y \in f(X)$ and any open neighborhood U of $f^{-1}(y)$, there exists an open neighborhood V of y such that $f^{-1}(V) \subseteq U$.*

For a proof see [55] (Theorem 11.2, p. 86). We leave the proofs of the next two result as an exercise

Proposition 10 *Let $f : Z \to Y$ and $g : X \to Z$ be functions between topological spaces. If f and g are closed (open), so is $f \circ g$. If $f \circ g$ is closed (open), and g is continuous and surjective, then f is closed (open). If $f \circ g$ is closed (open), and f is continuous and injective, then g is closed (open).*

Proposition 11 *Let O be an open subset of the topological product space $X \times Y$. For each $x \in X$, the section*

$$O_x \equiv \pi_2\left(O \cap \pi_1^{-1}(x)\right) = \{y \in Y : (x, y) \in O\}$$

is an open subset of Y.

3. Nonlinear programming

In this section, unless otherwise stated,

- X is an open and convex subset of $\mathbb{R}^n$ and
- f, g and h are continuous functions such that $f : X \to \mathbb{R}$, $g : X \to \mathbb{R}^m$ and $h : X \to \mathbb{R}^l$.

The goal of this section is to study the problem

$$\begin{array}{lll} \max_{x \in X} f(x) & & \\ \quad s.t. & g(x) \geq 0 & (1) \\ & h(x) = 0 & (2) \end{array} \tag{1.2}$$

f is called **objective function**; x **choice variable** vector; (1) and (2) in (1.2) **constraints**; g and h **constraint functions**;

$$C \equiv \{x \in X : g(x) \geq 0 \text{ and } h(x) = 0\}$$

is the **constraint set**;

$$\{x^* \in C : \forall x \in C,\ f(x^*) \geq f(x)\}$$

is called **solution set** to problem (1.2) and it is also denoted by arg max (1.2).

Under suitable assumptions on concavity (and differentiability) of the functions f, g, and h, arg max(1.2) basically coincides with the set of solutions to a well chosen system of equations and inequalities. For that reason, in Subsection 3.1 we present some concavity concepts in the case of C^0, differentiable and C^2 functions and then we study the relationship among those definitions. Although in both economic and optimization literature there are many classes of generalized concavity concepts, we consider only those which are used in this book. All the results listed here are proved in [7], [13] and [25].

In Subsection 3.2, we consider necessary and/or sufficient conditions for the existence of a solution to problem (1.2). Under suitable assumption, the solution set of problem (1.2) can be described by means of an appropriate system of equality and/or inequality conditions. Furthermore we give a sort of "simple recipe" which can help the reader in finding them.

3.1 Concavity and generalized concavity

3.1.1 Concavity

Definition 12 A C^0 function f is **concave** if $\forall\, x',x'' \in X, \forall\, \lambda \in [0,1]$,

$$f((1-\lambda)x' + \lambda x'') \geq (1-\lambda)f(x') + \lambda f(x'')$$

Proposition 13 *A differentiable function f is concave if and only if*

$$\forall x', x'' \in X, \quad f(x'') - f(x') \leq Df(x')(x'' - x')$$

Proposition 14 *Consider a concave, differentiable function f. If x^0 is a critical point, then it is a global maximum point.*

Proposition 15 *A C^2 function f is concave if and only if*

$$\forall x \in X, \quad D^2 f(x) \text{ is negative semidefinite}$$

Concavity does not guarantee the existence of a unique global maximum, which motivates the introduction of the following class of functions.

3.1.2 Strict concavity

Definition 16 A C^0 function f is **strictly concave** if $\forall\, x', x'' \in X$ such that $x' \neq x''$, $\forall\, \lambda \in (0,1)$,

$$f((1-\lambda)x' + \lambda x'') > (1-\lambda) f\left(x'\right) + \lambda f(x'')$$

Proposition 17 *A differentiable function f is strictly concave if and only if $\forall x', x'' \in X$ such that $x' \neq x''$,*

$$f(x'') - f(x') < Df(x')(x'' - x')$$

Proposition 18 *Consider a strictly concave, differentiable function f. If x^0 is a critical point, then it is a strict global maximum point, i.e., the unique global maximum point.*

Proposition 19 *Consider a C^2 function f. If*

$$\forall x \in X, \quad D^2 f(x) \text{ is negative definite} \tag{1.3}$$

then f is strictly concave.

The definitions contained in the next subsections try to give more general conditions under which the conclusions of Propositions 14 and 18 hold.

3.1.3 Quasiconcavity

Definition 20 A C^0 function f is **quasiconcave** if $\forall\, x', x'' \in X$, $\forall\, \lambda \in [0,1]$,

$$f((1-\lambda)x' + \lambda x'') \geq \min\left\{f\left(x'\right), f(x'')\right\}$$

Proposition 21 *A C^0 function f is quasiconcave if and only if*

$$\forall \alpha \in \mathbb{R}, \; B(\alpha) \equiv \{x \in X : \; f(x) \geq \alpha\} \text{ is convex}$$

Proposition 22 *A differentiable function f is quasiconcave if and only if*

$$\forall x', x'' \in X \;\; f(x'') - f(x') \geq 0 \;\Rightarrow\; Df(x')(x'' - x') \geq 0$$

Proposition 23 *If a C^2 function f is quasiconcave then*

$$\forall x \in X, \forall v \in \mathbb{R}^n \text{ such that } Df(x)\,v = 0, \quad vD^2 f(x)v \leq 0$$

When f is quasiconcave, necessary first order conditions fail to be sufficient. Moreover it is not the case that if x^0 is a local maximum point, then it is a global maximum point. On the other hand, the following property holds.

Proposition 24 *Consider a quasiconcave function f. If x^0 is a strict local maximum point, then it is a strict global maximum point.*

3.1.4 Strict quasiconcavity

Definition 25 A C^0 function f is **strictly quasiconcave** if $\forall\ x', x'' \in X$, such that $x' \neq x''$, and $\forall\ \lambda \in (0,1)$, we have that

$$f((1-\lambda)x' + \lambda x'') > \min\left\{f\left(x'\right), f(x'')\right\}$$

Proposition 26 *Consider a differentiable function f. If*

$$f(x'') - f(x') \geq 0 \;\Rightarrow\; Df(x')(x'' - x') > 0, \forall x', x'',\; x' \neq x'' \tag{1.4}$$

then f is strictly quasiconcave.

Definition 27 A differentiable function f is **differentiable strictly quasiconcave** if Condition (1.4) in Proposition 26 is verified.

Given a differentiable function, it is *not* the case that strict quasiconcavity implies differentiable strict quasiconcavity.

If we restrict the class of differentiable functions to whose with nonzero gradients everywhere in the domain, then differentiable strict quasiconcavity and strict quasiconcavity are equivalent.

Proposition 28 *Consider a C^2 function f. If*

$$Df(x)\,v = 0 \Rightarrow vD^2 f(x)v < 0 \;\; \forall x \in X, \forall v \in \mathbb{R}^n \backslash \{0\} \tag{1.5}$$

then f is differentiable strictly quasiconcave.

Definition 29 A C^2 function f is **differentiably strictly quasiconcave** if Condition (1.5) in Proposition 28 is verified.

By the use of bold character, we help the reader to distinguish the definition of differentia**bly** strict quasiconcavity from the definition of differentia**ble** strict quasiconcavity, since the concept names differ by just one letter.

Given a C^2 function, it is *not* the case that differentia**ble** strict quasiconcavity implies differentia**bly** strict quasi concavity.

Proposition 30 *Consider a differentiable strictly quasiccncave function* f. *If* $Df\left(x^0\right)=0$, *then* x^0 *is a strict global maximum point.*

The search for a more general definition of concavity for which critical points are global maxima motivates the following definition.

Definition 31 A differentiable function f is **pseudoconcave** if

$$\forall x', x'' \in X, \;\; f\left(x''\right) > f\left(x'\right) \Rightarrow Df\left(x'\right)\left(x''-x'\right) > 0$$

Proposition 32 *If* f *is a pseudoconcave function, then* x^0 *is a global maximum point if and only if* $Df\left(x^0\right)=0$.

We summarize the results of this subsection in the following tables.

Class of function	**Fundamental properties**		
	C ⇒ G max	L max ⇒ G max	Uniqueness of G max
Strictly concave	Yes	Yes	Yes
Concave	Yes	Yes	No
Diff.ably-str.-q.-conc.	Yes	Yes	Yes
Diff.ble-str.-q.-conc.	Yes	Yes	Yes
Pseudoconcave	Yes	Yes	No
Quasiconcave	No	No	No

where C stands for property of being a critical point, and L and G stand for local and global, respectively. Observe that the first, the second and the last row of the second column apply to the case of C^0 and not necessarily differentiable functions.

	C^0	(1)	Differentiable	(2)	C^2
Concavity	Def. 12	$\Leftrightarrow$	Prop. 13	$\Leftrightarrow$	Prop. 15
Strict concavity	Def. 16	$\Leftrightarrow$	Prop. 17	$\Leftarrow$ $\not\Rightarrow$	Prop. 19
Strictly quasiconcavity	Def. 25	$\Leftarrow$ $\not\Rightarrow$	Prop. 26	$\Leftarrow$ $\not\Rightarrow$	Prop. 28
Quasiconcavity	Def. 20 Prop. 21	$\Leftrightarrow$	Prop. 22	$\not\Leftarrow$ $\Rightarrow$	Prop. 23

where "Prop (Definition).... " should be read as "Condition stated in Proposition (Definition) ... ". In addition (1) and (2) mean "in the differentiable case" and "in the C^2 case", respectively.

3.1.5 Relationships among kinds of concavity

The relationships among different definitions of concavity in the case of differentiable functions (or C^2 for differentiably strictly quasi concave functions) are summarized in the following table.

		strict concavity		differentiably strict quasiconcavity
		$\Downarrow$	$\searrow$	$\Downarrow$
linearity	$\Rightarrow$	concavity		differentiable strict quasiconcavity
		$\Downarrow$	$\swarrow$	$\Downarrow$
		pseudoconcavity		strict quasiconcavity
		$\Downarrow$	$\swarrow$	
		quasiconcavity		

(1.6)

If a conditional connective is missing somewhere, the corresponding conditional statement is false.

3.2 Constrained optimization

3.2.1 The case of equality constraints: Lagrange theorems

Throughout the present subsection, $l \leq n$. Consider the problem

$$\max_{x \in X} \; f(x) \qquad s.t. \qquad h(x) = 0 \tag{1.7}$$

where f and h are differentiable functions. Define the associated lagrangean function as

$$\mathcal{L} : X \times \mathbb{R}^l \to \mathbb{R}, \quad (x, \lambda) \mapsto f(x) + \lambda h(x)$$

where λ is called the Lagrange multipliers vector associated with the constraint $h(x) = 0$.

Theorem 33 (Necessary conditions) *Consider problem (1.7) and assume that* $\text{rank}\,[Dh(x^*)] = l$. *If x^* is a local maximum point for (1.7), then there exists $\lambda^* \in \mathbb{R}^l$ such that*

$$\begin{cases} Df(x^*) + \lambda^* Dh(x^*) = 0 & (1) \\ h(x^*) = 0 & (2) \end{cases} \tag{1.8}$$

Remark 34 Observe that Conditions (1) and (2) in (1.8) can be respectively written as

$$\begin{array}{ll} D_x \mathcal{L}(x, \lambda) = 0 & (1) \\ D_\lambda \mathcal{L}(x, \lambda) = 0 & (2) \end{array}$$

The full rank condition in the above theorem is one of the several conditions, called constraint qualifications, which guarantee that Conditions (1.8) are necessary conditions for a solution to problem (1.7). For a discussion of other versions of constraint qualifications conditions, see for example [13] and [85]. When no constraint qualification holds, a solution to problem (1.7) verifies weaker necessary conditions, known as Fritz John necessary conditions.

Theorem 35 (Fritz John necessary conditions) *If x^* is a local maximum point for (1.7), then there exists $\mu^* \in \mathbb{R}^{l+1}$ such that*

$$\begin{cases} \mu^* \left[Df\left(x^*\right), Dh\left(x^*\right)\right] = 0 \\ h\left(x^*\right) = 0 \\ \mu^* \neq 0 \end{cases} \tag{1.9}$$

Theorem 36 (Sufficient conditions) *Consider problem (1.7). Assume that*

1. *f is pseudoconcave,*
2. *for $j = 1, ..., l$, h_j is quasiconcave,*
3. *there exist $(x^*, \lambda^*) \in X \times \mathbb{R}^l$ such that*
   - *(a)* $\lambda^* \geq 0$,
   - *(b)* $Df\left(x^*\right) + \lambda^* Dh\left(x^*\right) = 0$,
   - *(c)* $h\left(x^*\right) - 0$.

Then x^* solves (1.7).

Define

1. $K \equiv \{x \in X : \exists \lambda \in \mathbb{R}^l$ such that (x, λ) satisfies conditions (1.8)$\}$;
2. $K^+ \equiv \{x \in X : \exists \lambda \in \mathbb{R}^l_+$ such that (x, λ) satisfies conditions (1.8$\}$;
3. $M \equiv \arg\max$ (1.7).

Of course, $K^+ \subseteq K$. Theorem 33 says that if $\text{rank}\left[Dh\left(x^*\right)\right] = l$, then $M \subseteq K$. Theorem 36 says that whenever f is pseudoconcave and the g_i's are quasiconcave, $K^+ \subseteq M$. Therefore, in such a case, the conditions in (1.8) completely characterize the solutions to problem (1.7) if $K^+ = K$, that is, if for every (x, λ) which satisfies the conditions in (1.8), $\lambda \geq 0$.

3.2.2 The case of inequality constraints: Kuhn-Tucker theorems

Throughout the present subsection, $m \leq n$. Consider the following problem

$$\max_{x \in X} f(x) \qquad s.t. \qquad g(x) \geq 0 \tag{1.10}$$

where f and h are differentiable functions. Similarly to the case of equality constraints, we define the associated lagrangean function as

$$\mathcal{L} : X \times \mathbb{R}^m \to \mathbb{R}, \quad (x, \lambda) \mapsto f(x) + \lambda g(x)$$

where λ is still called the Lagrange multipliers vector associated with the constraint $g(x) = 0$.

Definition 37 Let $x^* \in X$. We say that j is a **binding constraint** if $g^j(x^*) = 0$. Let

$$J^* \equiv \left\{ j \in \{1, ..., m\} : g^j(x^*) = 0 \right\}$$

and

$$g^* \equiv \left(g^j\right)_{j \in J^*}, \quad \widehat{g} \equiv \left(g^j\right)_{j \notin J^*}$$

Theorem 38 (Kuhn-Tucker necessary conditions) *If x^* is a solution to problem (1.10), and one of the following constraint qualification conditions holds:*

1 *for $j = 1, ..., m$, g^j is pseudoconcave and there exists $x^{++} \in X$ such that $g(x^{++}) \gg 0$,*

2 $\operatorname{rank} Dg^*(x^*) = \#J^*$,

then there exists $\lambda^* \in \mathbb{R}^m$ such that (x^*, λ^*) solves the following conditions (known as Kuhn-Tucker conditions):

$$\begin{cases} Df(x) + \lambda Dg(x) & = 0 & & (1) \\ g^j(x) & \geq 0 & \text{for } j = 1, ..., m & (2) \\ \lambda^j & \geq 0 & \text{for } j = 1, ..., m & (3) \\ \lambda^j g^j(x) & = 0 & \text{for } j = 1, ..., m & (4) \end{cases} \tag{1.11}$$

Remark 39 Observe that Conditions (1) and (2) in (1.11) can be respectively written as

$$\begin{array}{ll} D_x \mathcal{L}(x, \lambda) = 0 & (1) \\ D_\lambda \mathcal{L}(x, \lambda) \geq 0 & (2) \end{array}$$

Theorem 40 (Kuhn-Tucker sufficient conditions) *Assume that f is pseudoconcave and for $j = 1, ..., m$ g^j is quasiconcave. If there exists $\lambda^* \in \mathbb{R}^m$ such that (x^*, λ^*) solves the Kuhn-Tucker conditions (1.11), then x^* solves problem (1.10).*

3.2.3 The case of both equality and inequality constraints: generalized Kuhn-Tucker theorems

Consider the problem

$$\begin{array}{lrcl} \max\limits_{x\in X} f(x) & & & \\ \quad s.t. & g(x) & \geq & 0 \\ & h(x) & = & 0 \end{array} \qquad (1.12)$$

and define the associated lagrangean function

$$\mathcal{L}(x,\lambda,\mu) = f(x) + \lambda g(x) + \mu h(x)$$

where f, g, and h are differentiable and $(\lambda,\mu) \in \mathbb{R}^m \times \mathbb{R}^l$ is the associated Lagrange multipliers vector.

Theorem 41 (Necessary conditions) *Consider problem (1.12) and assume that one of the following qualification constraint conditions holds:*

1 rank $\begin{bmatrix} Dg^*(x^*) \\ Dh(x^*) \end{bmatrix} = m^* + l,$

2 *g and h are affine.*

If x^* solves (1.12), then $\exists\,(x^*,\lambda^*,\mu^*) \in X \times \mathbb{R}^m \times \mathbb{R}^l$ which satisfies the following conditions (known as Kuhn-Tucker conditions)

$$\begin{cases} Df(x) + \lambda Dg(x) + \mu Dh(x) & = 0 & & (1) \\ g^j(x) & > 0 & for\ j = 1,...,m & (2) \\ \lambda^j & \geq 0 & for\ j = 1,...,m & (3) \\ \lambda^j g^j(x) & = 0 & for\ j = 1,...,m & (4) \\ h^k(x) & = 0 & for\ k = 1,...,l & (5) \end{cases} \qquad (1.13)$$

Remark 42 Observe that Conditions (1),(2) and (5) in (41) can be respectively written as

$$\begin{array}{ll} D_x\mathcal{L}(x,\lambda,\mu) = 0 & (1) \\ D_\lambda\mathcal{L}(x,\lambda,\mu) \geq 0 & (2) \\ D_\mu\mathcal{L}(x,\lambda,\mu) = 0 & (3) \end{array}$$

Theorem 43 (Sufficient conditions) *Assume that f is pseudoconcave, for $j = 1,...,m$, g^j is quasiconcave, and h is affine. If $(x^*,\lambda^*,\mu^*) \in X \times \mathbb{R}^m \times \mathbb{R}^l$ satisfies the Kuhn-Tucker conditions associated with (1.12), i.e., Condition (1.13), then x^* solves (1.12).*

Summarizing, we describe below the steps to be followed to try solve a maximization problem.

1 **The set X.**
Check that X is open and convex. If X is not open try to rewrite the problem in terms of an open set X' and added constraints.

2 **Existence.**
Apply the extreme value theorem ([90], Theorem 6.4, p.175). The constraint set has to be nonempty and compact. If the constraint set is not compact, it is sometimes possible to find a maximization problem with that property and with solution set contained in the solution set of the problem we are analyzing. If you are not able to show existence, but

(a) conditions hold, making the Kuhn-Tucker conditions sufficient for a solution,

(b) you are able to find a solution to the Kuhn-Tucker conditions,

then a solution exists.

3 **Canonical form.**
Write the problem in the (in fact, our definition of) canonical form (1.7) or (1.10) or (1.12).

4 **Uniqueness.**
Try to show that the objective function is strictly quasiconcave and the constraint set is convex, or the objective function is quasiconcave and the constraint set is strictly convex.

5 **Necessity of K-T.**
Check if conditions hold, making the Kuhn-Tucker conditions necessary for a solution.

6 **Sufficiency of K-T.**
Check if conditions hold, making the Kuhn-Tucker conditions sufficient for a solution.

7 **K-T conditions.**
Write the Lagrangian function and then the Kuhn-Tucker conditions.

8 **Solve the K-T conditions.**

Try to solve the system of Kuhn-Tucker conditions in the unknown variables (x, λ). To do that, either analyze all cases, or try to get a "good conjecture" and check if the conjecture is correct.

4. Our notation for rank computations

In this section, we present a procedure to compute the rank of a matrix which turns out to be very convenient for the general equilibrium models we analyze in this book (see Chapters 8-15). In the literature on these topics, the method is usually referred to as the *perturbation method.*

We start introducing the basic aspects of that method. For a better understanding of the procedure and in order to make the reader familiar with the notation, we compute ranks in very simple cases.

While the perturbation method is (obviously) equivalent to a set of well chosen elementary row and column operations, it is more suitable than that procedure when we have to compute the rank of a quite large matrix. In that case, the proposed method can be used to describe those computations in a relatively concise manner. This is actually done starting with (8.18) in Chapter 8 and at least once in almost all the following chapters.

To describe the perturbation method, we first present a very simple example to informally describe it; then we show a simple proposition which strengthens its power; finally, we describe a step-by-step procedure to be used in applying it.

4.1 Perturbing rows

Consider a matrix $A \in \mathcal{M}_{m,n}$ with $m \leq n$. The so called perturbations method builds upon the following very well known result: A has full row rank if and only if the systems $Ax = \mathbf{e}^1$, $Ax = \mathbf{e}^2$,...,$Ax = \mathbf{e}^m$ have a solution, where $\mathbf{e}^i$ is i-th elements of the canonical basis.

Example 44 Let A be the following matrix

$$A = \begin{bmatrix} 1 & 0 & 0 & 1 \\ 0 & 1 & 1 & 0 \\ 1 & 1 & 0 & 0 \end{bmatrix}$$

We want to show that A has full row rank, hence we want to show that the following three systems have solution:

$$\begin{bmatrix} 1 & 0 & 0 & 1 \\ 0 & 1 & 1 & 0 \\ 1 & 1 & 0 & 0 \end{bmatrix} \begin{bmatrix} x_1 \\ x_2 \\ x_3 \\ x_4 \end{bmatrix} = \begin{bmatrix} 1 \\ 0 \\ 0 \end{bmatrix} \tag{1.14}$$

$$\begin{bmatrix} 1 & 0 & 0 & 1 \\ 0 & 1 & 1 & 0 \\ 1 & 1 & 0 & 0 \end{bmatrix} \begin{bmatrix} x_1 \\ x_2 \\ x_3 \\ x_4 \end{bmatrix} = \begin{bmatrix} 0 \\ 1 \\ 0 \end{bmatrix} \tag{1.15}$$

and

$$\begin{bmatrix} 1 & 0 & 0 & 1 \\ 0 & 1 & 1 & 0 \\ 1 & 1 & 0 & 0 \end{bmatrix} \begin{bmatrix} x_1 \\ x_2 \\ x_3 \\ x_4 \end{bmatrix} = \begin{bmatrix} 0 \\ 0 \\ 1 \end{bmatrix} \tag{1.16}$$

Call b_i, with $i = 1, 2, 3$, the left hand side of systems (1.14), (1.15) and (1.16), respectively. To get a 1 as the first component of b_1, it is enough to pick just one nonzero element in the first row of A, say a_{1k}, and to choose $x_k = \frac{1}{a_{1k}}$, $x_j = 0$ with $j \neq k$. In the example, both a_{11} and a_{14} are equal to 1. On the other hand, we need to get 0 in all other components of b^1. In general, choosing some $x_k \neq 0$ implies that we could get non zero elements in the components of b^1 other than its first component if there are non zero elements in the k-th column of A. In the example, this is the case for a_{11}: there is 1 in the third position of the first column. To the contrary, $a_{14} = 1$ and there are no other elements different from zero in the fourth column. Therefore the vector $(0, 0, 0, x_4)$ with $x_4 = \frac{1}{a_{14}} = 1$ is the solution of system (1.14), i.e., it allows us to get 1 in first row, and 0 in the second and in the third one.

We can rephrase the above observations as follows. "We want to perturb the first equation, without disturbing any other equation; that can be done using x_4". We introduce the following notation to describe the above procedure:

$$(1) \leftarrow x_4$$

which we read as "use x_4 to perturb the first row or equation; no other row is disturbed"[3]. The solution to system (1.14) is therefore seen as a way of "perturbing the first row without disturbing the other rows".

The use of the word perturbation to describe the procedure may be misleading and requires some explanation. The method consists in finding a vector x^i such that $Ax^i = \alpha \mathbf{e}^i$ with $\alpha \in \mathbb{R} \setminus \{0\}$. A necessary step to do that is to find a nonzero element in the i-th row of A and use the corresponding variable to make the i-th component of Ax different from zero, that is, to "perturb $(Ax)_i$ away from zero". Shortly, and with abuse of language, we say that "we perturb the i-th row of A". Similar procedure can be used to find a solution to system (1.15). In this case, since we want to get 1 as the *second* component of b_2, we look for non zero elements in the *second* row of A. In the example, there is 1 in the third component of that row. Observe that all other components of column 3 are 0. Therefore, the choice of $x_3 = 1$ perturbs equation 2 without disturbing any other equation. Using the notation introduced above, we can write

$$(2) \leftarrow x_3$$

[3]This is equivalent to say that the fourth column of A is a multiple of the canonical vector $\mathbf{e}^1$.

Finally consider system (1.16). We can choose x_1, that is $x_1 = 1$, to perturb that row.

But since there is 1 in the first component of the first column of A, that choice does disturb equation (1). We have to look for a non zero element in the first row (but not in its first component) to adjust the disturbance caused by x_1. We can use x_4, that is $x_4 = -1$, to adjust the disturbance occurred in the first row. This does not disturb any other row. Therefore $(x_1, 0, 0, x_4)$, with $x_1 = 1$ and $x_4 = -1$, is the solution to system (1.16). We can summarize our way of solving system (1.16) as follows.

$$\begin{array}{lll} (3) \leftarrow x_1 & \rightsquigarrow (1) \\ (1) \leftarrow x_4 & \end{array}$$

i.e., we use x_1 to perturb the third row ; this disturbs the first row, which can be adjusted using x_4, which does not disturb further any row[4].

In the previous example we could have determined the exact value of x very easily. In some cases this task is more difficult. However what is really important is choosing the components to be used in order to perturb the rows of the matrix A. For this reason we simply refer to the variables which are used to perturb the rows, but we never specify their values. We consider the altered equations without computing these alterations.

Observe that it is convenient to write the name of the corresponding variable on top of the column of the coefficients associated with that variable, as done below for the matrix of Example 44

$$\begin{array}{cccc} x_1 & x_2 & x_3 & x_4 \end{array}$$
$$\begin{pmatrix} 1 & 0 & 0 & 1 \\ 0 & 1 & 1 & 0 \\ 1 & 1 & 0 & 0 \end{pmatrix}$$

4.2 Forgetting lower rows

In many general equilibrium models, we deal with matrices of the following form

$$M = \begin{bmatrix} A_{n\times(n+l)} \\ B_{k\times(n+l)} \end{bmatrix}_{(n+k)\times(n+l)}$$

where $k \leq l$ and $\operatorname{rank} A = n$, and we want to show that M has full row rank.

The following proposition allows us to compute the rank of M, by perturbing the last k rows, i.e., the rows of submatrix B.

[4] The reader can see that we can equivalently perturb the third row using x_2. This alters the second row that can be adjusted using x_3.

Proposition 45 *Consider the matrix* $M = \begin{bmatrix} A_{n\times(n+l)} \\ B_{k\times(n+l)} \end{bmatrix}$ *with* $k \le l$ *and assume that* $\operatorname{rank} A = n$. *Then, the following two statements are equivalent:*

1 $\operatorname{rank}\begin{bmatrix} A \\ B \end{bmatrix} = n + k$;

2 for $i = 1, ..., k$, $\mathbf{e}^i \in \mathbb{R}^k$, $\exists z^i \in \mathbb{R}^{n+l}$ *such that* $\begin{bmatrix} A \\ B \end{bmatrix} z^i = \begin{bmatrix} 0 \\ \mathbf{e}^i \end{bmatrix}$.

Proof. $[1 \Rightarrow 2]$ Since $\operatorname{rank}\begin{bmatrix} A \\ B \end{bmatrix} = n + k$, from Proposition 3, for every $b \in \mathbb{R}^{n+k}$, there exists $z \in \mathbb{R}^{n+l}$ such that $\begin{bmatrix} A \\ B \end{bmatrix} z = b$.

$[1 \Leftarrow 2]$ For each $j \in \{1, ..., n\}$, there exists $w^j \in \mathbb{R}^{n+l}$ such that $Aw^j = \mathbf{e}^j \in \mathbb{R}^n$. Let $Z \equiv \left[z^1, \ \ z^k\right]$ and $W \equiv \left[w^1 \ \ w^k\right]$. Then

$$M\,[W \ \ Z] = \begin{bmatrix} I_n & 0 \\ BW & I_k \end{bmatrix}$$

and M has full rank. ∎

Example 46 Consider the matrix

$$M = \begin{pmatrix} \overset{x_1}{1} & \overset{x_2}{0} & \overset{x_3}{0} & \overset{x_4}{1} & \overset{x_5}{0} & \overset{x_6}{2} & \overset{x_7}{0} \\ 0 & 1 & 1 & 0 & 0 & 0 & 1 \\ 1 & 1 & 0 & 0 & 0 & 0 & 2 \\ 0 & 0 & 0 & 2 & 0 & 1 & 3 \\ 3 & 4 & \sqrt{2} & 5 & 6 & -2 & 15 \end{pmatrix}$$

Observe that $M = \begin{bmatrix} A \\ B \end{bmatrix}$ with

$$A = \begin{bmatrix} 1 & 0 & 0 & 1 & 0 & 2 & 0 \\ 0 & 1 & 1 & 0 & 0 & 0 & 1 \\ 1 & 1 & 0 & 0 & 0 & 0 & 2 \end{bmatrix}$$

and

$$B = \begin{bmatrix} 0 & 0 & 0 & 2 & 0 & 1 & 3 \\ 3 & 4 & \sqrt{2} & 5 & 6 & -2 & 15 \end{bmatrix}$$

By Example 44 we know that the submatrix A has full row rank. Then from Proposition 45 and using our language, we are left with perturbing the last two rows of M, without disturbing the first 3 rows.

As a further application of Proposition 45, when we perturb row (4) we do not need to pay attention to what happens to row (5).

Perturbation of row (4)

$$\begin{array}{ll} (4) \leftarrow x_4 & \rightsquigarrow (1) \\ (1) \leftarrow x_1 & \rightsquigarrow (3) \\ (3) \leftarrow x_2 & \rightsquigarrow (2) \\ (2) \leftarrow x_3 & \end{array}$$

Perturbation of row (5)

$$(5) \quad \leftarrow \quad x_5$$

4.3 A step-by-step procedure

In terms of our perturbation technique, Proposition 45 can be used as follows. Consider an arbitrary matrix $C_{m \times n}$, with $m \leq n$.

Step 1. Consider the first row. Show that it has full rank 1, i.e., perturb it. You don't need to pay any attention to what happens to the rows from the 2-nd one to the m-th one (i.e., you can forget the lower rows).

Step 2. Consider the submatrix of C made up by its first two rows. To show that the submatrix has full row rank, using Proposition 45 and point 1 above, it is enough to show that you can perturb the second row without disturbing the first one. Of course, you can forget the rows from the 3-rd one to the m-th one.

Step i 3 $\leq$ i $\leq$ m. Consider the submatrix of C made up by its first i rows. It is enough to show that we can perturb the i-th row without disturbing the rows from the first one to the $(i-1)$-th one. Of course, you can forget the rows from the $(i+1)$-th one to the m-th one.

Clearly the main advantage of this procedure with respect to the one presented in the first part of this section is that when we want to perturb one row, we can "forget lower rows".

Example 47 Consider the matrix

$$A = \begin{array}{c} \begin{matrix} x_1 & x_2 & x_3 & x_4 \end{matrix} \\ \begin{pmatrix} 2 & 3 & 5 & 9 \\ 0 & 0 & 8 & 0 \\ 4 & 2 & 8 & 5 \\ 0 & 0 & 3 & 7 \end{pmatrix} \end{array}$$

Perturbation of row (1).

$$(1) \leftarrow x_1$$

Perturbation of row (2).

$$\begin{array}{ll} (2) \leftarrow x_3 & \rightsquigarrow (1) \\ (1) \leftarrow x_1 & \end{array}$$

Perturbation of row (3).

$$(3) \leftarrow x_2 \quad \rightsquigarrow (1)$$

$$\dots$$

Now to adjust the perturbation of row (1), we cannot use any x_i, for $i = 1, \dots 4$, otherwise the perturbation goes back to row 3: we have a sort of "short circuit" in our method. In this case, a way out is offered by resort to another established (well known) method (elementary row operations). Combining this method with ours, we can first group together the rows which caused the problem, i.e., row 1 and 3. Consider then the matrix where we moved row 3 immediately below row 1, i.e., we interchanged row 2 and 3. Clearly, this elementary operation does not change the rank of the matrix:

$$A' = \begin{bmatrix} 2 & 3 & 5 & 9 \\ 4 & 2 & 8 & 5 \\ 0 & 0 & 8 & 0 \\ 0 & 0 & 3 & 7 \end{bmatrix}$$

Now the 2 by 2 upper left submatrix of A has full rank, simply because its determinant is different from zero. Therefore we can rewrite A' as

$$A' = \begin{bmatrix} B & & C & \\ 0 & 0 & 8 & 0 \\ 0 & 0 & 3 & 7 \end{bmatrix}$$

where

$$B \equiv \begin{bmatrix} 2 & 3 \\ 4 & 2 \end{bmatrix} \text{ and } C \equiv \begin{bmatrix} 5 & 9 \\ 8 & 5 \end{bmatrix}$$

Now we can again proceed as we did in the previous examples, keeping into account that B has full rank and therefore we can perturb the first two rows of A', from now on the "first superrow" using (x_1, x_2).

We rename also the third and fourth row as second and third row of the new matrix.

Perturbation of superrow (1) **of** A'.

$$(1) \leftarrow (x_1, x_2)$$

Perturbation of row (2) **of** A'.

$$\begin{array}{ll} (2) \leftarrow x_3 & \rightsquigarrow (1) \\ (1) \leftarrow (x_1, x_2) & \end{array}$$

Perturbation of row (3) **of** A'.

$$\begin{array}{ll} (3) \leftarrow x_4 & \rightsquigarrow (1) \\ (1) \leftarrow (x_1, x_2) & \end{array}$$

The example above shows that the perturbation method can be also applied to partitioned matrices and their superrows. In that case, for each superrow it is enough to find a full row rank submatrix, whose corresponding variables can be used to perturb it.

Chapter 2

MANIFOLDS IN EUCLIDEAN SPACES

One perhaps reductive but certainly perspicuous way to describe the present book, if one looks at it from a completely abstract point of view, is to say that it is mainly devoted to the study of some nonlinear equation systems which the authors consider particularly important in applications to problems arising in general equilibrium models. To study a system of nonlinear equations means, among other things, to decide whether or not it has any solutions, and, in the affirmative case, to establish properties of its solution set. This is a much more difficult task than in the case of systems of linear equations.

For the latter type of systems, the existence issue is settled by straightforward computations (typically, a rank check); moreover, although these computations may be more heavy for some system than for some other, the ideas underlying them are completely general, in that they apply to every possible system in the same fashion. The situation is similar for the second issue (properties of solution sets): if the solution set is nonempty, its structure is always that of an affine subspace; the main feature distinguishing the solution set of a linear system from another being its dimension.

Nonlinear functions are extensively dealt with by means of the tools provided by the calculus; that is, many properties of nonlinear functions are established indirectly, by looking at the properties of their linear and higher order approximations, and by checking whether or not, and to what extent, the properties enjoyed by the approximations are shared by the original functions. A basic feature contrasting the study of nonlinear functions from the linear theory is a direct consequence of this approach, namely, the conclusions drawn are generally *local* or even *pointwise* in nature, at least on a first stage of analysis: since any given nonlinear function generally has different linear and higher order approximations at different elements of its domain, the properties established according to this method only hold at specific points, or in appropriate neighborhoods of them.

A typical example, and perhaps the most appropriate for our purposes, is the implicit function theorem. Let n and k be positive integers, and let $m \equiv n+k$. Suppose that f is any function, with domain D_f included in $\mathbb{R}^m$, and with values in $\mathbb{R}^n$; and consider the equation

$$f(x, z) = 0 \tag{2.1}$$

where $x \in \mathbb{R}^n$, $z \in \mathbb{R}^k$ and, for convenience, we have identified $\mathbb{R}^m$ with $\mathbb{R}^n \times \mathbb{R}^k$. When f is linear, the solution set S to (2.1) is always nonempty (it contains $(0, 0)$) and it can be given a convenient representation by inverting an appropriate submatrix of the $n \times m$ matrix representing f. When f is nonlinear, if we happen to know that the solution set S of (2.1) is nonempty and it contains a solution $(\overline{x}, \overline{z})$ belonging to the interior of D_f, and if some appropriate differentiability conditions are satisfied for f, then we can draw important conclusions concerning S. These conclusions are local, because they only concern $S \cap N$, where N is some neighborhood of $(\overline{x}, \overline{z})$ in $\mathbb{R}^m$. More precisely, according to the implicit function theorem (or, rather, some version of it), if the following assumptions hold:

a) $f(\overline{x}, \overline{z}) = 0$;

b) f is C^1 in some open neighborhood N of $(\overline{x}, \overline{z})$ contained in D_f;

c) the partial $(n \times n)$ jacobian matrix $D_x f(\overline{x}, \overline{z})$ is not singular;

then there exist open sets $U \subseteq \mathbb{R}^n$ and $W \subseteq \mathbb{R}^k$, and a function $\vartheta : W \to U$, with the following properties:

1 $(\overline{x}, \overline{z}) \in U \times W \subseteq N$;

2 $\forall (x, z) \in U \times W$, $f(x, z) = 0$ if and only if $x = \vartheta(z)$;

3 ϑ is C^1 on W.

Thus not all the solution set S, but at least $S \cap (U \times W)$, a subset of S consisting of elements which are sufficiently close to $(\overline{x}, \overline{z})$, is completely characterized as the graph of a C^1 function (here we only need to touch upon the fact that by assuming higher order differentiability of f, the differentiability result for ϑ is strengthened).

If we consider the function $(\vartheta, id_W) : W \to \mathbb{R}^m$, we realize that it maps W bijectively onto $S \cap (U \times W)$. Hence we have a C^1 bijection between W and $V \equiv S \cap (U \times W)$, a subset of $\mathbb{R}^m$. V is not generally an open set of $\mathbb{R}^m$, but it is an open subset of S as a topological subspace of $\mathbb{R}^m$; in fact, it is an open set of the most general type for the induced topology on S ([90], p. 89).

The best we can hope for the solution set S to our nonlinear system is to be able to replicate this construction at each of its elements. This is certainly possible if f is a C^1 function over all its domain, and its jacobian matrix $Df(x, z)$ has full rank everywhere. Or, even if S is not quite well known, if it is known that $Df(x, z)$ has full rank on S (this occurs fairly often). We could then rephrase the consequences of the implicit function theorem for S in the following way: under the assumptions made, if S is not empty, each element of S has a neighborhood in S which is the bijective image of some open subset of $\mathbb{R}^k$ under an appropriate C^1 function.

We may wonder whether all such functions can be seen as diffeomorphisms. To be sure, we cannot quite call a function like (ϑ, id_W) a diffeomorphism as yet, because its inverse is defined on V, and, from calculus in euclidean spaces, the concept of differentiability is not usually given for functions defined on arbitrary subsets of euclidean spaces. However, this concept can be meaningfully and conveniently defined; indeed, the first section of this chapter is devoted to introduce such a concept, and to discuss some of its properties. It turns out then, as we shall see in later chapters, that we can strengthen the conclusion just reached, and say that each element of S has a neighborhood in S which is diffeomorphic to an open subset of $\mathbb{R}^k$. In short, as we shall often say from now on, S is *locally diffeomorphic* to $\mathbb{R}^k$ around each of its points.

This example leads us to the concept of *manifold*: a subset M of an euclidean space $\mathbb{R}^l$ is a manifold of dimension k, with $k \leq l$, if it is locally diffeomorphic to $\mathbb{R}^k$. Even if a manifold M has not an affine subspace structure, it is at least possible to describe M in terms of a collection of local diffeomorphisms. Given any element x of M, an appropriate diffeomorphism associates one and only one point of an open subset of $\mathbb{R}^k$ to each element in a suitable neighborhood of x. Thus we can say that a system of *local coordinates* has been introduced on M.

Section 2 is devoted to a presentation of the definition of manifold, as well as to an attempt to explain the notion by several examples and applications.

In this chapter, and hence in the whole book, we have followed the approach, which from the very start looks at manifolds as subsets of finite dimensional euclidean spaces. We have thought it worthwhile to proceed in this way, in order to help the reader in better visualizing the concepts presented. We have been aware all the way long that this approach forces the reader into immediately coming to grasps with the somewhat sophisticated concept of induced topology, but we have considered that to be a necessary sacrifice. On the other hand, we urge to stress that presenting manifolds as subsets of finite dimensional euclidean spaces places severe, unnecessary restrictions on the scope and content of the theory of manifolds, and even tends to make the presentation (at least, our presentation) more cumbersome at times. Thus we could not resist enclosing an appendix, where we give a sketchy outline of the alternative approach, which looks at manifolds as abstract topological spaces endowed with an intrinsic differentiable structure. This has allowed us to give at least the statement of the celebrated Whitney's embedding theorem, which partially reconciles the two approaches (and which will be presented in Chapter 3).

1. Manifolds - Definition and examples

Definition 1 Consider two arbitrary sets $X \subseteq \mathbb{R}^m$, $Y \subseteq \mathbb{R}^n$ and $r \in \mathbb{N}^*$ (that is a positive integer of ∞). A function $f : X \to Y$ is said to be of **class** C^r if, for any $x \in X$, there exists an open neighborhood N of x in $\mathbb{R}^m$ and a C^r function $F : N \to \mathbb{R}^n$ such that, for any $x' \in N \cap X$, $f(x') = F(x')$. F is called a **local extension** of f (at x).

Remark 2 A consequence of our definition is the fact that a C^1 function $f : X \to Y$ is continuous. Indeed, given $x \in X$, a local C^1 extension $F : N \to \mathbb{R}^n$ of f, defined in a suitable neighborhood N of x in $\mathbb{R}^m$, is continuous. It follows that $f_{|N \cap X} = F_{|N \cap X}$ is continuous at x as a function from $N \cap X$ to Y, and then so is f.

Recall that here and in the rest of the book any subset X of an euclidean space $\mathbb{R}^m$ is a topological space endowed with the induced topology of $\mathbb{R}^m$. Therefore, while the notion of differential of a function $f : X \to Y$ at x - and consequently the notion that f is C^r - requires an extension of that function to an open neighborhood of x in $\mathbb{R}^m$, the continuity of f at x is an intrinsic (topological) concept, not requiring any extension of f to be defined.

If $f : X \to \mathbb{R}^n$ is a C^r function, it may have an infinite number of extensions, as the following example shows.

Example 3 Consider $S \equiv \{(x, y) \in \mathbb{R}^2 : y = 0\}$ and $f : S \to \mathbb{R}$, $(x, y) \mapsto 0$. Consider also $F, G : \mathbb{R}^2 \to \mathbb{R}$, $F : (x, y) \mapsto 0$ and $G : (x, y) \mapsto y$. Both F and G are C^∞ extensions (indeed, global extensions) of f. Observe also that for any $(x, y) \in \mathbb{R}^2$

$$DF(x, y) = (0, 0) \neq (0, 1) = DG(x, y)$$

Definition 4 Consider two sets $X \subseteq \mathbb{R}^m$, $Y \subseteq \mathbb{R}^n$. A function $f : X \to Y$ is a C^r **diffeomorphism** if

1 f is C^r,

2 f is bijective,

3 f^{-1} is C^r.

In this case X and Y are said to be C^r **diffeomorphic** (by f).

Exercise 5 Prove that two C^1 diffeomorphic sets are in particular homeomorphic (see Remark 2).

Example 6 The exponential function $\exp : \mathbb{R} \to \mathbb{R}_{++}$ is a C^∞ diffeomorphism. It is well known that exp is C^∞ and bijective; its inverse, $\ln : \mathbb{R}_{++} \to \mathbb{R}$, $x \mapsto \ln x$, is clearly C^∞.

Example 7 Any open interval (a, b) of $\mathbb{R}$ is C^∞ diffeomorphic to $\mathbb{R}$. To see this consider the tangent function

$$f : (-\pi/2, \pi/2) \to \mathbb{R}, \quad x \mapsto \tan x$$

f is a C^∞ diffeomorphism as it is well known. Then, let

$$g : (a, b) \to (-\pi/2, \pi/2), \quad x \mapsto \frac{\pi}{b-a}(x-a) - \pi/2$$

which is a C^∞ diffeomorphism between (a, b) and $(\pi/2, \pi/2)$. Finally, $f \circ g$ is a C^∞ diffeomorphism between (a, b) and $\mathbb{R}$.

Exercise 8 Let $L : \mathbb{R}^m \to \mathbb{R}^n$ be a linear (affine) function. Then, for any $X \subset \mathbb{R}^m$, $L_{|X} : X \to L(X)$ is C^∞.

Example 9 The subsets of $\mathbb{R}^2$

$$X \equiv [0, 1] \times [0, 1]$$
$$Y \equiv \{(x, y) \in \mathbb{R}^2 : y \in [0, 1],\ y < x < y + 1\}$$

are C^∞ diffeomorphic. A C^∞ diffeomorphism between X and Y is

$$f : X \to Y, \quad (x, y) \mapsto (x + y, y)$$

The check that f is actually a C^∞ diffeomorphism is left to the reader.

Example 10 Let $f : \mathbb{R} \to \mathbb{R}$, $x \mapsto x^3$. The function f is C^∞ and bijective. The inverse $f^{-1} : \mathbb{R} \to \mathbb{R}$, $x \mapsto \sqrt[3]{x}$, is not differentiable at 0 since $\lim_{h \to 0} \frac{\sqrt[3]{h}}{h} = +\infty$. Therefore, f is not a C^1 diffeomorphism.

Proposition 11 *Given any subset X of $\mathbb{R}^m$, the identity function $id_X : X \to X$ is a C^∞ diffeomorphism.*

Proof. id_X is C^∞ since it has the C^∞ (global) extension $id_{\mathbb{R}^m}$. Of course, the inverse of id_X is id_X. ■

Proposition 12 (Algebra of C^r functions) *Consider $X \subseteq \mathbb{R}^m$ and $Y \subseteq \mathbb{R}^n$. Let $f : X \to Y$ and $g : X \to Y$ be two C^r functions. Then*

1. *$f \pm g$ is C^r,*
2. *$f \cdot g$ is C^r,*
3. *f/g is C^r if $g(x) \neq 0$ for all $x \in X$.*

Moreover, consider $X \subseteq \mathbb{R}^m$, $Y \subseteq \mathbb{R}^n$ and $Z \subseteq \mathbb{R}^p$. Let $f : X \to Y$ and $g : Y \to Z$ be C^r. Then

4. *$g \circ f : X \to Z$ is C^r.*

Proof. The proofs of 1, 2, 3 are left as an exercise. To show that 4 holds, let $x \in X$, W be a neighborhood of x in $\mathbb{R}^m$ and V be a neighborhood of $f(x)$ in $\mathbb{R}^n$. Let also $F : W \to \mathbb{R}^n$ and $G : V \to \mathbb{R}^p$ be two C^r local extensions of f and g, respectively. Consider an open neighborhood U of x in $\mathbb{R}^m$ such that $F(U)$ is contained in V (U exists by continuity of F). Then, $G \circ F_{|U} : U \to \mathbb{R}^p$ is a local C^r extension of $g \circ f$ at x. ■

Proposition 13 *Consider $X \subseteq \mathbb{R}^m$, $Y \subseteq \mathbb{R}^n$ and $Z \subseteq \mathbb{R}^p$. If $f : X \to Y$ and $g : Y \to Z$ are C^r diffeomorphisms, then $g \circ f : X \to Z$ is a C^r diffeomorphism.*

Proof. By Proposition 12, $g \circ f$ is C^r. In addition, $g \circ f$ is invertible and its inverse, $f^{-1} \circ g^{-1}$, is C^r still by Proposition 12, since f^{-1} and g^{-1} are so. ■

The following result can be proved using the above proposition and the inverse function theorem ([3], Theorem 13.6, p. 372).

Proposition 14 *Let A and B be two C^1 diffeomorphic subsets of $\mathbb{R}^n$. If A is open in $\mathbb{R}^n$, then B is open in $\mathbb{R}^n$ as well.*

Proof. Let $f : A \to B$ a C^1 diffeomorphism. Consider $y_0 \in B$ and let $G : N \to A$ a local C^1 extension of $g \equiv f^{-1}$, where N is an open neighborhood of y_0 in $\mathbb{R}^n$. Let U be an open neighborhood of $x_0 \equiv f^{-1}(y_0)$ in A (and consequently in $\mathbb{R}^n$ since A is open) such that $f(U) \subseteq N$. Then $G \circ f_{|U} = id_U$ and

$$id_{\mathbb{R}^n} = d\left(G \circ f_{|U}\right)_{x_0} = dG_{y_0} \circ df_{x_0}$$

This implies that df_{x_0} is an isomorphism. Then, by the inverse function theorem for $\widetilde{f} : A \to \mathbb{R}^n$, $x \mapsto f(x)$, there exists an open neighborhood U' of x_0 such that $V' \equiv \widetilde{f}(U')$ is an open neighborhood of y_0 in $\mathbb{R}^n$. (In addition, the inverse function theorem states that U' and V' are diffeomorphic by $\widetilde{f}$.) Since V' is contained in B, as $f(U') = \widetilde{f}(U')$, we can conclude that any element of B admits an open neighborhood in $\mathbb{R}^n$, which is contained in B. Therefore, B turns out to be open in $\mathbb{R}^n$. ∎

The above proposition can appear trivially true since B is the inverse image of the open set A by the continuous function f^{-1}. Actually, since f^{-1} is defined on B, if we only use the continuity of this function, we simply prove that B is an open subset of itself, which is a not interesting result. A more general and classical result in topology, the invariance of domain theorem, states that if U is an open connected subset of $\mathbb{R}^n$ and $f : U \to \mathbb{R}^n$ is a one-to-one and continuous function, then $f(U)$ is open in $\mathbb{R}^n$ and connected. The proof of this theorem is not immediate. The reader can see for example [55]: Section 3 in Chapter 17.

Let us now give the definition of C^r manifold.

Definition 15 (Manifold) A set $M \subseteq \mathbb{R}^l$ is said to be **locally C^r diffeomorphic** to $\mathbb{R}^m$, or an **m-dimensional C^r manifold** of $\mathbb{R}^l$, if, for every $x \in M$, there exists a function $\phi : U \to V$, where

1 U is an open subset of $\mathbb{R}^m$,

2 V is an open neighborhood of x in M,

3 ϕ is a C^r diffeomorphism.

The triple (ϕ, U, V) is called a **local parametrization** of M around x. A family of local parametrizations $\{(\phi_\lambda, U_\lambda, V_\lambda)\}_{\lambda\in\Lambda}$ of M is called a **system of local parametrizations** of M provided that $M = \cup_{\lambda\in\Lambda} V_\lambda$.

Consider the inverse function $\psi : V \to U$ of ϕ. The triple (ψ, V, U) is called a **chart** of M at x. A family of charts $\{(\psi_\lambda, V_\lambda, U_\lambda)\}_{\lambda\in\Lambda}$, with $\cup_{\lambda\in\Lambda} V_\lambda = M$, is called an **atlas** of M.

In the sequel, the dimension of a manifold M will be denoted by the symbol $\dim M$. In addition, the term manifold will simply stand for a C^r manifold of $\mathbb{R}^l$ for some r and some l.

Remark 16 Consider two local parametrizations $(\phi_\lambda, U_\lambda, V_\lambda)$ and (ϕ_μ, U_μ, V_μ) for which $V_\lambda \cap V_\mu \neq \emptyset$. As a direct consequence of Proposition 13, the two open subsets $\widetilde{U}_\lambda \equiv \phi_\lambda^{-1}(V_\lambda \cap V_\mu)$ and $\widetilde{U}_\mu \equiv \phi_\mu^{-1}(V_\lambda \cap V_\mu)$ of $\mathbb{R}^m$ are mapped onto each other diffeomorphically by the composite functions $\phi_{\mu\lambda} \equiv \phi_\mu^{-1} \circ (\phi_\lambda)_{|\widetilde{U}_\lambda}$ and $\phi_{\lambda\mu} \equiv \phi_\lambda^{-1} \circ (\phi_\mu)_{|\widetilde{U}_\mu}$. Moreover, $\phi_{\mu\lambda} = (\phi_{\lambda\mu})^{-1}$.

According to our definition, if M is a C^r manifold of $\mathbb{R}^l$ of dimension m and (ϕ, U, V) is a local parametrization, ϕ is C^r in the ordinary sense if it is

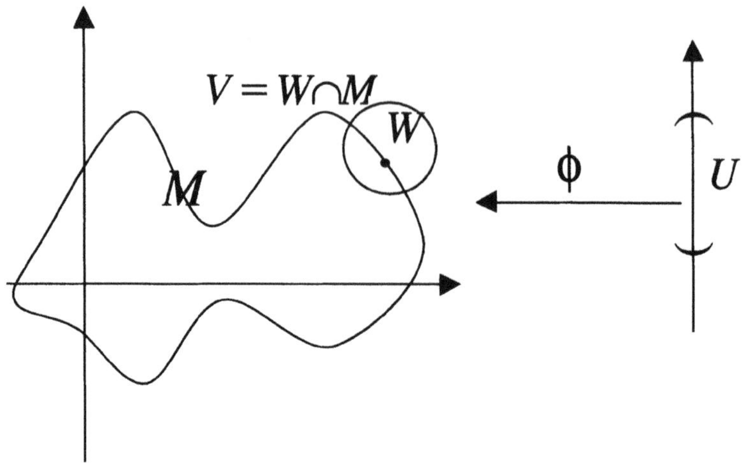

Figure 2.1. Local parametrization

looked at a function with values in $\mathbb{R}^l$. Concerning the chart (ψ, V, U), ψ is C^r since it has a local extension $\Psi_x : W_x \to \mathbb{R}^m$ at each $x \in V$, where W_x is a suitable neighborhood of x in $\mathbb{R}^l$.

Definition 17 Given a C^r manifold M and a subset N of M, if N is a C^r manifold, it is called a C^r **submanifold** of M. If M has dimension m and N has dimension n, $m - n$ is the **codimension** of N in M, denoted as codim N.

Clearly, any manifold M of $\mathbb{R}^l$ is a submanifold of $\mathbb{R}^l$.

Proposition 18 *Consider a C^r m-dimensional manifold M. An open subset S of M is a C^r m-dimensional submanifold of M.*

Proof. Let $x \in S$ and (ϕ, U, V) be a local parametrization of M around x. Since ϕ is a homeomorphism and $S \cap V$ is open in V, $U' \equiv \phi^{-1}(S \cap V)$ is an open subset of $\mathbb{R}^m$. It is immediate to see that $(\phi_{|U'}, U', S \cap V)$ is a local C^r parametrization of S at x. ■

The above proof can be adapted to show that if N is an open subset of M, and hence a submanifold, each atlas of M induces an atlas on N. More precisely, the atlas $\{(\psi_\lambda, V_\lambda, U_\lambda)\}_{\lambda \in \Lambda}$ induces the atlas

$$\{((\psi_\lambda)_{|V_\lambda \cap N}, V_\lambda \cap N, \psi_\lambda (V_\lambda \cap N))\}_{\lambda \in \Lambda'}$$

where Λ' is the set of those λ such that $V_\lambda \cap N \neq \emptyset$. On the other hand, this argument does not apply if N is a submanifold of M, which is not open in M.

Exercise 19 Let N be a submanifold of M. Show that $\dim N \leq \dim M$. In particular $\dim N = \dim M$ if and only if N is open in M.

Example 20 (The unit circle in $\mathbb{R}^2$) The unit circle in $\mathbb{R}^2$,

$$S^1 \equiv \left\{(x,y) \in \mathbb{R}^2 : x^2 + y^2 = 1\right\}$$

is a one-dimensional C^∞ manifold. Let us construct a system of four local parametrizations of S^1.

Consider $V_1 \equiv \left\{(x,y) \in S^1 : y > 0\right\}$ and $U \equiv (-1,1)$. V_1 is the upper semicircle, which is open in S^1. Take

$$\phi_1 : U \to V_1, \quad t \mapsto \left(t, \sqrt{1-t^2}\right)$$

ϕ_1 is C^∞ and bijective. In addition $\psi_1 \equiv \phi_1^{-1}$ is C^∞ as well since it is the restriction to V_1 of the projection

$$\pi_1 : \mathbb{R}^2 \to \mathbb{R}, \quad (x,y) \mapsto x$$

Then (ϕ_1, U, V_1) is a local parametrization around any point of V_1. The other three local parametrizations are analogous. Precisely we define (ϕ_2, U, V_2), (ϕ_3, U, V_3) and (ϕ_4, U, V_4) as follows:

$$\begin{array}{lll}
V_2 \equiv \left\{(x,y) \in S^1 : y < 0\right\} & \phi_2 : U \to V_2, & t \mapsto \left(t, -\sqrt{1-t^2}\right) \\
V_3 \equiv \left\{(x,y) \in S^1 : x > 0\right\} & \phi_3 : U \to V_3, & t \mapsto \left(\sqrt{1-t^2}, t\right) \\
V_4 \equiv \left\{(x,y) \in S^1 : x < 0\right\} & \phi_4 : U \to V_4, & t \mapsto \left(-\sqrt{1-t^2}, t\right)
\end{array}$$

Exercise 21 Check that (ϕ_2, U, V_2), (ϕ_3, U, V_3) and (ϕ_4, U, V_4) are actually three local parametrizations of S^1.

Example 22 (The unit circle in $\mathbb{R}^2$, continued) It is possible to obtain a system of two local parametrizations of S^1. This can be done e.g. by the use of polar coordinates.

Call P and Q respectively the two points of S^1 having coordinates $(1,0)$ and $(-1,0)$, and consider the two open subsets S^1, $V_1 \equiv S^1 \backslash \{Q\}$ and $V_2 \equiv S^1 \backslash \{P\}$. Then consider the intervals $U_1 \equiv (-\pi, \pi)$, $U_2 \equiv (0, 2\pi)$ and the functions

$$\begin{array}{ll}
\phi_1 : U_1 \to V_1, & t \mapsto (\cos t, \sin t) \\
\phi_2 : U_2 \to V_2, & t \mapsto (\cos t, \sin t)
\end{array}$$

$\{(\phi_1, U_1, V_1), (\phi_2, U_2, V_2)\}$ is a system of local parametrizations for S^1. It is clear that ϕ_1 and ϕ_2 are C^∞ and bijective functions. It remains to be seen that the inverses of ϕ_1 and ϕ_2 are C^∞; we show this for ϕ_1^{-1}; the check for ϕ_2^{-1} is completely analogous and is left as an exercise. The inverse of ϕ_1 is

$$\psi_1 : V_1 \to U_1, \quad (x,y) \mapsto \begin{cases} -\arccos x & \text{if } y<0 \\ 0 & \text{if } (x,y)=(1,0) \\ \arccos x & \text{if } y>0 \end{cases}$$

A global extension of ψ_1 is the function

$$\Psi_1 : \mathbb{R}^2 \backslash X_- \to (-\pi,\pi), \quad (x,y) \mapsto \text{sign } y \arccos \frac{x}{\sqrt{x^2+y^2}}$$

where $X_- \equiv (-\infty, 0] \times \{0\}$ is the nonpositive horizontal semiaxis in $\mathbb{R}^2$. Notice that Ψ_1 is positively homogeneous of degree zero, and continuous since the functions arccos and $(x,y) \mapsto \frac{x}{\sqrt{x^2+y^2}}$ are continuous on their domains, and the product sign y arccos $\frac{x}{\sqrt{x^2+y^2}}$ is equal to zero on $\mathbb{R}_{++} \times \{0\}$, even if the function $(x,y) \mapsto$ sign y is discontinuous there. After some computation, it is seen that Ψ_1 has the following first partial derivatives:

$$\frac{\partial \Psi_1}{\partial x}(x,y) = -\frac{y}{x^2+y^2} \qquad \frac{\partial \Psi_1}{\partial y}(x,y) = \frac{x}{x^2+y^2}$$

which are continuous on $\mathbb{R}^2 \backslash X$; moreover, a glance at the partial derivatives reveals that Ψ_1 has continuous partial derivatives of any order on its domain.

Exercise 23 Prove that no atlas of S^1 consists of only one chart. (Hint: show that S^1 is not homeomorphic to an open interval of $\mathbb{R}$).

Remark 24 Exercise 23 is probably the simplest example showing that manifolds are not, in general, globally diffeomorphic to some euclidean space.

Proposition 25 *The dimension of a manifold is uniquely determined, that is, if M is an m-dimensional and an n-dimensional manifold, then $m = n$.*

Proof. Let $x \in M$ and let $(\phi_1, U_1, V_1), (\phi_2, U_2, V_2)$ be two local parametrizations of M around x, with U_1 and U_2 open subsets of $\mathbb{R}^m$ and $\mathbb{R}^n$, respectively. By Remark 16, we have that

$$\phi_2^{-1} \circ \phi_1 : \phi_1^{-1}(V_1 \cap V_2) \to \phi_2^{-1}(V_1 \cap V_2)$$

is a C^r diffeomorphism for some r. Observe that $\phi_1^{-1}(V_1 \cap V_2)$ and $\phi_2^{-1}(V_1 \cap V_2)$ are open subsets of $\mathbb{R}^m$ and $\mathbb{R}^n$, respectively. Given any point $z \in \phi_1^{-1}(V_1 \cap V_2)$, the differential of $\phi_2^{-1} \circ \phi_1$ at z is a linear isomorphism between $\mathbb{R}^m$ and $\mathbb{R}^n$. It follows that $m = n$. ■

Proposition 26 *Let* X *and* Y *be two* C^r *manifolds and* $f : X \to Y$ *be a* C^r *function. Then*

$$\text{graph}\, f \equiv \{(x, y) \in X \times Y : x \in X \text{ and } y = f(x)\}$$

is a submanifold of $X \times Y$ *and it is* C^r *diffeomorphic to* X.

Proof. Define

$$F : X \to \text{graph}\, h,\ x \mapsto (x, f(x))$$

F is clearly one-to-one, onto and C^r. Moreover, $F^{-1} = \pi_{1_{|\text{graph}\, h}}$ is C^∞ (see Exercise 8). ■

We now present two examples of sets which are not manifolds.

Example 27 (The interval $[0, 1)$ **is not a manifold)** Since $(0, 1)$ is diffeomorphic to $\mathbb{R}$, $[0, 1)$ is locally diffeomorphic to $\mathbb{R}$ at each $x \neq 0$. Therefore, if $[0, 1)$ is a manifold, its dimension is one. Let us now show that this leads us to a contradiction, that is, $[0, 1)$ cannot be an one-dimensional manifold. This is done essentially by a topological argument.

Assume that there is a chart (ψ, V, U) of $[0, 1)$, with $0 \in V$. Without loss of generality, suppose that V is an interval of the form $[0, \alpha)$, with $0 < \alpha \leq 1$. As ψ is a diffeomorphism, U must be connected (see [90], p. 149) and, therefore, it is an open interval (a, b) in $\mathbb{R}$. In addition, $\psi_{|(0,\alpha)} : (0, \alpha) \to (a, b) \backslash \{\psi(0)\}$ is a diffeomorphism; but this is impossible since $(0, \alpha)$ is connected whereas $(a, b) \backslash \{\psi(0)\}$ is not.

Example 28 Another interesting example of a set which is not a manifold is given by the graph of the absolute value function, that is, the set $G \equiv \{(x, y) \in \mathbb{R}^2 : y = |x|\}$. Consider

$$G_- \equiv \{(x, y) \in G : x < 0\}$$
$$G_+ \equiv \{(x, y) \in G : x > 0\}$$

G_- and G_+ are both diffeomorphic to an open interval of $\mathbb{R}$. Therefore, if G is a manifold, it has necessarily dimension one. Assume that G is locally diffeomorphic to $\mathbb{R}$ around $(0, 0)$. Then there exists a local parametrization (ϕ, U, V) around $(0, 0)$. Without loss of generality assume that U is a neighborhood of 0 and that $\phi(0) = (0, 0)$. Denote $\phi = (\phi_1, \phi_2)$. We distinguish the two cases, $\phi_1'(0) \neq 0$ and $\phi_1'(0) = 0$.

Case a. Without loss of generality we can assume that $\phi_1'(0) > 0$. Then there exists $\delta \in \mathbb{R}_{++}$ such that $(-\delta, \delta) \subseteq U$ and ϕ_1 is strictly increasing on $(-\delta, \delta)$. Hence, using the definition of absolute value, we have that

$$h > 0 \Rightarrow \phi_1(h) > 0 \Rightarrow \phi_2(h) = \phi_1(h)$$
$$h < 0 \Rightarrow \phi_1(h) < 0 \Rightarrow \phi_2(h) = -\phi_1(h)$$

Then,

$$\lim_{h \to 0+} \frac{\phi_2(h) - \phi_2(0)}{h} = \lim_{h \to 0+} \frac{\phi_1(h) - 0}{h} = \phi_1'(0) > 0$$
$$\lim_{h \to 0^-} \frac{\phi_2(h) - \phi_2(0)}{h} = \lim_{h \to 0^-} \frac{-\phi_1(h) - 0}{h} = -\phi_1'(0) < 0$$

contradicting the fact that ϕ_2 is differentiable in 0.

Case b. Identical argument as in Case a shows that $\phi_2'(0) = 0$. If Ψ is any C^∞ extension of ϕ^{-1} to an open set of $\mathbb{R}^2$ containing V, we have that $\Psi_{|V} = \phi^{-1}$, $\Psi \circ \phi = id_U$. Then

$$(\Psi \circ \phi)'(0) = 1 \tag{2.2}$$

However

$$(\Psi \circ \phi)'(0) = \frac{\partial \Psi}{\partial x}(0) \cdot \phi_1'(0) + \frac{\partial \Psi}{\partial y}(0) \cdot \phi_2'(0) = 0$$

which contradicts (2.2).

Using the above example we can exhibit a manifold which is C^1, but not C^2.

Example 29 (A C^1 manifold which is not a C^2 manifold) Consider the following subset of $\mathbb{R}^2$

$$M \equiv \left\{(x, y) \in \mathbb{R}^2 : y = \text{sign } x \cdot x^2\right\}$$

In other words M is the graph of

$$f : \mathbb{R} \to \mathbb{R}, \quad x \mapsto \text{sign } x \cdot x^2$$

We prove that

1. M is a C^1 one-dimensional manifold;
2. M is not a C^2 manifold.

Proof of 1. M is (globally) C^1 diffeomorphic to $\mathbb{R}$. Indeed,

$$\phi : \mathbb{R} \to M, \quad t \mapsto (t, \text{sign } t \cdot t^2)$$

is C^1 and bijective; its inverse, the restriction to M of the projection $\pi_1 : \mathbb{R}^2 \to \mathbb{R}$, is C^∞. Therefore, M admits a system of local parametrizations consisting of the unique local parametrization $(\phi, \mathbb{R}, M)$.

Proof of 2. Suppose that M is an one-dimensional C^2 manifold and let (ϕ, U, V) be a local C^2 parametrization around $(0,0)$. Let $t_0 \equiv \phi^{-1}(0,0)$ and $\Psi : N \to U$ a local extension of ϕ^{-1} defined in an open neighborhood N of $(0,0)$ in $\mathbb{R}^2$. Denote $\phi = (\phi_1, \phi_2)$. There are two cases:

a) $\phi_1'(t_0) \neq 0$. By the inverse function theorem, ϕ_1 is invertible in a neighborhood of t_0 and its inverse ϕ_1^{-1} is C^2 in a neighborhood W of 0. Therefore, $\phi_2 \circ \left(\phi_1^{-1}\right)_{|W}$ is C^2. However, $\phi_2 \circ \left(\phi_1^{-1}\right)_{|W} = f_{|W}$ which is not C^2 at 0.

b) $\phi_1'(t_0) = 0$. Since $\phi_2(t) = f(\phi_1(t))$ and ϕ_1 is one-to-one, it follows

$$\phi_2'(0) = \lim_{t \to t_0} \frac{\phi_2(t) - \phi_2(t_0)}{t - t_0} =$$

$$\lim_{t \to t_0} \left(\frac{f(\phi_1(t)) - f(\phi_1(t_0))}{\phi_1(t) - \phi_1(t_0)} \cdot \frac{\phi_1(t) - \phi_1(t_0)}{t - t_0} \right) = f'(0) \cdot \phi_1'(0) = 0$$

Hence, arguing as in Case b of Example 29, ϕ cannot be a diffeomorphism.

Exercise 30 Let $f : \mathbb{R} \to \mathbb{R}$ be a continuous function and x_0 be a real number such that

1 f is C^1 in $\mathbb{R} \backslash \{x_0\}$;

2 f admits no derivative at x_0;

3 $\dfrac{f(x) - f(x_0)}{x - x_0}$ is bounded in a neighborhood of x_0.

Prove that the graph of f is not a manifold.

The following exercise shows the existence of functions which are not C^∞ and with a C^∞ manifold as graph.

Exercise 31 Let

$$f : \mathbb{R} \to \mathbb{R}, \quad x \mapsto \sqrt[3]{x}$$

Prove that the graph of f is a C^∞ manifold.

Exercise 32 Prove that the set $G \equiv \left\{(x,y) \in \mathbb{R}^2 : y^2\left(x^2 + y^2\right) = y^2 - x^2\right\}$, i.e. "the figure 8", is not a manifold. (Hint: proceed as in Example 27, that is, at first prove that if G is a manifold, it has necessarily dimension one, then show, by topological arguments, that it is not a one-dimensional manifold.)

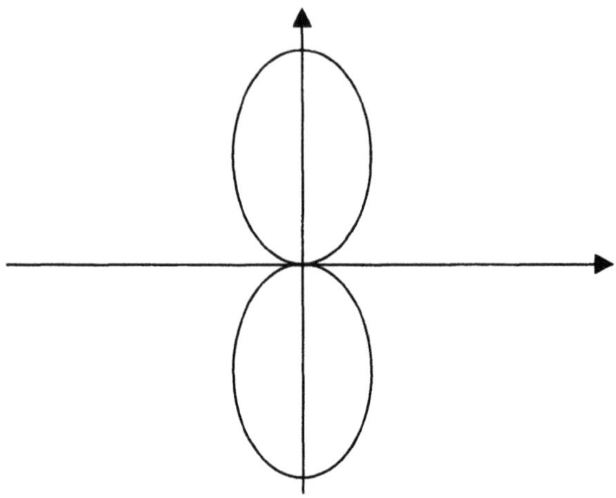

Figure 2.2. A set which is not a manifold

Example 33 Another example of a subset of $\mathbb{R}^2$ which is not a manifold is

$$G \equiv ((-\infty, 0] \times \{-1\}) \cup S^1$$

To prove that G is not a manifold we can follow the topological argument of Example 27. From a different point of view it is interesting to observe that G is the image of

$$\phi : (-\infty, 2\pi) \to \mathbb{R}^2, \quad t \mapsto \begin{cases} (t, -1) & \text{if } t \leq 0 \\ (\cos(t - \pi/2), \sin(t - \pi/2)) & \text{if } t \in (0, 2\pi) \end{cases}$$

The function ϕ is C^1 and bijective, but it is not a diffeomorphism (the proof that ϕ^{-1} is not C^1 is left as an exercise).

Proposition 34 *A set M is a zero-dimensional manifold indexManifold!zero-dimensional if and only if it is a set of isolated points.*

Proof. 1) If $m = 0$, then $\mathbb{R}^m = \{0\}$. Therefore, if M is a zero-dimensional manifold, for each $x \in M$, there exists a neighborhood V of x in M which is diffeomorphic to $\{0\}$. Thus V must be a singleton and then x is an isolated point in M. In addition, as any function between two singletons is C^∞, we have that a zero-dimensional manifold is C^∞.

2) If M is a set of isolated points, given $x \in M$, $\{x\}$ is an open neighborhood of x in M which is clearly C^∞ diffeomorphic to $\{0\} = \mathbb{R}^0$. Then M is a zero-dimensional C^∞ manifold. ■

Example 35 By the previous proposition it is easy to verify that

i) the set $\{1/n\}_{n\in\mathbb{N}}$ is a zero-dimensional manifold,

ii) the set $\{1/n\}_{n\in\mathbb{N}} \cup \{0\}$ is a not a zero-dimensional manifold. More generally, it is not a manifold.

Example 36 (The two-dimensional torus) Let $T^2 \equiv S^1 \times S^1$. Recalling the definition of S^1 (see Example 20), T^2 is a manifold of $\mathbb{R}^2 \times \mathbb{R}^2$ (see also Proposition 40 below). By the identification of $\mathbb{R}^2 \times \mathbb{R}^2$ with $\mathbb{R}^4$, we have

$$T^2 = \left\{(u, v, w, t) \in \mathbb{R}^4 : u^2 + v^2 = w^2 + t^2 = 1\right\}$$

An atlas for T^2 consisting of four charts is obtained as a "product atlas" by a suitable and straightforward modification of the definitions given in Example 22. Recalling the atlas $\{(\psi_\alpha, V_\alpha, U_\alpha)\}_{\alpha\in\{1,2\}}$ in that example, for each couple (α, β) in $\{1, 2\} \times \{1, 2\}$ let

$$V_{\alpha,\beta} \equiv V_\alpha \times V_\beta \quad U_{\alpha,\beta} \equiv U_\alpha \times U_\beta \tag{2.3}$$

$$\psi_{\alpha,\beta} : (u, v, w, t) \mapsto \begin{pmatrix} \alpha' + (\arccos u - \alpha') \operatorname{sign} v \\ \beta' + (\arccos w - \beta') \operatorname{sign} t \end{pmatrix} \tag{2.4}$$

with $\alpha' \equiv (\alpha - 1)\,\pi$, $\beta' \equiv (\beta - 1)\,\pi$. A traditional and perhaps more intuitive description of the two-dimensional torus makes use of its representation as a subset of $\mathbb{R}^3$; more precisely, T^2 is described by means of its image under a C^∞ diffeomorphism. To this purpose, let R and r be real numbers such that $0 < r < R$, and consider the function

$$f : \mathbb{R}^4 \to \mathbb{R}^3, \quad \begin{pmatrix} u \\ v \\ w \\ t \end{pmatrix} \mapsto \begin{pmatrix} (R + rw)\, u \\ (R + rw)\, v \\ rt \end{pmatrix}$$

The reader is invited to check that f is C^∞ with a full rank jacobian matrix on each point of T^2, that its restriction to T^2 is a bijection of T^2 onto $\mathcal{T}^2 \equiv f\left(T^2\right)$, and that the inverse of this restriction

$$(x, y, z) \mapsto \left(\frac{x}{\sqrt{x^2 + y^2}}, \frac{y}{\sqrt{x^2 + y^2}}, \frac{\sqrt{x^2 + y^2} - R}{r}, \frac{z}{r}\right) \tag{2.5}$$

extends to $\mathbb{R}^3 \backslash Z$ as a C^∞ function, where Z denotes the Z-axis. In conclusion, $f_{|T^2} : T^2 \to \mathcal{T}^2$ is a C^∞ diffeomorphism. We shall refer to $\mathcal{T}^2$ as "the torus of $\mathbb{R}^3$" from now on. In order to better visualize the "doughnut" shape of $\mathcal{T}^2$, notice what follows:

1 the equation

$$\left(\sqrt{x^2+y^2}-R\right)^2+z^2=r^2 \tag{2.6}$$

completely characterizes T^2. Thus T^2 is the surface obtained by rotating around the Z-axis the circle C of the XZ-plane with center at $(R,0,0)$ and radius r.

2 For each $(u,v)\in S^1$, $f\left[\{(u,v)\}\times S^1\right]$ is the circle with center $(Ru, Rv, 0)$ and radius r, belonging to the vertical plane of equation

$$vx-uy=0$$

Each such circle corresponds to a position assumed by C in the rotation; it is called "meridian circle", or simply "meridian"; each vertical plane through the origin contains two (opposite) meridians. The centers of the meridians build up a second, horizontal circle $\widetilde{C}$; it has center at the origin and radius R. Thus T^2 can be looked at as providing a three dimensional visualization of the cartesian product $S^1\times S^1$ by means of the two circles C and $\widetilde{C}$, a "copy" of C being associated to each element of $\widetilde{C}$.

3 Similarly, for each $(w,t)\in S^1$, $f\left[S^1\times\{(w,t)\}\right]$ is the circle with center at $(0,0,t)$ and radius $R+rw$, belonging to the horizontal plane of equation

$$z=rt$$

Each such circle is called "parallel circle" or just "parallel"; there are two parallels at a given height greater than $-r$ and smaller than r, and just one parallel at height r or $-r$.

To see directly that T^2 is a two-dimensional manifold, it simply suffices to compose with f each element in the system of local parametrizations of T^2 implicit in the atlas (2.3)-(2.4): for each (α,β) in $\{1,2\}\times\{1,2\}$,

$$U_{\alpha,\beta}\equiv U_\alpha\times U_\beta$$

$$\phi_{\alpha,\beta}:\begin{pmatrix}\theta\\ \xi\end{pmatrix}\mapsto\begin{pmatrix}(R+r\cos\xi)\cos\theta\\ (R+r\cos\xi)\sin\theta\\ r\sin\xi\end{pmatrix}$$

$$F_{\alpha,\beta}\equiv\phi_{\alpha,\beta}\left(U_{\alpha,\beta}\right)$$

Exercise 37 Check that each $F_{\alpha,\beta}$ is indeed an open set in T^2.

We conclude the section with some basic properties of manifolds. We limit ourselves to a few results which will be useful throughout the book.

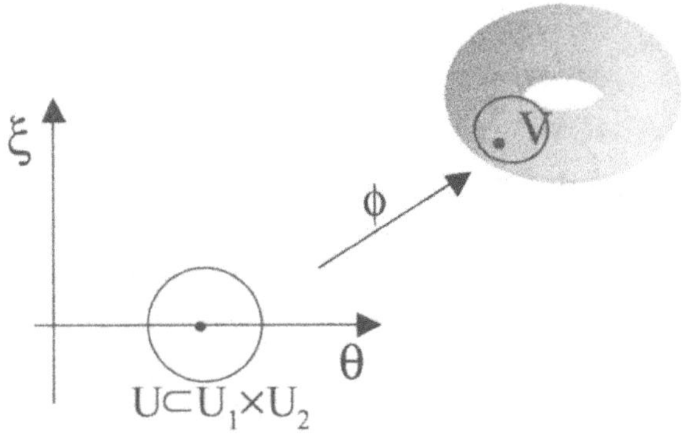

Figure 2.3. The torus in $\mathbb{R}^3$

Proposition 38 *A manifold is a locally compact topological space.*

Proof. Let M be an m-dimensional manifold. Given $x \in M$, consider a local parametrization (ϕ, U, V) of M around x. Let U' be an open neighborhood of $\phi^{-1}(x)$ in $\mathbb{R}^m$ with compact closure contained in U. By the continuity of ϕ, $\phi(\overline{U}')$ is a compact neighborhood of x in M. ■

According to a classical result in topology (see [90], p. 256), any subset of $\mathbb{R}^l$ is paracompact as topological subspace of $\mathbb{R}^l$. A straightforward consequence of this property is the following proposition.

Proposition 39 *Any manifold is a paracompact topological space.*

Let us prove another useful result.

Proposition 40 *If M and N are C^r manifolds, then*

1 *$M \times N$ is a C^r manifold,*

2 $\dim M \times N = \dim M + \dim N$.

Proof. Let $m \equiv \dim M$ and $n \equiv \dim N$. Let $(x, y) \in M \times N$. To prove that $M \times N$ is a C^r manifold, with dimension $m+n$, we need to show that there exists an open neighborhood of (x, y) in $M \times N$ which is C^r diffeomorphic to an open subset of $\mathbb{R}^{m+n}$.

Let (ϕ_1, U_1, V_1) and (ϕ_2, U_2, V_2) be, respectively, two local parametrizations of M around x and of N around y. $V_1 \times V_2$ is an open neighborhood of (x, y) in $M \times N$ and $U_1 \times U_2$ is an open subset of $\mathbb{R}^m \times \mathbb{R}^n$. It is straightforward to see that $\Phi \equiv \phi_1 \times \phi_2$ is a diffeomorphism. ■

Given two manifolds M and N and two corresponding atlases, the above proof provides a method to obtain an atlas of the product manifold $M \times N$. The reader can try it as an exercise on products of manifolds already presented in the previous examples.

2. Appendix - Abstract manifolds

In this appendix we briefly introduce the notion of abstract manifold. An abstract manifold is a topological space which is locally diffeomorphic to an euclidean space. The notion of (local) diffeomorphism between a topological space X and an euclidean space is not immediate, since X does not generally have a linear or metric structure. Hence, the definition of manifold (Definition 15) does not generally apply. However, as we will see below, in some particular cases a concept of local diffeomorphism between X and a suitable euclidean space can be defined. This extends the concept given in Definition 15 and it allows us to use the word "manifold" for these special topological spaces, while the term "abstract" specifies that we deal with objects which are not necessarily contained in euclidean spaces. In other word, a manifold which is a subset of an euclidean space can be seen as an abstract manifold, but the converse is not true.

Abstract manifolds are certainly more difficult for the reader's understanding, but, at the same time, they turn out to be crucial in solving a large number of problems from calculus. In addition, a lot of topics in differential topology, which are out of the scope of this book, can be fully developed only if the limitation to manifolds as submanifolds of euclidean spaces is dropped.

For our purposes, some problems arising in the second part of the book will be tackled using properties of grassmannian manifolds, which are abstract manifolds of a particular type, to be defined and reviewed there.

Definition 41 (Topological manifold) A topological space M is a said to be an **m-dimensional topological manifold** if there exists a collection of triples $(\phi_i, U_i, V_i)_{i\in I}$ such that

1 $\{V_i\}_{i\in I}$ is an open covering of M;

2 for each i, U_i is an open subset of $\mathbb{R}^m$;

3 $\phi_i : U_i \to V_i$ is a homeomorphism.

Every triple (ϕ_i, U_i, V_i) is called a **local parametrization** of M and the set of local parametrizations is called a **system of local parametrizations** . Conversely, (ψ_i, V_i, U_i), where $\psi_i \equiv \phi_i^{-1}$, is called a **chart** of M and $\{(\psi_i, V_i, U_i)\}_{i\in I}$ is called an **atlas** of M.

Definition 42 Let M be an m-dimensional topological manifold. Consider an atlas of M $\{(\psi_i, V_i, U_i)\}_{i \in I}$, and let $r \in \mathbb{N}^*$. M is said to be an **m-dimensional C^r manifold** if, for every $i, j \in I$ such that $V_i \cap V_j \neq \emptyset$,

$$\psi_j \circ \left(\psi_i^{-1}\right)_{|\psi_i(V_i \cap V_j)} : \psi_i(V_i \cap V_j) \to \psi_j(V_i \cap V_j)$$

is a C^r diffeomorphism.

The family $\{(\psi_i, V_i, U_i)\}_{i \in I}$ is called a C^r **atlas** of M and every triple (ψ_i, V_i, U_i) is a C^r **chart**. The corresponding parametrizations (ϕ_i, U_i, V_i) are called C^r **local parametrizations**.

We stress that the term manifold is used to denote two distinct notions expressed by different definitions. However, the two concepts are not completely independent. First of all, it is not difficult to see that a subset of an euclidean space satisfying Definition 15, that is, a submanifold of an euclidean space, satisfies Definition 42; this can be proved recalling Remark 16. Of course the converse is not true since a manifold according to Definition 42 is not necessarily a subset of some euclidean space.

By a classical theorem due to Whitney, which will be recalled in the next chapter (see Theorem 73), any abstract manifold, which is a paracompact Hausdorff topological space is diffeomorphic to a submanifold of an euclidean space. This requires a more general notion of diffeomorphism which will be introduced in the next chapter.

Example 43 (The projective space) Consider the two-dimensional unit sphere in $\mathbb{R}^3$,

$$S^2 \equiv \left\{(x, y, z) \in \mathbb{R}^3 : x^2 + y^2 + z^2 = 1\right\}$$

Then, induce on S^2 the equivalence relation "$\sim$" which identifies the antipodal points, that is, $(x, y, z) \sim (x', y', z')$ if $(x, y, z) = \pm(x', y', z')$.

The quotient set is called **two-dimensional real projective space**, and denoted by the symbol $\mathbb{P}^2(\mathbb{R})$. It is a topological space with the quotient topology (see [90], Section 11 in Chapter 2).

For any $(x, y, z) \in S^2$, we denote by $[(x, y, z)]$ the corresponding element of $\mathbb{P}^2(\mathbb{R})$, that is, $[(x, y, z)] \equiv \{(x, y, z), -(x, y, z)\}$. In addition, let $p : S^2 \to \mathbb{P}^2(\mathbb{R})$ be the projection associated with the equivalence relation.

We prove that $\mathbb{P}^2(\mathbb{R})$ is a C^∞ two-dimensional (abstract) manifold in terms of Definition 42 by constructing a system of three local parametrizations. Let

$$U \equiv \left\{(s, t) \in \mathbb{R}^2 : s^2 + t^2 < 1\right\}$$

and

$$
\begin{aligned}
&V_1 \equiv \{[(x,y,z)] \in \mathbb{P}^2(\mathbb{R}) : x \neq 0\} \\
&\phi_1 : U \to V_1, \ \ (s,t) \mapsto \left[\left(\sqrt{1-s^2-t^2}, s, t\right)\right] \\
&V_2 \equiv \{[(x,y,z)] \in \mathbb{P}^2(\mathbb{R}) : y \neq 0\} \\
&\phi_2 : U \to V_2, \ \ (s,t) \mapsto \left[\left(s, \sqrt{1-s^2-t^2}, t\right)\right] \\
&V_3 \equiv \{[(x,y,z)] \in \mathbb{P}^2(\mathbb{R}) : z \neq 0\} \\
&\phi_3 : U \to V_3, \ \ (s,t) \mapsto \left[\left(s, t, \sqrt{1-s^2-t^2}\right)\right]
\end{aligned}
$$

Observe that the $V_i's$ are open in $\mathbb{P}^2(\mathbb{R})$ by definition of quotient topology, since they are the images by p of open hemispheres in $\mathbb{R}^3$; for example V_3 is obtained by the projection on $\mathbb{P}^2(\mathbb{R})$ of the set $\left\{(s,t,\sqrt{1-s^2-t^2}) : (s,t) \in U\right\}$, which is the open upper hemisphere of S^2. In addition, $V_1 \cup V_2 \cup V_3 = \mathbb{P}^2(\mathbb{R})$. Consider now $V_1 \cap V_2$. We have

$$U_{1,2} \equiv \phi_1^{-1}(V_1 \cap V_2) = \phi_2^{-1}(V_1 \cap V_2) = \{(s,t) \in U : s > 0\}$$

and

$$\phi_2^{-1} \circ \phi_1 : U_{1,2} \to U_{1,2}, \quad (s,t) \mapsto \left(\sqrt{1-s^2-t^2}, t\right)$$

which is clearly C^∞. The cases for $V_1 \cap V_3$ and $V_2 \cap V_3$ are completely analogous.

Chapter 3

DIFFERENTIALS

1. Introduction

This chapter is devoted to the introduction of a few basic tools making the reader acquainted, we hope, with the development of elementary calculus on manifolds. As we argued in the introduction of Chapter 2, one of the main objects of elementary calculus is the representation of the local behavior of nonlinear functions mapping open subsets of $\mathbb{R}^m$ into $\mathbb{R}^n$ by means of their linear and higher order approximations. Similarly, one of the main purposes of elementary calculus on manifolds is the development of appropriate objects allowing the construction of linear and higher order approximations for the local behavior of functions mapping manifolds into manifolds.

Consider a C^r function $f : U \to \mathbb{R}^n$, with U an open subset of $\mathbb{R}^m$ and $r \in \mathbb{N}^*$, and an element x of U. The first order approximation of f at x is the differential df_x. The differential being a linear function, its domain and value set are linear spaces; indeed, $df_x : \mathbb{R}^m \to \mathbb{R}^n$. Higher order approximations involve higher order differentials. On the other hand, let $f : M \to N$ be a function mapping a C^r manifold M of $\mathbb{R}^l$ into a C^r manifold N of $\mathbb{R}^p$, of dimensions m and n respectively, and let x be an element of M. Going back to Section 1 of Chapter 2, to say that f is a C^r function at x has already a definite meaning, by resort (Definition 1) to some local C^r extension F of f at x (if existing). But how could the differential df_x of f at x be defined?

Even to begin talking about the differential of f at x as a linear function, one has to clarify what the domain and value set of such an object should be. Certainly the two manifolds M and N cannot play that role, since quite generally they are not linear subspaces of $\mathbb{R}^l$ and $\mathbb{R}^p$ (there would be no need to study manifolds if they were). One may consider resorting to dF_x, the differential of the local extension; this is a function from $\mathbb{R}^l$ to $\mathbb{R}^p$. To define df_x plainly as dF_x

would be meaningless, since there is quite an amount of arbitrariness in locally extending f at x into a C^r function defined on a full open neighborhood of x in $\mathbb{R}^l$ with values in $\mathbb{R}^p$; and we have already observed (Example 3 of Chapter 2) that different local extensions of f at x may have different differentials at x. In particular, the differential of any chart of an m-manifold of $\mathbb{R}^l$ is not defined as yet if $m < l$ (as opposed to the corresponding local parametrization); this is going to be overcome in Proposition 32.

It will turn out that a consistent definition of df_x, even if it might not look like that at first sight, can be seen as obtained by restricting the differential dF_x of any local extension F of f at x to appropriate linear subspaces of $\mathbb{R}^l$ and $\mathbb{R}^p$. These subspaces are called *tangent spaces* of M and N (*at* x and *at* $f(x)$ respectively); they provide a translated image of the first order approximation of M at x and of N at $f(x)$. Section 2 introduces the concept of tangent space to a manifold at one of its points, and proceeds to the derived concepts of tangent bundle and normal bundle.

The definition of differential for functions between manifolds is given in Section 3. For any $x \in M$, the definition uses two arbitrarily chosen charts (ψ, V, U) (at x) and (ξ, T, S) (at $f(x)$) in the atlases of the two manifolds, and builds upon the differential dg_u of $g \equiv \xi \circ f \circ \psi^{-1}$ at $u = \psi(x)$ (if existing). Such a function g is called a local representation of f at x, because it is an ordinary function between open subsets of $\mathbb{R}^m$ and $\mathbb{R}^n$, which keeps track of the behavior of f in a neighborhood of x (in M) in terms of local coordinates. It is then argued that the definition is consistent, in that the object defined does not depend on the choice of ψ and ξ. In addition, some simple properties of differentials are established.

As it happens in ordinary calculus, the differential of a function at a point provides important information concerning the behavior of that function in some appropriate neighborhood of that point. This is the object of Section 4. The main, basic classification enucleates functions for which the differential as a linear map is an isomorphism, or just injective, or just surjective. For the first class of functions we state and prove an inverse function theorem for manifolds. The functions in the other two classes are called immersive, resp. submersive. Two important theorems characterize immersive and submersive functions as locally representable by means of canonical immersions and projections. We conclude the chapter giving sufficient conditions under which the image of a manifold via a C^r function is in fact a manifold itself. That entails defining the concept of an embedding which is a particular case of an immersive function, and allows us to give the statement of the celebrated Whitney's embedding theorem .

2. Tangent spaces

Definition 1 Let M be a manifold. A (continuously differentiable) **curve** in M is a C^1 function $\gamma : I \to M$ where $I = (a, b)$ is an open real interval. The curve γ is **through** $x \in M$ **at** $t \in I$ if $x = \gamma(t)$.

According to the above definition, a curve γ in a manifold of $\mathbb{R}^l$ is just a curve in $\mathbb{R}^l$ for which $\operatorname{Im}\gamma$ is contained in the manifold. Notice that we do not assume regularity, a common hypothesis requiring $d\gamma_t \neq 0$ for each $t \in I$. Recall that, if γ is a curve in $\mathbb{R}^l$ and $d\gamma_t \neq 0$, γ has a tangent line L_t at the point $\gamma(t)$, namely, the one-dimensional affine subspace of $\mathbb{R}^l$

$$L_t \equiv \left\{ z \in \mathbb{R}^l : \exists s \in \mathbb{R},\ z = \gamma(t) + s\gamma'(t) \right\}$$

where $\gamma'(t) \equiv d\gamma_t(1)$.

Definition 2 Let M be a manifold of $\mathbb{R}^l$, and let x be an element of M. An element v of $\mathbb{R}^l$ is a **tangent vector** to M at x if there exists a curve γ through x at t such that $\gamma'(t) = v$. The **tangent space** to M at x is the set T_xM of all tangent vectors to M at x.

Notice that for every $x \in M$ the null vector (of $\mathbb{R}^l$) belongs to T_xM, because any constant function γ with $\operatorname{Im}\gamma = \{x\}$ is a curve in M through x at each t, according to Definition 1. We now prove that the tangent space has indeed a vector subspace structure.

Proposition 3 *Let M be an m-dimensional manifold of $\mathbb{R}^l$, and let x be an element of M. For any local parametrization (ϕ, U, V) of M around x such that $x = \phi(u)$,*

$$T_xM = \operatorname{Im} d\phi_u$$

In particular, T_xM is an m-dimensional subspace of $\mathbb{R}^l$, and $d\phi_u : \mathbb{R}^m \to T_xM$ is a linear isomorphism.

Remark 4 It must be pointed out right away that to write $d\phi_u : \mathbb{R}^m \to T_xM$, as we have done, is a slight abuse of notation. Strictly speaking, the differential at u of $\phi : U \to V \subseteq \mathbb{R}^l$ is a linear function $d\phi_u : \mathbb{R}^m \to \mathbb{R}^l$, and rather generally it is not invertible because not surjective (unless $m = l$). However, we are going to need the *restriction* $d\phi_{u|\mathbb{R}^m, T_xM} : \mathbb{R}^m \to T_xM$ (e.g., in defining differentials), and we have chosen to save notational weight. Thus from now on $d\phi_u$ shall denote $d\phi_{u|\mathbb{R}^m, T_xM}$; in this sense, $d\phi_u$ is a linear isomorphism and $(d\phi_u)^{-1}$ is well defined.

Proof of Proposition 3. Let first v be an element of $\operatorname{Im} d\phi_u$, with $v = d\phi_u(w)$. Take $\delta > 0$ such that the segment $(u - \delta w, u + \delta w)$ is included in

U. The function $\gamma : (-\delta, \delta) \to M,\ t \mapsto \phi(u + tw)$ is a curve in M through x at 0, since ϕ is a C^r function with $r \in \mathbb{N}^*$. Moreover, $\gamma'(0) = v$ by the chain rule ([3], Theorem 12.7, p. 352).

Conversely, let v be an element of $T_x M$, that is, let $x = \gamma(t)$ and $v = \gamma'(t)$ for some curve γ in M. Let (ψ, V, U) the chart associated with (ϕ, U, V), and let $\Psi : W \to \mathbb{R}^m$ be a local extension of ψ at x, defined on open neighborhood W of x in $\mathbb{R}^l$. Without loss of generality[1] we shall assume that $W \cap M = V$. Ψ is differentiable at x, and $d\Psi_x$ is a linear function of $\mathbb{R}^l$ into $\mathbb{R}^m$. If $w \equiv d\Psi_x(v)$, $d\phi_u(w)$ is well defined. Consider $\widetilde{\gamma} \equiv \Psi \circ \gamma$, a curve in U. Then

$$\widetilde{\gamma}'(t) = d\Psi_x(\gamma'(t)) = w$$
$$\phi \circ \widetilde{\gamma} = \gamma$$
$$v = (\phi \circ \widetilde{\gamma})'(t) = d\phi_u(w)$$

i.e., v is an element of $\operatorname{Im} d\phi_u$.

The last two statements are immediate, because $d\phi_u$ is linear, and $\Psi \circ \phi = id_U$ implies (by the chain rule) $d\Psi_x \circ d\phi_u = d(\Psi \circ \phi)_u = id_{\mathbb{R}^m}$; hence $\operatorname{Im} d\phi_u$ cannot be less than m (rank of a product, Proposition 6, Chapter 1), nor more (the domain of $d\phi_u$ is $\mathbb{R}^m$, Lemma 4 there). ■

One step of the above proof deserves autonomous consideration; we state it as a corollary.

Corollary 5 *According to the notational change discussed in Remark 4, for any local extension Ψ at x of the chart (ψ, V, U) associated with (ϕ, U, V)*

$$(d\phi_u)^{-1} = (d\Psi_x)_{|T_x M} \tag{3.1}$$

The statement of the corollary may be seen as the differential restriction placed on an arbitrary function from $\mathbb{R}^l$ to $\mathbb{R}^m$ by the requirement that it be a local extension of the chart ψ. More geometrically, we may say that the jacobian matrix of a local extension of ψ is arbitrary, in as much as it yields some prescribed directional derivatives along directions which are tangent to M. If $v = d\phi_u(w) \in T_x M$, the directional derivative of Ψ along v at x, $D\Psi_x(v)$, must be equal to w by (3.1). In algebraic terms, if X is the $m \times l$ ("low") jacobian matrix $D\Psi(x)$, and A is the $l \times m$ ("high") jacobian matrix $D\phi(u)$, X has to be a solution of the matrix equation

$$XA = I_m$$

Remark 6 The first order approximation of M at x is the set $\{x\} + T_x M$, an m-dimensional affine subspace of $\mathbb{R}^l$. This generalizes the concepts of tangent line for regular curves in $\mathbb{R}^l$ and tangent plane for regular surfaces in $\mathbb{R}^3$.

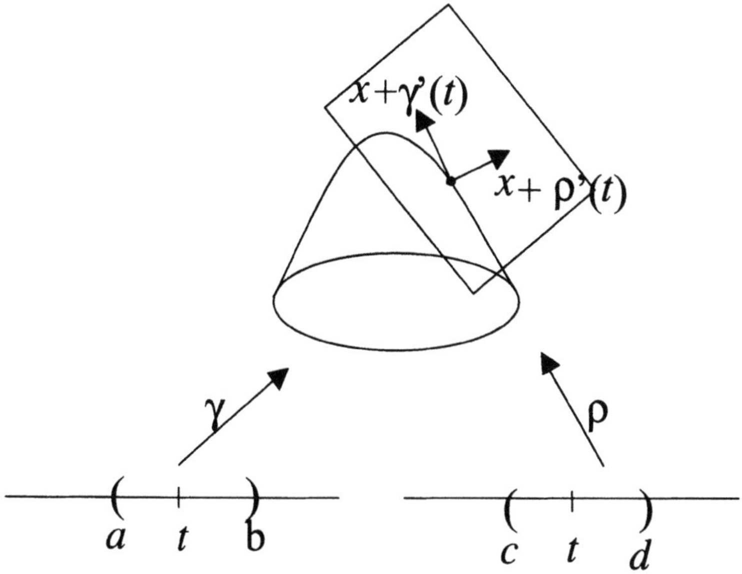

Figure 3.1. The (translated) tangent space

Exercise 7 If M is an open subset of $\mathbb{R}^l$, then $T_xM = \mathbb{R}^l$ for each $x \in M$.

Example 8 If S is an open subset of a manifold M, then for any $x \in S$, $T_xS = T_xM$. Indeed, arguing as in the proof of Proposition 18, Chapter 2, and in the same notation, $d\phi_u = d\left(\phi_{|U'}\right)_u$.

Proposition 9 *If M is a submanifold of N and $x \in M$, then $T_xM \subseteq T_xN$.*

Proof. This follows directly from Definition 2, since any curve in M through x is a curve in N as well. ■

Example 10 If M is a one-dimensional manifold, any local parametrization (ϕ, U, V) around $x \in M$, with U connected, defines a curve in M through x (just forget V). Thus T_xM is the linear subspace generated by $\phi'(\psi(x))$.

Exercise 11 Let X be a point of the circle S^1 (Examples 20 and 22 of Chapter 2). Then T_XS^1 is the line through the origin O having the radius OX as normal unit vector.

[1] It suffices to restrict F if $V \subsetneq W \cap M$, and to extend it if $W \cap M \subsetneq V$

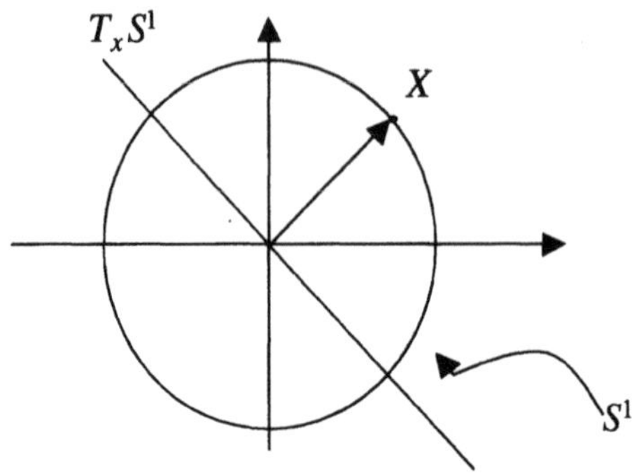

Figure 3.2. The tangent space is orthogonal to the radius

Proposition 12 *Let M and Q be two manifolds, of dimension m and q respectively. For each $(x, z) \in M \times Q$,*

$$T_{(x,z)}(M \times Q) = T_x M \times T_z Q$$

up to the canonical linear isomorphism of $\mathbb{R}^m \times \mathbb{R}^q$ with $\mathbb{R}^{m+q}$.

Proof. In proving that $M \times Q$ is a manifold of dimension $m+q$ (Proposition 40, Chapter 2), it was seen that, up to the canonical linear isomorphism of $\mathbb{R}^m \times \mathbb{R}^q$ with $\mathbb{R}^{m+q}$, a local parametrization of $M \times Q$ around (x, z) is

$$(\phi \times \alpha, U \times A, V \times B)$$

whenever (ϕ, U, V) and (α, A, B) are two local parametrizations of M and Q around x and z, respectively. If, say, $x = \phi(u)$ and $z = \alpha(a)$, the claim follows from Proposition 3 and from the fact that for any two differentiable functions (Proposition 2, Chapter 1)

$$d(\phi \times \alpha)_{(u,a)} = d\phi_u \times d\alpha_a$$

■

Exercise 13 Let P be a point of the torus T^2 ((Example 36 of Chapter 2). Describe the tangent space to T^2 at P.

Exercise 14 Let T^2 be the torus of $\mathbb{R}^3$. Show that meridian and parallel circles are one-dimensional submanifolds of T^2. Check that $T_{(x,y,z)}T^2 = T_{(x,y,z)}M + T_{(x,y,z)}P$, where M and P are the meridian and parallel circles through (x, y, z).

Up to now, a specific tangent space has been associated with each element of the manifold M. The corresponding global object associated with M as a whole is called tangent bundle. Tangent bundles are naturally endowed with a manifold structure, as we are about to see. Moreover, the concept of tangent bundle is going to be essential in next section, in order to extend the notion of differential df_x of a function $f : M \to N$ at a point x into a global notion of derivative df of that function. Finally, the concept of normal bundle shall be used extensively in Chapter 7.

Definition 15 Let M be a manifold. The **tangent bundle** of M, denoted by TM, is the set

$$TM \equiv \left\{(x, v) \in \mathbb{R}^l \times \mathbb{R}^l : x \in M \text{ and } v \in T_x M\right\}$$

Rephrasing the definition in a different notation,

$$TM = \bigcup_{x \in M} (\{x\} \times T_x M) \tag{3.2}$$

In order to prove the main result of our simple presentation of tangent bundles, we state two of their basic properties.

Lemma 16 *Let M be a manifold, and V be an open set of M. Then*

$$TV = TM \cap \left(V \times \mathbb{R}^l\right) \tag{3.3}$$

Proof. The claim follows directly from Example 8 and (3.2). ■

Exercise 17 Consider the canonical projection of the tangent bundle of a manifold on the manifold itself

$$p : TM \to M, \ (x, v) \mapsto x$$

and assume TM endowed with the topology induced by $\mathbb{R}^{2l}$. The projection p is continuous.

Hint. Look at Lemma 16.

Proposition 18 *Let M be an m-dimensional C^r manifold of $\mathbb{R}^l$. The tangent bundle TM is a $2m$-dimensional C^{r-1} manifold of $\mathbb{R}^{2l}$.*

Proof. Let (ϕ, U, V) be a local parametrization of M around $x = \phi(u)$, and let

$$d\phi : U \times \mathbb{R}^m \to TV \subseteq \mathbb{R}^{2l}, \ (w, z) \mapsto (\phi(w), d\phi_w(z))$$

Then $(d\phi, U \times \mathbb{R}^m, TV)$ is a local parametrization of TM around (x, v) for any $v \in T_x M$. Indeed,

a) $U \times \mathbb{R}^m$ is open in $\mathbb{R}^{2m}$,

b) TV is open in TM by the lemma above,

c) $d\phi$ is well defined, because $d\phi_w(z) \in T_{\phi(w)}M$ for each $w \in M$ and each $z \in \mathbb{R}^m$, and it is onto, because $\phi(U) = V$ and $\operatorname{Im} d\phi_w = T_{\phi(w)}M$ for each $w \in U$ (consider (3.2)),

d) $d\phi$ is one to one because ϕ is so, and $\{r\} \times T_r M$ is disjoint from $\{s\} \times T_s M$ if $r \neq s$,

e) $d\phi$ is C^{r-1} because it is C^{r-1} in w for each $z \in \mathbb{R}^m$ (ϕ being C^r), and it is linear in z for each $w \in U$,

f) $d\phi^{-1}$ is C^{r-1} because it is equal to $d\Psi_{|TV}$ for any C^r local extension Ψ of the chart (ψ, V, U) corresponding to (ϕ, U, V) (by Corollary 5).

■

Definition 19 The tangent bundle TM of an m-dimensional manifold M is **trivial** if there exists a homeomorphism $h : M \times \mathbb{R}^m \to TM$ such that, for every $x \in M$, the restriction $h_{|\{x\}\times\mathbb{R}^m}$ is a linear isomorphism of $\{x\} \times \mathbb{R}^m$ with $T_x M$.

Exercise 20 (Every tangent bundle is **locally trivial**) Let M be a manifold. Each element of M has a neighborhood V such that TV is trivial.

Hint. For any $x \in M$, let (ϕ, U, V) be a local parametrization of M around $x = \phi(u)$, with (ψ, V, U) the corresponding chart; define

$$h : V \times \mathbb{R}^m \to TV, \quad (v, z) \mapsto \left(v, d\phi_{\psi(v)}(z)\right)$$

and notice that, for each $(v, z) \in V \times \mathbb{R}^m$, $h(v, z) = d\phi(\psi(v), z)$, with $d\phi$ as in the proof of Proposition 18.

Remark 21 The reader may think that Proposition 12 (concerning tangent spaces of manifolds M, N, $M \times N$ at given points x, y, (x, y) respectively) directly extends into a general statement concerning tangent bundles, namely, the following:

$$T(M \times N) = TM \times TN \tag{3.4}$$

It must be observed, however, that such a statement makes sense *only* up to the following identification:

$$\forall x \in M, \ \forall y \in N, \ \forall v \in T_x M, \ \forall w \in T_y N$$

$$((x, y), (v, w)) \leftrightarrow ((x, v), (y, w))$$

In the sequel, we shall take the above identification for granted, and we will make use of (3.4) whenever convenient.

Definition 22 Let M be a manifold of $\mathbb{R}^l$, and let $x \in M$. The **normal space** to M at x, denoted $N_x M$, is the orthogonal complement of the tangent space to M at x

$$N_x M \equiv \left\{ n \in \mathbb{R}^l : \forall v \in T_x M,\ vn = 0 \right\}$$

Definition 23 Let M be a manifold of $\mathbb{R}^l$. The **normal bundle** of M, denoted NM, is the set

$$NM \equiv \left\{ (x, n) \in \mathbb{R}^l \times \mathbb{R}^l : x \in M \text{ and } v \in N_x M \right\}$$

Like the tangent bundle, the normal bundle is a manifold of appropriate dimension. This will be proved later on (Proposition 18, Chapter 4).

3. Definition of differential

Throughout next two sections,

1 M is a C^r manifold of $\mathbb{R}^l$ of dimension m,

2 N is a C^r manifold of $\mathbb{R}^p$ of dimension n,

3 $f : M \to N$ is a continuous function (to begin with)

4 (ϕ, U, V) is a local parametrization of M around x with $x = \phi(u)$, and (ψ, V, U) is the corresponding chart,

5 (χ, S, T) is a local parametrization of N around $y = f(x)$ with $y = \chi(s)$, and (ξ, T, S) is the corresponding chart,

6 $f(V) \subseteq T$.

Occasionally, e.g. when stating important results which are likely to be read again in isolation, some of these assumptions will be repeated.

In the introduction of Chapter 2 we have suggested that the actual definition of a manifold might be rather involved (for example, M and N may arise as solution sets to some complicated but sufficiently regular nonlinear systems of equations). Therefore, a function f mapping M into N may be well defined in some abstract way, yet it may be quite hard to work out an ordinary closed formula corresponding to its definition[2]. At any rate, it seems reasonable to rely on systems of local coordinates in order to have a concrete representation (that is, in terms of vectors of real numbers) of a restriction of f. This is achieved

[2] This is even the more true (in fact, exactly true) if one takes the point of view which we have sketched in the appendix of Chapter 2, and considers manifolds as abstract topological spaces, endowed with some differential structure but "living outside" euclidean spaces.

by a so called "local representation" of f, an object which is best illustrated by the following table:

$$\begin{array}{ccc} x \in V \subseteq M & \xrightarrow{\quad f \quad} & y \in T \subseteq N \\ \uparrow \phi & & \downarrow \xi \\ u \in U \subseteq \mathbb{R}^m & \xrightarrow[\xi \circ f \circ \phi]{} & s \in S \subseteq \mathbb{R}^n \end{array}$$

Definition 24 The function $g \equiv \xi \circ f \circ \phi$ is called a **local representation** of f at x, **induced** by the parametrizations (ϕ, U, V) and (χ, S, T) (with $\xi = \chi^{-1}$).

As the charts vary in the atlases of the two manifolds, the local representation of f changes, in a consistent way. The differential theory for functions between manifolds builds up on the system of local representations induced for them by the existing atlases, according to the above definition. The first item is the differential of a function at a given point.

Definition 25 $f : M \to N$ is said to be **differentiable** at x if some local representation g, induced by the parametrizations (ϕ, U, V) around x and (χ, S, T) around $f(x)$, is differentiable at $u = \psi(x)$. The **differential** df_x of f at x is

$$df_x \equiv d\chi_s \circ dg_u \circ (d\phi_u)^{-1} : T_x M \to T_{f(x)} N \tag{3.5}$$

The definition is illustrated by the following picture and table:

$$\begin{array}{ccc} T_x M & \xrightarrow{\quad df_x \quad} & T_y N \\ (d\phi_u)^{-1} \downarrow & & \uparrow d\chi_s \\ \mathbb{R}^m & \xrightarrow[dg_u = d(\xi \circ f \circ \phi)_u]{} & \mathbb{R}^n \end{array}$$

Observe that $s = g(u)$, and that reference to the inverse function $(d\phi_u)^{-1}$ is unavoidable, since the differential $d\psi_x$ of $\psi : V \to U$ at x remains undefined as yet, because V is an open subset of a manifold (consider again Corollary 5, however).

The definition makes df_x a linear function, as it should be. In several occasions, the most important piece of information needed to analyze problems involving a given function f is the rank of its differential. Since $d\chi_s : \mathbb{R}^n \to T_{f(x)}M$ and $(d\phi_u)^{-1} : T_x M \to \mathbb{R}^m$ are linear isomorphisms, Proposition 6 of Chapter 1 yields directly the characterization of rank df_x in terms of the local representation g.

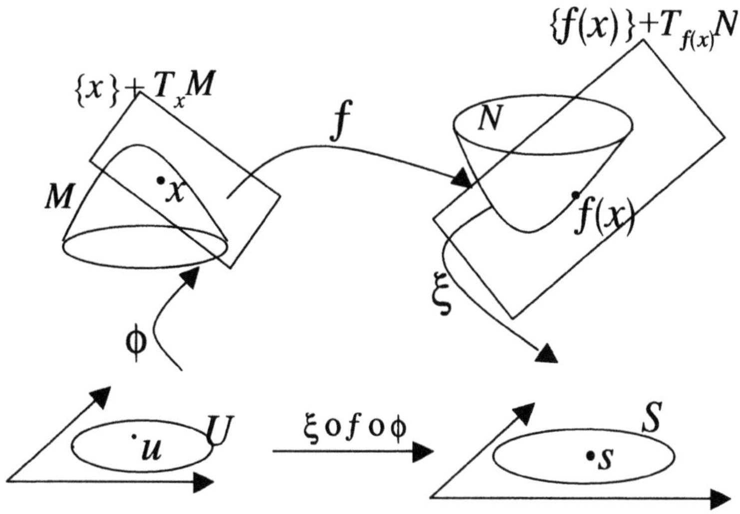

Figure 3.3. The differential of f in terms of its local representation

Proposition 26 *Let f be differentiable at x; then*

$$\operatorname{rank} df_x = \operatorname{rank} dg_u$$

Equivalently, $\dim \ker df_x = \dim \ker dg_u$*; hence df_x is surjective (injective) if and only if dg_u is so.*

The definition given, however, still needs to be justified, since it is given in terms of an arbitrarily chosen local representation of f. Moreover, it must be carefully compared with the differentiability concept introduced at the beginning of Chapter 2, which is based on the existence of differentiable extensions of f. This is the object of next three propositions.

Proposition 27 *Definition 25 is consistent, i.e., the object defined in (3.5) does not depend on the choice of the parametrizations (ϕ, U, V), (χ, S, T) inducing the local representation g.*

Proof. Let $\left(\hat{\phi}, \hat{U}, \hat{V}\right)$ and $\left(\hat{\chi}, \hat{S}, \hat{T}\right)$ be some other parametrizations, around $x = \hat{\phi}\left(\hat{u}\right)$ and $y = \hat{\chi}\left(\hat{s}\right)$ respectively, with $\hat{\psi}$ and $\hat{\xi}$ as inverses, and let $\hat{g}$ be the induced local representation of f. Without loss of generality we shall assume $\hat{V} = V$ and $\hat{T} = T$. Then

$$\begin{aligned} \hat{g} &= \hat{\xi} \circ f \circ \hat{\phi} \\ &= \left(\hat{\xi} \circ \chi\right) \circ \left(\xi \circ f \circ \phi\right) \circ \left(\psi \circ \hat{\phi}\right) \\ &= \left(\hat{\xi} \circ \chi\right) \circ g \circ \left(\psi \circ \hat{\phi}\right) \end{aligned}$$

and $\hat{\xi} \circ \chi$, $\psi \circ \hat{\phi}$ are C^r functions; hence, $\hat{g}$ is differentiable at $\hat{u}$ if g is differentiable at u. Moreover, applying the chain rule (three times), the inverse differential rule, and the inverse composition formula ([3], Theorem 12.7, p. 352; [3], Note p. 373), and taking into account that $u = \left(\psi \circ \hat{\phi}\right)(\hat{u})$ and $\hat{s} = \left(\hat{\xi} \circ \chi\right)(s)$,

$$\begin{aligned} d\hat{\chi}_{\hat{s}} \circ d\hat{g}_{\hat{u}} \circ \left(d\hat{\phi}_{\hat{u}}\right)^{-1} &= d\hat{\chi}_{\hat{s}} \circ \left[d\left(\hat{\xi} \circ \chi\right)_s \circ dg_u \circ d\left(\psi \circ \hat{\phi}\right)_{\hat{u}}\right] \circ \left(d\hat{\phi}_{\hat{u}}\right)^{-1} \\ &= \left[d\hat{\chi}_{\hat{s}} \circ d\left(\hat{\xi} \circ \chi\right)_s\right] \circ dg_u \circ \left[d\left(\hat{\psi} \circ \phi\right)_u\right]^{-1} \circ \left(d\hat{\phi}_{\hat{u}}\right)^{-1} \\ &= d\left(\hat{\chi} \circ \hat{\xi} \circ \chi\right)_s \circ dg_u \circ \left[d\left(\hat{\phi} \circ \hat{\psi} \circ \phi\right)_u\right]^{-1} \\ &= d\chi_s \circ dg_u \circ (d\phi_u)^{-1} \end{aligned}$$

■

Exercise 28 Consider the two atlases of the circle S^1 given in Chapter 2, and the function $f : S^1 \to \mathbb{R}$, $(x, y) \mapsto x^3 + y^2$. Check that in either atlas Definition 25 yields $df_{\left(\frac{\sqrt{3}}{2}, \frac{1}{2}\right)} : v \mapsto \frac{13}{4} v$.

If f is differentiable at x for each $x \in M$, we obtain a global notion of derivative for f, as a mapping between tangent bundles.

Definition 29 Let f be differentiable at every $x \in M$. The **derivative** of f, denoted df, is the function

$$df : TM \to TN, \ (x, v) \mapsto (f(x), df_x(v))$$

Comparing with Definition 25, and taking into account Proposition 27, it is easily seen that the following statements are equivalent:

1 f has a derivative df;

2 for every x in M, every local parametrization (ϕ, U, V) of M around x, and every local parametrization (χ, S, T) of N around $y \equiv f(x)$ (with (ψ, V, U), (ξ, T, S) the corresponding charts), the induced local representation $g = \xi \circ f \circ \phi : U \to S$ is differentiable at every $u \in U$.

More generally, differentiability classes of any order for functions between manifolds, which have been already defined in Chapter 2 (Definition 1), can be characterized on the model of Definition 29.

Proposition 30 *Let M and N be C^r manifolds, and let $f : M \to N$. f is of class C^r if and only if for every x in M, every local parametrization (ϕ, U, V) of M around x, and every local parametrization (χ, S, T) of N around $y \equiv f(x)$ such that $f(V) \subseteq T$ (with (ψ, V, U), (ξ, T, S) the corresponding charts), the induced local representation $g = \xi \circ f \circ \phi : U \to S$ is of class C^r.*

Proof. 1. Let $x \in M$ be given. If f is C^r at x, there exist an open neighborhood W of x in $\mathbb{R}^l$ and a C^r function $F : W \to \mathbb{R}^p$ such that $F(z) = f(z)$ for any $z \in W \cap M$. Let now (ϕ, U, V)-(ψ, V, U) and (χ, S, T)-(ξ, T, S) be arbitrary couples parametrization-chart at x and y (respectively) such that $f(V) \subseteq T$ as in the theorem's statement. Consider the couples parametrization-chart $\left(\hat{\phi}, \hat{U}, \hat{V}\right)$-$\left(\hat{\psi}, \hat{V}, \hat{U}\right)$ and $\left(\hat{\chi}, \hat{T}, \hat{S}\right)$-$\left(\hat{\xi}, \hat{T}, \hat{S}\right)$ obtained by appropriate restriction, that is, where

$$\begin{array}{llll} \hat{V} \equiv V \cap (W \cap M) & \hat{U} \equiv \psi\left(\hat{V}\right) & \hat{\psi} \equiv \psi_{|\hat{V},\hat{U}} & \hat{\phi} \equiv \phi_{|\hat{U},\hat{V}} \\ \hat{T} \equiv T \cap f(W \cap M) & \hat{S} \equiv \xi\left(\hat{T}\right) & \hat{\xi} \equiv \xi_{|\hat{T},\hat{S}} & \hat{\chi} \equiv \chi_{|\hat{S},\hat{T}} \end{array}$$

It is clear that $\hat{\phi}$, $\hat{\psi}$, $\hat{\chi}$, and $\hat{\xi}$ are of class C^r, and that $f\left(\hat{V}\right) \subseteq \hat{T}$. By the ordinary chain rule, the local representation

$$\hat{\xi} \circ f \circ \hat{\phi} = \hat{\xi} \circ F \circ \hat{\phi} : \hat{U} \to \hat{S}$$

is C^r on $\hat{U}$. If u is any element of $U \backslash \hat{U}$, the above argument can be repeated for $\phi(u)$ in place of x.

2. Let $x \in M$ be given, and let (ϕ, U, V), (ψ, V, U) and (χ, S, T), (ξ, T, S) be arbitrary couples parametrization-chart at x and y (respectively) such that the induced local representation

$$g = \xi \circ f \circ \phi : U \to S$$

is of class C^r. Since M is a C^r manifold, ψ is C^r at x, and there exist an open neighborhood W of x in $\mathbb{R}^l$, and a C^r extension $\Psi : W \to \mathbb{R}^m$ of ψ. U being open and Ψ being continuous, we can assume without loss of generality that $\Psi(W) \subseteq U$. Hence the function

$$F \equiv \chi \circ g \circ \Psi : W \to T$$

is well defined and of class C^r (because χ is C^r too, N being a C^r manifold). It is straightforward to check that $F(z) = f(z)$ for every $z \in W \cap V$, and this concludes the proof that f is C^r at x. ■

Proposition 31 *Let F be a local C^1 extension of $f : M \to N$ at x, and let $y \equiv f(x)$. Then*

$$df_x = (dF_x)_{|T_x M, T_y N} \tag{3.6}$$

Proof. Let (ϕ, U, V)-(ψ, V, U) and (χ, S, T)-(ξ, T, S) be arbitrary couples parametrization-chart at x and y (respectively) such that $f(V) \subseteq T$ and V is contained in the neighborhood of x where F is defined. From previous Proposition 30, the local representation $\xi \circ f \circ \phi$ is of class C^r on U, and df_x is well defined. We show now that $dF_x(T_xM) \subseteq T_yN$. For any v in T_xM, according to Definition 2, $v = \gamma'(t)$ for some curve γ in M through x at t. Then $f \circ \gamma = F \circ \gamma$ is a curve in N through y at t, and

$$(F \circ \gamma)'(t) = dF_x \circ \gamma'(t) = dF_x(v)$$

is a tangent vector to N at y. To prove (3.6), it is enough to take into account the chain rule and Corollary 5: for every $z \in \mathbb{R}^l$, and for any local extension Ξ of ξ at y,

$$\begin{aligned} dg_u &= d(\xi \circ f \circ \phi)_u = d(\Xi \circ F \circ \phi)_u \\ &= d\Xi_y \circ dF_x \circ d\phi_u \\ &= (d\chi_s)^{-1} \circ (dF_x)_{|T_xM,T_yN} \circ d\phi_u \\ df_x &= d\chi_s \circ dg_u \circ (d\phi_u)^{-1} \\ &= (dF_x)_{|T_xM,T_yN} \end{aligned}$$

■

The following four results are a more or less immediate consequences of Proposition 31. Their proofs are left to the reader as an exercise.

Corollary 32 *Let M be a manifold and let $x \in M$. For any chart (ψ, V, U) of M at x, ψ is differentiable at x, and $d\psi_x = (d\phi_u)^{-1}$, where (ϕ, U, V) is the parametrization corresponding to ψ.*

Corollary 33 *Let $L : \mathbb{R}^l \to \mathbb{R}^p$ be a linear function such that $L(M) \subseteq N$, and let*

$$f \equiv L_{|M,N} : M \to N, \; x \mapsto L(x)$$

be the restriction of L obtained by taking M as domain and N as value set. Then the derivative of f is the restriction of L having TM as domain and TN as value set:

$$df = L_{|TM,TN} : TM \to TN \tag{3.7}$$

We recall that, given any two sets X and Y such that $X \subseteq Y$, the insertion of X into Y, denoted in short $i_{X,Y}$ or just i when X and Y are clear from the context, is the restriction of id_Y to X.

Corollary 34 *Let M be a submanifold of N. The derivative of the insertion $i_{M,N}$ is the insertion of TM in TN, i.e.,*

$$\textit{if} \qquad i_{M,N} : M \to N, \ x \mapsto x$$

$$\textit{then} \quad d(i_{M,N}) = i_{TM,TN} : TM \to TN, \ (x,v) \mapsto (x,v)$$

Corollary 35 *Let y be an element of $\mathbb{R}^j$ for some $j \in \mathbb{N}$. Consider the function $ci : M \to M \times \{y\}, \ x \mapsto (x,y)$. For each $x \in M$,*

$$d(ci)_x : T_xM \to T_xM \times \{0\}, \ v \mapsto (v,0)$$

We conclude this section with a few properties of differentials, many of which extend standard results in ordinary calculus to calculus on manifolds.

Exercise 36 If f is of class C^1, then df is a continuous function.

Exercise 37 Let f be constant. Then, for every $x \in M$, df_x is the zero function.

Hint. Look at any local representation.

Proposition 38 *Consider the cartesian product $M \times N$ of two manifolds. The derivative of either projection in $M \times N$ is the corresponding projection in $T(M \times N) = TM \times TN$ (recall Remark 21), i.e., taking for example the second projection,*

$$\textit{if } \pi_2 : M \times N \to N, \ (x,y) \mapsto y$$

$$\textit{then } d\pi_2 : T(M \times N) \to TN, \ ((x,y),(v,w)) \mapsto (y,w)$$

Proof. For every $y \in N$, the partial function $(\pi_2)_y$ is constant, and its differential is zero. Now it is enough to apply Proposition 40 to π_2 for every $(x,y) \in M \times N$. Why can't you use directly Exercise 33? ■

Proposition 39 (Chain rule) *Let Q, M, and N be manifolds, let $h : Q \to M$ be differentiable at z, and let $f : M \to N$ be differentiable at $x \equiv h(z)$. Then $f \circ h$ is differentiable at z and*

$$d(f \circ h)_z = df_x \circ dh_z$$

Proof. Let (α, A, B), (β, B, A) be a couple parametrization-chart for Q around z, with A an open subset of $\mathbb{R}^q$, B an open subset of Q, $a = \beta(z)$, and such that $h(B) \subseteq V$; and let the couples parametrization-chart for M at x and for N at $f(x)$ be as usual. Then the local representation of h is

$$k \equiv \psi \circ h \circ \alpha$$

and the local representation of $f \circ h$ is

$$\begin{aligned} \xi \circ (f \circ h) \circ \alpha &= (\xi \circ f) \circ (h \circ \alpha) \\ &= (\xi \circ f) \circ (\phi \circ \psi) \circ (h \circ \alpha) \\ &= (\xi \circ f \circ \phi) \circ (\psi \circ h \circ \alpha) \\ &= g \circ k \end{aligned}$$

By assumption, g and k are ordinary differentiable functions (at u and a, respectively). Taking into account the ordinary chain rule ([3], Theorem 12.7, p. 352),

$$\begin{aligned} d(f \circ h)_z &= d\chi_s \circ d(g \circ k)_a \circ (d\alpha_a)^{-1} \\ &= d\chi_s \circ (dg_u \circ dk_a) \circ (d\alpha_a)^{-1} \\ &= \left[d\chi_s \circ dg_u \circ (d\phi_u)^{-1}\right] \circ \left[d\phi_u \circ dk_a \circ (d\alpha_a)^{-1}\right] \\ &= df_x \circ dh_z \end{aligned}$$

■

The following table may help the reader in keeping track of the purely algebraic arguments contained in the above proof.

$$\begin{array}{ccccc} B \subseteq Q & \xrightarrow{\quad h \quad} & V \subseteq M & \xrightarrow{\quad f \quad} & T \subseteq N \\ \alpha \uparrow \; \downarrow \beta & & \phi \uparrow \; \downarrow \psi & & \chi \uparrow \; \downarrow \xi \\ A \subseteq \mathbb{R}^q & \xrightarrow[k = \psi \circ h \circ \alpha]{} & U \subseteq \mathbb{R}^m & \xrightarrow[g = \xi \circ f \circ \phi]{} & S \subseteq \mathbb{R}^n \end{array}$$

$$\xrightarrow[\xi \circ (f \circ h) \circ \alpha = g \circ k]{}$$

Proposition 40 (Total differential formula) *Let M, Q, and N be manifolds, and let $p : M \times Q \to N$ be differentiable at (x, z). Consider the partial functions $f \equiv p_z : M \to N$, $x \mapsto p(x, z)$ and $h \equiv p_x : Q \to N$, $z \mapsto p(x, z)$. Then*

1 *f is differentiable at x and h is differentiable at z,*

2 *for each $(v, w) \in T_{(x,z)}(M \times Q)$,*

$$dp_{(x,z)}(v, w) = df_x(v) + dh_z(w)$$

Proof. With the same notation used in the proofs of the chain rule and of Proposition 12 (however, notice that h takes values in N now), consider couples parametrization-chart (ϕ, U, V)-(ψ, V, U) for M around $x = \phi(u)$, (α, A, B)-(β, B, A) for Q around $z = \alpha(a)$, and $(\phi \times \alpha, U \times A, V \times B)$-$(\psi \times \beta, V \times B, U \times A)$ for $M \times Q$ around (x, z). Local representations for f, h and p are now the following:

$$\begin{aligned} g &\equiv \xi \circ f \circ \phi \\ k &\equiv \xi \circ h \circ \alpha \\ q &\equiv \xi \circ p \circ (\phi \times \alpha) \end{aligned}$$

which means, in particular, that g and k are partial functions for q, i.e., $g = q_a$ and $k = q_u$. The first claim is proved, because the differentiability assumption on p is in fact a differentiability assumption for q, and this in turn implies differentiability of the partial functions g and k, and hence of f and h.

Concerning the second claim, the definition of differential and Proposition 2 of Chapter 1 (taking into account that $(\phi \times \alpha)^{-1} = \phi^{-1} \times \alpha^{-1}$ for any two invertible functions ϕ and α) give

$$\begin{aligned} dp_{(x,z)} &= d\chi_s \circ dq_{(u,a)} \circ \left[d(\phi \times \alpha)_{(u,a)}\right]^{-1} = \\ &= d\chi_s \circ dq_{(u,a)} \circ \left[(d\phi_u)^{-1} \times (d\alpha_a)^{-1}\right] \end{aligned}$$

Then, by [3], Theorem 12.5, p. 347, applied to q, and a little algebra,

$$\begin{aligned} dp_{(x,z)}(v, w) &= \left(d\chi_s \circ dq_{(u,a)}\right) \left[(d\phi_u)^{-1}(v), (d\alpha_a)^{-1}(w)\right] = \\ &= d\chi_s \left(dg_u \left[(d\phi_u)^{-1}(v)\right] + dk_a \left[(d\alpha_a)^{-1}(w)\right]\right) = \\ &= \left(d\chi_s \circ dg_u \circ [d\phi_u]^{-1}\right)(v) + \left(d\chi_s \circ dk_a \circ [d\alpha_a]^{-1}\right)(w) = \\ &= df_x(v) + dh_z(w) \end{aligned}$$

■

Exercise 41 Let Q, M, and N be manifolds, let $f : M \to N$ be differentiable at x, and let $h : Q \to N$ be differentiable at z. Then

$$d(f \times h)_{(x,z)} = df_x \times dh_z$$

Hint. Still keeping the notation just used (dropping p), a local representation for $f \times h : M \times Q \to N \times N$ is $g \times k = (\xi \circ f \circ \phi) \times (\xi \circ h \circ \alpha)$. The claim follows from Proposition 2, Chapter 1, and the definition of differential.

Exercise 42 Let M and N be manifolds, let Q be a submanifold of M, and let $f : M \to N$ be differentiable at $x \in Q$. Then,

1 the restriction $f_{|Q}$ is differentiable at x as well,

2 the differential of $f_{|Q}$ is equal to the restriction of df_x to the tangent space to Q at x, that is,

$$\left(df_{|Q}\right)_x = \left(df_x\right)_{|T_x Q}$$

Hint. Compare the definition of differential at x for f and $f_{|Q}$, using a couple parametrization-chart for $f_{|Q}$ obtained by restriction (this has been discussed after Proposition 18, Chapter 2).

4. Differentials and local behavior

4.1 Preliminaries

A good reason to study differentials is explained by the aphorism: "knowing just a little about f' tells us a lot about f" ([113], page 173). This extends to manifolds. In a few words, we have that for any differentiable function $f : M \to N$

1 if df_x is a linear isomorphism, then f is a local diffeomorphism at x;

2 if df_x is injective, then f is locally injective at x;

3 if df_x is surjective, then f is essentially surjective at x.

These results are formally described and proved in the remainder of this section. Local diffeomorphisms, locally injective and essentially surjective functions are formally presented in Definitions 44, 51 and 62, respectively. The precise statements of 1, 2 and 3 above are given in Theorems 48, 52 and 66, respectively.

Most of the arguments rely heavily on the inverse function theorem for euclidean spaces, to the point that we have thought it better to restate the theorem here.

Theorem 43 (Inverse function theorem) *Let U be an open subset of $\mathbb{R}^n$, let $g : U \to \mathbb{R}^n$ be a function of class C^r, and let dg_u be a linear isomorphism. Then there exist an open neighborhood W of u in U, and an open neighborhood Z of $s \equiv g(u)$, such that $Z = g(W)$, and $g_{|W,Z}$ is a C^r diffeomorphism.*

4.2 The inverse function theorem for manifolds

We shall now state and prove the extension to manifolds of Theorem 43. In order to do that, we need to give the appropriate notion of C^r diffeomorphism for functions of manifolds into manifolds.

Definition 44 For given manifolds M and N, $f : M \to N$ is a **local** C^r **diffeomorphism**at $x \in M$ if there exist an open neighborhood X of x in M, and an open neighborhood Y of $y \equiv f(x)$ in N, such that $f_{|X} : X \to Y$ is a C^r diffeomorphism (according to Definition 4, Chapter 2). f is a **local** C^r **diffeomorphism** if it is a local C^r diffeomorphism at x for every $x \in M$.

Example 45 A canonical example of a local C^∞ diffeomorphism which is not a (global) diffeomorphism is the function $(\cos, \sin) : \mathbb{R} \to S^1$. Another one is the function $f : \mathbb{R}^2 \to \mathbb{R}^2$, $(x, y) \mapsto (e^x \cos y, e^x \sin y)$. A third example is given by the functions $\tan$ and $\cot$ (the reader probably knows their domains) or by removing from the domain of the functions sin and cos all points where their derivative equals zero.

Remark 46 Notice that the statement of the inverse function theorem can be rephrased in short as point 1 above, under the restricted interpretation that M is an open subset of $\mathbb{R}^n$ and $N = \mathbb{R}^n$.

Next proposition is a simple converse of the first item in the list presented introducing the section.

Proposition 47 *If $f : M \to N$ is a local C^r diffeomorphism, then df_x is a linear isomorphism for every $x \in M$.*

Proof. Given $x \in M$, assume without loss of generality (by appropriate restrictions) that the parametrization (ϕ, U, V) of M around x is such that $V = X$ (as in Definition 44). Then $\left(f_{|V} \circ \phi, U, Y\right)$ is a parametrization of N around y, which implies, by Definition 15 and Proposition 25, Chapter 2, that n is equal to m. Thus $T_x M$ and $T_y N$ are vector spaces of the same dimension, and $\left(df_{|V}^{-1}\right)_y \circ \left(df_{|V}\right)_x = id_{T_x M}$ implies ([78], p. 90 and Corollary p. 106) that $df_x = \left(df_{|V}\right)_x$ is a linear isomorphism. ■

We conclude the subsection with its main result.

Theorem 48 (Inverse function theorem for manifolds) *Let M, N be C^r manifolds, and let $f : M \to N$ be a function of class C^r such that df_x is a linear isomorphism for some $x \in M$. Then f is a local C^r diffeomorphism at x. More precisely, there exist charts $\left(\overline{\psi}, X, W\right)$ of M at x and $\left(\overline{\xi}, Y, W\right)$ of N at y inducing a local representation of f which is equal to the identity*

$$\overline{\xi} \circ f \circ \overline{\psi}^{-1} : W \to W, \quad w \mapsto w$$

Proof. The argument proceeds in three steps: i) starting with arbitrary[3] parametrizations, we discuss the properties of the local representation they

[3] That is, within the conditions 4-6 at the beginning of Section 3.

induce; ii) the latter being a function between open subsets of euclidean spaces (of the same dimension), we are in the position to invoke the ordinary inverse function theorem; iii) this allows us to exhibit the required local representation, which in turn leads to the conclusion concerning f.

Since $df_x : T_xM \to T_yN$ is a linear isomorphism, T_xM and T_yN have the same dimension, and the same is true for M and N. Recalling Proposition 26, dg_u is a linear isomorphism as well (of $\mathbb{R}^n$ with itself). Therefore, the local representation $g : U \to S$, which is of class C^r by Proposition 30, satisfies the assumptions of Theorem 43, and there exist an open neighborhood W of u in U, and an open neighborhood Z of $g(u)$ in S, such that

1 $Z = g(W) = \xi(f[\phi(W)])$,

2 $g_{|W,Z}$ is a C^r diffeomorphism.

Then, let $h \equiv (g_{|W,Z})^{-1}$ and let $X \equiv \phi(W)$, an open neighborhood of x in M; observe that $Y \equiv \chi(Z) = f(X)$ is an open neighborhood of y in N; and conclude that the triples $(\phi_{|W,X}, W, X)$ and $(\chi \circ g_{|W,Z}, W, Y)$ are local parametrizations of M and N around x and y respectively. With the positions $\overline{\psi} \equiv \psi_{|X,W}$ and $\overline{\xi} \equiv h \circ \xi_{|Y,Z}$, the corresponding charts are $(\overline{\psi}, X, W)$ and $(\overline{\xi}, Y, W)$. The local representation of f at x induced by these parametrizations is

$$\begin{aligned}\overline{\xi} \circ f \circ \overline{\psi}^{-1} &= h \circ (\xi_{|Y,Z} \circ f \circ \phi_{|W,X}) \\ &= h \circ (\xi \circ f \circ \phi_{|W}) \\ &= h \circ g_{|W} \\ &= id_W\end{aligned}$$

Finally, composition with $\overline{\psi}$ on the right and with $\overline{\xi}^{-1}$ on the left gives

$$f_{|X,Y} = \chi \circ g_{|W,Z} \circ \psi_{|X,W}$$

showing that $f_{|X,Y}$ is a C^r diffeomorphism by Proposition 30. ∎

4.3 Immersions and embeddings

In next two sections we weaken somewhat the assumption made in the statement of the implicit function theorem for manifolds (Theorem 48), according to which df_x is a linear isomorphism. This is done in two opposite directions: either we retain the requirement that df_x is injective (here), or that it is surjective (later). In both cases we state and prove a theorem characterizing the local behavior of f, which is formulated in order to be seen as an extension of Theorem 48, yielding that theorem as a particular case.

Definition 49 For any positive integers m and q, the **canonical immersion** of $\mathbb{R}^m$ into $\mathbb{R}^{m+q}$ (up to isomorphism with $\mathbb{R}^m \times \mathbb{R}^q$), denoted $ci_{m,q}$ and in short ci (when m and q are clear from the context), is defined as follows:

$$ci_{m,q} : \mathbb{R}^m \to \mathbb{R}^m \times \mathbb{R}^q, \quad x \mapsto (x, 0)$$

$ci_{m,q}$ is a linear isomorphism of $\mathbb{R}^m$ with $\mathbb{R}^m \times \{0\}$.

Definition 50 $f : M \to N$ is C^r **immersive** at $x \in M$ if it is of class C^r and df_x is injective; f is a C^r **immersion** if it is C^r immersive at x for every $x \in M$.

It is clear from Proposition 47 that f is immersive at x whenever it is a local diffeomorphism at x.

Definition 51 $f : M \to N$ is **locally injective** at $x \in M$ if there exists an open neighborhood X of x in M, such that $f_{|X}$ is injective.

Theorem 52 *Let M and N be C^r manifolds of dimension m and n respectively, and let $f : M \to N$ be a function which is C^r immersive at x. Then f is locally injective at x. More precisely, $q \equiv n - m$ is nonnegative, and*

1 either q is zero, and there exist charts $(\overline{\psi}, X, W)$ of M at x and $(\overline{\xi}, Y, W)$ of N at y inducing a local representation of f which is equal to id_W;

2 or q is positive, and there exist charts $(\overline{\psi}, X, W)$ at x and $(\overline{\xi}, Y, W \times H)$ at y inducing a local representationof f which is equal to the restriction to W of the canonical immersion of $\mathbb{R}^m$ into $\mathbb{R}^{m+q}$:

$$\overline{\xi} \circ f \circ \overline{\psi}^{-1} = (ci)_{|W} : W \to W \times H, \quad w \mapsto (w, 0)$$

where H is an open neighborhood of 0 in $\mathbb{R}^q$.

In order to proceed, we shall need the following lemma, which is a consequence of Theorem 43, and plays in the proof of Theorem 52 the same role played by Theorem 43 in the proof of Theorem 48.

Lemma 53 *Let $g : U \subseteq \mathbb{R}^m \to \mathbb{R}^{m+q}$ be a function of class C^r on the open set U, let $u \in U$, and let dg_u be of full rank. Then there exists a C^r diffeomorphism $G : W \times H \to Z$, such that:*

1 $g(W) \subseteq Z$

2 $G \circ (ci_{m,q})_{|W} = g_{|W,Z}$

where W is an open neighborhood of u in U, Z is an open neighborhood of $g(u)$ in $\mathbb{R}^{m+q}$, and H is an open neighborhood of 0 in $\mathbb{R}^q$.

$$\begin{array}{ccc} W \subseteq U \subseteq \mathbb{R}^m & \overset{g_{|W,Z}}{\longrightarrow} & Z \subseteq g(U) \subseteq \mathbb{R}^{m+q} \\ & ci \searrow & \uparrow G \\ & & W \times H \subseteq U \times \mathbb{R}^q \subseteq \mathbb{R}^m \times \mathbb{R}^q \end{array}$$

Proof. By assumption, $\operatorname{rank} dg_u = m$. Let K be any complementary subspace to $\operatorname{Im} dg_u$ in $\mathbb{R}^{m+q}$ ([78], Theorem 8, p. 56); thus $\dim K = q$, and there exists a linear isomorphism $k : \mathbb{R}^q \to K$. Consider the C^r function

$$F \equiv g \circ \pi_1 + k \circ \pi_2 : U \times \mathbb{R}^q \to \mathbb{R}^{m+q} : (w, p) \mapsto g(w) + k(p)$$

Then, by the total differential formula (Proposition 40), and by linearity of k,

$$\operatorname{Im} dF_{(u,0)} = \operatorname{Im} dg_u + \operatorname{Im} dk_0 = \operatorname{Im} dg_u + \operatorname{Im} k = \mathbb{R}^{m+q}$$

and $dF_{(u,0)}$ has full rank. The implicit function theorem applies, and there exist

1 an open neighborhood of $(u, 0)$ in $U \times \mathbb{R}^q$, which can be taken of the form $W \times H$ with W open neighborhood of u in U and H open neighborhood of 0 in $\mathbb{R}^q$,

2 an open neighborhood Z of $g(u) = F(u, 0)$,

such that $Z = F(W \times H)$, and $G \equiv F_{|W \times H, Z}$ is a C^r diffeomorphism. Then, for each $w \in W$, $G(w, 0) = g(w)$, which proves both claims. ■

Proof (of the theorem). Since $df_x : T_x M \to T_y N$ is injective, by Theorem 3, p. 99 in [78], $\dim T_x M \leq \dim T_y N$, i.e., $m \leq n$. If $m = n$, df_x is an isomorphism, Theorem 48 applies, and the first claim is proved. Let then $m < n$. By Proposition 26, the differential $dg_u : \mathbb{R}^m \to \mathbb{R}^{m+q}$ of the local representation of f at x is injective, too. Therefore, $g : U \to S$ satisfies the assumptions of Lemma 53, and there exist an open neighborhood W of u in U, an open neighborhood Z of $s = g(u)$, an open neighborhood H of 0 in $\mathbb{R}^q$, and a C^r diffeomorphism $G : W \times H \to Z$, such that

1 $g(W) = \xi(f[\phi(W)]) \subseteq Z$,

2 $G \circ (ci_{m,q})_{|W} = g_{|W}$.

Let then $X \equiv \phi(W)$, an open neighborhood of x in M; observe that $Y \equiv \chi(Z)$ is an open neighborhood of y in N, and conclude that the triples $(\phi_{|W}, W, X)$ and $(\chi \circ G, W \times H, Y)$ are local parametrizations of M and

N around x and y respectively; moreover, $f(X) \subseteq Y$. With the positions $\overline{\psi} \equiv \psi_{|X}$ and $\overline{\xi} \equiv G^{-1} \circ \xi_{|Y}$, the corresponding charts are $(\overline{\psi}, X, W)$ and $(\overline{\xi}, Y, W \times H)$. The induced local representation $\overline{g}$ of f at x is

$$\begin{aligned} \overline{g} &= \overline{\xi} \circ f \circ \phi_{|W} \\ &= G^{-1} \circ (\xi_{|Y} \circ f \circ \phi_{|W}) \\ &= G^{-1} \circ g_{|W} \\ &= (ci_{m,q})_{|W} \end{aligned}$$

■

It is certainly not the case that the image of a manifold via a C^r function f is a manifold. In fact, we do not get that result even if f is an immersion (see example 59 below): an additional topological property is needed. This is the motivation for the following definition.

Definition 54 $f : M \to N$ is a C^r **embedding** if

1 f is a C^r immersion;

2 the restriction $f : M \to f(M)$ is a homeomorphism.

Remark 55 It must be understood that in 2 above $f(M)$ is considered as a topological subspace of N, i.e., it is endowed with the topology induced by N.

Proposition 56 *Let M be a submanifold of N. Then the canonical insertion of M into N is a C^∞ embedding.*

Proof. It is clear that $i_{M,N}$ is C^∞ immersive at every $x \in M$, and that it is injective. Let V be any open set of M (that is, V is an element of the topology induced on M from $\mathbb{R}^p$); then $V = M \cap W$ for some open set W of $\mathbb{R}^p$. Since

$$M \cap W = (M \cap N) \cap W = M \cap (N \cap W)$$

and $N \cap W$ is open in N, $i_{M,N}(V) = V$ is open in $i_{M,N}(M) = M$ as well (that is, V is an element of the topology induced on M from N). ■

Proposition 57 *Let $f : M \to N$ be a C^r embedding. Then $f(M)$ is a submanifold of N, and the restriction $f : M \to f(M)$ is a C^r diffeomorphism.*

Proof. Suppose first that m is less than n. Let y be any element of $f(M)$, let $x \equiv f^{-1}(y)$, and let $(\overline{\psi}, X, W)$ and $(\overline{\xi}, Y, W \times H)$ be charts of M at x and of N at y, respectively, as in the statement of Theorem 52. Then $f(X)$ is an open subset of $f(M)$ by the second condition of Definition 54, and there exists an open neighborhood Z of y in N, which without loss of generality we assume contained in Y, such that $f(X) = f(M) \cap Z$. Since $\overline{\xi}(Z)$ is open in $W \times H$, and $(\overline{\xi} \circ f)(X) = W \times \{0\}$, the section

$$V \equiv \left(\pi_1 \circ \overline{\xi} \circ f\right)(X) = \overline{\xi}(Z)_0$$

is open in W (Proposition 11, Chapter 1), and the triple $\left(\pi_1 \circ \overline{\xi}_{|f(X)}, f(X), V\right)$ is a C^r chart of $f(M)$ at y. Thus an atlas for $f(M)$ is obtained from the given atlas of N by appropriate restrictions.

If, on the other hand, m is equal to n, the argument is entirely analogous, and actually simpler, with $W \times H$ replaced by W, and no projection involved. ■

Proposition 58 *Let $f : M \to N$ be a C^r immersion which is injective, and let M be compact. Then f is a C^r embedding.*

Proof. This is a consequence of a standard result in general topology ([90], Theorem 5.6, p. 167), according to which any continuous invertible function of a compact set onto a Hausdorff space is a homeomorphism. ■

Example 59 Consider the curve in $\mathbb{R}^2$ through the origin

$$\gamma : \mathbb{R} \to \mathbb{R}^2, \quad t \mapsto \frac{(t^2, t)}{(1+t^2)^2}$$

which is of class C^∞; its tangent vector at t

$$\gamma'(t) = \frac{\left(2t\left(1-t^2\right), 1-3t^2\right)}{(1+t^2)^3}$$

is different from zero for every $t \in \mathbb{R}$. Thus γ is a C^∞ immersion; moreover, γ is (globally) injective, since any line through the origin of slope $\frac{1}{m}$ contains the image under γ of exactly one real number (namely, m) and the two axes, too, contain just the image of 0 (the origin). However, γ is not an embedding of $\gamma(\mathbb{R})$ in $\mathbb{R}^2$, because it is not a homeomorphism. Indeed,

$$\lim_{t \to +\infty} \gamma(t) = (0,0)$$
$$\lim_{t \to +\infty} \gamma(t) = (0,0)$$

so that every sufficiently small[4] open neighborhood of $(0,0)$ in $\gamma(\mathbb{R})$ is the image under γ of the disjoint union

$$(-\infty, a) \cup (b, c) \cup (d, +\infty)$$

[4]For example, $B_\varepsilon \cap \operatorname{Im} \gamma$ for any $\varepsilon < \frac{1}{3}$.

for some real numbers a, b, c, d such that $a < b < 0 < c < d$. Hence the image of every bounded open neighborhood of 0 in $\mathbb{R}$ is not an open set of $\gamma(\mathbb{R})$.

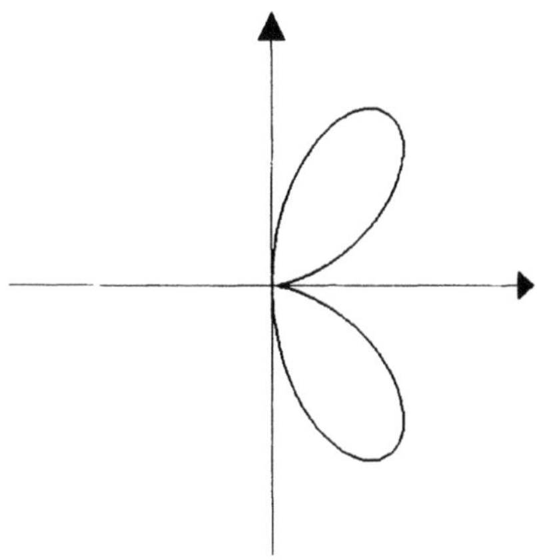

Figure 3.4. The image of γ

4.4 Submersions

Definition 60 For any positive integers n and q, the **canonical projection** of $\mathbb{R}^{n+q}$ onto $\mathbb{R}^n$, up to the canonical isomorphism of $\mathbb{R}^{n+q}$ with $\mathbb{R}^n \times \mathbb{R}^q$, is just the first projection π_1 of $\mathbb{R}^n \times \mathbb{R}^q$ onto $\mathbb{R}^n$

$$\pi_1 : \mathbb{R}^n \times \mathbb{R}^q \to \mathbb{R}^n, \ (x, y) \mapsto x$$

π_1 is a surjective linear function.

Definition 61 $f : M \to N$ is C^r **submersive at** $x \in M$ if it is of class C^r and df_x is surjective; f is a C^r **submersion** if it is C^r submersive at x for every $x \in M$.

It is clear from Proposition 47 that f is submersive at x whenever it is a local diffeomorphism at x.

Definition 62 $f : M \to N$ is **essentially surjective** at $x \in M$ if the image of each open neighborhood of x in M contains an open neighborhood of $f(x)$ in N

$$\forall O \subseteq M, \quad \left.\begin{array}{c} O \text{ open} \\ x \in O \end{array}\right\} \Rightarrow \exists P \subseteq N, \ \left\{\begin{array}{c} P \text{ open} \\ f(x) \in P \\ P \subseteq f(O) \end{array}\right.$$

Proposition 63 *Let M, N be topological spaces, and let $f : M \to N$. f is open if and only if it is essentially surjective at every element of M.*

Proof. Let f be essentially surjective at x for each $x \in M$, and let O be open in M. For each $y \in f(O)$, let x_y be any element of $f^{-1}(y)$. Since f is essentially surjective at x_y, there exists an open neighborhood P_y of y in N such that $P_y \subseteq f(O)$. Thus $f(O)$ is open in N. The converse statement is trivial. ■

Remark 64 The reader is entitled to an explanation for our denomination "essentially surjective". On one hand, the property defined under that denomination is of some importance in the sequel. On the other, uniformity with the last two subsections would have suggested to talk about "locally surjective" functions at an element of their domain. However, established usage of the word "locally" seems to require an existential rather than universal quantifier for neighborhoods where a property is assumed to hold. The corresponding modification of Definition 62 would have yielded a weaker property than required by our developments; moreover, the correct denomination in that case should have probably been "locally open". Since we are unaware of alternative denominations, we have decided to create our own.

As in the last two sections, in order to prove the main theorem concerning submersive functions between manifolds (Theorem 66) we need a corresponding lemma for submersive function between euclidean spaces.

Lemma 65 *Let $g : U \subseteq \mathbb{R}^{n+q} \to \mathbb{R}^n$ be a function of class C^r on the open set U, let $u \in U$, and let dg_u be of full rank. Then there exists a C^r diffeomorphism $G : Z \times H \to W$, such that*

1 $g(W) = Z$

2 $g_{|W} \circ G = (\pi_1)_{|Z \times H}$

where W is an open neighborhood of u in U, Z is an open neighborhood of $s \equiv g(u)$ in $\mathbb{R}^n$, and H is an open subset H of $\mathbb{R}^q$.

$$\begin{array}{ccc} Z \times H \subseteq \mathbb{R}^n \times \mathbb{R}^q & \overset{G}{\to} & W \subseteq U \subseteq \mathbb{R}^{n+q} \\ & \pi_1 \searrow & \downarrow g_{|W} \\ & & Z \subseteq \mathbb{R}^n \end{array}$$

Proof. The argument proceeds in three steps. First we check that the claim is true for linear functions; this amounts to a change of basis in $\mathbb{R}^{n+q}$. Then we prove that the claim is true for any g such that $dg_u = \pi_1$; this amounts to finding a local inverse of the function $g \times (\pi_2)_{|U}$, by means of the inverse function theorem. Finally, we consider an arbitrary g, and we draw on the previous results in order to conclude.

Let $L : \mathbb{R}^{n+q} \to \mathbb{R}^n$ be linear. To say that $dL_u = L$ is of full rank means that $\operatorname{Im} L = \mathbb{R}^n$, and in turn this yields that $\dim \ker L = q$. Thus define a set $\mathcal{V}$ of $n+q$ vectors in $\mathbb{R}^{n+q}$ as follows: first, for each canonical basis vector $\mathbf{e}^j$ of $\mathbb{R}^n$, choose v_j in $\mathbb{R}^{n+q}$ such that $L(v_j) = \mathbf{e}^j$; second, let $\{v_j\}_{j=n+1}^{n+q}$ be a basis of $\ker L$. $\mathcal{V}$ is a linearly independent set[5]. If A is the $(n+q) \times (n+q)$ matrix with the vectors in $\mathcal{V}$ as columns, the function

$$G : \mathbb{R}^{n+q} \to \mathbb{R}^{n+q}, u \mapsto Au$$

is a linear isomorphism; moreover, for each canonical base vector $\overline{\mathbf{e}}^j$ of $\mathbb{R}^{n+q}$, $G\left(\overline{\mathbf{e}}^j\right) = v_j$. Then

$$(L \circ G)(u) = (L \circ G)\left(\sum_{j=1}^{n+q} u_j \overline{\mathbf{e}}^j\right) = L\left(\sum_{j=1}^{n+q} u_j v_j\right) = \sum_{j=1}^{n} u_j \mathbf{e}^j$$

which shows that $L \circ G = \pi_1$. Thus the lemma holds for L, and the stated diffeomorphism is the linear automorphism of $\mathbb{R}^{n+q}$ transforming coordinates with respect to the base $\mathcal{V}$ into canonical coordinates.

Let now g be as in the statement of the lemma, and assume that $dg_u = \pi_1$. Consider the function

$$F \equiv g \times (\pi_2)_{|U} : U \subseteq \mathbb{R}^{n+q} \to \mathbb{R}^{n+q}$$

then

$$\pi_1 \circ F = g$$
$$dF_u = \pi_1 \times \pi_2 = id_{\mathbb{R}^{n+q}}$$

and Theorem 43 applies to F. There exist an open neighborhood W of u in U, and an open neighborhood of $F(u) = (g(u), u_2)$ in $\mathbb{R}^{n+q}$, which can be taken of the form $Z \times H$ with Z open neighborhood of s in $\mathbb{R}^n$ and H open in $\mathbb{R}^q$, such that $Z \times H = F(W)$, and $F_{|W}$ is a C^r diffeomorphism. Let

$$G \equiv \left(F_{|W}\right)^{-1} : Z \times H \to W$$

Then

$$g(W) = \pi_1(F(W)) = Z$$
$$g_{|W} \circ G = \left(\pi_1 \circ F_{|W}\right) \circ \left(F_{|W}\right)^{-1} = \pi_1 \circ id_{Z \times H} = (\pi_1)_{|Z \times H}$$

[5]The argument for this often appears in proofs of the dimension theorem ([78], Theorem 3, p.69); in short, $\sum_{j=1}^{n+q} \alpha_j v_j = 0$ implies $0 = L\left(\sum_{j=1}^{n+q} \alpha_j v_j\right) = \sum_{j=1}^{n} \alpha_j \mathbf{e}^j$ and hence $\alpha_j = 0$ for $j \leq n$; then $\sum_{j=n+1}^{n+q} \alpha_j v_j = 0$ gives $\alpha_j = 0$ for $j > n$.

Finally, remove the provisional assumption that $dg_u = \pi_1$. By the first part of the argument, there exists a linear automorphism E of $\mathbb{R}^{n+q}$ such that $dg_u \circ E = \pi_1$; let $\overline{U} \equiv E^{-1}(U)$ and $\overline{u} \equiv E^{-1}(u)$. The function $\overline{g} \equiv g \circ E : \overline{U} \to \mathbb{R}^n$ is differentiable at $\overline{u}$, and $d\overline{g}_{\overline{u}}$ is equal to π_1; notice that $g = \overline{g} \circ \left(E^{-1}\right)_{|U}$ and $s = \overline{g}(\overline{u})$. By the second part of the argument, there exists a local C^r inverse $\overline{G}$ of $\overline{F} \equiv \overline{g} \times (\pi_2)_{|\overline{U}}$, defined on some neighborhood $Z \times H$ of $\overline{F}(\overline{u}) = (s, \overline{u}_2)$, with Z open neighborhood of s in $\mathbb{R}^n$ and H as above, with $\operatorname{Im} \overline{G} = \overline{W}$ for some neighborhood $\overline{W}$ of $\overline{u}$ in $\overline{U}$, and such that $\overline{g}\left(\overline{W}\right) = Z$ and $\overline{g}_{|\overline{W}} \circ \overline{G} = (\pi_1)_{|Z \times H}$. Then $W \equiv E\left(\overline{W}\right)$ is an open neighborhood of u in U, and $G \equiv E \circ \overline{G}$ satisfies the conditions in the statement of the lemma. Indeed,

$$G(Y \times H) = E\left(\overline{W}\right) = W$$

$$g(W) = \left(\overline{g} \circ E^{-1}_{|U}\right)(W) = \left(\overline{g} \circ E^{-1}_{|U}\right)\left(E\left(\overline{W}\right)\right) = \overline{g}\left(\overline{W}\right) = Z$$

$$g_{|W} \circ G = g_{|W} \circ \left(E \circ \overline{G}\right) = \left(\overline{g} \circ \left(E^{-1}\right)_{|W}\right) \circ \left(E_{|\overline{W}} \circ \overline{G}\right)$$

$$= \overline{g} \circ id_{\overline{W}} \circ \overline{G} = \overline{g}_{|\overline{W}} \circ \overline{G} = (\pi_1)_{|Z \times H}$$

■

Theorem 66 *Let M and N be C^r manifolds of dimension m and n respectively, and let $f : M \to N$ be a function which is C^r submersive at x. Then f is essentially surjective at x. More precisely, $q \equiv m - n$ is nonnegative, and*

1 either q is zero, and there exist charts $\left(\overline{\psi}, X, W\right)$ at x and $\left(\overline{\xi}, Y, W\right)$ at y inducing a local representation of f which is equal to id_W;

2 or q is positive, and there exist charts $\left(\overline{\psi}, X, Z \times H\right)$ at x and $\left(\overline{\xi}, Y, Z\right)$ at y inducing a local representation of f which is equal to the restriction to $Z \times H$ of the canonical projection of $\mathbb{R}^{n+q}$ into $\mathbb{R}^n$:

$$\overline{\xi} \circ f \circ \overline{\psi}^{-1} = (\pi_1)_{|Z \times H} : Z \times H \to Z, \quad (z, h) \mapsto z$$

where H is an open set of $\mathbb{R}^q$.

Proof. Since $df_x : T_xM \to T_yN$ is surjective, $\dim T_xM \geq \dim T_yN$, i.e., $m \geq n$. If $m = n$, df_x is an isomorphism, Theorem 48 applies, and once again the first claim is proved. Let then $m > n$. By Proposition 26, the differential $dg_u : \mathbb{R}^{n+q} \to \mathbb{R}^n$ of the local representation of f at x is surjective, too. The assumptions of Lemma 53 are satisfied by $g : U \to S$, and there exist an open neighborhood W of u in U, an open neighborhood Z of $s \equiv g(u)$, an open subset H of $\mathbb{R}^q$, and a C^r diffeomorphism $G : Z \times H \to W$, such that

1 $g(W) = Z$, and

2 $g_{|W} \circ G = (\pi_1)_{|Z \times H}$.

Let then $X \equiv \phi(W)$, an open neighborhood of x in M, and let $Y \equiv \chi(Z)$, an open neighborhood of y in N. Since $\psi(X) = W$, $f(X) = (\chi \circ g \circ \psi)(X) = Y$. The triples $(\phi_{|W} \circ G, Z \times H, X)$ and $(\chi_{|Z}, Z, Y)$ are local parametrizations of M and N around x and y respectively. With the positions $\overline{\psi} \equiv G^{-1} \circ \psi_{|X}$ and $\overline{\xi} \equiv \xi_{|Z}$, the corresponding charts are $(\overline{\psi}, X, Z \times H)$ and $(\overline{\xi}, Y, Z)$. The induced local representation $\overline{g}$ of f at x is

$$\begin{aligned}\overline{g} &= \overline{\xi} \circ f \circ (\phi_{|W} \circ G) \\ &= (\xi_{|Z} \circ f \circ \phi_{|W}) \circ G \\ &= g_{|W} \circ G \\ &= (\pi_1)_{|Z \times H}\end{aligned}$$

Finally, $\overline{g}$ is essentially surjective at every element of $Z \times H$, because every projection is an open function ([90], Exercise 4, p. 91); and the same is true for $f_{|X} = \chi_{|Z} \circ \overline{g} \circ \overline{\psi}$, because $\chi_{|Z}$ and $\overline{\psi}$ are diffeomorphisms. ∎

5. Appendix - C^r functions on abstract manifolds and Whitney's theorem

This appendix is a continuation to the appendix of Chapter 2. It contains two things: a sketch of the concept of C^r function for abstract manifolds, together with a short list of properties which will be needed in the sequel, and a statement of Whitney's theorem in a form which is suitable to our purposes. All the technical terms used refer to definitions contained in the appendix to Chapter 2. The interested reader is invited to consult, e.g., [18] or [71].

Definition 67 Let M and N be C^r manifolds with C^r atlases $\{(\psi_i, V_i, U_i)\}_{i \in I}$ and $\{(\xi_j, T_j, S_j)\}_{j \in J}$, respectively. A C^r **function**, or a **function of class** C^r, of M into N is a function $f : M \to N$ such that, for each $x \in M$, there exist charts (ψ_i, V_i, U_i) and (ξ_j, T_j, S_j) with $x \in V_i$, $f(x) \in T_j$, and $f(V_i) \subseteq T_j$, for which the local representation

$$g_{ij} \equiv \xi_j \circ f \circ \psi_i^{-1} : U_i \to S_j$$

is of class C^r.

It follows from the above definition and Definition 42 that if $f : M \to N$ is a C^r function, x is any element of M, and (ψ_k, V_k, U_k) and (ξ_l, T_l, S_l) are any charts with $x \in V_k$, $f(x) \in T_l$, and $f(V_k) \subseteq T_l$, the local representation $g_{kl} \equiv \xi_l \circ f \circ \psi_k^{-1}$ is of class C^r.

Definition 68 A C^r **diffeomorphism** of the C^r manifolds M and N is a bijective function of M into N which is of class C^r together with its inverse.

Proposition 69 *Let Q, M, and N be C^r manifolds, and let $h : Q \to M$ and $f : M \to N$ be C^r functions. Then $f \circ h$ is of class C^r.*

Remark 70 If M is a C^r manifold, and $f : M \to \mathbb{R}^n$, $h : M \to \mathbb{R}^n$, and $k : M \to \mathbb{R}$ are C^r functions, the functions $f + h$, fh, and kf are of class C^r.

Proposition 71 *Let Q, M, and N be C^r manifolds such that Q is included in M, and let f be a C^r diffeomorphism of M into N. Then $f(Q)$ is a C^r manifold included in N.*

Remark 72 The reader should notice that we have carefully avoided the word submanifold in the statement of the above proposition. For abstract manifolds the notion of C^r submanifold is more restrictive than the simple notion of included manifold.

Theorem 73 *Let M be a paracompactand Hausdorff topological space which is an m-dimensional C^r manifold. Then there exists a C^r diffeomorphism $f : M \to \widetilde{M}$, where $\widetilde{M}$ is an m-dimensional C^r manifold $\widetilde{M}$ of $\mathbb{R}^{2m}$.*

Chapter 4

REGULAR VALUES

1. Introduction

Many interesting and important properties of C^r functions of manifolds into manifolds, as in the case of functions of euclidean spaces into euclidean spaces, do not hold in full generality, but only under appropriate restrictions. For example, most, but not all, the level sets of a C^r function exhibit a manifold structure (if they are not empty). The concept of regular value provides in this case the appropriate dividing line, as stated by the regular value theorem. This is the object of the first section, which contains the definition of regular and critical values, followed by a discussion of some their properties, the theorem's statement and proof, and an interpretation of the result in terms of nonlinear systems of equations. In the same spirit, the concept of proper function is introduced and discussed in the second section. The assumption of properness allows us to sharpen our understanding the structure of solution sets to nonlinear systems of equations. Even when such an assumption is not granted, some of the conclusions achieved in the properness case can be seen to hold; this is the object of the third section. The final section contains a few results based on regularity and properness which are heavily used in the second part of the book.

2. Definitions

As usual, and according to the definitions given, in this chapter M and N are C^r manifolds of dimension m and n respectively, and f is a C^r function from M to N. Occasionally, e.g. when stating particularly important results which are likely to be read again in isolation, these assumptions will be repeated or specified.

Definition 1 An element $x \in M$ is a **regular point** for f if df_x is surjective, and it is a **critical point** for f otherwise. An element $y \in N$ is a **regular value** for f if every x in $f^{-1}(y)$ is regular, and it is a **critical value** otherwise.

Remark 2 Notice that $y \in N$ is regular if $y \notin f(M)$; in particular, when $m < n$ $y \in N$ is a regular value if and only if $y \notin f(M)$. On the other hand, $y \in N$ is a critical value if and only if it is the image of a critical point.

The set of all the regular points for f and the set of all the critical points for f are called RP_f and CP_f respectively; the set of all the regular values for f and the set of all the critical values for f are called R_f and C_f respectively.

Remark 3 The equality $C_f = f(CP_f)$ is always true, whereas the equality $R_f = f(RP_f)$ is generally false: first, because any $y \notin \operatorname{Im} f$ is regular; second, because $f^{-1}(y)$ may contain regular as well as critical points. Thus $R_f \cap \operatorname{Im} f = f(RP_f)$ if and only if, for each y in N, the elements of $f^{-1}(y)$ are either all regular or all critical.

Exercise 4 What is (if any) the inclusion holding between the two sets R_f and $f(RP_f)$?

Example 5 Let $f : \mathbb{R} \to \mathbb{R} : x \mapsto x^4 - 2x^2$. Observe that $CP_f = \{-1, 0, 1\}$, hence $\sqrt{2}$ and $-\sqrt{2}$ are regular points; however, $0 = f(\sqrt{2}) = f(-\sqrt{2})$ is a critical value.

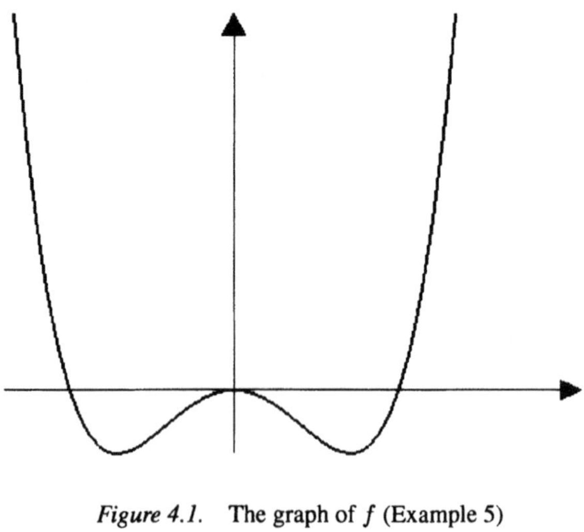

Figure 4.1. The graph of f (Example 5)

Example 6 Consider $f : \mathbb{R}^3 \to \mathbb{R}^2$, $(x, y, z) \mapsto (x^3 - 2xy - z, x^5 - 2z - 4)$. Then

$$Df(x, y, z) = \begin{bmatrix} 3x^2 - 2y & -2x & -1 \\ 5x^4 & 0 & -2 \end{bmatrix}$$

and the set CP_f of critical points of f is

$$\{(x, y, z) \in \mathbb{R}^3 : \operatorname{rank} Df(x, y, z) < 2\}$$

i.e., critical points for f are characterized by proportionality of the rows of the jacobian matrix. This yields

$$CP_f = \{(x, y, z) \in \mathbb{R}^3 : x = y = 0\}$$

i.e., every point not on the z-axis is regular. The set of critical values is

$$C_f = f(CP_f) = \{(u, v) \in \mathbb{R}^2 : v = 2u - 4\}$$

and the image of the set of regular points is

$$f(RP_f) = \mathbb{R}^2$$

For example, for each $(u, v) \in \mathbb{R}^2$,

$$(x, y, z) \equiv \left(1, \frac{-2u + v + 5}{4}, -\frac{v + 3}{2}\right)$$

is a regular point and

$$f(x, y, z) = (u, v)$$

In this example, too, the set R_f of regular values is strictly included in $f(RP_f)$.

Example 7 Let M and N be both equal to S^1, and let

$$f : S^1 \to S^1, \ (x, y) \mapsto (x^2 + y^2)(\cos \pi y, \sin \pi y)$$

We are going to work with the atlas $\{(\psi_\lambda, V_\lambda, U_\lambda)\}_{\lambda \in \{1,2\}}$ of Example 22, Chapter 2. Call P, Q, R, S the distinguished elements of S^1 having coordinates $(1, 0)$, $(-1, 0)$, $(0, 1)$, $(0, -1)$ respectively. Observe that

$$f(P) = f(Q) = P \qquad f(R) = f(S) = Q$$

and that f is bijective ("making a full round on $S^1 \setminus \{Q\}$") when restricted to the open arcs $S\overset{\frown}{P}R$, $R\overset{\frown}{Q}S$. On the other hand, f is not bijective ("making a half round back and forth with turn at Q") when restricted to the open arcs $P\overset{\frown}{R}Q$, $Q\overset{\frown}{S}P$. These four arcs are the maximal ones having image under f which is fully represented by means of a single chart (ψ_1, V_1, U_1) or (ψ_2, V_2, U_2). In other words, as the variable point X runs through S^1 with a single counterclockwise rotation (starting at R or at S), its image point $f(X)$ covers S^1 twice (starting at Q), once counterclockwise and once clockwise. Thus Q is a "turning point" for the "orbit" $f(X)$, and from this intuitive description it is perhaps already clear that Q is a critical value (in fact, the only critical value) of f.

In order to discuss the matter in detail by means of the local representations of f, we need a few preliminaries from elementary calculus. Consider the following four functions:

$$\begin{aligned} s &\equiv \text{sign} \circ \sin \\ k &\equiv \arccos \circ \cos \\ \gamma &\equiv sk \\ \eta &\equiv sk + \pi(1 - s) \end{aligned}$$

Then s is piecewise constant, with $s(t) = (-1)^{\lfloor t/\pi \rfloor}$ for every $t \in \mathbb{R}$; k is continuous, piecewise linear, with corners at $t = z\pi$, $z \in \mathbb{Z}$, and derivative $k' = s$ where defined; γ is piecewise linear, with jumps at $t = (2z+1)\pi$, $z \in \mathbb{Z}$, and unitary slope; and η is piecewise linear, with jumps at $t = 2z\pi$, $z \in \mathbb{Z}$, and unitary slope.

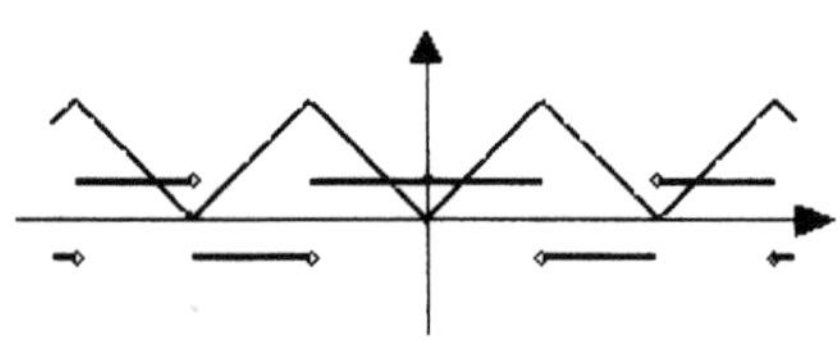

Figure 4.2. The graph of k

In particular, γ restricts to the identity on $(-\pi, \pi)$, and η restricts to the identity on $(0, 2\pi)$ and to the function $t \mapsto t + 2\pi$ on $(-2\pi, 0)$.

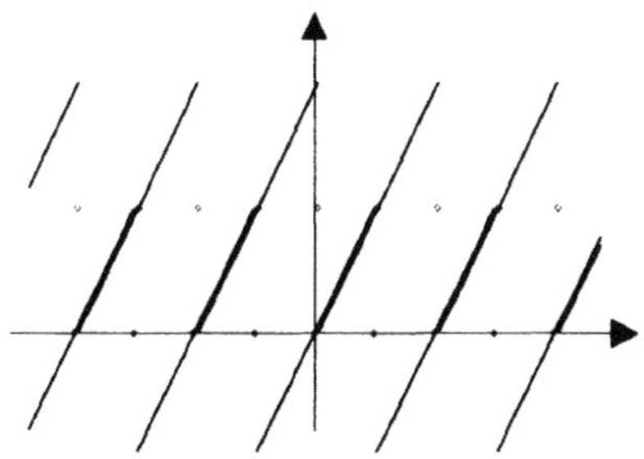

Figure 4.3. The graph of γ and η

The local representations for the restrictions of f to the arcs introduced above are as follows:

$$\widehat{SPR} \quad (\psi_1 \circ f \circ \phi_1)_{|(-\frac{\pi}{2},\frac{\pi}{2})} : t \to \gamma(\pi \sin t) = \pi \sin t \tag{4.1}$$

$$\widehat{RQS} \quad (\psi_1 \circ f \circ \phi_2)_{|(\frac{\pi}{2},\frac{3\pi}{2})} : t \to \gamma(\pi \sin t) = \pi \sin t \tag{4.2}$$

$$\widehat{QSP} \quad (\psi_2 \circ f \circ \phi_1)_{|(-\pi,0)} : t \to \eta(\pi \sin t) = \pi \sin t + 2\pi \tag{4.3}$$

$$\widehat{PRQ} \quad (\psi_2 \circ f \circ \phi_1)_{|(0,\pi)} : t \to \eta(\pi \sin t) = \pi \sin t \tag{4.4}$$

It is clear that the right-hand sides of (4.1)-(4.4) are C^∞. Computation of the derivative is now easy, and it shows that the critical points for f are characterized by the condition $\cos t = 0$, which is satisfied for $t = -\frac{\pi}{2}$ in (4.3), for $t = \frac{\pi}{2}$ in (4.4), and nowhere else. R and S are critical points but regular values for f, P is both a regular point and a regular value for f, and Q is a regular point but the unique critical value of f. Hence in this example

$$R_f = S^1 \setminus \{P\} \qquad f(RP_f) = S^1$$

We end up this section with a simple property characterizing regularity.

Lemma 8 *If $f : M \to N$ is of class C^1, the set of regular points of f is open in M.*

Proof. Let $x \in RP_f$, and let $g : U \to S$, $u \mapsto (\xi \circ f \circ \phi)(u)$ be some local representation of f, with (ϕ, U, V) local parametrization of M around x, $x = \phi(w)$, and (ξ, T, S) chart of N at $f(x)$, as usual. Since df_x is surjective, so is dg_w, and hence there exists a nonsingular $n \times n$ submatrix $Dg(w)$. Since g is a C^1 function as f, for every u in some open subset U' of U containing w, the corresponding submatrix of $Dg(u)$ is nonsingular too. Thus dg_u is onto $\forall u \in U'$, and so is df_z for any z in $V' \equiv \phi(U')$, which is an open neighborhood of x in M. ■

3. Inverse images of regular values

3.1 The regular value theorem

The implicit function theorem ([3], Theorem 13.7, p. 374) asserts some regularity of the set X of solutions to an equation

$$f(x) = y \tag{4.5}$$

when $f : D_f \subseteq \mathbb{R}^{n+k} \to \mathbb{R}^n$ is a sufficiently regular function. Namely, if $x \in X$ is such a solution, f is of class C^1 in a neighborhood of x, and the differential of f at x is surjective, then there exists a neighborhood U of x such that $X \cap U$ is the graph of a uniquely determined C^1 function.

The following theorem has a central role in the study of manifolds, and extends the implicit function theorem to functions defined between manifolds. According to the theorem, whenever y is a regular value for f, the set of solutions to equation (4.5) is either empty or a C^1 manifold.

Theorem 9 (Regular value theorem) *Let M and N be C^r manifolds of dimensions m and n respectively, let $f : M \to N$ be a C^r function, and let y be a regular value for f. If $y \in \operatorname{Im} f$, $f^{-1}(y)$ is a C^r manifold of dimension $m - n$.*

Proof. Let

$$Q \equiv f^{-1}(y) \subseteq M$$

let $q \equiv m - n \geq 0$ (recall Remark 2), and let x be any element of Q. By assumption, f is C^r submersive at x, and Theorem 66 of Chapter 3 applies. We must distinguish two cases.

$(m > n)$ There exist charts (ψ, X, U) for M at x and (ξ, Y, Z) for N at y, and corresponding parametrizations (ϕ, U, X) and (χ, Z, Y), with Z open subset of $\mathbb{R}^n$ and $U = Z \times H$ for some open subset H of $\mathbb{R}^q$, such that the induced local representation of f is the restriction to U of the canonical projection of $\mathbb{R}^m$ into $\mathbb{R}^n$

$$g \equiv \xi \circ f \circ \phi = (\pi_1)_{|U} : U \to Z, \quad (z, h) \mapsto z$$

Let now $w \equiv \xi(y)$. For each (z, h) in U, $\phi(z, h)$ is in Q if and only if $g(z, h) = w$, that is, $z = w$. Thus

$$Q \cap X = \phi(\{w\} \times H)$$

or, equivalently,

$$\psi(Q \cap X) = \{w\} \times H$$

It follows that $\left((\pi_2)_{|\{w\} \times H} \circ \psi, Q \cap X, H\right)$ is a C^r chart for Q at x, since $(\pi_2)_{|\{w\} \times H}$ is a linear isomorphism of $\{w\} \times H$ with H, $Q \cap X$ is open in Q (because X is open in M), and H is open in $\mathbb{R}^q$. The corresponding parametrization is $(\phi \circ \varrho, H, Q \cap X)$, where $\varrho : h \mapsto (w, h)$ is the inverse of $(\pi_2)_{|\{w\}\times H}$. The set of all triples $\left((\pi_2)_{|\{w\}\times H} \circ \psi, Q \cap X, H\right)$ obtained as x varies in Q is an atlas for Q, making it a C^r manifold of dimension q.

($m = n$) There exist charts (ψ, X, W) at x and (ξ, Y, W) at y inducing a local representation of f which is equal to id_W. In particular, $f_{|X} : X \to Y$ is a bijection, and

$$Q \cap X = \{x\}$$

Thus Q is a set of isolated points of M, and hence a C^∞ manifold of dimension 0 (Remark 34 of Chapter 2). ■

Definition 10 For each $n \in \mathbb{N}$, and each $x_0 \in \mathbb{R}^n$, let

$$S_r^{n-1}(x_0) \equiv \{x \in \mathbb{R}^n : \|x - x_0\| = r\}$$

be the **sphere** of $\mathbb{R}^n$ of radius r centered at x_0. Let S_r^{n-1} denote the sphere centered at 0, and let S^{n-1} denote the sphere of $\mathbb{R}^n$ centered at 0 and of radius 1, in short the **unit sphere** of $\mathbb{R}^n$.

Example 11 Consider $f : \mathbb{R}^n \to \mathbb{R}$, $x \to \|x\|^2 = xx$. Then, $Df(x) = 2x$. Therefore, df_x is surjective unless $x = 0$; $CP_f = \{0\}$, $RP_f = \mathbb{R}^n \setminus \{0\}$, $C_f = \{0\}$, $R_f = \mathbb{R} \setminus \{0\}$. Thus $S_r^{n-1} = f^{-1}(r^2)$ is an $(n-1)$-dimensional submanifold of $\mathbb{R}^n$. Needless to say, $S_r^{n-1}(x_0)$ is an $(n-1)$-dimensional submanifold of $\mathbb{R}^n$, too.

Example 12 Consider $f : \mathbb{R}^3 \to \mathbb{R}$, $(x, y, z) \to x^2 + y^2 + z^2 + R^2 - r^2 - 2R\sqrt{x^2 + y^2}$, where R and r are positive real numbers such that $R > r$. Then, for $(x, y) \neq (0, 0)$ and any z,

$$Df(x, y, z) = 2\left[\begin{array}{ccc} x\left(1 - \frac{R}{\sqrt{x^2+y^2}}\right) & y\left(1 - \frac{R}{\sqrt{x^2+y^2}}\right) & z \end{array}\right]$$

whereas f is not differentiable at any point of the z-axis (Z from now on). Let then g be the restriction of f to $\mathbb{R}^3 \setminus Z$. Since

$$CP_g = \left\{(x, y, z) \in \mathbb{R}^3 : x^2 + y^2 = R^2 \text{ and } z = 0\right\}$$

if (x, y, z) is a critical point of g, the value of g at (x, y, z) is $-r^2 < 0$. It follows that 0 is a regular value for g. This shows (in a third way; see example 36 of Chapter 2) that $\mathcal{T}^2$, the torus of $\mathbb{R}^3$, is a two-dimensional manifold.

Proposition 13 *Let $m > n$, $f : M \to N$ be a C^r function, let y be a regular value for f, and let x be any element of the manifold $Q \equiv f^{-1}(y)$. Then*

$$T_x Q = \ker df_x$$

Proof. Consider the restriction $f_{|Q} : Q \to N$, a constant function. From Exercise 37 of Chapter 3, $d\left(f_{|Q}\right)_x : T_xQ \to T_yN$ is the zero linear function. From Exercise 42 of the same chapter, $(df_x)_{|T_xQ}$ is the zero linear function as well. Thus

$$T_xQ \subseteq \ker df_x$$

Since df_x is surjective by assumption, the dimension theorem ([78], Theorem 3, p. 99), gives

$$\dim \ker df_x = m - n$$

Since the dimension of Q has just been proved to be $m - n$, Proposition 3 of Chapter 3 gives

$$\dim T_xQ = m - n$$

and the conclusion follows (from [78], Corollary 1, p. 53). ∎

Example 14 Let x be an element of S_r^{n-1}; then $T_xS_r^{n-1} = \{v \in \mathbb{R}^n : xv = 0\}$. More generally, if x is an element of $S_r^{n-1}(x_0)$, then

$$T_xS_r^{n-1}(x_0) = \{v \in \mathbb{R}^n : (x - x_0)\,v = 0\}$$

Thus a normal vector for the tangent space to the sphere at a point x is the radius connecting x with the center of the sphere.

Example 15 Let $P = (x, y, z)$ be an element of $\mathcal{T}^2$, the torus of $\mathbb{R}^3$ (Example36, Chapter 2); then, taking into account Example 12,

$$T_P\mathcal{T}^2 =$$
$$\left\{(u, v, w) \in \mathbb{R}^3 : \left(x - \frac{x}{\sqrt{x^2+y^2}}R\right)u + \left(y - \frac{y}{\sqrt{x^2+y^2}}R\right)v + zw = 0\right\}$$

Thus a normal vector for the tangent space to the torus of $\mathbb{R}^3$ at a point P is the radius connecting P with the center of the meridian circle containing P.

Corollary 16 *Let M and N be C^r manifolds of the same dimension, respectively, let $f : M \to N$ be a C^r function, and let y be a regular value for f such that $f^{-1}(y)$ is compact. Then $f^{-1}(y)$ is a finite subset of M.*

Proof. From the second part of the proof of Theorem 9, $f^{-1}(y)$ is a set of isolated points. Then, for each $x \in f^{-1}(y)$, there exists an open set W_x (in $\mathbb{R}^l$) such that $W_x \cap f^{-1}(y) = \{x\}$. The family $\mathcal{W} \equiv \{W_x\}_{x \in f^{-1}(y)}$ is an open covering of $f^{-1}(y)$, and, $f^{-1}(y)$ being compact, there exists a finite subcovering $\mathcal{W}' = \{W_{x_i}\}_{i=1}^r$. Thus

$$f^{-1}(y) = f^{-1}(y) \cap \left(\cup_{i=1}^r W_{x_i}\right) = \cup_{i=1}^r f^{-1}(y) \cap W_{x_i} = \{x_i\}_{i=1}^r$$

∎

3.2 Regular values and nonlinear systems of equations

A summary of the conclusions reached by means of the regular value theorem is given in the following table:

For a regular value y of $f : M \to N$

1.	if $\dim M > \dim N$	either $f^{-1}(y) = \emptyset$, or $f^{-1}(y)$ is a submanifold of M of dimension $m - n$	
2.	if $\dim M = \dim N$	either $f^{-1}(y) = \emptyset$, or $f^{-1}(y)$ is set of isolated points of M	(4.6)
3.	if $\dim M = \dim N$ and $f^{-1}(y)$ is compact	either $f^{-1}(y) = \emptyset$, or $f^{-1}(y)$ is a finite subset of M	
4.	if $\dim M < \dim N$	$f^{-1}(y) = \emptyset$	

The content of the table can be given a very intuitive meaning. Consider, for fixed y, a nonlinear system of equations

$$f(x) = y \tag{4.7}$$

where $f : M \to N$ is a C^1 function which is defined on an open subset of $\mathbb{R}^m$ and takes values in $\mathbb{R}^n$, or, more generally, where f is defined on an m-dimensional manifold of $\mathbb{R}^l$ and takes values in an n-dimensional manifold of $\mathbb{R}^q$. In the latter case, system (4.7) must be considered as decomposed in several systems of the form

$$g(u) = s \tag{4.8}$$

where $g : U \to S$ is the local representation of f obtained by a specific choice of charts in the atlases of M and N, and hence U and S are open sets of $\mathbb{R}^m$ and $\mathbb{R}^m$ respectively, and s corresponds to y in the chosen chart at y (in the former case, just set $g \equiv f, U \equiv M$, and so on). Thus in all cases m is the number of variables appearing in the equation system under study, and n is the number of equations.

Suppose that, for some reason or another, y is known to be a regular value for f; equivalently, suppose that s is known to be a regular value for g, no matter what representation g has been chosen. Then we can draw the following preliminary conclusions concerning the set of solutions to (4.7):

1 when the number of variables is bigger than the number of equations, there can be solutions only in infinite number; in such a case, a description of the solution set entails some arbitrary parameters, their number $m - n$ being equal to the number of "exceeding" variables (the so called degree of indeterminacy of the system);

2 when the number of variables is equal to the number of equations, a solution (if any) has to be locally unique; in particular, if the domain of f is compact, or if all solutions are known to belong to a compact set, then there is at most a finite number of them.

3 when the number of variables is smaller than the number of equations, there are no solutions.

The actual scope of the above conclusions needs to be clarified. They all depend on the regularity assumption. But how could we know that s is a regular value for every g? In the first two cases, that is, when $n \leq m$, according to Definition 1 dg_u may have been found to be surjective by a rank check on its jacobian matrix $Dg(u)$ (an $n \times m$ matrix). Precisely, for each g, $Dg(u)$ may be a matrix of full row rank n no matter how u is chosen in U, or at least it may be known to have full row rank if u is chosen in the set of solutions to (4.8).

On the other hand, our third statement doesn't seem to be conclusive; in fact, it would appear rather circular. If $m < n$, $Dg(u)$ cannot have full row rank; indeed, we have already noticed (Remark 2) that s can be regular only if $g^{-1}(s)$ is empty, i.e., if system (4.8) has no solutions. Thus point 4 in the table is a pure tautology, as of now. However, the transversality theorem (Theorem 26, Chapter 6) will make sense of it. As a matter of fact, point 4 of table (4.6) and the transversality theorem turn out to be very useful tools in order to analyze the solution set to a system of nonlinear equations.

3.3 A partial converse of the regular value theorem and the normal bundle

Once established that nonempty inverse images of regular values are manifolds, the question naturally arises, whether or not all manifolds can be seen as inverse images of regular values for appropriately chosen functions. The answer is that this is true only locally, as stated in the first proposition of this section. This is enough to yield a simple proof of the fact that the normal bundle of a manifold has itself a manifold structure, as stated in the second proposition of the section.

Proposition 17 *Let N be a C^r manifold of $\mathbb{R}^p$ of dimension n, and let M be a C^r submanifold of N of dimension $m \leq n$. Then, for any $x \in M$, there exist an open neighborhood T of x in N, and a C^r function $k : T \to \mathbb{R}^{n-m}$, such that:*

1 0 *is a regular value for* k;

2 $k^{-1}(0) = M \cap T$.

Proof. If $m = n$, the statement of the proposition is trivial, because every chart (ψ, V, U) of M at x is a chart of N at x as well, and there exists only one function $k : V \to \mathbb{R}^0 = \{0\}$, which is of course constant; k has a (constant) surjective differential everywhere on V, hence the unique value of k is regular, and $k^{-1}(0)$ equals $V = M \cap V$.

Let then $q \equiv n - m$ be positive, and consider the insertion $i_{M,N}$ of M in N. By Corollary 34 of Chapter 3, $i_{M,N}$ is C^r immersive at x. By Theorem 52, there exist charts $(\overline{\psi}, X, W)$ and $(\overline{\xi}, Y, W \times H)$ at x for M and N respectively, with $(\overline{\phi}, W, X)$ and $(\overline{\chi}, W \times H, Y)$ corresponding parametrizations, such that the induced local representation of $i_{M,N}$ is the restriction to W of the canonical immersion of $\mathbb{R}^m$ into $\mathbb{R}^{m+q}$:

$$\overline{\xi} \circ i_{M,N} \circ \overline{\phi} = (ci)_{|W} : W \to W \times H, \quad w \mapsto (w, 0) \tag{4.9}$$

where H is an open neighborhood of 0 in $\mathbb{R}^q$ (in particular, $i_{M,N}(X) \subseteq Y$, that is, $X \subseteq Y$). Since X is open in M, there exists an open neighborhood $\hat{T}$ of x in N such that $X = M \cap \hat{T}$. Let now

$$\begin{aligned} T &\equiv Y \cap \hat{T} \\ S &\equiv \overline{\xi}(T) \\ \xi &\equiv \overline{\xi}_{|T,S} \\ k &\equiv \pi_2 \circ \xi = (\pi_2)_{|S} \circ \xi \end{aligned}$$

and notice that T is an open neighborhood of x in N, S is an open neighborhood of s in $W \times H$, $k : T \to \mathbb{R}^q$ is of class C^r, and $M \cap T = X \cap Y = X$. Thus $X \subseteq T$, ξ is well defined on $i_{M,N}(X) = X$, and

$$\begin{aligned} \xi(X) &= \left(\overline{\xi} \circ i_{M,N} \circ \overline{\phi}\right)(W) \\ &= (ci)_{|W}(W) \\ &= W \times \{0\} \end{aligned}$$

It follows that $W \times \{0\} \subseteq S$, and that for each element $(w, 0)$ of S there exists an element z of X such that $\xi(x) = (w, 0)$. Since ξ is injective and $X \subseteq T$, $z = \xi^{-1}((w, 0))$, and hence

$$\begin{aligned} k^{-1}(0) &= \left(\xi^{-1} \circ (\pi_2)_{|S}^{-1}\right)(0) \\ &= \xi^{-1}(W \times \{0\}) \\ &= X \\ &= M \cap T \end{aligned}$$

Finally, for each $t \in T$,

$$dh_t = \pi_2 \circ d\xi_t$$

has full rank q because $d\xi_t$ is a linear isomorphism (see Proposition 6, Chapter 1), and hence every $w \in W$ is a regular value for h. ∎

Proposition 18 *Let M be an m-dimensional C^r manifold of $\mathbb{R}^l$. The normal bundle NM is an l-dimensional C^{r-1} manifold of $\mathbb{R}^{2l}$.*

Proof. If $m = l$, $NM = M \times \{0\}$, and the conclusion follows directly from Proposition 40, Chapter 2.

Let then $q \equiv l - m$ be positive, and let x be any element of M. By the last proposition, there exist an open neighborhood T of x in $\mathbb{R}^l$, and a C^r function $k : T \to \mathbb{R}^q$, such that 0 is a regular value for k, and $k^{-1}(0) = M \cap T$: As a matter of fact, in the proof of the proposition it has been shown that $k^{-1}(0) = X$ for some local parametrization $(\overline{\phi}, W, X)$ of M around x, and that every $t \in T$ is a regular point for k, that is, rank $dk_t = q$ and the rows of the jacobian matrix $Dk(t)$ are linearly independent. Moreover, by Proposition 13, for each $t \in X$, $T_tX = \ker dk_t$, that is, $Dk(t)v = 0$ for each $v \in T_tX$, and the rows of the jacobian matrix $Dk(t)$ belong to N_tM. With the position

$$\nu : W \times \mathbb{R}^q \to NX \subseteq \mathbb{R}^{2l}, \quad (w, z) \mapsto \left(\overline{\phi}(w), zDh\left(\overline{\phi}(w)\right)\right)$$

the triple $(\nu, W \times \mathbb{R}^q, NX)$ is a local parametrization of NM around (t, n) for each $(t, n) \in NX$. ■

4. Proper functions

As we have seen in the Corollary 16, if y is a regular value for f and $\dim M = \dim N$, the solution set to the equation $f(x) = y$ is finite if it is compact. The compactness of M is sufficient for that property to hold. A weaker but still sufficient condition for that is about to be defined. Up to now, and as it usually happens in the second part of the book, we are mainly concerned with functions having subsets of euclidean spaces as domain and value set. Compactness of the domain is usually not assumed. However, in view of applications contained in the last chapter of the book, where the relevant vector spaces are infinite dimensional, with (utility) functions as elements, we give a formulation in terms of arbitrary metric spaces ([90], p. 119) rather than standard euclidean spaces. The following theorem provides a foundation to the main definition of the section.

Theorem 19 *Let (X, d) and (Y, d') be metric spaces, and let $f : X \to Y$ be continuous. The following conditions are equivalent:*

1. *f is closed and $f^{-1}(y)$ is compact for each $y \in Y$;*
2. *$f^{-1}(K)$ is a compact subset of X for each compact subset K of Y;*
3. *every sequence $(x_n)_{n\in\mathbb{N}}$ in X such that $(f(x_n))_{n\in\mathbb{N}}$ converges has a converging subsequence $(x_{n_k})_{k\in\mathbb{N}}$.*

Proof. ($1 \Rightarrow 3$) Let $(x_n)_{n\in\mathbb{N}}$ be a sequence in X such that $(f(x_n))_{n\in\mathbb{N}}$ converges. For each $m \in \mathbb{N}_0$, let $S_m \equiv \{x_{n+m}\}_{n\in\mathbb{N}}$; since the sequence $(f(x_{n+m}))_{n\in\mathbb{N}}$ is a subsequence of $(f(x_n))_{n\in\mathbb{N}}$, if y is the limit of $(f(x_n))_{n\in\mathbb{N}}$, it is the limit of $(f(x_{n+m}))_{n\in\mathbb{N}}$ as well ([3], Theorem 4.5, p. 72). Thus y belongs to $\overline{f(S_m)}$ for each $m \in \mathbb{N}_0$. ([90], Lemma 10.2, p. 128). Since f is continuous, $f(\overline{S}_m) \subseteq \overline{f(S_m)}$ ([90], Theorem 7.1, p. 103); since f is closed, $f(\overline{S}_m)$ is a closed set containing $f(S_m)$. It follows that $f(\overline{S}_m) = \overline{f(S_m)}$ for each $m \in \mathbb{N}_0$. Summing up, for each $m \in \mathbb{N}_0$, y belongs to $f(\overline{S}_m)$, that is, the set $C_m \equiv \overline{S}_m \cap f^{-1}(y)$ is nonempty, and closed (by the first condition, $f^{-1}(y)$ is compact, hence closed as well, [90], Theorem 5.3, p. 166 and Theorem 6.1, p. 93). Since the sequence $(C_m)_{m\in\mathbb{N}_0}$ is nonincreasing (w.r. to inclusion), every finite subfamily of $\{C_m\}_{m\in\mathbb{N}_0}$ has nonempty intersection. By compactness of $f^{-1}(y)$, there exists some x belonging to $\bigcap_{m\in\mathbb{N}_0} C_m = \left(\bigcap_{m\in\mathbb{N}_0} \overline{S}_m\right) \cap f^{-1}(y)$ ([90], Theorem 5.9, p. 170). In particular, $f(x) = y$, and every neighborhood of x meets every S_m ([90], Theorem 6.5, p. 95). For each $k \in \mathbb{N}$, there exists some $m_k \in \mathbb{N}_0$ such that x_{m_k} is in $S_{m_{k-1}}$, and $d(x_{m_k}, x) < \frac{1}{k}$. The sequence $(m_k)_{k\in\mathbb{N}}$ is strictly increasing, and the subsequence $(x_{m_k})_{k\in\mathbb{N}}$ converges to x.

($3 \Rightarrow 2$) Let K be compact in Y, and hence closed and sequentially compact ([90], Theorem 7.4, p. 181), and let $(x_n)_{n\in\mathbb{N}}$ be a sequence in $f^{-1}(K)$, which is a closed subset of X because f is continuous. The sequence $(f(x_n))_{n\in\mathbb{N}}$ is in K, and there exists a subsequence $(f(x_{n_k}))_{k\in\mathbb{N}}$ converging to some $y \in K$. For each $k \in \mathbb{N}$, let $z_k \equiv x_{n_k}$. Thus $(f(z_k))_{k\in\mathbb{N}}$ converges to y, and by the third condition $(z_k)_{k\in\mathbb{N}}$ has a converging subsequence $(z_{k_h})_{h\in\mathbb{N}}$, which is a subsequence of $(x_n)_{n\in\mathbb{N}}$ as well. The limit of $(z_{k_h})_{h\in\mathbb{N}}$ is in $f^{-1}(K)$, the latter being a closed set. This proves that $f^{-1}(K)$ is a sequentially compact subset of X, and hence compact.

($2 \Rightarrow 1$) Since every singleton in a topological space is a compact set, it must only be proved that f is closed. Let C be a closed subset of X, and let $(y_n)_{n\in\mathbb{N}}$ be a sequence in $f(C)$ converging to some $y \in Y$. This means that there exists a sequence $(x_n)_{n\in\mathbb{N}}$ in C with $y_n = f(x_n)$ for each $n \in \mathbb{N}$. The set $K \equiv \{y_n\}_{n\in\mathbb{N}} \cup \{y\}$ is compact because in any open covering of K the set containing y contains all but finitely many y_n's. By the second condition, $f^{-1}(K)$ is compact too. Since $(x_n)_{n\in\mathbb{N}}$ is a sequence in $f^{-1}(K)$, there exists a subsequence $(x_{n_k})_{k\in\mathbb{N}}$ which is convergent. Let $x \equiv \lim_k x_{n_k}$; then $x \in C$ because C is closed and, by continuity of f, $f(x) = y$ ([90], Theorem 10.3, p. 128). This shows that $y \in f(C)$, and hence that $f(C)$ is closed ([90], Lemma 10.2, p. 128). ∎

Definition 20 Let (X, d) and (Y, d') be metric spaces. A function $f : X \to Y$ is **proper** if it is continuous and one among conditions 1-3 of Theorem 19 holds.

Remark 21 The reader should perhaps be given notice that when the sets X, Y involved in the above definition fail to satisfy the first countability axiom ([90], p. 190), the third condition of Theorem 19 loses most of its importance; similarly, the second condition is not any more equivalent to the first when Y is not locally compact ([90], p. 182), or X is not Hausdorff ([90], p. 98). In a general topological setting, the first condition alone is often used in the definition of properness.

Remark 22 Notice that if f is a continuous injective function, f is proper if and only if it is closed, or if and only if it is a homeomorphism of X with $f(X)$.

Example 23 (Proper and non proper elementary functions) Any constant function with a non compact domain is not proper. The functions sin and cos, as any periodic function, are not proper. The function $\exp : \mathbb{R} \to \mathbb{R}_{++}$ (a homeomorphism) is proper. The function $f : \mathbb{R} \to \mathbb{R},\ x \mapsto e^x$ is not proper. The function $\exp : \mathbb{C} \to \mathbb{C}$ is not proper. The function $\ln : \mathbb{R}_{++} \to \mathbb{R}$ is proper. The power functions $\{f_n\}_{n\in\mathbb{N}}$, $f_n : \mathbb{R} \to \mathbb{R},\ x \mapsto x^n$, are proper.

Proposition 24 *If (X, d) and (Y, d') are metric spaces, X is compact, and $f : X \to Y$ is continuous, then f is proper.*

Proof. If $K \subseteq Y$ is compact, it is closed; then $f^{-1}(K)$ is closed, and hence compact ([90], Theorem 5.2, p. 165), and the second condition of Theorem 19 is satisfied. ■

Exercise 25 Let (X, d) be a metric space, and let C be closed in X. The insertion $i_{C,X}$ of C in X is proper.

Hint. Recall that $\left(C, d_{|C\times C}\right)$ is a metric space, where the topology induced by the distance coincides with the subspace topology ([90], Exercise 1, p. 135).

Proposition 26 *Let (X, d), (Z, d'), and (Y, d'') be metric spaces, and let $f : Z \to Y$, $g : X \to Z$ be continuous functions. Then*

1 if f and g are proper, so is $f \circ g$;

2 if $f \circ g$ is proper and g is surjective, then f is proper;

3 if $f \circ g$ is proper, so is g.

Proof. 1. Left as an exercise.

2. Let K be compact, hence closed in Y, let $J \equiv (f \circ g)^{-1}(K)$, and let $H \equiv g(J)$. Then J is compact in X by properness of $f \circ g$, and H is compact in Z by continuity of g. Since $f^{-1}(K) = H$ by surjectivity of g, f satisfies the second condition of Theorem 19.

3. Let K be compact hence closed in Z, let $J \equiv f(K)$, and let $H \equiv f^{-1}(J)$. Then J is compact in Y by continuity of f, and $(f \circ g)^{-1}(J)$ is compact by properness of $f \circ g$. Moreover, K is included in H, and $g^{-1}(K)$, a closed set by continuity of g, is included in $g^{-1}(H)$. Since $g^{-1}(H) = (f \circ g)^{-1}(J)$, $g^{-1}(K)$ is compact, and g satisfies the second condition of Theorem 19. ■

Corollary 27 *If (X, d) and (Y, d') are metric spaces, $f : X \to Y$ is proper, and C is closed in X, then the restriction $f_{|C} : C \to Y$ is proper.*

Proof. Just notice that $f_{|C} = f \circ i_{C,X}$, and consider Proposition 26. ■

Corollary 28 *If (X, d) and (Y, d') are metric spaces, $f : X \to Y$ is proper, T is closed in Y, and $S \equiv f^{-1}(T)$, then the restriction $f_{|S,T} : S \to T$ is proper.*

Proof. Since S is closed by continuity of f, $f_{|S,Y}$ is proper by the previous corollary. Since $f_{|S,Y} = i_{|T,Y} \circ f_{|S,T}$, $f_{|S,T}$ is proper by the third point of Proposition 26. ■

Corollary 29 *Let (X, d) be a topological space, let $\{(Y_\lambda, d_\lambda)\}_{\lambda \in \Lambda}$ be a family of metric spaces, and for each $\lambda \in \Lambda$ let $f_\lambda : X \to Y_\lambda$ be a proper function. Then the vector function*

$$(f_\lambda)_{\lambda \in \Lambda} : X \to \times_{\lambda \in \Lambda} Y_\lambda$$

is proper.

Proof. Let $f \equiv (f_\lambda)_{\lambda \in \Lambda}$; then $f_\lambda = \pi_\lambda \circ f$, and the claim follows from Proposition 26, point 3. ■

Corollary 30 *Let $\{(X_\lambda, d_\lambda)\}_{\lambda \in \Lambda}$ and $\{(Y_\lambda, d'_\lambda)\}_{\lambda \in \Lambda}$ be families of metric spaces, and for each $\lambda \in \Lambda$ let $f_\lambda : X_\lambda \to Y_\lambda$ be a proper function. Then the product function*

$$\times_{\lambda \in \Lambda} f_\lambda : \times_{\lambda \in \Lambda} X_\lambda \to \times_{\lambda \in \Lambda} Y_\lambda$$

is proper.

Proof. Let $f \equiv \times_{\lambda \in \Lambda} f_\lambda$, let K be a compact subset of $\times_{\lambda \in \Lambda} Y_\lambda$, and, for each $\lambda \in \Lambda$, let $K_\lambda \equiv \pi_\lambda(K)$. Then, for each $\lambda \in \Lambda$, K_λ is a compact subset of Y_λ, and $f_\lambda^{-1}(K)$ is a compact subset of X_λ; moreover, $f^{-1}(K)$ is closed, and contained in $H \equiv \times_{\lambda \in \Lambda} f_\lambda^{-1}(K_\lambda)$. Since H is a compact subset of $\times_{\lambda \in \Lambda} X_\lambda$ by Tychonov's theorem ([90], Theorem 1.1, p. 232), and $\times_{\lambda \in \Lambda} X_\lambda$ with the product topology is a Hausdorff space ([90], Theorem 1.1, p. 197), f satisfies the second condition of Theorem 19. ■

Corollary 31 *If (X, d) and (Y, d') are metric spaces, and X is compact, the projection $\pi_2 : X \times Y \rightarrow Y$ is proper.*

Proof. As any projection, π_2 is continuous ([90], proof of Theorem 8.5, p. 115). If $K \subseteq Y$ is compact, $\pi_2^{-1}(K) = X \times K$ is the product of compact sets, hence compact ([90], Theorem 5.7, p. 167). ■

Lemma 32 *If $f : M \rightarrow N$ is a proper C^1 function, the set of regular values for f is open in N.*

Proof. From Lemma 8, CP_f is closed in M, and so is $C_f = f(CP_f)$ in N (first condition of Theorem 19). The statement follows. ■

From now on, we shall be using the concept of properness for functions of manifolds into manifolds. Although we do not define it explicitly, it must be understood that all the manifolds we deal with are naturally endowed with the distance induced from the euclidean spaces in which they are included.

Proposition 33 *Let $f : M \rightarrow N$ be a C^r immersion which is injective. Then f is a C^r embedding if and only if it is proper.*

Proof. Immediate from Remark 22 and Definition 54 in Chapter 3. ■

If f is a proper C^1 function, m is equal to n, and we consider a sufficiently small neighborhood Y of a regular value y belonging to $\mathrm{Im}\, f$, the structure of $f^{-1}(Y)$ is pretty nice. It looks like a "stack of records" (see [66], page 26).

Proposition 34 *Let $f : M \rightarrow N$, be a C^1 function which is proper, let $\dim M = \dim N$, and let $y \in \mathrm{Im}\, f$ be a regular value for f. Then the following holds:*

1 $f^{-1}(y)$ is a finite set, namely, $f^{-1}(y) = \{x_1, ..., x_r\}$ for some $r \in \mathbb{N}$;

2 there exists an open neighborhood Y of y in N, and for each i an open neighborhood U_i of x_i in M, such that:

(a) $U_j \cap U_k = \emptyset$ if $j \neq k$;

(b) $f^{-1}(Y) = \cup_{i=1}^{r} U_i$;

(c) $f_{|U_i} : U_i \rightarrow Y$ is a diffeomorphism.

Proof. 1. This is just the content of Corollary 16.

2. Since $\dim M = \dim N$ and y is a regular value, $df_{x_i} : TM_{x_i} \rightarrow TN_y$ is an isomorphism for $i = 1, ..., r$. By the inverse function theorem (Theorem 48, Chapter 3), for each i there exist an open neighborhood Y_i of y in N, and an open neighborhood W_i of x_i in M, such that $f_{|W_i} : W_i \rightarrow Y_i$ is a diffeomorphism.

Moreover, since the x_i's are finite in number, the W_i's can be taken so that $W_j \cap W_k = \emptyset$ if $j \neq k$. We now construct the neighborhoods Y and $\{U_i\}_{i=1}^r$ with the required properties. By Proposition 9 of Chapter 1, since $\cup_{i=1}^r W_i$ is an open neighborhood of $f^{-1}(y)$ in M and f is closed, there exists an open neighborhood Y_0 of y in N such that

$$f^{-1}(Y_0) \subseteq \cup_{i=1}^r W_i$$

and $f^{-1}(Y_0)$ is an open neighborhood of $f^{-1}(y)$ in M, too (and thus of each x_i). Let then

$$\begin{aligned} Y &\equiv \cap_{i=0}^r Y_i \\ U_i &\equiv f^{-1}(Y) \cap W_i \qquad (i = 1, \ldots, r) \end{aligned}$$

It follows that Y is an open neighborhood of y in N, that for each i U_i is an open neighborhood of x_i in M, and that $U_j \cap U_k = \emptyset$ if $j \neq k$. Moreover,

$$f^{-1}(Y) \subseteq f^{-1}(Y_0) \subseteq \cup_{i=1}^r W_i$$

and hence

$$\begin{aligned} f^{-1}(Y) &= f^{-1}(Y) \cap [\cup_{i=1}^r W_i] \\ &= \cup_{i=1}^r [f^{-1}(Y) \cap W_i] \\ &= \cup_{i=1}^r U_i \end{aligned}$$

Finally, $f_{|U_i} : U_i \to f(U_i)$ is a restriction of the diffeomorphism $f_{|W_i}$, thus a diffeomorphism itself, and

$$\begin{aligned} f(U_i) &= f\left[f^{-1}(Y) \cap W_i\right] \\ &= f\left[f^{-1}(Y)\right] \cap f(W_i) \\ &= Y \cap Y_i \\ &= Y \end{aligned}$$

(the second equality holds by Proposition 1 of Chapter 1, and the third because $Y \subseteq f(M)$). ■

Remark 35 Concerning the second part of the proof, the reader may have thought that a ready candidate for the set Y of the statement of the Proposition were given by $\cap_{i=1}^r Y_i$ (with associated well chosen restrictions of the W_i's). However, that choice does not always work, as the following picture shows. In the picture, the set (c, d) is not a neighborhood of x_1, nor of x_2. In general, the problem arises because $f^{-1}(\cap_{i=1}^r Y_i)$ may strictly include $\cup_{i=1}^r W_i$. Hence a further restriction of $\cap_{i=1}^r Y_i$ has been unavoidable.

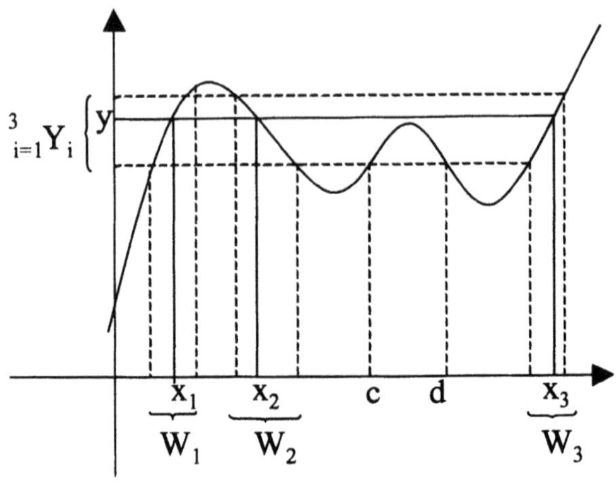

Figure 4.4. Isolated regular points

5. Local properness

When the map f fails to be proper, to be a regular point for f is not anymore an open condition (recall lemma 32). This is illustrated in the example below.

Example 36 A nonproper map with a regular value which is an accumulation point of critical values.

Let

$$f : \mathbb{R} \to \mathbb{R}, \; x \mapsto \begin{cases} \frac{x}{1+x^2} + \frac{\sin^2(x)}{x} & \text{if } \; x \neq 0 \\ 0 & \text{if } \; x = 0 \end{cases}$$

f is (at least) of class C^1, and 0 is a regular point for f. Indeed,

$$f'(x) = \frac{1-x^2}{(1+x^2)^2} + \frac{2x \sin x \cos x - \sin^2 x}{x^2} \qquad (x \neq 0)$$

$$f'(0) = 2$$

Observe moreover that the following is true:

$$x > 0 \Rightarrow f(x) > \frac{x}{1+x^2} > 0$$

$$x < 0 \Rightarrow f(x) < \frac{x}{1+x^2} < 0$$

so that $f^{-1}(0) = \{0\}$ and 0 is a regular value for f. As the plotted graph suggests, there is an increasing unbounded sequence $\{x_n\}_{n \in \mathbb{N}}$ of positive numbers

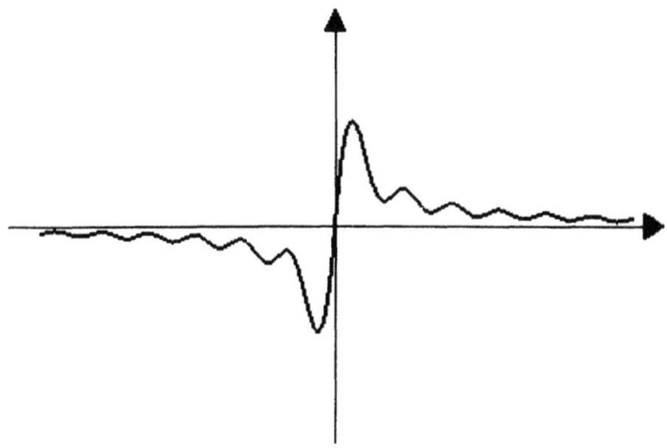

Figure 4.5. The regular value 0 is a limit point of critical values

such that:

$$\forall n \in \mathbb{N}, \quad f'(x_n) = 0 \qquad \text{and} \qquad \lim_{n \to +\infty} f(x_n) = 0 \tag{4.10}$$

Thus if $y_n = f(x_n)$, every y_n is a critical value for f, and every neighborhood of the regular value 0 contains all but a finite number of the y_n's. In order to convince yourself that an increasing unbounded sequence with properties (4.10) exists, let

$$u_n \equiv \frac{\pi}{4} + 2n\pi \qquad v_n \equiv \frac{3\pi}{4} + 2n\pi$$

and observe that for each n

$$f'(u_n) = \frac{1 - u_n^2}{(1 + u_n^2)^2} + \frac{u_n - \frac{1}{2}}{u_n^2} > \frac{-u_n^2}{u_n^4} + \frac{u_n}{2u_n^2} = \frac{u_n - 2}{2u_n^2} > 0$$

$$f'(v_n) = \frac{1 - v_n^2}{(1 + v_n^2)^2} + \frac{-v_n - \frac{1}{2}}{v_n^2} < 0$$

Then, for each n, the existence of $x_n \in (u_n, v_n)$ with the assumed properties is a consequence of a well known theorem in calculus ([3], Theorem 4.32, p. 85) for the continuous function f'. It is immediate that f is not proper.

In spite of this state of affairs, it is still possible to obtain a weaker but important result for maps which are not proper. To begin with, the following corollary establishes a local version of the results of Proposition 34 for functions between manifolds of the same dimension which satisfy a local properness condition.

Corollary 37 *Let:*

1 *M and N be two manifolds of the same dimension;*

2 *$f : M \to N$ be a C^1 function;*

3 *$y \in \operatorname{Im} f$ be a regular value for f;*

4 *Ω be an open neighborhood of $f^{-1}(y)$ such that f is proper on $\overline{\Omega}$.*

Then the following holds:

1 *$f^{-1}(y)$ is a finite set, namely, $f^{-1}(y) = \{x_1, ..., x_i, ..., x_r\}$ for some $r \in \mathbb{N}$;*

2 *there exist an open neighborhood Y of y in N, and for each i an open neighborhood U_i of x_i in Ω, such that:*

   (a) $U_j \cap U_k = \emptyset$ if $j \neq k$;
   (b) $f^{-1}(Y) \cap \overline{\Omega} = \cup_{i=1}^r U_i$;
   (c) $f_{|U_i} : U_i \to Y$ is a diffeomorphism.

Proof. 1. Let $g \equiv f|_\Omega$, $\overline{g} \equiv f|_{\overline{\Omega}}$, and notice that $f^{-1}(y) = g^{-1}(y) = \overline{g}^{-1}(y)$, hence y is a regular value for g as it is for f; moreover, Ω is a submanifold of M of the same dimension (by Proposition 18, Chapter 2). Point 3 of Table 4.6 applies to g, and we conclude that $g^{-1}(y)$ is finite if it is compact, which it is, because it coincides with $\overline{g}^{-1}(y)$ and $\overline{g}$ is proper.

2. Closely replicating the proof of Proposition 34, from the inverse function theorem we obtain r open neighborhoods $\{Y_i\}$ of y in N, and r disjoint open neighborhoods $\{W_i\}$ of the x_i's in Ω, such that $f_{|W_i} = \overline{g}_{|W_i} = g_{|W_i} : W_i \to Y_i$ is a diffeomorphism. By Proposition 9 of Chapter 1, since $\cup_{i=1}^r W_i$ is an open neighborhood of $\overline{g}^{-1}(y) = g^{-1}(y)$ in Ω and $\overline{g}$ is closed, there exists an open neighborhood Y_0 of y in N such that:

$$\overline{g}^{-1}(Y_0) \subseteq \cup_{i=1}^r W_i$$

and hence, since $\cup_{i=1}^r W_i \subseteq \Omega$,

$$g^{-1}(Y_0) = \overline{g}^{-1}(Y_0)$$

Then again, letting

$$Y \equiv \cap_{i=0}^r Y_i$$
$$U_i \equiv \overline{g}^{-1}(Y) \cap W_i$$

Y is an open neighborhood of y in N, the U_i's are disjoint open neighborhoods of the x_i's in Ω, and

$$\begin{aligned} \overline{g}^{-1}(Y) &= \overline{g}^{-1}(Y) \cap [\cup_{i=1}^r W_i] \\ &= \cup_{i=1}^r [\overline{g}^{-1}(Y) \cap W_i] \\ &= \cup_{i=1}^r U_i \end{aligned}$$

Thus, for each i, $f_{|U_i} = \overline{g}_{|U_i} : U_i \to \overline{g}(U_i)$ is a diffeomorphism, and

$$\begin{aligned} \overline{g}(U_i) &= \overline{g}[\overline{g}^{-1}(Y) \cap W_i] \\ &= \overline{g}[\overline{g}^{-1}(Y)] \cap \overline{g}(W_i) \\ &= Y \cap Y_i \\ &= Y \end{aligned}$$

■

Extending Lemma 32 to functions between manifolds of the same dimension which satisfy a local properness condition, Corollary 37 leads to establish that for such functions regularity is a "locally open" condition: if y is a regular value for f, every y' close enough to y is a regular value for the restriction of f to some open neighborhood of $f^{-1}(y)$.

Corollary 38 *Under the same hypotheses of Corollary 37, there exists an open neighborhood Y of y in N such that:*

1 $f(\partial\Omega) \cap Y = \emptyset$;

2 for each $y' \in Y$, y' is regular for $f_{|\Omega}$ and $\# f^{-1}(y') = \# f^{-1}(y)$.

Proof. By Corollary 37, for some $r \in \mathbb{N}$,

$$f^{-1}(y) = \{x_1, ..., x_i, ..., x_r\}$$

there exist an open neighborhood Y of y in N and disjoint open subsets U_i of Ω such that $f^{-1}(Y) \cap \overline{\Omega} = \cup_{i=1}^r U_i$, and, for $i \in \{1, \ldots, r\}$, x_i belongs to U_i, $f_{|U_i} : U_i \to Y$ is a diffeomorphism. Then every $u \in U_i$ is a regular point for $f_{|U_i}$, and hence every $y' \in Y$ is a regular value for $f_{|U_i}$. Finally,

$$\cup_{i=1}^r U_i = f^{-1}(Y) \cap \overline{\Omega} = f^{-1}(Y) \cap \Omega$$

(because $\forall i$, $U_i \subseteq \Omega$), so that

$$f^{-1}(Y) \cap \partial\Omega = \emptyset$$

and

$$\begin{aligned}\emptyset &= f\left[f^{-1}(Y)\cap\partial\Omega\right]\\ &= f\left[f^{-1}(Y)\right]\cap f(\partial\Omega)\\ &= Y\cap f(\partial\Omega)\end{aligned}$$

(by Proposition 1 of Chapter 1). Thus every $y' \in Y$ is regular value for $f_{|\Omega}$. ■

The final result of this section is a consequence of the simple fact that, by local compactness, every map between manifolds of the same dimension satisfies, at a regular value with a finite inverse image, the local properness condition referred to in the previous two corollaries.

Theorem 39 *Let $f : M \to N$ be any C^1 function between two manifolds M and N such that* $\dim M = \dim N$, *and let y be a regular value for f such that $f^{-1}(y)$ is finite (possibly empty). Then there exist an open neighborhood V of y in N and an open neighborhood U of $f^{-1}(y)$ in M, such that, for each $y' \in V$, y' is regular for $f_{|U}$. More precisely,*

1 if $f^{-1}(y) = \{x_i\}_{i=1}^r$ for some $r \in \mathbb{N}$, then $U = \cup_{i=1}^r U_i$, where

(a) $\{U_i\}_{i=1}^r$ is a family of pairwise disjoint open subsets of M,

(b) for each i, U_i is a neighborhood of x_i,

(c) for each i, $f_{|U_i} : U_i \to V$ is a diffeomorphism,

(d) $f(\partial U) \cap V = \emptyset$;

2 if $f^{-1}(y)$ is empty, so is $f^{-1}(y') \cap U$.

Proof. 1. It suffices to observe that in the proof of Corollary 37 the neighborhoods $\{W_i\}_{i=1}^r$ obtained by application of the inverse function theorem can be chosen to have compact closure, because M, being a manifold, is locally compact (Theorem 38 of Chapter 2). Then, if $\Omega \equiv \cup_{i=1}^r W_i$, Ω is an open neighborhood of $f^{-1}(y)$, $f_{|\overline{\Omega}}$ is proper because $\overline{\Omega}$ is compact, and Corollaries 37 and 38 apply. By taking as V the neighborhood Y in their proofs, and restricting to $U \equiv \cup_{i=1}^r U_i$ the theses of the corollaries, the claim is proved.

2. Let U be any open neighborhood of $f^{-1}(y)$, that is, any open subset of M, such that $\overline{U}$ is compact. Then $f(\overline{U})$ is compact, hence closed, and does not contain y; hence $V \equiv N \backslash f(\overline{U})$ is an open neighborhood of y such that $f^{-1}(V) \cap \overline{U}$ is empty. ■

6. Two applications

6.1 Submersiveness of functions related to the equilibrium manifold

We now present a theorem[1] which is going to be used several times in the second part of the book.

The starting point of the analysis is a function $f : \Xi \times \Pi \to \mathbb{R}^n$, where Ξ and Π are open subsets of $\mathbb{R}^n$ and $\mathbb{R}^m$, respectively. In short, each element π of Π describes the parameters characterizing an "economy", and each element ξ of Ξ is a vector of "endogenous" variables describing agents' behavior (consumption levels, prices, ...)[2]. The zeros of the function f, called "equilibria", are the main object under investigation. In fact, since it is shown that 0 is a regular value for f, the set $f^{-1}(0)$ is a manifold, the so called "equilibrium" manifold. Theorem 40 provides a simple way of addressing two different but connected problems which are related to the equilibrium manifold $f^{-1}(0)$.

The first problem concerns some function h, which is defined on $f^{-1}(0)$. It is essential to know whether or not h is submersive. As we are going to see in Chapters 8 and 13, a widely spread approach to equilibrium analysis, the so called "reduced approach", studies the properties of h directly. Under the approach presented in this book, the primary concern is on studying the properties of f. Theorem 40 links the two approaches, relating submersiveness of h with submersiveness of f.

The second problem, which is the object of Chapter 15, is the comparative static analysis of the intervention of an economic policy institution on the economy, as described by the equilibrium manifold $f^{-1}(0)$. Such an institution is assumed to control some policy tools, described by a vector τ, which influence the endogenous variables ξ via a function η. In turn, this influences some goal variables, described by a vector v, which are assumed to depend on ξ via a function g. If the differential of $g \circ \eta$ is surjective, we can conclude that, by a well chosen local change of τ, a local change in v along any direction can be achieved (see Definition 62 and Theorem 66 in Chapter 3 on this respect).

Theorem 40 shows that the key step to address both problems described above requires only the computation of the ranks of the two matrices $D_\xi f(\xi, \pi)$ and $D(f \times g)(\xi, \pi)$. In order to make future reference to the theorem statement handier, we have phrased it in a way that includes the content of the implicit function theorem ([3], [3] Theorem 13.7, pp. 374-5) among the conclusions. We acknowledge that this is rather strange, and we apologize right away.

[1] The main ideas of the present subsections are contained in [26].

[2] Accurate definitions of the above terms are given in the early chapters of second part of the book.

[3] In fact [3] states the implicit function theorem for C^1 function. A C^r version of the theorem can be found in [106] pp. 70-71.

Theorem 40 *Let $X \subseteq \mathbb{R}^n$ and $Z \subseteq \mathbb{R}^m$ be open sets, let $f : X \times Z \to \mathbb{R}^n$, $g : X \times Z \to \mathbb{R}^l$ be functions, where $l \leq m$, let $(\overline{x}, \overline{z})$ be any element of $X \times Z$, and let y be an element of $\mathbb{R}^n$.*
Assume:

1 *f and g are of class C^r,*

2 *$f(\overline{x}, \overline{z}) = y$,*

3 *the partial jacobian matrix $D_x f(\overline{x}, \overline{z})$ is nonsingular.*

Consider the restriction

$$h \equiv g_{|f^{-1}(y)} : f^{-1}(y) \to \mathbb{R}^l$$

Then

1 *there exist an open neighborhood U of $\overline{z}$ in Z and a unique C^r function $\eta : U \to \mathbb{R}^m$, such that $\overline{x} = \eta(\overline{z})$ and*

$$\forall z \in U, \qquad f(\eta(z), z) = y \tag{4.11}$$

2 *given the composite function*

$$k \equiv g \circ (\eta, id_U) : U \to \mathbb{R}^l, \quad z \mapsto g(\eta(z), z)$$

the following two statements are equivalent:

(a) *$D(f, g)(\overline{x}, \overline{z})$ has full rank $n + l$,*

(b) *$Dk(\overline{z})$ has full rank l;*

3 *if y is a regular value for f, then (a) and (b) are equivalent to the statement*

(c) *h is submersive at $(\overline{x}, \overline{z})$.*

Proof. 1. As said, the existence of the function η is the statement of the implicit function theorem.

2. Since f is of class C^r, $D_x f(x, z)$ is nonsingular in some open neighborhood of $(\overline{x}, \overline{z})$ in $X \times Z$, which without loss of generality we assume to be of the form $T \times U$, for some open neighborhood T of $\overline{x}$ in X. The jacobian matrix of (f, g) is

$$D(f, g) = \begin{pmatrix} D_x f & D_z f \\ D_x g & D_z g \end{pmatrix}$$

where, for the sake of readability, we drop for the moment the indication that full and partial jacobian matrices are evaluated at $(x, z) \in T \times U$. For each such (x, z), since $D_x f(x, z)$ is nonsingular, gaussian elimination in block form (Lemma 8, Chapter 1) gives

$$\begin{aligned} \operatorname{rank} D(f, g) &= \operatorname{rank} \begin{pmatrix} D_x f & D_z f \\ 0 & D_z g - D_x g (D_x f)^{-1} D_z f \end{pmatrix} \\ &= n + \operatorname{rank} \left(D_z g - D_x g (D_x f)^{-1} D_z f \right) \end{aligned}$$

Moreover, since by (4.11) the function $f \circ (\eta, id_U)$ is constant on its domain U, the chain rule (Proposition 39, Chapter 3) gives

$$df_{(x,z)} \circ (d\eta_z, id_{\mathbb{R}^m}) = 0$$

In terms of the jacobian matrices, taking into account the total differential formula (Proposition 40, Chapter 3), we have

$$\begin{bmatrix} D_x f & D_z f \end{bmatrix} \begin{bmatrix} D\eta \\ I \end{bmatrix} = 0$$

and hence

$$D\eta = -(D_x f)^{-1} D_z f$$

The rank of $Dk(z)$ can now be evaluated along similar lines:

$$\begin{aligned} \operatorname{rank} Dk(z) &= \operatorname{rank} dk_z \\ &= \operatorname{rank} dg_{(x,z)} \circ (d\eta_z, id_{\mathbb{R}^m}) \\ &= \operatorname{rank} \begin{bmatrix} D_x g & D_z g \end{bmatrix} \begin{bmatrix} D\eta \\ I \end{bmatrix} \\ &= \operatorname{rank} \left(-D_x g (D_x f)^{-1} D_z f + D_z g \right) \end{aligned}$$

which proves that conditions (a) and (b) are equivalent (as a matter of fact, for each (x, z) in $T \times U$, not just for $(\overline{x}, \overline{z})$).

3. If y is a regular value for f, $f^{-1}(y)$ is a C^r manifold (Theorem 9), and the stated properties of η imply that the triple $(\eta, id_U, U, \operatorname{graph} \eta)$ is a parametrization of $f^{-1}(y)$ around $(\overline{x}, \overline{z})$. Moreover, k is just the local representation of h induced by this parametrization (and by the trivial parametrization of $\mathbb{R}^l$). Hence the conclusion follows directly from Definition 25 and Proposition 26 of Chapter 3. ■

Corollary 41 *Let X, Z, f, g, y, and h be as in Theorem 40. If y is a regular value for f, h is a submersion if and only if Condition 2.(a) in the theorem's statement holds at every element of the manifold $f^{-1}(y)$. Under such an assumption, y is a regular value for h.*

Proof. It is only necessary to make clear that when (x, z) varies in $f^{-1}(y)$, regularity of y requires some $n \times n$ submatrix of the jacobian matrix $Df(x, z)$ (not necessarily $D_x f$) to be nonsingular. A construction similar to the one carried through in the theorem's statement and proof can be replicated, albeit under a different identification of $\mathbb{R}^n \times \mathbb{R}^m$ with the direct sum of two appropriate subspaces of dimensions n and m. ■

Example 42 Consider

$$\begin{aligned} f &: \mathbb{R}^2 \to \mathbb{R}, (x, z) \mapsto x^2 + z^2 - 1 \\ g &: \mathbb{R}^2 \to \mathbb{R}, (x, z) \mapsto \sin x + \cos z \end{aligned}$$

and h as in the theorem's statement, with $y = 0$. Clearly, $f^{-1}(0) = S^1$. Then $(1, 0)$ and $(0, 1)$ are a critical point and a regular point for h, respectively. Indeed,

$$D(f, g) = \begin{pmatrix} 2x & 2z \\ \cos x & -\sin z \end{pmatrix}$$

Moreover,

$$\begin{aligned} D_x f(1, 0) &= D_z f(0, 1) = 2 \\ \det D(f, g)(x, z) &= -(2x \sin z + 2z \cos x) \\ \det D(f, g)(1, 0) &= 0 \\ \det D(f, g)(0, 1) &= -2 \cos 1 \neq 0 \end{aligned}$$

Two other simple corollaries of Theorem 40 arise when $l = m$ and the function g is just the second projection π_2; this occurs frequently in the second part of the book.

Corollary 43 *Let X, Z, f, and y be as in Theorem 40, let pr be the restriction of π_2 to $f^{-1}(y)$*

$$pr : f^{-1}(y) \to Z, \quad (x, z) \mapsto z$$

and let z be any element of Z. If

1 y is a regular value for f;

2 y is a regular value for the partial function f_z;

then z is a regular value for pr.

Proof. From Theorem 9 and the first assumption, $f^{-1}(y)$ is a C^1 submanifold of $X \times Z$ of dimension m. If $pr^{-1}(z)$ is empty, we are done. Otherwise, let x be any element of $pr^{-1}(z)$, and let

$$g \equiv (\pi_2)_{|X \times Z}$$

The jacobian matrix of (f,g) at (x,z) is

$$D(f,g)(x,z) = \begin{pmatrix} D_x f(x,z) & D_z(x,z) \\ 0 & I \end{pmatrix}$$

and, by second assumption, $D_x f(x,z) = Df_z(x)$ is nonsingular, no matter how x has been chosen in $pr^{-1}(z)$. All the assumptions of Theorem 40 are satisfied, and it is enough to observe that $pr = g_{|f^{-1}(y)}$ in order to conclude that (x,z) is a regular point for pr. ∎

Corollary 44 *Let X, Z, f, pr, and y be as in Corollary 43, and let z be any element of Z. If*

1 *y is a regular value for f;*

2 *y is a regular value for the partial function f_z;*

3 *pr is surjective;*

4 *pr is proper;*

then

1 *z is a regular value for pr;*

2 *$pr^{-1}(z)$ is a finite set, i.e., $pr^{-1}(z) = \{(x^i, z)\}_{i=1}^{r}$ for some $r \in \mathbb{N}$;*

3 *there exist an open neighborhood V of z in Z, and for each i an open neighborhood U_i of (x^i, z) in $f^{-1}(y)$, such that*

   (a) $U_j \cap U_k = \emptyset$ if $j \neq k$;
   (b) $pr^{-1}(V) = \cup_{i=1}^{r} U_i$;
   (c) $pr_{|U_i} : U_i \to V$ is a diffeomorphism.

Proof. From Corollary 43 and the first two assumptions, the first claim is true. By the third assumption, $pr^{-1}(z)$ is nonempty. Together with the fourth assumption, this implies that Proposition 34 applies to the function pr, and last two claims follow. ∎

6.2 The Lagrange multiplier rule

Next theorem has already been stated - without proof - in an almost equivalent form in the chapter on mathematical prerequisites (Theorem 33, Chapter 1). Here we simply want to show how its statement translates in the language of differentiable manifolds, thereby allowing for a clean and direct rephrasing of its proof.

Theorem 45 *Let $Z \subseteq \mathbb{R}^m$ be open, and let $f : Z \to \mathbb{R}$, $g : Z \to \mathbb{R}^n$ be C^1 functions, where $m \geq n$. Let $y \in \mathbb{R}^n$ be an element of $\operatorname{Im} g$ and a regular value for g, let $X \equiv g^{-1}(y)$, and consider the restriction*

$$h \equiv g_{|X} : X \to \mathbb{R}$$

If z is a local maximum point of h, i.e., if there exists an open neighborhood W of z in X such that z solves the problem

$$\max_{x \in W} \ f(x) \quad s.t. \quad g(x) = y$$

then there exist a vector $\lambda \in \mathbb{R}^n$ such that z is a critical point of the function

$$L \equiv f - \lambda g$$

Proof. By the regular value theorem (Theorem 9), X is a C^1 manifold of $\mathbb{R}^m$ of dimension $m - n$. Let v be any element of $T_z X$, and let γ (Definition 2) be a curve in X through x at t, such that $\gamma(0) = z$, $\gamma(t) \in W$, and $\gamma'(t) = v$. The composite function $\eta \equiv f \circ \gamma$ achieves a maximum at 0, and hence

$$0 = \eta'(0) = df_z(v) = Df(z)\,v$$

Thus the gradient $Df(z)$ is orthogonal to every element of $T_z X$, that is,

$$Df(z) \in N_z X \tag{4.12}$$

By Proposition 13, $T_z X = \ker dg_z$, that is, for each $v \in T_z X$

$$Dg(z)\,v = 0$$

In other words, for each $i \in \{1, \ldots, n\}$,

$$Dg_i(z) \in N_z X$$

Moreover, the gradients $\{Dg_i(z)\}_{i=1}^n$ form a basis of $N_z X$ (see the proof of Proposition 18). This means that (4.12) is equivalent to

$$Df(z) = \sum_{i=1}^{n} \lambda_i Dg_i(z) = \lambda Dg(z)$$

■

Remark 46 The reader can easily check that nothing really changes in the theorem's statement and proof if the lagrangean function is defined as $f + \lambda g$ instead of $f - \lambda g$.

Chapter 5

MANIFOLDS WITH BOUNDARY

1. Manifolds with boundary

According to our definition of C^r manifold, given in Chapter 2, we cannot use our "differential topology tool kit" for many interesting sets such as a closed unit ball, a solid torus, a compact cylindrical surface. Those sets fail to be C^r manifolds because of points they have on their "boundaries". In fact, around those points they look like closed euclidean halfspaces or even convex cones. The definition of manifold with boundary deals with this issue, encompassing a wider class of objects which includes some of the examples we have just referred to.

Manifolds with boundary play a crucial role in the degree theory which will be presented in Chapter 7. As we will see, the domain of a homotopy function is $M \times [0, 1]$ where M is a manifold without boundary, that is the type of manifolds we have been studying up to now; since $[0, 1]$ is in fact a manifold with boundary, $M \times [0, 1]$ is such a manifold too, as it will be shown in Proposition 27.

Up to now, we have been talking about "boundary" and "interior" of subsets of topological spaces according to the common usage of general topology (see [90] p. 95 and p. 101). In this chapter, following the tradition of differential topology, we are going to define new meanings for the same two words. Which of the two meanings is relevant should be clear from the context; at any rate, we will be using different symbols, adding at times the qualification "differential" or "topological", in order to make ourselves clear.

Definition 1 A **closed euclidean halfspace** of dimension $m > 0$ is

$$H^m \equiv \{x \in \mathbb{R}^m : a(x) \geq 0\}$$

where $a : \mathbb{R}^m \to \mathbb{R}$ is a nontrivial linear function.
The topological boundary of H^m in $\mathbb{R}^m$ is

$$\delta H^m = \{x \in \mathbb{R}^m : a(x) = 0\} = \ker\ a$$

A standard choice of the function a is $\pi_1 : x \mapsto x_1$. In that case,

$$H^m = \{(x_1, x_2, ..., x_m) \in \mathbb{R}^m : x_1 \geq 0\}$$

and

$$\delta H^m = \{x \in \mathbb{R}^m : x_1 = 0\} = \{0\} \times \mathbb{R}^{m-1}$$

where the second equality holds up to the canonical linear isomorphism of $\{0\} \times \mathbb{R}^{m-1}$ with $\mathbb{R}^{m-1}$.

We now introduce a new and more general definition of manifold, which has the definition given in Chapter 2 as a special case.

Definition 2 A set $M \subseteq \mathbb{R}^l$ is a C^r m-**dimensional manifold** of $\mathbb{R}^l$ if, for any $x \in M$, there exists a function $\phi : U \to V$ where

1 U is an open subset of H^m,

2 V is an open neighborhood of x in M,

3 ϕ is a C^r diffeomorphism.

The triple (ϕ, U, V) is called a **local parametrization** of M around x. A family of local parametrizations $\{(\phi_\lambda, U_\lambda, V_\lambda)\}_{\lambda \in \Lambda}$ is called a **system of local parametrizations** of M provided that $M = \cup_{\lambda \in \Lambda} V_\lambda$.

The triple (ψ, V, U), where $\psi : V \to U$ is the inverse function of ϕ, is called a **chart** of M. A family of charts $\{(\psi_\lambda, V_\lambda, U_\lambda)\}_{\lambda \in \Lambda}$ such that $\cup_{\lambda \in \Lambda} V_\lambda = M$ is called an **atlas** of M.

Definition 3 Given a C^r manifold M and a subset N of M, if N is a C^r manifold, it is called a C^r **submanifold** of M. If M has dimension m and N has dimension n, $m - n$ is the **codimension** of N in M, denoted as codim N.

Remark 4 Recalling how the induced topology acts on H^m from $\mathbb{R}^m$, in order to prove that a subset M of $\mathbb{R}^l$ is a C^r manifold of dimension m, we can show that, for any $x \in M$, there exists a function $\phi : U \to V$ where

1 $U \equiv U' \cap H^m$ with U' is an open set in $\mathbb{R}^m$,

2 $V \equiv W \cap M$ with W is an open set in $\mathbb{R}^l$,

3 ϕ is a C^r diffeomorphism.

Definition 5 Let $\{(\phi_\lambda, U_\lambda, V_\lambda)\}_{\lambda\in\Lambda}$ be a system of local parametrizations of M. The **differential boundary** of M is

$$\partial M \equiv \cup_{\lambda\in\Lambda}\phi_\lambda\left(U_\lambda \cap \delta H^m_\lambda\right)$$

and the **differential interior** of M is the difference $M\backslash\partial M$.

Definition 6 A manifold M is a **manifold with boundary** if ∂M is nonempty, and it is a **manifold without boundary**, or **boundaryless manifold**, otherwise.

Remark 7 In the above definition, ∂M can be empty if and only if, for any parametrization, $U_\lambda \cap \delta H^m_\lambda$ is empty for each $\lambda \in \Lambda$; that is, if each U_λ is included in the interior of H^m_λ, and hence it is an open subset of $\mathbb{R}^m$. This shows that the present definition of manifold without boundary is just the definition we have been using in Chapters 2-4. On the other hand, if, for some λ, $U_\lambda \cap \delta H^m_\lambda$ is nonempty, then U_λ is open in H^m_λ but not in $\mathbb{R}^m$, and the relevant definition for ϕ being of class C^r is Definition 1 of Chapter 2.

Since the only linear function defined on a 0-dimensional vector space is the zero function, Definition 1 does not apply to the case $m = 0$. For our purposes, we can deal with the issue by simply convening that 0-dimensional manifolds are boundaryless.

The concept of differential boundary of a manifold is well defined, in the sense that it does not depend on the choice of the system of local parametrizations. Before proving this result in Proposition 17, we present some examples.

Example 8 A closed euclidean halfspace H^m is a C^∞ m-dimensional manifold; its boundary ∂H^m is an $(m-1)$-dimensional manifold without boundary, which is diffeomorphic to $\mathbb{R}^{m-1}$. To show that H^m is a C^∞ m-dimensional manifold, it is sufficient to observe that it has an atlas with a single chart $(\psi, V, U) \equiv (id_{H^m}, H^m, H^m)$. It is then immediate to conclude that $\partial H^m = \delta H^m$.

Example 9 $M \equiv [0,1]$ is a C^∞ one-dimensional manifold with boundary. Consider the halfspace $H^1 \equiv \{x \in \mathbb{R} : x \geq 0\}$ and take the following two parametrizations:

a. $U \equiv V \equiv [0,1)$ and $\phi \equiv id_{[0,1)}$;

b. $U \equiv [0,1)$, $V \equiv (0,1]$, and $\phi : [0,1) \to (0,1]$, $t \mapsto 1-t$.

In case b, since there is no open set U' in $\mathbb{R}$ such that $U' \cap H^1 = (0,1]$, we cannot make the simple, and wrong, choice of ϕ as $id_{(0,1]}$, unless we consider the halfspace $-H^1 = \{x \in \mathbb{R} : x \leq 0\}$.

When the dimension m is sufficiently low, m-dimensional manifolds can be classified. A full (and simple) classification of one-dimensional manifolds is given in the following theorem, which is going to be used in Chapter 7. We simply state it without proof[1].

Theorem 10 (Classification of compact one-dimensional manifolds) *Every connected compact one-dimensional manifold with boundary is diffeomorphic either to a circle or to a bounded closed interval.*

Remark 11 From Theorem 10 any connected compact one-dimensional manifold with boundary M is "equal to" either a circle or a closed interval via a diffeomorphism. Since the boundary of any bounded closed interval has two points and the boundary of circle is the empty set, Theorem 10 implies that the boundary of any compact one-dimensional manifold with boundary M has an even numbers of points.

Example 12 Let $M \equiv \overline{B}^2$, the closed unit ball in $\mathbb{R}^2$. We want to verify that M is a C^∞ two-dimensional manifold with boundary. Let us see a system of five local parametrizations.

a. Consider the halfspace $H^2 \equiv \{(x_1, x_2) \in \mathbb{R}^2 : x_2 \geq 0\}$. The first local parametrization is $(\phi, B^2(0,1), B^2)$, where $B^2(0,1)$ is the unit ball with center $(0,1)$ and $\phi(x_1, x_2) \equiv (x_1, x_2 - 1)$.

b. Consider $V \equiv \{(x_1, x_2) \in M \cap H^2 : x_2 > 0\}$.

Let $g : V \to \mathbb{R}^2,\ (x_1, x_2) \mapsto (x_1, 1 - x_1^2 - x_2^2)$. Observe that the jacobian matrix

$$Dg(x) = \begin{bmatrix} 1 & 0 \\ -2x_1 & -2x_2 \end{bmatrix}$$

has full rank if and only if $x_2 \neq 0$, as we assumed. As a consequence of the inverse function theorem (Theorem 43, Chapter 3), for any $x \in V$, there exists an open neighborhood $V' \subseteq V$ of x such that $g_{V'}$ is a diffeomorphism. Since g is one-to-one on V, as it is not difficult to prove, it follows that g is a diffeomorphism. Call $U \equiv g(V)$ and $\phi \equiv g^{-1}$; we obtain the local parametrization (ϕ, U, V). Notice that U is an open subset of H^2 containing $(-1,1) \times \{0\}$

c. Consider $V \equiv M \backslash H^2$ and proceed as in the previous case.

d. Take $H^2 \equiv \{x \in \mathbb{R}^2 : x_1 \geq 0\}$.

Consider $V \equiv \{(x_1, x_2) \in M \cap H^2 : x_1 > 0\}$. Let $g : V \to \mathbb{R}^2,\ (x_1, x_2) \mapsto (1 - x_1^2 - x_2^2, x_2)$. Let $U \equiv g(V)$. U is an open subset of H^2 containing $\{0\} \times (-1,1)$. Call $\phi \equiv g^{-1}$; we obtain the local parametrization (ϕ, U, V).

[1] A proof of the theorem can be found in [66], Appendix 2, or in [89], Appendix.

e. Take $V \equiv M \backslash H^2$ (being H^2 the halfspace of case d) and proceeds as in the previous cases.

Example 13 (An atlas for $\overline{B}^2$ with two charts) Consider the closed unit disk $\overline{B}^2$; its boundary is the unit circle S^1. The set obtained by removing one point of S^1 from $\overline{B}^2$ is open in $\overline{B}^2$; from the point of view of differential topology, such a set is equivalent to H^2 in the sense that they are diffeomorphic. This is shown by an atlas with two charts for $\overline{B}^2$ that we are about to construct. Let $P \equiv (1, 0)$ and $Q \equiv (-1, 0)$. The first chart maps $\overline{B}^2 \backslash \{P\}$ diffeomorphically on the infinite semiclosed stripe $[0, 1) \times \mathbb{R}$, and it is not difficult to see that the latter is diffeomorphic to H^2 (the reader is invited to construct such a diffeomorphism). The second chart does the same for $\overline{B}^2 \backslash \{Q\}$ and $(-1, 0] \times \mathbb{R}$. The idea underlying the construction of the first chart is simple. Consider the family of all the circles which are tangent in P to the vertical line through P, and contained in $\overline{B}^2$; each circle in the family has its center on the x-axis, with abscissa in $[0, 1)$, and each point of $\overline{B}^2$ which is distinct from P belongs to exactly one circle in the family. The set obtained by removing P from each such circle is diffeomorphic to a straight-line, and indeed our chart projects it from P onto the vertical line through its center. Here below we present the details for the first chart; the construction of the second chart is left as an exercise.

Let $V \equiv \overline{B}^2 \backslash \{P\}$, $U \equiv [0, 1) \times \mathbb{R}$, and define

$$\psi : V \to U, \quad (x, y) \mapsto (u(x, y), v(x, y))$$

$$u(x, y) \equiv \frac{1 - (x^2 + y^2)}{2(1 - x)}, \quad v(x, y) = \frac{y}{2}\left(1 + \frac{y^2}{(1 - x)^2}\right)$$

For each (x, y) in V, (ψ, V, U) is a chart at (x, y). This is seen as follows:

1 ψ is of class C^∞, since so are u and v;

2 ψ is a local diffeomorphism, since its jacobian matrix

$$D\psi(x, y) = \frac{1}{2(1 - x)^3}\begin{pmatrix} (1 - x)^3 - (1 - x)y^2 & -2(1 - x)^2 y \\ 2y^3 & (1 - x)^3 + 3(1 - x)y^2 \end{pmatrix}$$

is nonsingular on V

$$4 \det D\psi(x, y) = \left(1 + \frac{y^2}{(1 - x)^2}\right)^2$$

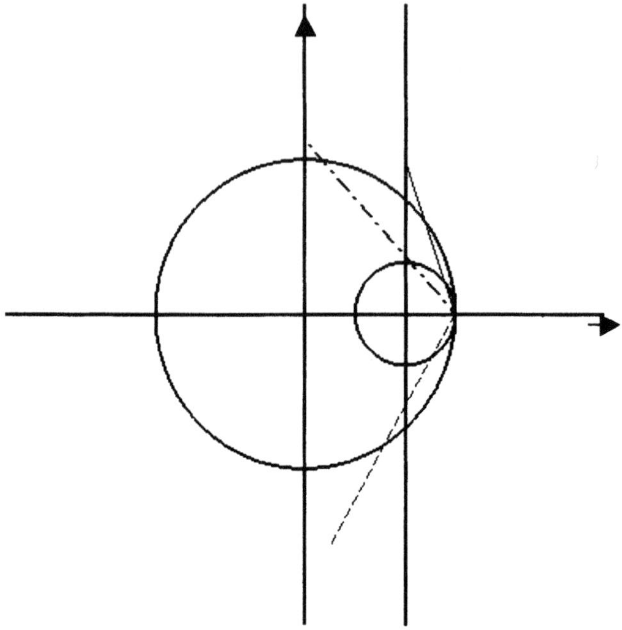

Figure 5.1. $S^1\backslash\{P\}$ is diffeomorphic to $\mathbb{R}$

3 ψ is injective on V because it projects from P each circle which is (contained in V) and tangent to $x = 1$ in P onto the line of equation $x = u$; the reader can check by straightforward computation that

$$\begin{cases} (x-u)^2 + y^2 = (1-u)^2 \\ \dfrac{y}{1-x} = \dfrac{v}{1-u} \end{cases}$$

Example 14 $M \equiv \{x \in \mathbb{R}^2 : x_1^2 + x_2^2 \leq 1 \text{ and } x_2 \geq 0\}$ is not a manifold with boundary, because of the points $(-1, 0)$ and $(1, 0)$. Suppose otherwise that M is a manifolds. This implies that any neighborhood of $(-1, 0)$ or $(1, 0)$ in δM is diffeomorphic to an open real interval (see also Proposition 25 below). However, this is not possible and it can be seen along the lines of the Example 28 of Chapter 2.

Our first three examples exhibit topological and differential boundaries which are the same. In our next examples the opposite is true, i.e., $\delta M \neq \partial M$.

Example 15 Let

$$M \equiv \{x \in \mathbb{R}^3 : x_1^2 + x_2^2 + x_3^2 \leq 1 \text{ and } x_3 = 0\}$$

Then,

$$\partial M = \left\{x \in \mathbb{R}^3 : x_1^2 + x_2^2 + x_3^2 = 1 \text{ and } x_3 = 0\right\}$$

while $\delta M = M$.

Example 16 Let $O \equiv (0,0)$, $H^2 \equiv \left\{(x_1, x_2) \in \mathbb{R}^2 : x_2 \geq 0\right\}$ and

$$M \equiv H^2 \backslash \{O\}$$

Then

$$\partial M = (\{0\} \times \mathbb{R}) \backslash \{O\}$$

and

$$\delta M = \{0\} \times \mathbb{R}$$

Proposition 17 *The definition of differential boundary of a manifold M (Definition 5) is consistent in the following sense: for any two local parametrizations* $(\phi_\lambda, U_\lambda, V_\lambda)$ *and* (ϕ_μ, U_μ, V_μ) *such that* $V_\lambda \cap V_\mu$ *is nonempty, then*

$$x \in \phi_\lambda (U_\lambda \cap \delta H_\lambda^m) \Leftrightarrow x \in \phi_\mu \left(U_\mu \cap \delta H_\mu^m\right)$$

In other words, the proposition states that the definition of differential boundary does not depend on the choice of the system of local parametrizations.

Proof. Let $(\phi_\lambda, U_\lambda, V_\lambda)$ and (ϕ_μ, U_μ, V_μ) be two local parametrizations of M around $x \in V_\lambda \cap V_\mu$, with $x = \phi_\lambda(u_\lambda) = \phi_\mu(u_\mu)$. Suppose that $x \in \phi_\lambda (U_\lambda \cap \delta H_\lambda^m)$ and let us show that $u_\mu \in U_\mu \cap \delta H_\mu^m$. To this purpose, consider the following sets and mappings (we call as usual ψ_λ and ψ_μ the inverse functions of ϕ_λ and ϕ_μ):

$$\begin{array}{ll} V \equiv V_\lambda \cap V_\mu & \\ S_\lambda \equiv \psi_\lambda (V) & S_\mu \equiv \psi_\mu (V) \\ T_\lambda \equiv S_\lambda \cap \delta H_\lambda^m & T_\mu \equiv S_\mu \cap \delta H_\mu^m \\ g \equiv \psi_\mu \circ (\phi_\lambda)_{|S_\lambda} & \end{array}$$

Observe that $x \in V$, $u_\lambda \in T_\lambda$, $u_\mu \in S_\mu$, $u_\mu = g(u_\lambda)$, and g is a diffeomorphism. It must be shown that indeed $u_\mu \in T_\mu$. Since $H_\mu^m \backslash \delta H_\mu^m$ is open in $\mathbb{R}^m$, it follows that $S_\mu \backslash T_\mu = S_\mu \cap (H_\mu^m \backslash \delta H_\mu^m)$ is open in $\mathbb{R}^m$. An identical conclusion holds for $S_\lambda \backslash T_\lambda$. Then from Proposition 14 in Chapter 2, $g(S_\lambda \backslash T_\lambda)$ is open in $\mathbb{R}^m$ too, and hence its intersection with δH_μ^m is empty. It follows that $g(S_\lambda \backslash T_\lambda) \subseteq S_\mu \backslash T_\mu$. Since the argument can be reversed, $g(S_\mu \backslash T_\mu) = S_\lambda \backslash T_\lambda$ and finally $g(T_\lambda) = T_\mu$ and in particular $u_\mu = g(u_\lambda) \in T_\mu$ ■

Since in the discussion of Example 8 we pointed out that $\delta H^m = \partial H^m$, from now we will drop the symbol δH^m and always use ∂H^m for the boundary of H^m (in either sense).

We need to develop now the notion of tangent space at an element of a manifolds with boundary. In order to extend the scope of Definition 2, Chapter 3, which was given for boundaryless manifolds, we start with a slight extension of the notion of curve, given in Chapter 3 (in particular concerning the domain).

Definition 18 Let M be a manifold. A C^1 **curve** in M is a C^1 function $\gamma : I \to M$ where $I = [a, b]$ is a closed real interval. The curve γ is **through** $x \in M$ **at** $t \in I$ if $x = \gamma(t)$.

According to the above definition, a curve γ in a manifold of $\mathbb{R}^l$ is just a curve in $\mathbb{R}^l$ for which $\operatorname{Im} \gamma$ is contained in the manifold. Notice that we allow $x = \gamma(t)$ may belong to the boundary of M. The Definition 2, Chapter 3, can be reformulated for manifolds with boundary.

Definition 19 Let M be a manifold of $\mathbb{R}^l$, and let x be an element of M. An element v of $\mathbb{R}^l$ is a **tangent vector** to M at x if there exists a curve in γ through x at t such that $\gamma'(t) = v$. Whenever $x = \gamma(a)$ or $x = \gamma(b)$, $\gamma'(a)$ and $\gamma'(b)$ are the derivative of any C^1 extension of γ. The **tangent space** to M at x is the set $T_x M$ of all tangent vectors to M at x.

Remark 20 By a straightforward application of the above definition, given a closed euclidean halfspace H^m and a point x in H^m, the **tangent space** to H^m at x is $T_x H^m = \mathbb{R}^m$.

The following proposition gives a characterization of a tangent space to a manifold. In particular it shows that the tangent space is actually an m-dimensional vector space.

Proposition 21 *Let M be an m-dimensional manifold of $\mathbb{R}^l$, and let x be an element of M. For any local parametrization (ϕ, U, V) of M around x such that $x = \phi(u)$,*

$$T_x M = \operatorname{Im} d\phi_u$$

Proof. If $x \in M \backslash \partial M$ the statement has been proved in Proposition 3 in Chapter 3. Therefore consider $x \in \partial M$. Suppose first that v is an element of $\operatorname{Im} d\phi_u$. There exists $w \in \mathbb{R}^m$ such that $v = d\phi_u(w)$. Suppose that the segment $[u, u + \delta w]$ is included in U for $\delta > 0$ small enough. The function

$$\gamma : [0, \delta] \to M, \quad t \mapsto \phi(u + tw)$$

is a curve in M through x at 0, since ϕ is C^1. Moreover, $\gamma'(0) = v$ by the chain rule ([3], Theorem 12.7, p. 352). If, otherwise, the segment $[u, u + \delta w]$ is not included in U for any $\delta > 0$, then there exists $\delta > 0$ such that the segment $[u - \delta w, u]$ is contained in U. Analogously to the above case,

$$\gamma : [-\delta, 0] \to M, \quad t \mapsto \phi(u + tw)$$

is a C^1 curve through x at 0 with $\gamma'(0) = v$.

Suppose, conversely, that $x = \gamma(t)$ and $v = \gamma'(t)$ for some curve γ in M. Let (ψ, V, U) the chart associated with (ϕ, U, V), and let $\Psi : W \to \mathbb{R}^m$ be a local extension of ψ at x, defined on an open neighborhood W of x in $\mathbb{R}^l$. Without loss of generality (see also proof of Proposition 3 in Chapter 3) we shall assume that $W \cap M = V$. Ψ is C^1, and $d\Psi_x$ is a linear function between $\mathbb{R}^l$ and $\mathbb{R}^m$. If $w \equiv d\Psi_x(v)$, $d\phi_u(w)$ is well defined. Consider $\widetilde{\gamma} \equiv \Psi \circ \gamma$. By the chain rule, $\widetilde{\gamma}'(t) = w$ and hence $d\phi_u(w) = v$. ∎

Remark 22 As an immediate consequence of the proposition, the tangent space to an m-dimensional manifold at any point is an m-dimensional vector space (being the image of a linear function).

As we have already pointed out in Chapter 3, from now on $d\phi_u$ shall denote $d\phi_{u|\mathbb{R}^m, T_xM} : \mathbb{R}^m \to T_xM$; we recall that in this sense, $d\phi_u$ is a linear isomorphism and $d\psi_x$ is well defined.

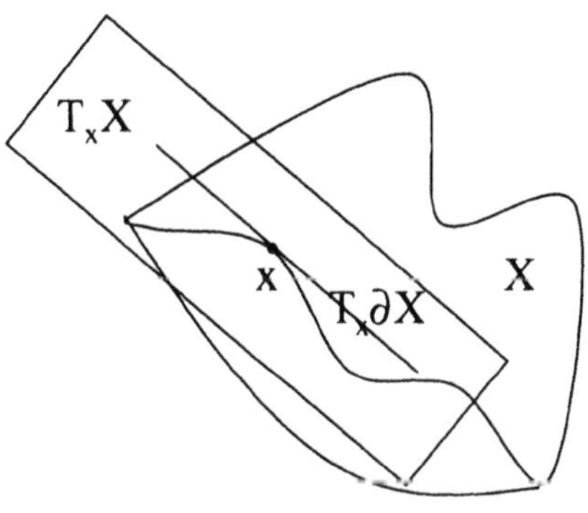

Figure 5.2. The (translated) tangent space

Remark 23 It may be thought that when x is on the boundary of M, so that u belongs to ∂H^m, and ϕ is defined on a set U which is not an open neighborhood of u in $\mathbb{R}^m$, the value of $d\phi_u$ along directions pointing outside U is to some extent arbitrary. However, the differentiability of ϕ at u hinges upon the differentiability at u of an appropriate extension of ϕ to a full neighborhood of u in $\mathbb{R}^m$; according to Definition 1 of Chapter 2, all such extensions have the same differential at u. Indeed let $\widetilde{\phi}$ be a differentiable extension of ϕ at u, and suppose that $u + tv$ does not belong to U for $t > 0$. It follows that $u + t(-v)$ is in U, t positive and small enough. The directional derivative of $\widetilde{\phi}$ at u, along $-v$ is opposite to the directional derivative of ϕ at u, along v.

Proposition 24 *Let M be a C^r m-dimensional manifold with boundary.*

1 $M\backslash\partial M$ is a C^r m-dimensional manifold without boundary.

2 Let O be an open subset of M. O is a C^r m-dimensional manifold. Moreover, O is a manifold with boundary if and only if $O \cap \partial M \neq \emptyset$.

Proof. 1. Consider an arbitrary system of local parametrizations of M $\{(\phi_\lambda, U_\lambda, V_\lambda)\}_{\lambda\in\Lambda}$. By definition of ∂M, for any $x \in M\backslash\partial M$ there exists at least a local parametrization (ϕ, U, V) around x such that $u = \phi^{-1}(x)$, $u \in H^m\backslash\partial H^m$. Observe that $H^m\backslash\partial H^m$ is open in $\mathbb{R}^m$, hence there exists an open neighborhood $\widetilde{U}$ of u in $\mathbb{R}^m$ such that $\widetilde{U} \subseteq U \cap (H^m\backslash\partial H^m)$. Thus $\left(\phi, \widetilde{U}, \phi(\widetilde{U})\right)$ is a local parametrization around x of $M\backslash\partial M$.

2. The first part of the statement follows from a verbatim replica of the proof of Proposition 18 in Chapter 2, under the current new interpretation; the second part is a direct consequence of 1. ∎

Proposition 25 *If M is a C^r m-dimensional manifold with boundary, then ∂M is a C^r $(m-1)$-dimensional manifold without boundary.*

Proof. We are going to check the definition of C^r boundaryless manifold on ∂M. Consider $x \in \partial M$, a local parametrization (ϕ, U, V) of M with $\phi(u) = x$. In Example 8 we have seen that ∂H^m is C^∞ diffeomorphic to $\mathbb{R}^{m-1}$; moreover, $U \cap \partial H^m$ is open in the relative topology of ∂H^m. Then $U \cap \partial H^m$ is diffeomorphic to an open subset U' of $\mathbb{R}^{m-1}$, i.e., there exists a C^∞ diffeomorphism $g : U' \to U \cap \partial H^m$. Take $V' \equiv V \cap \partial M$; then the function

$$\sigma : U' \to V', \quad t \mapsto (\phi \circ g)(t)$$

is a C^r diffeomorphism, with $\sigma(U') = \phi(U \cap \partial H^m) = V'$ and $\sigma(t) = \phi(u) = x$. The last equality follows from the fact that $x \in V'$ is equivalent to $\phi^{-1}(x) \in U \cap \partial H^m = g(U')$. ∎

As a consequence of the proof in Proposition 25, the tangent space at $x \in \partial M$, $T_x(\partial M)$, is a vector subspace of dimension $(m-1)$ of T_xM. More precisely

$$T_x(\partial M) = d\phi_u(\partial H^m)$$

Indeed,

$$T_x(\partial M) = \operatorname{Im} d\sigma_t = d\phi_u(\operatorname{Im} dg_t)$$

dg_t takes values in ∂H^m (a subspace of $\mathbb{R}^m$) because g takes values in $U \cap \partial H^m$, and dg_t is onto because g is a diffeomorphism.

Example 26 Let U be an open subset of $\mathbb{R}^m$ such that $U \cap \partial H^m \neq \emptyset$. From the above, Example 8, Propositions 24 and 25, $U \cap H^m$ is manifold with boundary and $U \cap \partial H^m$ is its boundary. Moreover, at any element of the boundary the tangent space to $\partial(U \cap \partial H^m)$ is ∂H^m.

Proposition 27 *Assume that*

1 M *is a* C^r m*-dimensional manifold without boundary, and*

2 N *is a* C^r n*-dimensional manifold with boundary.*

Then

1 $M \times N$ *is a* C^r $(m+n)$*-dimensional manifold with boundary, and*

2 $\partial(M \times N) = M \times \partial N$.

Proof. 1. Take $z \in M \times N$, $z \equiv (x, y)$. Given two local parametrizations (ϕ, U, V) of M around x and (ϕ', U', V') of N around y, respectively, we want to show that

$$((\phi \times \phi'), U \times U', V \times V')$$

is a local parametrization of $M \times N$ around z.

First of all observe that $V \times V'$ is an open neighborhood of z in $M \times N$. In addition, $(\phi \times \phi')$ is a C^r diffeomorphism between $U \times U'$ and $V \times V'$. Therefore, it remains to prove that $U \times U'$ is open in H^{m+n}. To see this, identify $\mathbb{R}^m \times \mathbb{R}^n$ with $\mathbb{R}^{m+n}$, so $U \times U' \subseteq \mathbb{R}^{m+n}$. As U is open in $\mathbb{R}^m$ and U' is open in H^n, then $U \times U'$ is open in $\mathbb{R}^m \times H^n$. Now it is immediate to see that $\mathbb{R}^m \times H^n$ is a closed euclidean halfspace in $\mathbb{R}^{m+n}$ and this completes the proof.

2. Let a be a linear function as in Definition 1. For any $z \in M \times N$, $z \in \partial(M \times N)$ if and only if $a(y) = 0$, that is if and only if $y \in \partial N$. ■

Example 28 (The solid torus) Let $ST^2 \equiv S^1 \times \overline{B}^2$. Here $\partial\overline{B}^2 = S^1$ and $\partial ST^2 = S^1 \times S^1$ is the (ordinary) torus T^2, discussed as Example 36 in Chapter 2. A representation of ST^2 as a subset of $\mathbb{R}^3$ can be given by the same construction presented there for T^2. Indeed, observe that the restriction to ST^2 of the map f defined in that example is still a diffeomorphism, since it holds true that $w \in [-1, 1]$ for each $(uv, w, t) \in ST^2$, hence the two conditions

$$\sqrt{x^2 + y^2} = R + rw$$

$$0 < R - r \leq R + rw \leq R + r$$

remain satisfied. The jacobian matrix of f still is of full rank, and the inverse of $f_{|ST^2}$ keeps having a C^∞ extension. The diffeomorphic image of ST^2 obtained by means of the map $f_{|ST^2}$, which is going to be denoted $\mathcal{ST}^2$, is the set of solutions to the inequality

$$\left(\sqrt{x^2 + y^2} - R\right)^2 + z^2 \leq r^2$$

Indeed,

$$\overline{B}^2 = \bigcup_{\sigma\in[0,1]} S^1_\sigma$$

$$ST^2 = \bigcup_{\sigma\in[0,1]} S^1 \times S^1_\sigma$$

$$\mathcal{S}\mathcal{T}^2 = \bigcup_{\sigma\in[0,1]} \mathcal{T}^2_\sigma$$

where each set $\mathcal{T}^2_\sigma$ appearing in the last union is the solution set to the equation

$$\left(\sqrt{x^2+y^2} - R\right)^2 + z^2 = \sigma r^2$$

and is the (diffeomorphic) image of $S^1 \times S^1_\sigma$ under the map f defined in that example. The topological and differential boundaries of $\mathcal{S}\mathcal{T}^2$ coincide:

$$\partial\mathcal{S}\mathcal{T}^2 = \mathcal{T}^2 = f\left(S^1 \times S^1\right) = f\left(S^1 \times \partial\overline{B}^1\right)$$

Exercise 29 Describe the image under f of $\overline{B}^2 \times S^1$

Notice that when M and N are both manifolds with boundary, $M \times N$ is never a manifold. Concerning this, consider the following example.

Example 30 $[0,1] \times [0,1]$ is not a manifold of $\mathbb{R}^2$. Indeed, none of its four vertices has a neighborhood in $[0,1] \times [0,1]$ which is diffeomorphic to a an euclidean halfspace (see again Example 28 of Chapter 2).

Our next aim is to state and prove some results which are similar in flavor to the regular value theorem (Theorem 9 in Chapter 4), though applying in the present, more general context. These results establish that, under specific regularity assumptions, the inverse images of well behaved sets are manifolds with boundary. The first result we present concerns real valued functions and inverse images of halflines; later on we will be dealing with functions between manifolds and inverse images of submanifolds.

Proposition 31 (Cutting lemma) *Let M be a C^r m-dimensional manifold without boundary and let $f : M \to \mathbb{R}$ be a C^r function . If $\bar{y} \in f(M)$ is a regular value for f, then*

1. *$X \equiv \{x \in M : f(x) \geq \bar{y}\}$ is a C^r manifold with boundary of dimension m;*
2. *$\partial X = f^{-1}(\bar{y})$.*

Proof. By considering the function $x \mapsto f(x) - \bar{y}$ we can take $\bar{y} = 0$ without loss of generality.

Consider first the set $Z \equiv f^{-1}((0, +\infty))$ which is an open subset of M. Z is a boundaryless m-submanifold of M (see Proposition 18). Any local parametrization of Z around x is now a local parametrization of X around x.

Consider then $x \in f^{-1}(0)$; since 0 is a regular value for f, there exists a local representation of f which is equal to the canonical projection $\pi_1 : U \to \mathbb{R}$, $u \to u_1$, (see Theorem 66, Chapter 3) and we get the following diagram

$$\begin{array}{ccc} M & \xrightarrow{\quad f \quad} & \mathbb{R} \\ \uparrow \phi & & \downarrow id \\ U \subseteq \mathbb{R}^m & \xrightarrow[\pi_1 = id \circ f \circ \phi]{} & S \subseteq \mathbb{R} \end{array}$$

Notice that, $\phi^{-1}(x) = (0, w_2, ..., w_m)$; moreover, for any $z \in X \cap \phi(U)$ we get $\phi^{-1}(z) = (u_1, u_2, ..., u_m)$ with $u_1 \geq 0$. Conversely, for any $u \in U$ such that $u_1 \geq 0$ we get $\pi_1(u) = u_1 \geq 0$, hence $\phi(u) \in X$. Then define $\widehat{U} \equiv U \cap H^m$ and $\widehat{V} \equiv V \cap X$ so that $(\phi_{\widehat{U}}, \widehat{U}, \widehat{V})$ is a local parametrization of X around x. Thus the first claim is proved. By the previous construction, $\phi_{\widehat{U}}(\widehat{U} \cap \partial H^m) = \widehat{V} \cap f^{-1}(0)$, which proves that $\partial X = f^{-1}(0)$. ■

Corollary 32 *Let M be a boundaryless manifold of $\mathbb{R}^l$, and let $a : \mathbb{R}^l \to \mathbb{R}$ be a nontrivial linear map. If*

1 $(\partial H^l_a) \cap M$ *is nonempty;*

2 0 *is a regular value for the restriction* $a_{|M}$*;*

then

1 $H^l_a \cap M = \{x \in M : a(x) \geq 0\}$ *is a manifold with boundary of dimension* $\dim M$*;*

2 $\partial (H^l_a \cap M) = (\partial H^l_a) \cap M = \{x \in M : a(x) = 0\}$.

Proof. Immediate. Just take $f \equiv a_{|M} : M \to \mathbb{R}$ in the statement of the previous theorem. ■

Exercise 33 Consider the function

$$f : \mathbb{R}^2 \to \mathbb{R}, \;\; (x, y) \mapsto xy(1 - y)(1 - x)$$

The set $X \equiv \{x \in \mathbb{R}^2 : f(x) \geq 0\}$ is not a manifold with boundary. Why?

Example 34 Using Corollary 32 we can prove that any closed ball $\overline{B_r^m}$ is a manifold with boundary, and $\partial\overline{B_r^m} = S_r^{m-1}$.[2] Let $f : \mathbb{R}^m \to \mathbb{R}$, $x \mapsto r^2 - \|x\|^2 = r^2 - (xx)$. Then $df_x : v \mapsto -2(xv)$, the only critical point of f is 0, with critical value r^2. Moreover, $\{x \in \mathbb{R}^m : f(x) \geq 0\} = \overline{B_r^m}$ and $f^{-1}(0) = S_r^{m-1}$.

2. Intersection and inverse images of manifolds: the notion of transversality

The present section addresses two issues which turn out to be considerably related[3].

As indicated by the regular value theorem (Theorem 9, Chapter 4), a vast amount of the manifolds we consider arises as solution sets to single equations or systems of equations with well behaved left hand sides.

It is common practice to study systems of equations by splitting them into components (single equations or subsystems) and investigating the properties of each individual component separately. The problem naturally arises, then, to determine whether or not the set of solutions to a single, large system of equations obtained by combination of different subsystems has a manifold structure, when each of its components is sufficiently regular to have a manifold as solution set. In somewhat more general terms, our first issue is to determine whether or not the intersection of two (or more) manifolds is a manifold itself. More appropriately, whether or not the intersection of two submanifolds of a given manifold, is again a submanifold of that manifold.

Another problem arises if, in considering a C^r function $f : M \to N$, one tries to extend the regular value theorem by investigating the structure of inverse images, according to f, of subsets rather than of single elements. When dealing with single elements, which in themselves can be looked at as (zero-dimensional) manifolds, the requirement has been for them to be regular values. Our second issue consists in specifying, when dealing with subsets, a specific and appropriate regularity condition to be required on these sets in order for their inverse images to have a manifold structure; in addition, of course, to the easily expected restriction that they be submanifolds of N. A perhaps obvious but unfortunately far too restrictive condition would be to assume that such submanifolds entirely consist of regular values.

In both issues the intervening generalization of the notion of regularity is called transversality. For a regular value, the tangent space of the manifold N has to be "covered" by the image of the differential of f (at all the elements of its inverse image). A corresponding sort of covering is required for two

[2] The reader is advised to compare this example with Example 11.

[3] The reader is advised to compare the usage of the word transversality, which is specific of differential topology, with the one of Chapter 6 and Part II, more popular among the general equilibrium theorists.

submanifolds Y and Z of a given manifold N to be transverse, or for a function $f : M \to N$ to be transverse to a submanifold Y of N. In the first case, at any intersection point, the tangent space of N must be spanned (as a vector sum) by the tangent spaces of Y and Z *together*. In the second case, the tangent space of N at y must be spanned by the image of the differential df_x and the tangent space of Y at y *together*, whenever $f(x) = y$.

It may be already clear that the relationship between the two notions of transversality is quite strict. In Remark 42 it will be shown that either one can be derived from the other.

Both issues are going to be taken up allowing for one (but not two!) of the involved manifolds to have a nonempty boundary, giving rise to a fair, but not overwhelming, amount of complication if compared to the fully boundaryless case.

2.1 Intersecting manifolds

The intersection of two manifolds is not necessarily a manifold.

Example 35 Consider the function

$$f : \mathbb{R} \to \mathbb{R}, \quad x \mapsto \begin{cases} x^4 \cos \frac{1}{x} & \text{for } x \neq 0 \\ 0 & \text{for } x = 0 \end{cases}$$

and let $Y \equiv \operatorname{graph} f$, $Z \equiv x$-axis. Y and Z are manifolds but $Z \cap Y$ is not a manifold. See Example 35 of Chapter 2.

Example 36 Consider the function $f : \mathbb{R}^2 \to \mathbb{R}$, $(x, y) \mapsto xy$, $Y \equiv \operatorname{graph} f$ and let Z be the xy-plane in $\mathbb{R}^3$. Y and Z are clearly manifolds but $Z \cap Y$ is not a manifold.

Example 37 Consider the following two real functions of three real variables:

$$f : (x, y, z) \mapsto x^2 + z^2 \qquad\qquad g : (x, y, z) \mapsto y^2 + z^2$$

and let P and Q be the solution sets, respectively, to the equations

$$f(x, y, z) = 1 \qquad\qquad g(x, y, z) = 1$$

P and Q are closed circular unbounded cylinders. Since $Df(x, y, z) = 2(x, 0, z)$ and $Dg(x, y, z) = 2(0, y, z)$, the only critical value of both f and g is 0, and hence P and Q are 2-dimensional boundaryless manifolds of $\mathbb{R}^3$ by the regular value theorem (Theorem 9, Chapter 4). The intersection $P \cap Q$ of the two manifolds is the solution set to the system

$$\begin{cases} x^2 + z^2 = 1 \\ y^2 + z^2 = 1 \end{cases}$$

so that for any element $(\overline{x}, \overline{y}, \overline{z})$ in $P \cap Q$ the condition

$$\overline{x}^2 = \overline{y}^2$$

must hold. Thus $P \cap Q$ is completely contained in the union of the two planes π, ρ described (respectively) by the equations

$$x - y = 0 \qquad\qquad x + y = 0$$

$P \cap Q$ is not a manifold. Indeed, $P \cap Q$ is the union of two ellipses μ and ν, $\mu \subseteq \pi$ and $\nu \subseteq \rho$, joined together at the two points $p = (0,0,1)$ and $q = (0,0,-1)$. Thus $P \cap Q$ is locally diffeomorphic to $\mathbb{R}$ around each of its points which is different from p and q, but not around p or q: each sufficiently small neighborhood of p (q) in $\mathbb{R}^3$ intersects $P \cap Q$ along two elliptical arcs, which are subarcs of μ and ν, respectively, and meet at p (q).

2.2 Transversality

Definition 38 Let Y and Z be submanifolds of a manifold N. Y and Z are said to be **transverse** in N, which is denoted by $Y \pitchfork Z$, if, $\forall\, x \in Y \cap Z$,

$$T_x Y + T_x Z = T_x N$$

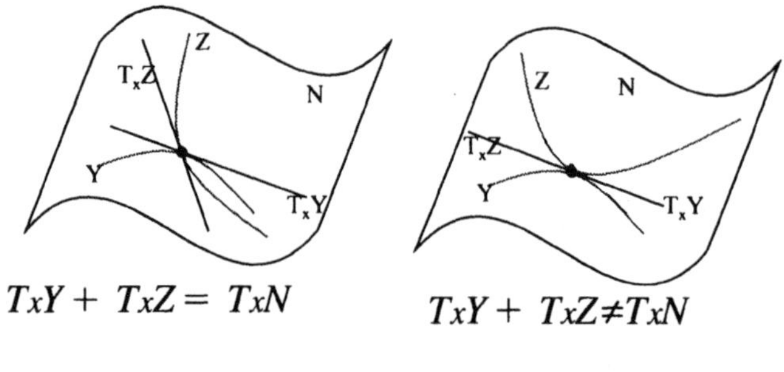

Figure 5.3. Transversality for submanifolds

Remark 39 Any disjoint submanifolds of N are transverse.

Example 40 $\mathbb{R}^n$ and $\{0\}$ are transverse in $\mathbb{R}^n$. More generally, two linear subspaces V and W of $\mathbb{R}^l$ such that $V + W = \mathbb{R}^l$ are transverse in $\mathbb{R}^l$, since for each x, and, in particular, for $x \in V \cap W$, $T_x V = V$ and $T_x W = W$.

Definition 41 Let M, N, Y be C^r manifolds, Y submanifold of N and let $f : M \to N$ be a C^r function. f is said to be **transverse** to Y, in symbols $f \pitchfork Y$, if, $\forall\, x \in f^{-1}(Y)$,

$$df_x (T_x M) + T_{f(x)} Y = T_{f(x)} N \tag{5.1}$$

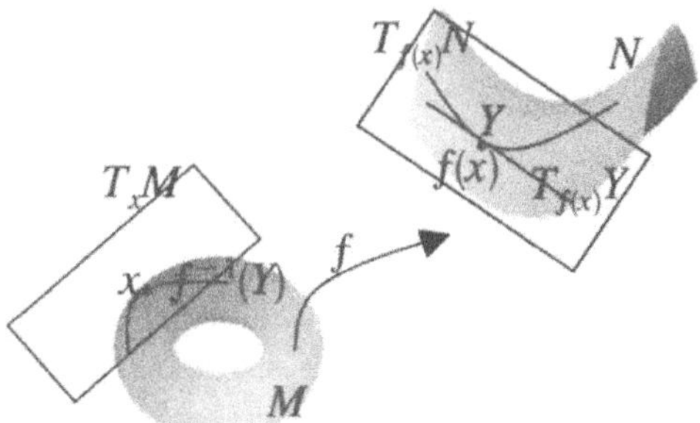

Figure 5.4. f is transverse to Y

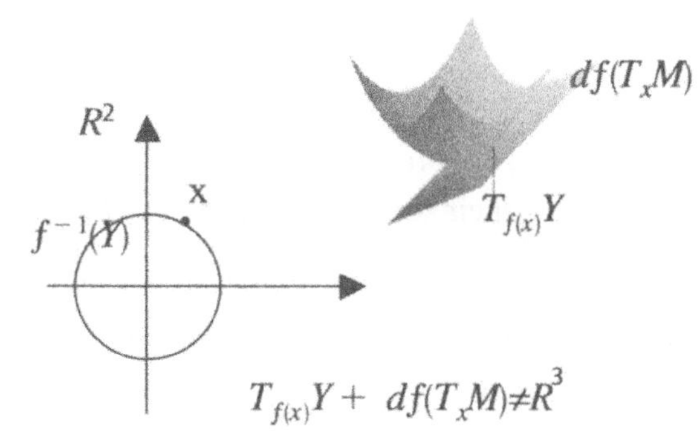

Figure 5.5. f is not transverse to Y

Remark 42 As with Definition 38, f and Y are transverse when Im $f \cap Y = \emptyset$. Moreover, Definition 41 specializes in Definition 38 when $M = Z \subseteq N$ and $f = i : Z \to N$ is the canonical insertion. Conversely, Definition 38 specializes in Definition 41 when $Z = \text{Im } f$ for a function f having a submanifold of N as its image (see, e.g., Proposition 57 of Chapter 3).

Example 43 Let $f : \mathbb{R} \to \mathbb{R}$, $x \mapsto x^2$, and let $Y \equiv \mathbb{R} \times \{0\}$, $Z \equiv \text{graph } f$. Then

$$T_{(0,0)}Y + T_{(0,0)}Z = Y + Y = Y \neq \mathbb{R}^2$$

that is, Y and Z are not transverse in $\mathbb{R}^2$. More generally, this is true for any C^∞ real functions of one real variable which are tangent in a point (or more).

Example 44 Let $N \equiv \mathbb{R}^3$, $L \equiv S^2$, and $M \equiv xy$-plane. Then L and M are transverse in N. Indeed, if $A \equiv (x, y, z)$ is in $L \cap M$, then $x^2 + y^2 = 1$, $z = 0$, and hence

$$T_A L = \{(u, v, w) \in \mathbb{R}^3 : xu + yv = 0\}$$
$$T_A M = \{(u, v, w) \in \mathbb{R}^3 : z = 0\}$$

A basis for $T_A N = \mathbb{R}^3$ is provided, e.g., by the vectors

$$(y, -x, 0) \qquad (0, 0, 1) \qquad (x, y, 0)$$

Example 45 Recalling the cylinders P and Q defined in Example 37, if $p \equiv (p_1, p_2, p_3)$ is an element of P, and $q \equiv (q_1, q_2, q_3)$ is an element of Q, according to Corollary 13 in Chapter 4, the tangent spaces of P at p and of Q at q can be expressed as solution sets, respectively, to the equations

$$Df(p)\,(x, y, z) = 0 \qquad Dg(q)\,(x, y, z) = 0$$

that is,

$$p_1 x + p_3 z = 0 \qquad q_2 y + q_3 z = 0$$

It is quickly seen that when $p \equiv q \equiv (0, 0, 1)$ or $p \equiv q \equiv (0, 0, -1)$ the tangent spaces $T_q Q$ and $T_p P$ coincide, their sum is strictly included in $\mathbb{R}^3$ and hence P and Q are not transverse in $\mathbb{R}^3$.

Proposition 46 *Under the assumptions of Definition 41, if* $f \pitchfork Y$ *and* $\dim M + \dim Y < \dim N$, *then* $f(M) \cap Y = \emptyset$.

Proof. If $f^{-1}(Y) = \emptyset$ there is nothing to prove, otherwise let y be an element of $f(M) \cap Y$. Then

$$\begin{aligned} \dim\left(df_x(T_x M) + T_{f(x)}Y\right) &\leq \dim df_x(T_x M) + \dim T_{f(x)}Y \\ &\leq \dim M + \dim Y < \dim N \\ &= \dim T_{f(x)}N \end{aligned}$$

contradicting (5.1) of Definition 41. ■

Remark 47 Definition 41 can be seen as a generalization of the notion of regularity given in Chapter 4, in the sense that when the manifold Y consists of a single element, $Y = \{y\}$, then $f \pitchfork \{y\}$ just means that y is a regular value for f. This is a direct consequence of the simple fact that $T_y\{y\} = \{0\}$, so that condition (5.1) of Definition 41 reduces to Definition 1 of Chapter 4.

Definition 48 Let $f : M \to N$ where M is a manifold with boundary. The function

$$\partial f : \partial M \to N$$

is the restriction of f to ∂M.

Of course, since ∂M is manifold, we can apply Definition 41 to ∂f as well as the statement of Proposition 46 holds for ∂f. Transversality of both f and ∂f to a submanifold Y of N plays a fundamental role in the following theorem which generalizes the regular value theorem of Chapter 4.

Theorem 49 (General preimage theorem) *Let*

1 M, N, Y be C^r manifolds of dimensions m, n, l, respectively, such that $Y \subseteq N$ and $\partial N = \partial Y = \emptyset$;

2 $f : M \to N$ be a C^r function such that $f \pitchfork Y$ and $\partial f \pitchfork Y$;

3 $f^{-1}(Y) \neq \emptyset$.

Then

1 $f^{-1}(Y)$ is a submanifold of M;

2 codim $f^{-1}(Y)$ in M = codim Y in N, i.e., $\dim f^{-1}(Y) = m - n + l$;

3 $\partial\left(f^{-1}(Y)\right) = f^{-1}(Y) \cap \partial M$.

Proof. First of all observe that the assumptions imply $m - n + l \geq 0$ since otherwise from Proposition 46 $f^{-1}(Y) = \emptyset$ and that contradicts the referred assumptions.

Consider $\overline{x} \in f^{-1}(Y)$ with $\overline{y} \equiv f(\overline{x})$. From Proposition 17 in Chapter 4, there exist an open neighborhood O of $\overline{y}$ in N and a C^r function $h : O \to \mathbb{R}^{n-l}$ such that 0 is a regular value for h and $Y \cap O = h^{-1}(0)$. Since f is continuous, $f^{-1}(O)$ is open in M. Let (ϕ, U, V) be a local parametrization of M around $\overline{x}$, with $\overline{x} = \phi(\overline{u})$, such that $V \subseteq f^{-1}(O)$ and let (χ, S, T) be a local parametrization of N around $\overline{y}$.

Our argument is going to proceed much as in proof of the regular value theorem, with one basic difference. There we have used a local representation of f, namely the map $\xi \circ f \circ \phi$, which had $\overline{u}$ as a regular point. Since there is no reason to expect $\overline{u}$ to be a regular point for $\xi \circ f \circ \phi$ now, we shall compose $f \circ \phi$ with the function h rather than with ξ. Notice that $f^{-1}(Y \cap O) = (h \circ f)^{-1}(0)$.

$$
\begin{array}{ccccc}
V \subseteq f^{-1}(O) \subseteq M & \xrightarrow{\quad f \quad} & O \subseteq \mathbb{R}^n & & \\
\uparrow \phi & & \uparrow \xi & h \searrow & \\
U \subseteq \mathbb{R}^m & \xrightarrow{\xi \circ f \circ \phi} & S \subseteq \mathbb{R}^n & & \mathbb{R}^{n-l} \\
 & & \xrightarrow{\qquad g \equiv h \circ f \circ \phi \qquad} & &
\end{array}
$$

Our first task is to prove that, if $\overline{x}$ does not belong to the boundary of M, the composite map $g \equiv h \circ f \circ \phi$ (which takes values in $\mathbb{R}^{n-l}$) has 0 as a regular value. The transversality of f to Y is essential here. The result leads us to directly deduce from the regular value theorem that $f^{-1}(Y)$ is locally diffeomorphic to $\mathbb{R}^{m-n+l}$ around $\overline{x}$.

Our second task is to prove that, if $\overline{x}$ belongs to the boundary of M, then necessarily $m - n + l > 0$ and $f^{-1}(Y)$ is locally diffeomorphic to H^{m-n+l} around $\overline{x}$. In the first place, this requires consideration of an extension $\overline{g}$ of g, defined on a full neighborhood of $\overline{u}$ in $\mathbb{R}^m$. Replicating for $\overline{g}$, under an appropriate restriction of its domain, the argument already given for g, we obtain that $\overline{g}^{-1}(0)$ is a boundaryless submanifold of $\mathbb{R}^m$ of dimension $m - n + l$. In the second place, we prove that $\overline{g}^{-1}(0) \cap H^m$ is a manifold with boundary of the same dimension of $\overline{g}^{-1}(0)$. According to Corollary 32, this amounts to showing that 0 is a regular value for the restriction to $\overline{g}^{-1}(0)$ of the linear map a defining H^m. Transversality of ∂f to Y comes in at this point, as it is going to be seen later on. Let

$$g \equiv h \circ f \circ \phi : U \to \mathbb{R}^{n-l}$$

and suppose first that $\overline{x} \in M \backslash \partial M$; then, for any u in U, an open subset of $\mathbb{R}^m$,

$$dg_u = dh_y \circ df_x \circ d\phi_u : \mathbb{R}^m \to \mathbb{R}^{n-l} \tag{5.2a}$$

$$\operatorname{Im} dg_u = (dh_y \circ df_x)(T_x M) \tag{5.2b}$$

where $x = \phi(u) \in V$, and $y = f(x) \in O$. If, moreover, y belongs to Y, so that $h(y) = 0$, by evaluating dh_y on both sides of the equality defining transversality of f with Y, taking into account the regularity of value 0 for h,

Corollary 13 of Chapter 4 and (5.2b), we get

$$\begin{aligned}\mathbb{R}^{n-l} &= dh_y\left(T_yN\right) = dh_y\left(df_x\left(T_xM\right) + T_yY\right) \\ &= \operatorname{Im} dg_u + dh_y\left(T_yY\right) = \operatorname{Im} dg_u\end{aligned} \tag{5.3}$$

Thus every u in $g^{-1}(0)$ is a regular point for g, that is, g has 0 as a regular value; by the regular value theorem, $g^{-1}(0)$ is a boundaryless manifold of $\mathbb{R}^m$ of dimension $m - n + l$, and so is in $\mathbb{R}^l$ its C^r diffeomorphic image[4]

$$\phi\left(g^{-1}(0)\right) = f^{-1}\left(h^{-1}(0)\right) \cap V = f^{-1}(Y \cap O) \cap V = f^{-1}(Y) \cap V$$

Since $f^{-1}(Y) \cap V$ is a neighborhood of $\overline{x}$ in $f^{-1}(Y)$, $f^{-1}(Y)$ is locally C^r diffeomorphic to $\mathbb{R}^{m-n+l}$ around $\overline{x}$.

Consider now $\overline{x} \in \partial M$. Suppose that $m - n + l = 0$; recalling that $\partial f \pitchfork Y$ and $\dim(\partial M) = m - 1$, applying Proposition 46 to the function ∂f,

$$(\partial f)^{-1}(Y) = f^{-1}(Y) \cap \partial M = \emptyset$$

Therefore, if $\overline{x} \in \partial M$ then we necessarily have $m - n + l > 0$. Then $U = U' \cap H^m$ for some open subset U' of $\mathbb{R}^m$, and $\partial U = U' \cap \partial H^m$ is open in ∂H^m, hence diffeomorphic to an open subset of $\mathbb{R}^{m-1}$. Since g is C^r there exists a C^r extension $\overline{g}$ of g, defined on U'', some open subset of $\mathbb{R}^m$, such that $U = U'' \cap H^m$, and there is no loss of generality in assuming that $U'' = U'$. Since g is defined in $\overline{u}$, (5.2a), (5.2b) and (5.3) apply to $\overline{u}$; recalling Remark 23, $d\overline{g}_{\overline{u}} = dg_{\overline{u}}$, and $\overline{u}$ is a regular point for $\overline{g}$.

By Lemma 8 of Chapter 4, there is a neighborhood U''' of $\overline{u}$ in U' entirely consisting of regular points for $\overline{g}$, and once more there is no loss of generality in assuming that $U''' = U'$. Again from the regular value theorem, $\overline{g}^{-1}(0)$ is a boundaryless submanifold of U' (and of $\mathbb{R}^m$) of dimension $m - n + l$. Moreover, since $\overline{g}$ and g agree on U,

$$\overline{g}^{-1}(0) \cap U = g^{-1}(0) \cap U \qquad \overline{g}^{-1}(0) \cap \partial U = g^{-1}(0) \cap \partial U \tag{5.4a}$$

$$\phi\left(\overline{g}^{-1}(0) \cap U\right) = f^{-1}\left(h^{-1}(0)\right) \cap V = f^{-1}(Y) \cap V \tag{5.4b}$$

$$\phi\left(\overline{g}^{-1}(0) \cap \partial U\right) = f^{-1}\left(h^{-1}(0)\right) \cap \partial V = f^{-1}(Y) \cap V \cap \partial M \tag{5.4c}$$

By Corollary 32, we can conclude that $\overline{g}^{-1}(0) \cap U = \overline{g}^{-1}(0) \cap H^m$ is a submanifold of U (and $\mathbb{R}^m$) containing $\overline{u}$, of the same dimension, with boundary

$$\partial\left(\overline{g}^{-1}(0) \cap U\right) = \overline{g}^{-1}(0) \cap \partial U$$

if we can show that 0 is a regular value of the map $a_{|\overline{g}^{-1}(0)}$, where a is the linear function defining H^m (for all u in U). By (5.4a)-(5.4c), this will also prove that

[4]The first equality comes from the fact that while ϕ is onto, f is defined on all M. Also, remember that $f(V) \subset O$ implies $V \subset f^{-1}[f(V)] \subset f^{-1}(O)$.

$f^{-1}(Y) \cap V$ is a submanifold of $\mathbb{R}^m$ of dimension $m - n + l$, with boundary equal to $f^{-1}(Y) \cap V \cap \partial M$, or that $f^{-1}(Y)$ is locally C^r diffeomorphic to H^{m-n+l} around $\overline{x}$.

Thus we are left to show 0 is a regular value of the map $a_{|\overline{g}^{-1}(0)}$.

To this purpose, let u be any element of $\overline{g}^{-1}(0) \cap \partial H^m$, and notice that (5.3) holds for $d\overline{g}_u$; moreover, taking into account transversality of ∂f with Y, the argument can be replicated for $d(\partial g)_u = d(\partial \overline{g})_u = d\left(g_{|\partial U}\right)_u$ in the following way:

$$\begin{aligned} \mathbb{R}^{n-l} &= dh_y[T_y N] = dh_y\left[d(\partial f)_x[T_x \partial M] + T_y Y\right] \\ &= \operatorname{Im} d(\partial g)_u + dh_y(T_y Y) = \operatorname{Im} d(\partial g)_u \end{aligned} \tag{5.5}$$

Thus, invoking theorem on the dimension of the kernel and the image for the functions $d\overline{g}_u$ and $d(\partial g)_u$ (Theorem 3 p. 99 [78]),

$$\dim \ker d(\partial g)_u = \dim \ker d\overline{g}_u - 1$$

and hence, recalling that $T_u \partial U$ is a subspace of $T_u U$,

$$\ker d\overline{g}_u = \ker d(\partial g)_u + L[v]$$

for some $v \in T_u U \backslash T_u \partial U$ such that $d\overline{g}_u(v) = 0$, where $L[v]$ is the subspace of $\mathbb{R}^m$ generated by v. Notice that $v \notin T_u \partial U$ means $v \notin \partial H^m$ (Example 26) and hence $a(v) \neq 0$.

On the other hand, since a is linear and Proposition 13 of Chapter 4 holds,

$$\begin{aligned} d\left(a_{|\overline{g}^{-1}(0)}\right)_u \left(\overline{g}^{-1}(0)\right) &= a\left(T_u \overline{g}^{-1}(0)\right) = a\left[\ker d\overline{g}_u\right] \\ &= a\left(\ker d(\partial g)_u + L[v]\right) = \{0\} + L[a(v)] = \mathbb{R} \end{aligned}$$

This completes the regularity check for $a_{|\overline{g}^{-1}(0)} : \overline{g}^{-1}(0) \to \mathbb{R}$. ∎

Remark 50 Consider the insertion function $i : H^2 \to \mathbb{R}^2$. The differential of i at any point of H^2 is just $id_{\mathbb{R}^2}$. Then the differential of $\partial i : \partial H^2 \to \mathbb{R}^2$ equals everywhere the canonical immersion $i : \mathbb{R} \to \mathbb{R}^2$, $x \mapsto (x, 0)$, which is not onto. Thus a regular value for f is not necessarily a regular value for ∂f.

Example 51 Consider $M = \bar{B}^3$, $N = \mathbb{R}^4$, $Y = \{(x, y, z, t) \in \mathbb{R}^4 : 2x - z = 0\}$ and $f : M \to N$, $(x, y, z, t) \mapsto (2x, 3y, z, 1)$. In this case the assumptions of Theorem 49 are verified and therefore we get the following conclusions: $f^{-1}(Y) = \{(x, y, z) \in \bar{B}^3 : x = z\}$ is a 2-dimensional manifold and $\partial f^{-1}(Y) = \{(x, y, z) \in \bar{B}^3 : x^2 + y^2 + z^2 = 1,\ x = z\}$; codim$Y$ in $N = 4 - 3 = 1$, codim $f^{-1}(Y)$ in $M = 3 - 2 = 1$, $\partial f^{-1}(Y) = f^{-1}(Y) \cap \partial M$.

Example 52 Consider $M = [-10, 10]$, $N = \mathbb{R}^2$, $Y = \{(x, y) \in \mathbb{R}^2 : y = 0\}$ and $f : M \to N$, $x \mapsto (x, x^2 + 6x + 5)$. In this case the assumptions of Theorem 49 are verified and therefore we get the following conclusions: $f^{-1}(Y) = \{(-1, 0), (-5, 0)\}$ is a 0-dimensional manifold (so that $\partial f^{-1}(Y) = \emptyset$), codim$Y$ in $N = 2 - 1 = 1$, and codim $f^{-1}(Y)$ in $M = 1 - 0 = 1$.

As a direct consequence of Theorem 49 we get the following corollaries. The proof of the first one is immediate.

Corollary 53 (Regular value theorem for manifolds with boundary) *Consider two C^r manifolds M and N with $\partial N = \emptyset$ and a C^r function $f : M \to N$. If $y \in f(M)$ is a regular value for f and ∂f, then*

1. *$f^{-1}(y)$ is a C^r submanifold of M;*
2. *$\dim f^{-1}(y) = m - n$;*
3. *$\partial\left(f^{-1}(y)\right) = f^{-1}(y) \cap \partial M$.*

Corollary 54 *Consider three C^r manifolds N, Z, Y, with $\partial N = \emptyset$, $\partial Y = \emptyset$ and $Z \subseteq N$, $Y \subseteq N$.*

If Y is transverse both to Z and ∂Z in N then

1. *$Y \cap Z$ is a submanifold of N;*
2. $$\begin{array}{lcll} \text{codim}\,(Y \cap Z) \text{ in } Z & = & \text{codim}\,Y \text{ in } N, & \text{i.e.,} \\ \text{codim}\,(Y \cap Z) \text{ in } N & = & \text{codim}\,Y \text{ in } N + \text{codim}\,Z \text{ in } N & \text{i.e.,} \\ \dim\,(Y \cap Z) & = & \dim Y + \dim Z - \dim N & \end{array}$$
3. *$\partial(Y \cap Z) = (Y \cap Z) \cap \partial Z = Y \cap \partial Z$.*

Proof. As in Remark 41, the above conclusions 1 and 3, and the first equality in 2 follow simply identifying the function f of Theorem 49 with the insertion function. The second and third equality of 2 is the result of the following trivial computation

$$\begin{array}{l} \dim N - \dim\,(Y \cap Z) = \dim N - \dim Z + \dim Z - \dim\,(Y \cap Z) = \\ \text{codim}\,Z \text{ in } N + \text{codim}\,Y \text{ in } N = \dim N - \dim Z + \dim N - \dim Y \end{array}$$

■

Example 55 Consider $L = \left\{(x, y, z) \in \mathbb{R}^3 : z = 0\right\}$ and $K = \overline{B}^3$. The intersection $K \cap L$ is the image of $\overline{B}^2$ under the canonical immersion from $\mathbb{R}^2$ in $\mathbb{R}^3$. Observe that codim$(L \cap K)$ in K and codim$L$ in $N = \mathbb{R}^3$ are both equal to 1, and that $\partial(L \cap K)$, which is equal to the image of S^1 under the same immersion, coincides with the intersection of ∂K,(i.e., S^2) with L.

The following examples point out that the assumptions of Corollary 54 are only sufficient conditions and not necessary for the conclusions.

Example 56 Let Y, Z_1, Z_2, Z_3, Z_4 be subsets of $\mathbb{R}^3$ defined as solution sets to the following equations or systems of equations:

$$(Y) \qquad x^2 + y^2 + z^2 = 25$$

$$(Z_1) \qquad x^2 + y^2 = 25$$

$$(Z_2) \qquad \begin{cases} x^2 + y^2 = 16 \\ z \geq 3 \end{cases}$$

$$(Z_3) \qquad \begin{cases} x^2 + y^2 = 16 \\ 4z \geq 3x \end{cases}$$

Y is a sphere centered at the origin; all the Z_i's are circular, upwards unbounded cylinders, centered on the z-axis, Z_1 is downwards unbounded as well, whereas Z_2, Z_3 are downwards bounded by specific planes. For each i, Corollary 54 is not relevant for the intersection $Y \cap Z_i$, since one or another assumption fails; yet in all cases, except when $i = 3$, $Y \cap Z_i$ is a manifold, with or without boundary. This is summarized in the following tables:

	$i = 1$	$i = 2$	$i = 3$
$Y \cap Z_i$	$C_5 \cap P_0$	$C_4 \cap P_3$	$\{(-4, 0, 3)\} \bigcup (C_4 \cap P_3)$
∂Z_i	$\emptyset$	$C_4 \cap P_3$	$C_4 \cap P_2$
$Y \cap \partial Z_i$	$\emptyset$	$C_4 \cap P_3$	$\{(4, 0, 3), (-4, 0, 3)\}$

where $C_4 \equiv \{(x, y, z) \in \mathbb{R}^3 : x^2 + y^2 = 16\}$, $C_5 \equiv \{(x, y, z) \in \mathbb{R}^3 : x^2 + y^2 = 25\}$, $P_0 \equiv \{(x, y, z) \in \mathbb{R}^3 : z = 0\}$, $P_2 \equiv \{(x, y, z) \in \mathbb{R}^3 : 4z = 3x\}$, $P_3 \equiv \{(x, y, z) \in \mathbb{R}^3 : z = 3\}$.
Consequently regarding the Assumption of the Corollary 54 we get

Assumption	$i = 1$	$i = 2$	$i = 3$
$Y \pitchfork Z_i$	no	yes	yes
$Y \pitchfork \partial Z_i$	yes	no	no

while the following table is about the conclusions of the Corollary 54

Conclusion	$i = 1$	$i = 2$	$i = 3$
1	yes	yes	no
2	yes	no	no
3	yes	yes	no

where "yes" and "no" means the assumptions (conclusions) do or not hold true respectively.

Example 57 Let Y, Z be subsets of $\mathbb{R}^2$ defined as follows:

$$(Y) \qquad y = x^5$$

$$(Z) \qquad y \geq x^3$$

Then

$$\begin{array}{ll} Y \cap Z & \left\{(x,y) \in \mathbb{R}^2 : -1 \leq x \leq 0, y = x^5\right\} \cup \left\{(x,y) \in \mathbb{R}^2 : x \geq 1, y = x^5\right\} \\ \partial Z & \left\{(x,y) \in \mathbb{R}^2 : y = x^3\right\} \\ Y \cap \partial Z & \{(-1,-1), (0,0), (1,1)\} \\ \partial (Y \cap Z) & \{(-1,-1), (0,0), (1,1)\} \end{array}$$

Consequently no assumption of Corollary 54 (that is $Y \pitchfork Z$ and $Y \pitchfork \partial Z$) is verified but $Y \cap Z$ is 1-dimensional manifold with boundary, $\dim (Y \cap Z) = 1+2-2 = 1$ and $Y \cap \partial Z = \partial (Y \cap Z)$, i.e., all the conclusions of the corollary hold.

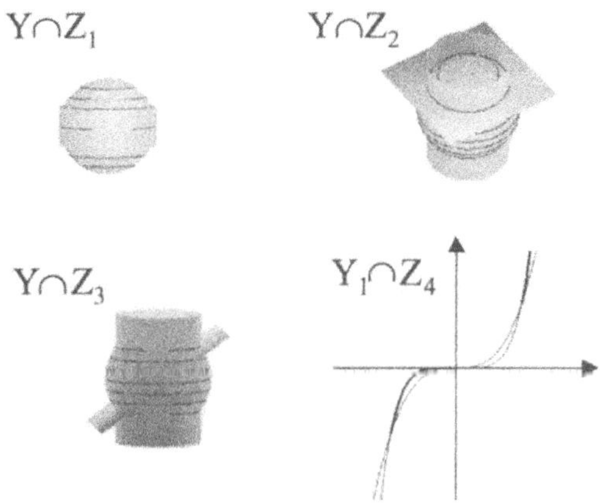

Figure 5.6. Intersection of manifolds

Remark 58 The intersection of the two manifolds in the Example 36 fails to be a manifold and indeed $T_{(0,0)}Y + T_{(0,0)}Z \neq \mathbb{R}^3$.

The following Example shows that f transverse to Y is a sufficient, but not a necessary condition for $f^{-1}(Y)$ to be a manifold. It also gives some motivation for the use of the word "transverse", i.e., "not tangent".

Example 59 Consider $S^2 \subseteq \mathbb{R}^3$ and $Z_c \equiv \left\{(x,y,z) \in \mathbb{R}^3 : z = c\right\}$.
a) $Z_c \cap S^2 = \emptyset$ for $c \in \mathbb{R} \backslash [-1,1]$

b) $Z_c \cap S^2$ is a one-dimensional manifold for $c \in (-1, 1)$. Indeed, at any element $P = (x, y, z)$ of $Z_c \cap S^2$, taking into account Corollary 13 of Chapter 4,

$$T_P Z_c = \{(u, v, w) \in \mathbb{R}^3 : w = 0\}$$
$$T_P S^2 = \{(u, v, w) \in \mathbb{R}^3 : xu + yv + zw = 0\}$$

Thus $\mathcal{B} = \{(1, 0, 0), (0, 1, 0)\}$ can be chosen as a basis for $T_P Z_c$ and $\mathcal{C} = \{(xz, yz, c^2 - 1), (y, -x, 0)\}$ can be chosen as a basis for $T_P S^2$ (notice that $\mathcal{C}$ is a linearly independent set because $x^2 + y^2 = -(c^2 - 1) > 0$). It follows that Z_c and S^2 are transverse in $\mathbb{R}^3$, since $T_P Z_c + T_P S^2$ contains

$$\{(1, 0, 0), (0, 1, 0), (xz, yx, c^2 - 1)\}$$

which is a basis for $\mathbb{R}^3$. Our statement is now a consequence of Corollary 54.
c) $Z_1 \cap S^2 = \{(0, 0, 1)\}$ and $Z_{-1} \cap S^2 = \{(0, 0, -1)\}$ are zero-dimensional manifolds, even if Corollary 54 does not apply,.because Z_1 and S^2 (resp. Z_{-1} and S^2) are not transverse:

$$T_{(0,0,1)} S^2 + T_{(0,0,1)} Z_1 = T_{(0,0,-1)} S^2 + T_{(0,0,-1)} Z_{-1} = Z_0 + Z_0 \subsetneq \mathbb{R}^3$$

Chapter 6

SARD'S THEOREM AND TRANSVERSALITY

1. Introduction

In many problems of calculus on manifolds and in differential topology, it is crucial to have some knowledge about the distribution of regular values of a C^r function $f : M \to N$, with M and N manifolds. For example, the construction of degree (Chapter 7) is based on the fact that the set of regular values of f is dense in N.

In this chapter we present a classical and very important theorem due to Morse and Sard. In the particular context of functions between euclidean spaces, this result states that

- *if $f : \mathbb{R}^m \to \mathbb{R}^n$ is C^r function, with*

$$r > \max\{m - n, 0\}$$

then the set of critical values of f has (Lebesgue) measure zero in $\mathbb{R}^n$.

Roughly speaking, Sard's theorem says that the number of critical values of f is really small from the point of view of Lebesgue measure.

The theorem has many applications and Section 3 of the chapter provides one of them, concerning the study of nonlinear systems of equations depending on a set of parameters.

2. Sard's theorem

This section is divided in three parts: first, in order to make the reader better understand the meaning of Sard's theorem and its applications, we recall the definition of (Lebesgue) measure zero set, and discuss some properties of that definition; secondly, we present Sard's theorem for functions between euclidean spaces, and some of its proofs (there are indeed several somewhat different proofs, depending on the dimensions of the euclidean spaces involved); finally, in Subsection 2.3, we extend the notion of zero measure to subsets of manifolds, and we give a more general version of Sard's theorem, concerning functions between manifolds.

We urge to point out that our discussion about sets of measure zero and their properties covers just a particular aspect of the general theory of Lebesgue measure. That theory starts defining measurable sets, and proceeds to assign to each measurable set a real nonnegative number, the measure of that set. We will not follow such an approach, since a discussion of Lebesgue measure theory is out of the scope of the book. In Definition 4 below we give a notion of set of measure zero which turns out, in our construction, to be completely independent of the existence of a general definition of measure. Nevertheless, one can prove that our notion of measure zero is equivalent to the corresponding one in the Lebesgue measure theory. The reader interested in Lebesgue measure theory can see for instance [3], Chapters 10 and 15.

2.1 Sets of (Lebesgue) measure zero

Definition 1 Let $a = (a_1, ..., a_n)$, $b = (b_1, ..., b_n)$ be two points in $\mathbb{R}^n$, such that $a \ll b$, and let

$$I^n(a,b) \equiv [a_1, b_1] \times ... \times [a_n, b_n]$$
$$J^n(a,b) \equiv (a_1, b_1) \times ... \times (a_n, b_n)$$

$I^n(a,b)$ and $J^n(a,b)$ are called (nondegenerate) **closed n-interval** and **open n-interval,** respectively. The points of the set $\{a_1, b_1\} \times \cdots \times \{a_n, b_n\}$ are called **vertices** of $I^n(a,b)$ and of $J^n(a,b)$. The (Lebesgue) measure of n-intervals of both types, denoted $\mu(I^n(a,b))$ and $\mu(J^n(a,b))$, is defined by the product of the side lengths

$$\prod_{i=1}^{n}(b_i - a_i)$$

Closed and open n-intervals play slightly different roles in the constructions below, because of their different topological properties, and we found it convenient to have both classes of intervals available.

Remark 2 In the sequel, the word n-interval will stand for closed n-interval.

Definition 3 An **n-cube** $C \subseteq \mathbb{R}^n$ of side $\lambda > 0$ is the cartesian product of n closed intervals of length λ.

Definition 4 A subset $S \subseteq \mathbb{R}^n$ has **measure zero** in $\mathbb{R}^n$ or it is a **null set** in $\mathbb{R}^n$ if for each $\varepsilon > 0$ there exists a countable family $\{I_s\}_{s \in \mathbb{N}}$ of n-intervals such that

1 $S \subseteq \bigcup_{s=1}^{\infty} I_s$,

2 $\sum_{s=1}^{\infty} \mu(I_s) < \varepsilon$.

There is no contradiction between Definitions 1 and 4 in the sense that any n-interval, which has positive measure according to Definition 1, has not measure zero in $\mathbb{R}^n$ according to Definition 4.

Proposition 5 *Let* $I \equiv I^n(a, b)$ *be an* n*-interval and* $\{I_s\}_{s\in\mathbb{N}}$ *a countable family of* n*-intervals such that*

$$I \subseteq \bigcup_{s=1}^{\infty} I_s$$

Then

$$\mu(I) \leq \sum_{s=1}^{\infty} \mu(I_s)$$

Proof. First of all we show that the conclusion holds for every countable family $\{J_s\}_{s\in\mathbb{N}}$ of *open* n-intervals covering I. Since I is compact, there is a finite open subcovering of I , say $\{J_1, \ldots, J_m\}$. For each $s \in \{1, \ldots, m\}$, denote

$$I_s \equiv (a^s, b^s) \quad a^s \equiv (a_i^s)_{i=1}^n \quad b^s \equiv (b_i^s)_{i=1}^n$$

The argument relies on an accurate estimate of the numbers $N_0, N_1, \ldots, N_m$ of integer lattice points of $\mathbb{R}^n$ (i.e., points in $\mathbb{R}^n$ with integer coordinates) contained in $I, J_1, \ldots, J_m$, respectively. For any two given real numbers α and β such that $\alpha < \beta$, the number of integers belonging to $[\alpha, \beta]$ cannot exceed $\beta - \alpha + 1$ and cannot fall short of $\beta - \alpha - 1$, (nor of 0, to be sure). Hence for each $s \in \{0, 1, \ldots, m\}$

$$\prod_{i=1}^{n} (b_i^s - a_i^s + 1) \leq N_s \leq \prod_{i=1}^{n} (b_i^s - a_i^s + 1)$$

where $a^0 \equiv a$ and $b^0 = b$ for the sake of uniformity in the notation. Since $\{J_s\}_{s=1}^m$ covers I,

$$\prod_{i=1}^{n} (b_i^0 - a_i^0 - 1) \leq \sum_{s=1}^{m} \prod_{i=1}^{n} (b_i^s - a_i^s + 1)$$

For each $\lambda \in \mathbb{R}_{++}$, the above argument can be replicated with respect to $I(\lambda a^0, \lambda b^0)$ and $\{J(\lambda a^s, \lambda b^s)\}_{s=1}^m$, the latter being an open covering of the former, yielding

$$\prod_{i=1}^{n} (\lambda b_i^0 - \lambda a_i^0 - 1) \leq \sum_{s=1}^{m} \prod_{i=1}^{n} (\lambda b_i^s - \lambda a_i^s + 1)$$

Dividing through λ^n on both sides, and taking limits for $\lambda \to +\infty$, we get

$$\mu(I) = \prod_{i=1}^{n}(b_i - a_i) \leq \sum_{s=1}^{m}\prod_{i=1}^{n}(b_i^s - a_i^s) = \sum_{s=1}^{m}\mu(J_s)$$

Let now $\{I_s\}_{s\in\mathbb{N}}$ be a covering of I by means of n-intervals, and let $\varepsilon > 0$ be given. For each $s \in \mathbb{N}$, there exists an open n-interval J_s^ε, containing I_s, such that [1]

$$\mu(J_s^\varepsilon) < \mu(I_s) + \frac{\varepsilon}{2^n}$$

It follows that $\{J_s^\varepsilon\}_{s\in\mathbb{N}}$ is a fortiori a covering of I, and therefore

$$\mu(I) \leq \sum_{s=1}^{\infty}\mu(J_s^\varepsilon) < \sum_{s=1}^{\infty}\mu(I_s) + \varepsilon$$

Since ε is arbitrary,

$$\mu(I) \leq \sum_{s=1}^{\infty}\mu(I_s)$$

■

As immediate consequence of Definition 4 and Proposition 5 we have the following two corollaries (the proofs are left as exercises)

Corollary 6 *Any subset of a null set is a null set.*

Corollary 7 *Any nonempty open set in $\mathbb{R}^n$ is not a null set.*

The reader can observe that the concept of measure zero for a set is relative to the euclidean space containing it. Consider for example the two squares $C \equiv [0,1] \times [0,1] \subseteq \mathbb{R}^2$ and $D \equiv [0,1] \times [0,1] \times \{0\} \subseteq \mathbb{R}^3$. By Definitions 1 and 4 it is clear that C has not measure zero in $\mathbb{R}^2$, while D has measure zero in $\mathbb{R}^3$. This should be not surprising because C and D are two different sets, even if under a wrong perception they sometimes appear as the same set, the square of side 1.

Remark 8 Given an n-cube C of side λ, for any x, y in C, $||x - y|| < \lambda\sqrt{n}$. Moreover, consider a nonempty open subset A of $\mathbb{R}^n$. If, for any x, y in A, $||x - y|| < r$, then there exists an n-cube of side r containing A.

In some cases it is worthwhile to use cubes to verify that a set has measure zero. The next proposition gives a sufficient and necessary condition for a set to have measure zero.

[1] it suffices to take, e.g., $J_s^\varepsilon \equiv J(a^s - \delta\mathbf{1}, b^s - \delta\mathbf{1})$ with $\delta < \frac{\min_i(b_i^s - a_i^s)}{2n}\ln\left(1 + \frac{\varepsilon}{2^n\mu(I_s)}\right)$

Proposition 9 *A subset $S \subseteq \mathbb{R}^n$ has measure zero in $\mathbb{R}^n$ if and only if for each $\varepsilon > 0$ there exists a countable covering $\{C_s\}_{s\in\mathbb{N}}$ of S of n-cubes such that*

$$\sum_{s=1}^{\infty} \mu(C_s) < \varepsilon$$

Proof. (1. $\Rightarrow$) Let $I(a,b)$ an interval. For each $i \in \{1,\ldots,n\}$, and each $\delta > 0$, let

$$t_i \equiv b_i - a_i \qquad r_i \equiv \left\lceil \frac{t_i}{\delta} \right\rceil$$

where $\lceil x \rceil$, the ceiling of x, is the minimum integer which is greater than or equal to x. Then for each $i \in \{1,\ldots,n\}$, we have

$$0 < t_i \leq \delta r_i < t_i + \delta$$

and hence

$$b \leq a + \delta r$$

It follows that the interval $I(a, a+\delta r)$ contains I. Moreover, $I(a, a+\delta r)$ is the union of exactly $\rho \equiv \prod_{i=1}^{n} r_i$ adjacent n-cubes of side δ, with vertices in the grid

$$\{a + \delta g\}_{g\in\mathcal{G}}$$

where $\mathcal{G}$ is the set of all points with integer coordinates contained in $I(0,r)$. If $\{C_l\}_{l=1}^{\rho}$ is the family of those cubes, we have

$$\sum\nolimits_{l=1}^{\rho} \mu(C_l) = \rho\delta^n$$

Given a positive ε and choosing any δ which is less than $\sqrt[n]{\frac{\mu(I)+\varepsilon}{\rho}}$, we have

$$\sum\nolimits_{l=1}^{\rho} \mu(C_l) < \mu(I) + \varepsilon$$

Now, let then $\varepsilon > 0$ be given, let $\{I_s\}_{s\in\mathbb{N}}$ be a countable covering of S by means of n-intervals such that

$$\sum_{s\in\mathbb{N}} \mu(I_s) < \varepsilon$$

and let

$$h \equiv \varepsilon - \sum_{s\in\mathbb{N}} \mu(I_s)$$

For each $s \in \mathbb{N}$, I_s can be covered by a finite family $\{C_l^s\}_{l=1}^{\rho_s}$ of n-cubes such that

$$\sum\nolimits_{l=1}^{\rho_s} \mu(C_l^s) < \mu(I_s) + \frac{h}{2^s}$$

Then the family

$$\bigcup_{s\in\mathbb{N}} \{C_l^s\}_{l=1}^{\rho_s}$$

is a countable covering of I such that

$$\sum_{s\in\mathbb{N}} \left(\sum\nolimits_{l=1}^{\rho_s} \mu\left(C_l^s\right)\right) < \sum_{s\in\mathbb{N}} \mu\left(I_s\right) + \sum_{s\in\mathbb{N}} \frac{h}{2^s} = \varepsilon$$

(2. $\Leftarrow$) Immediate. ■

Proposition 10 *The union of a countable family of subsets of $\mathbb{R}^n$ of measure zero has measure zero.*

Proof. Let $\{S_m\}_{m\in\mathbb{N}}$ be a countable family of subsets of $\mathbb{R}^n$ of measure zero. Let $0 < \varepsilon < 1$ and, for each $m \in \mathbb{N}$, consider a family $\{I_{sm}\}_{s\in\mathbb{N}}$ of n-intervals such that

i) $S_m \subseteq \bigcup_{s=1}^{\infty} I_{sm}$,

ii) $\sum_{s=1}^{\infty} \mu\left(I_{sm}\right) < \varepsilon/2^{m+1}$.

Since the union of countable sets is countable, then $\{I_{sm}\}_{s,m\in\mathbb{N}}$ is a countable family of intervals such that

i) $S \subseteq \bigcup_{m=1}^{\infty} \left(\bigcup_{s=1}^{\infty} I_{sm}\right)$,

ii) $\sum_{m=1}^{\infty} \sum_{s=1}^{\infty} \mu\left(I_{sm}\right) \leq \sum_{m=1}^{\infty} \varepsilon/2^{m+1} = \varepsilon/2$

and the proposition is proved. ■

Proposition 11 *The $(n-1)$-dimensional subspace $S \equiv \{x \in \mathbb{R}^n : x_1 = 0\}$ of $\mathbb{R}^n$ has measure zero in $\mathbb{R}^n$.*

Proof. Let $\mathcal{C}$ be the family of $(n-1)$-cubes having side 1 and all the vertices with integer coordinates. Clearly, $\mathcal{C} \equiv \{C_j\}_{j\in\mathbb{N}}$ is countable and covers $\mathbb{R}^{n-1}$.

Let $0 < \varepsilon < 1$ and

$$D_j \equiv \left[-\frac{\varepsilon}{2^{j+2}}, \frac{\varepsilon}{2^{j+2}}\right] \times C_j$$

Clearly, the union of all the D_j's contains S. Moreover, for any $j \in \mathbb{N}$, $\mu(D_j) = \varepsilon/2^{j+1}$, and hence

$$\sum_{j=1}^{\infty} \mu(D_j) = \sum_{j=1}^{\infty} \varepsilon/2^{j+1} = \varepsilon/2$$

■

Remark 12 As an easy generalization it can be proved that, for $k < m$, the k-dimensional subspace $S \equiv \{x \in \mathbb{R}^m : x_1 = ... = x_{m-k} = 0\}$ of $\mathbb{R}^m$ has measure zero.

By Proposition 11 (and Remark 12) we observe that the property of a set to have measure zero is not invariant under diffeomorphisms. For example, any $(n-1)$-dimensional subspace S of $\mathbb{R}^n$ and $\mathbb{R}^{n-1}$ are linearly isomorphic but S has measure zero (in $\mathbb{R}^n$), while $\mathbb{R}^{n-1}$ has not measure zero in itself. However, if we consider subsets of the same euclidean space the measure zero is invariant under diffeomorphisms (see Lemma 15).

Proposition 13 *A subset S of $\mathbb{R}^n$ has measure zero if for any $x \in S$ there exists a neighborhood U_x of x such that $U_x \cap S$ has measure zero.*

Proof. By Lindelöf's theorem ([90], Section 11 in Chapter 4), any open covering of $\mathbb{R}^n$ has a countable subcovering and, therefore, this holds for any subset of $\mathbb{R}^n$. For any $x \in S$, let U_x be an open neighborhood of x such that $U_x \cap S$ has measure zero. The open covering $\{U_x \cap S\}_{x \in S}$ of S admits a countable subcovering $\{U_{x_k} \cap S\}_{k \in \mathbb{N}}$. Hence, the claim follows from Proposition 10. ∎

Lemma 14 *Let U be an open subset of $\mathbb{R}^m$ and $f : U \to \mathbb{R}^n$ be any function. If for any $x \in S$ there exists a neighborhood W_x of x such that $f(W_x \cap S)$ has measure zero in $\mathbb{R}^n$, then $f(S)$ has measure zero in $\mathbb{R}^n$.*

Proof. As the previous result, this one is a consequence of the above cited Lindelöf's theorem. Indeed, given any $x \in S$, let W_x be an open neighborhood of x in U such that

$$\mu (f (W_x \cap S)) = 0$$

and let $\{W_{x_j}\}_{j \in \mathbb{N}}$ be a countable subfamily of $\{W_x\}_{x \in S}$ such that

$$S \subseteq \bigcup_{j \in \mathbb{N}} W_{x_j}$$

Hence

$$f (S) \subseteq \bigcup_{j \in \mathbb{N}} f (W_{x_j} \cap S)$$

By Proposition 10, it follows that $f (S)$ has measure zero. ∎

Lemma 15 *Consider an open set $U \subseteq \mathbb{R}^n$ and a C^1 function $f : U \to \mathbb{R}^n$. If $S \subseteq U$ is a set of measure zero, then $f (S)$ has measure zero as well.*

Proof. Applying the above lemma, it is sufficient to prove that for every $x \in S$ there exists a neighborhood W_x of x such that $f(W_x \cap S)$ has measure zero in $\mathbb{R}^n$. Given $x \in S$, let W_x be an open neighborhood of x such that $\overline{W}_x$ is contained in the interior of a compact subset K_x of U. As K_x is compact and f is C^1, there exists k such that

$$||df_z|| < k, \quad \forall z \in K_x$$

Let $\varepsilon > 0$. Since $S_x \equiv W_x \cap S$ has measure zero, there exists a family $\{I_s\}_{s\in\mathbb{N}}$ of n-intervals, $I_s \equiv I_s(a^s, b^s)$ such that

$$S_x \subseteq \bigcup_{s\in\mathbb{N}} I_s \subseteq K_x$$

and that

$$\sum_{s\in\mathbb{N}} \mu(I_s) < \frac{\varepsilon}{(2k)^n}$$

For every $s \in \mathbb{N}$, and any $x', x'' \in I_s$, by the mean value theorem (see [3] p. 355) applied to f along the line segment $\overline{x'x''}$, we have

$$||f(x') - f(x'')|| < k||x' - x''||$$

In particular $||f(x') - f(a^s)|| < k||x' - a^s||$. This implies that $f(I_s)$ is contained in the n-interval $J_s \equiv J_s(\alpha^s, \beta^s)$, where

$$\alpha_s = f(a^s) - k(b^s - a^s), \quad \beta_s = f(a^s) + k(b^s - a^s)$$

The reader can prove as an exercise that $\mu(J_s) = (2k)^n \mu(I_s)$. Since

$$f(S_x) \subseteq \bigcup_{s\in\mathbb{N}} f(I_s) \subseteq \bigcup_{s\in\mathbb{N}} J_s$$

and

$$\sum_{s\in\mathbb{N}} \mu(J_s) = (2k)^n \sum_{s\in\mathbb{N}} \mu(I_s) < \varepsilon$$

we obtain that $f(S_x)$ has measure zero. ■

In spite of previous lemma, it is not true without a smoothness assumption that the image of a null set is a null set[2]. In addition, the reader should notice that it is crucial in the lemma that domain and value set have the same dimension.

We close this subsection with another result which will be used in the proof of Sard's theorem.

[2]The Peano curve is a celebrated function $p : [0,1] \to [0,1] \times [0,1]$ which turns out to be surjective. Consider the function $f : [0,1] \times [0,1] \to [0,1] \times [0,1]$, $(x, y) \mapsto p(x)$. Then, for any $y \in [0,1]$, $f([0,1] \times \{y\}) = [0,1] \times [0,1]$.

Proposition 16 (Fubini's theorem for null sets) *Let S be a closed subset of $\mathbb{R}^n$. Assume that, for each $t \in \mathbb{R}$, the subset of $\mathbb{R}^{n-1}$*

$$S_t \equiv \left\{x \in \mathbb{R}^{n-1} : (t, x) \in S\right\}$$

has measure zero in $\mathbb{R}^{n-1}$. Then S has measure zero in $\mathbb{R}^n$.

Proof. The argument closely reflects the proof given in [66] Since S is closed and hence the intersection of S with any compact subset of $\mathbb{R}^n$ is compact, S can be written as a countable union of compact sets. Hence, by Proposition 10, it is enough to give the proof in the case when S is compact. and, consequently, contained in $[a, b] \times \mathbb{R}^{n-1}$, for some real numbers a, b.

The following two properties hold:

1 given a finite open covering of intervals $\{J_1, ..., J_m\}$ of $[a, b]$, there exists a finite covering of intervals $\{J'_1, ..., J'_m\}$ of $[a, b]$ such that each J'_j is contained in some J_i and $\sum \mu(J'_j) < 2(b - a)$.

2 Let K be a compact subset of $\mathbb{R}^n$. Let W be an open subset of $\mathbb{R}^{n-1}$ containing the t-section K_t for some t. Then, there is a real interval U, containing t, such that W contains $K_U \equiv \cup_{s \in U} K_s$.

The proof of the first point is left as an exercise. If point 2 is not true, there exists a sequence $((t_l, x_l))_{l \in \mathbb{N}}$ of points of K, where t_l tends to t and $x_l \notin W$. Up to a subsequence, x_l tends to a point of K_t and then belonging to W.

Let now $\varepsilon > 0$ be given. For any $t \in [a, b]$, let $\{I_i^t\}_{i \in \mathbb{N}}$ an open covering of S_t by $(n-1)$-intervals such that the sum of their measures is less than $\frac{\varepsilon}{2(b-a)}$. By the compactness of S_t consider a finite subcovering $\left\{I_1^t, ..., I_{m(t)}^t\right\}$. Let U_t be an open interval of $\mathbb{R}$, containing t, such that $\left\{U_t \times I_1^t, ..., U_t \times I_{m(t)}^t\right\}$ covers $S_{U_t} \equiv \cup_{s \in U_t} S_s$. Clearly,

$$\sum_{i=1}^{m(t)} \mu\left(U_t \times I_i^t\right) < \mu\left(U_t\right) \frac{\varepsilon}{2(b-a)}$$

We obtain an open covering of intervals of S, that is

$$S \subseteq \cup_{t \in [a,b]} \left(\cup_{i=1}^{m(t)} U_t \times I_i^t\right)$$

By the previous points 1 and 2, there is a finite covering $\{U'_1, ..., U'_{m'}\}$ of $[a, b]$ such that

i) each U'_i is contained in U_{t_i}, for a suitable $t_i \in [a, b]$,

ii) $\sum_{i=1}^{m'} \mu\left(U'_i\right) < 2(b - a)$;

iii) $S \subseteq \cup_{i=1}^{m'} \left(\cup_{j=1}^{m(t_i)} U_i' \times I_j^{t_i} \right)$.

We have obtained a finite covering of S. Collecting the above three conditions

$$\sum_{i=1}^{m'} \left(\sum_{j=1}^{m(t_i)} \mu \left(U_i' \times I_j^{t_i} \right) \right) < \varepsilon$$

■

2.2 Sard's theorem for functions between euclidean spaces

We present here Sard's theorem for functions between euclidean spaces. The strategy of the proof is taken from [71] and [14].

Theorem 17 (Sard's theorem in euclidean spaces) *Let U be an open subset of $\mathbb{R}^m$ and $f : U \to \mathbb{R}^n$ be a C^r function, with*

$$r > \max\{m - n, 0\}$$

Then, the set CP_f of critical values of f has measure zero in $\mathbb{R}^n$.

Proof. If $m - n < 0$ the proof is a consequence of Lemma 15 and Remark 12. Indeed, let $k \equiv n - m$ and consider the C^r function

$$F : \mathbb{R}^k \times U \to \mathbb{R}^n, \quad (x, y) \mapsto f(y)$$

By Remark 12, $\{0\} \times U$ has measure zero in $\mathbb{R}^k \times \mathbb{R}^m$. In addition, $F(\{0\} \times U) = f(U)$ and, by Lemma 15, $f(U)$ has measure zero in $\mathbb{R}^n$. Then, the set of critical values of f has measure zero, being contained in $f(U)$.

If $m \geq n$ the proof is quite hard. We will give it in the following three cases:

a) $m = n$,

b) $m > n > 1$,

c) $m > n = 1$ and $r > m$.

In other words we do not discuss[3] the case $m > n = 1$ and $r = m$. The proof of Sard's theorem in this particular case can be found in [50].

Proof of case a). By Lemma 14, it is enough to verify that, for any $x \in CP_f$, there exists a cube C contained in U and containing x such that $f(CP_f \cap C)$ is a null set. Let C be a cube of side λ, contained in U and containing x.

[3]As the reader will see in the proof of cases b) and c), the classification on m, n are r arises to the fact that the inequality $m - nr < 0$ is used.

Since f is C^1, there exist a nonnegative number k and a nonnegative function α defined on the real interval $(0, \lambda\sqrt{n})$, such that:

i)
$$k = \max_{z \in C} \|df_z\|$$

ii) for any z in C,
$$\|f(z) - f(x)\| \leq k \|z - x\|$$

iii) for any two elements z, z' of C,
$$\|f(z') - f(z) - df_z(z' - z)\| \leq \alpha(\|z' - z\|) \|z' - z\|$$

iv) α is nondecreasing,

v)
$$\lim_{t \to 0} \alpha(t) = 0$$

Observe that i) holds because df is continuous on the compact set C, and ii) is obtained by the continuity of df on C and the application of the mean value theorem along any direction. Moreover,

$$\begin{aligned}
\|f(z') - f(z) - df_z(z' - z)\| &= \left\| \int_0^1 \frac{d}{ds} \left(f\left[z + s(z' - z)\right]\right) - df_z(z' - z)\, ds \right\| = \\
&= \left\| \int_0^1 df_{z+s(z'-z)}(z' - z) - df_z(z' - z)\, ds \right\| \leq \\
&\leq \int_0^1 \left\| \left(df_{z+s(z'-z)} - df_z\right)(z' - z) \right\| ds
\end{aligned}$$

Let then

$$\alpha(t) \equiv \sup_{s \in [0,1],\, z \in C,\, z' \in C,\, \|z' - z\| = t} \left\| df_{z+s(z'-z)} - df_z \right\|$$

The function α is well defined, because of i). Finally, iii) and iv) are immediate from the definition, and v) follows from the fact that the continuity of df_z (with respect to z) is uniform given that C is compact.

Let $A \equiv \{f(x)\} + \operatorname{Im} df_x$, which is an affine subspace of $\mathbb{R}^n$, and let H be any hyperplane containing A. The existence of H follows from the fact that $\operatorname{Im} df_x$ is a proper subspace of $\mathbb{R}^n$, because x is a critical point. We now check how close $f(C)$ is to H. For each $z \in C$, the distance $d(f(z), H)$ of $f(z)$ from H does not exceed $\|f(z) - f(x) - df_x(z - x)\|$, since $f(x) + df_x(z - x)$ is an element of A. The latter in turn does not exceed $\alpha(\|z-x\|)\|z-x\|$, by iii). Therefore,

$$\forall z \in C, \ \|z-x\| < \delta \Rightarrow d(f(z), H) \leq \alpha(\delta)\delta$$

At the same time, from ii),

$$\forall z \in C, \ \|z-x\| < \delta \Rightarrow \|f(z) - f(x)\| \leq k\delta$$

Thus the image of any z with distance strictly less than δ from x belongs to $\overline{B}^n_{k\delta}(f(x)) \cap L$, where $\overline{B}^n_{k\delta}(f(x))$ is the closed n-dimensional ball with center $f(x)$ and radius $k\delta$, and L is the "layer" thick $2\alpha(\delta)\delta$ around H.

Perform an orthogonal transformation of $\mathbb{R}^n$ mapping the normal unit vector of H into, say, the first canonical vector e_1, and after that perform a translation bringing $f(x)$ to the origin. H is mapped onto

$$K \equiv \{x \in \mathbb{R}^n : x_1 = 0\}$$

L is mapped onto

$$P \equiv \{x \in \mathbb{R}^n : x_1 \in [-\alpha(\delta)\delta, \alpha(\delta)\delta]\}$$

and $\overline{B}^n_{k\delta}(f(x))$ onto

$$B \equiv \overline{B}^n_{k\delta}$$

the closed ball centered at the origin of radius $k\delta$. It should be clear that

$$B \cap K = \{0\} \times \overline{B}^{n-1}_{k\delta} \subset \{0\} \times C^{n-1}_{2k\delta}$$

where $C^{n-1}_{2k\delta}$ is the $(n-1)$-cube of side $2k\delta$ centered at the origin (of $\mathbb{R}^{n-1}$), and that

$$P \cap B \subseteq [-\alpha(\delta)\delta, \alpha(\delta)\delta] \times C^{n-1}_{2k\delta}$$

In addition the measure of the n-interval $[-\alpha(\delta)\delta, \alpha(\delta)\delta] \times C^{n-1}_{2k\delta}$ is

$$\mu\left([-\alpha(\delta)\delta, \alpha(\delta)\delta] \times C^{n-1}_{2k\delta}\right) = 2\alpha(\delta)\delta(2k\delta)^{n-1} = 2^n\delta^n k^{n-1}\alpha(\delta)$$

The orthogonal transformation we have mentioned is a C^∞ function (with C^∞ inverse), say T, of $\mathbb{R}^n$ into itself. By Lemma 15, to prove that the set regular values of f in $f(C)$ has measure zero it is sufficient to prove that the set of regular values of $T \circ f$ in $T(f(C))$ has measure zero. By the above construction, $f(C)$ is contained in $\overline{B}^n_{k\delta}(f(x)) \cap L$ and the latter set is transformed by T in a set contained in P. Therefore, without loss of generality we can assume that $f(C) \subseteq P$.

We are now ready to run through a subdivision argument. For each $s \in \mathbb{N}$, divide each real interval defining C (as a product) in s pairwise adjacent

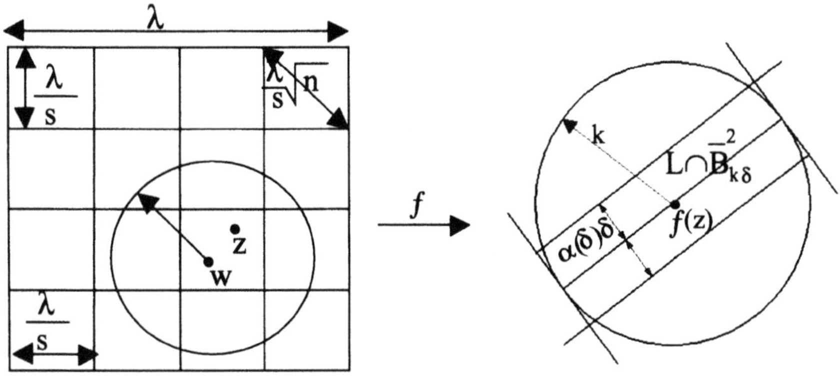

Figure 6.1. The layer around H

subintervals of equal length λ/s, and obtain in this way a family $\{C_i\}_{i=1}^{s^n}$ of n-cubes of side λ/s exactly covering C. Every critical point w of f in C has to belong to one (at least) of these cubes. Consider the subfamily $\{C_{i_j}\}_{j=1}^{l}$ of the cubes actually containing some critical point - in short, denote $\{C_j\}_{j=1}^{l}$ this subfamily. For each j, $f(C_j)$ is contained in an n-interval D_j of the form

$$\left[-\alpha\left(\frac{\lambda}{s}\sqrt{n}\right)\frac{\lambda}{s}\sqrt{n},\ \alpha\left(\frac{\lambda}{s}\sqrt{n}\right)\frac{\lambda}{s}\sqrt{n}\right]\times D'_j$$

where D'_j is an $(n-1)$-cube of side $2k\lambda\sqrt{n}/s$. Then

$$f(CP_f\cap C)\subseteq\bigcup_{j=1}^{l}D_j$$

and [4]

$$\begin{aligned}\sum_{j=1}^{l}\mu(D_j) &\leq s^n\,2\,\alpha\left(\frac{\lambda}{s}\sqrt{n}\right)\frac{\lambda}{s}\sqrt{n}\,2^{n-1}\,k^{n-1}\left(\frac{\lambda}{s}\sqrt{n}\right)^{n-1}\\ &= 2^n\,k^{n-1}\,\lambda^n\,\sqrt{n}^n\,\alpha\left(\frac{\lambda}{s}\sqrt{n}\right)\end{aligned}$$

[4] If, contrary to the assumption, we suppose $m>n$, the second and the third terms of the formula become $s^m\,2\,\alpha\left(\frac{\lambda}{s}\sqrt{n}\right)\frac{\lambda}{s}\sqrt{n}\,2^{n-1}\,k^{n-1}\left(\frac{\lambda}{s}\sqrt{n}\right)^{n-1}$ and $s^{m-n}2^n\,k^{n-1}\,\lambda^n\,\sqrt{n}^n\,\alpha\left(\frac{\lambda}{s}\sqrt{n}\right)$, which do not converge to zero as n tends to ∞.

Finally, by v), for each $\varepsilon > 0$ there exists $\delta > 0$ such that

$$0 < t < \delta \Rightarrow \alpha(t) < \frac{\varepsilon}{2^n k^{n-1} \lambda^n \sqrt{n^n}}$$

and it suffices to consider some $s > \frac{\lambda\sqrt{n}}{\delta}$ in order to conclude that

$$\sum_{j=1}^{l} \mu(D_j) < \varepsilon$$

and finally that $f(CP_f \cap C)$ is a null set.

Proof of cases b) and c) Assume that the conditions of case b) or c) hold. The strategy of the proof is the same for both cases. Decompose the set CP_f of critical points of f in the disjoint union of the three subsets A, B and C defined as follows:

* A is the set of points $x \in CP_f$ in which, for any component function f_i of f, $i = 1, ..., n$, all partial derivatives of f_i of each order $k \leq r$ vanish; that is,

  $$\frac{\partial^k f_i}{\partial x_1^{k_1} \partial x_2^{k_2} ... \partial x_n^{k_n}}(x) = 0$$

  where $k_1 + k_2 + ... + k_n = k$.

* B is the set of points $x \in CP_f$ such that

  $$\frac{\partial f_i}{\partial x_j}(x) \neq 0$$

  for some i, j.

* C is the set of points $x \in CP_f$ in which the Jacobian matrix vanishes and there exist a component f_i of f and an integer k, with $2 \leq k \leq r$, such that

  $$\frac{\partial^k f_i}{\partial x_1^{k_1} \partial x_2^{k_2} ... \partial x_n^{k_n}}(x) \neq 0$$

  where $k_1 + k_2 + ... + k_n = k$.

Notice that C is well defined since f is at least C^2. In addition B is empty if $n = 1$.

The proof that $f(CP_f)$ has measure zero is done in three steps, namely, we prove that

1. $f(A)$ is a null set,

2. $f(B)$ is a null set,

3. $f(C)$ is a null set.

Step 1. We proceed by an argument which is similar to that used in the proof of case a). For any $x \in A$, by the Taylor expansion of order r of f at x, there exist a neighborhood W_x of x in U and $\beta \geq 0$ such that, for any $y \in W_x$,

$$||f(x) - f(y)|| \leq \beta ||x - y||^r \tag{6.1}$$

Without loss of generality assume that W_x is a cube of side λ. Let $s \in \mathbb{N}$. Divide W_x in the union of s^m cubes of side λ/s. Let also $\{C_1, ..., C_l\}$ the set of those cubes having nonempty intersection with A. Given any C_j and any two elements $x', x'' \in C_j$, we have

$$||x' - x''|| \leq \sqrt{m}\lambda/s$$

Hence, $f(C_j)$ is contained in a cube D_j, the side of which not larger than $2\beta\left(\sqrt{m}\frac{\lambda}{s}\right)^r$, by (6.1) and the triangle inequality of the euclidean norm in $\mathbb{R}^n$. Then, the sum of the measures of these cubes is

$$g(s) \equiv \sum_{j=1}^{l} \mu(D_j) \leq s^m (2\beta)^n \left(\sqrt{m}\frac{\lambda}{s}\right)^{rn} = s^{m-rn} (2\beta)^n (\sqrt{m}\lambda)^{rn}$$

Now, observe that the assumptions defining the present case - case b) - imply that $m - rn < 0$. Then

$$\lim_{s \to +\infty} g(s) = 0$$

which implies that $f(W_x \cap A)$ has measure zero. Therefore, by Lemma 14, $f(A)$ has measure zero and step 1 is done.

Step 2. We prove that $f(B)$ has measure zero by induction on the dimension m $(\geq n)$ of the domain of f. If $m = n$, then $f(B)$ has measure zero by the proof of case a). In addition, if $n = 1$, then B is empty and it has of course measure zero. Let $m > n > 1$ and assume that the statement holds for functions defined in an open subset of $\mathbb{R}^p$ with $p < m$.

Let $\overline{x} \in B$. For suitable i, j

$$\frac{\partial f_i}{\partial x_j}(\overline{x}) \neq 0$$

For the sake of simplicity, put $i = n$ and $j = m$. Consider the C^r function

$$h : U \to \mathbb{R}^m, \ (x_1, ..., x_{m-1}, x_m) \mapsto (x_1, ..., x_{m-1}, f_n(x_1, ..., x_{m-1}, x_m))$$

Since the differential of h at $\overline{x}$ is an isomorphism, by the inverse function theorem, there exists an open neighborhood Ω of $\overline{x}$, such that $h_{|\Omega} : \Omega \to h(\Omega)$

is a C^r diffeomorphism. Without loss of generality, let Ω be such that $h(\Omega) = V \times W$, where V is an open subset of $\mathbb{R}^{m-1}$ and W is an open subset of $\mathbb{R}$. Then, $f_n \circ (h_{|\Omega})^{-1}$ is the n-th component of $h \circ (h_{|\Omega})^{-1} = id_{h_{|\Omega}}$ and hence

$$f_n \circ (h_{|\Omega})^{-1} : V \times W \to \mathbb{R}, \quad (x_1, ..., x_{m-1}, x_m) \mapsto x_m$$

Now, consider the C^r function

$$\widetilde{f} : V \times W \to \mathbb{R}^n, \quad \widetilde{f} \equiv f \circ (h_{|\Omega})^{-1}$$

$\widetilde{f}$ is of the form

$$\widetilde{f}\,(x_1, ..., x_{m-1}, x_m) = \left(g\,(x_1, ..., x_{m-1}, x_m)\,, x_m\right)$$

Since $h_{|\Omega}$ is a diffeomorphism, $x \in \Omega$ is a critical point of f if and only if $h(x) \in V \times W$ is a critical point of $\widetilde{f}$. Hence the sets of critical values of $\widetilde{f}$ and $f_{|\Omega}$ coincide.

Computing the jacobian matrix of $\widetilde{f}$, it is not difficult to check that any $(x_1, ..., x_{m-1}, x_m)$ is a critical point of $\widetilde{f}$ if and only if $(x_1, ..., x_{m-1})$ is a critical point for the partial function

$$(g_{x_m})_{|V} : V \to \mathbb{R}^{n-1}, \quad (x_1, ..., x_{m-1}) \mapsto g\,(x_1, ..., x_{m-1}, x_m)$$

which is C^r and depends on x_m, as a parameter. Therefore,

$$CP_{\widetilde{f}} \cap (V \times W) = \bigcup_{x_m \in W} CP_{g_{x_m}} \times \{x_m\}$$

Applying the induction hypothesis on m to each function g_{x_m}, which is defined on $V \subseteq \mathbb{R}^{m-1}$, we have that $g_{x_m}\left(CP_{g_{x_m}}\right)$ has measure zero in $\mathbb{R}^{n-1}$. By the Fubini's theorem for null sets the claim follows.

Step 3. The claim is proved by induction on the dimension m of the domain of f. If $m = n$, then $f(C)$ has measure zero by the proof of case a). Let $m > n$ and assume that the statement holds for functions defined in an open subset of $\mathbb{R}^p$ with $p < m$.

Let $x \in C$. By definition of C, there exist i, j, k such that

$$\begin{aligned} &\frac{\partial^{k-1} f_i}{\partial x_1^{k_1} ... \partial x_j^{k_j - 1} ... \partial x_n^{k_n}}(x) = 0 \\ \frac{\partial}{\partial x_j} \frac{\partial^{k-1} f_i}{\partial x_1^{k_1} ... \partial x_j^{k_j - 1} ... \partial x_n^{k_n}}(x) = &\frac{\partial^{k} f_i}{\partial x_1^{k_1} ... \partial x_j^{k_j} ... \partial x_n^{k_n}}(x) \neq 0 \end{aligned} \tag{6.2}$$

where $k_1 + k_2 + ... + k_n = k$.

Fix i, j and k, and let X the set of points of U verifying formula (6.2), for i, j and k. Since all possible combinations of i, j and k are finite, it is sufficient to prove that $f(X)$ has measure zero and, to obtain this, by Lemma 14, it is sufficient to verify that any point x of X has a neighborhood W_x such that $f(W_x)$ is a null set.

Define

$$g : U \to \mathbb{R}, \quad x \mapsto \frac{\partial^{k-1} f_i}{\partial x_1^{k_1} ... \partial x_j^{k_j - 1} ... \partial x_n^{k_n}}(x)$$

For any $x \in X$, there exists an open neighborhood V_x of x in U such that

$$\frac{\partial g}{\partial x_j}(x') \neq 0 \tag{6.3}$$

for all x' in V_x. By (6.3), $0 \in \mathbb{R}$ is a regular value for $g_{|V_x}$.

As f_i is C^r ($r \geq 2$), then g is C^{r-1}. Hence $X = g^{-1}(0) \cap V_x$ is a C^{r-1} submanifold of U of dimension $m - 1$.

Let W_x be an open neighborhood of x contained in V_x and C^{r-1} diffeomorphic to an open subset U_x of $\mathbb{R}^{m-1}$. Denote by $\psi : W_x \to U_x$ the diffeomorphism. The set of critical points of $f \circ \psi^{-1}$ coincides with the set of critical points of $f_{W_x \cap X}$. Now, applying the induction hypothesis, $CP_{f \circ \psi^{-1}}$ has measure zero (recall that the domain U_x of $f \circ \psi^{-1}$ has dimension $m-1$). Therefore $CP_{f_{X \cap W_x}}$ has measure zero and then $f\,(CP_f \cap X)$ has measure zero. ∎

The regularity assumption on f, that is, the fact that f is at least of class C^{m-n+1} (if $m \geq n$) may sound a little strange. The reader may suspect that the theorem holds under weaker assumptions (for example if f is C^1), perhaps with a harder proof. Actually, the assumption that f is C^{m-n+1} cannot be weakened. In fact, in a paper of 1935 (see [127]), Whitney has defined a C^1 function $f : \mathbb{R}^2 \to \mathbb{R}$ with an interval of critical values.

2.3 Sard's theorem for functions between manifolds

Let us now extend the above version of Sard's theorem to functions between manifolds. First of all we have to define the concept of measure zero for a subset of a manifold.

Definition 18 Let M be an m-dimensional manifold. A subset S of M has **measure zero** in M or S is a **null set** in M if for each $x \in S$ there is a chart (ψ, V, U) of M around x such that $\psi(V \cap S)$ has measure zero in $\mathbb{R}^m$.

Definition 19 A subset S of a manifold M is called a **full measure** subset of M if $M \backslash S$ has measure zero in M.

Remark 20 By Lemma 15 and Proposition 13 it possible to prove that if a subset S of M has **measure zero** in M, then, for *any* chart (ψ, V, U) of M, $\psi(V \cap S)$ has measure zero in $\mathbb{R}^m$.

The reader can prove as an exercise that an m-dimensional manifold M of $\mathbb{R}^l$ has measure zero in $\mathbb{R}^l$ if $m < l$. In this case, every subset S of M has measure zero in $\mathbb{R}^l$ as well. On the other hand, any nonempty open subset of M has not measure zero in M according to Definition 18. This does not mean that there is a contradiction between the two Definitions 4 and 18. Indeed, a subset S of a manifold M of $\mathbb{R}^l$, which has measure zero in $\mathbb{R}^l$, need not have measure zero in M.

As in the previous subsection, before introducing Sard's theorem, we need the following two lemmas which generalize Lemmas 14 and 15, respectively. The proofs are left as an exercise since they can be done in the same way.

Lemma 21 *Let $f : M \to N$ be a function between manifolds M and N. Let $S \subseteq M$ be such that for any $x \in S$ there exists a neighborhood W_x of x such that $f(W_x \cap S)$ has measure zero in N. Then, $f(S)$ has measure zero in N.*

Lemma 22 *Let M and N be manifolds of the same dimension, and $f : M \to N$ be a C^1 function. If $S \subseteq M$ has measure zero, then $f(S)$ is a set of measure zero as well.*

Theorem 23 (Sard's theorem for manifolds) *Let M and N be C^r manifolds of dimensions m, n and $f : M \to N$ be a C^r function. If*

$$r > \max\{m - n, 0\}$$

then the set of critical values of f has measure zero in N.

Proof. By Lemma 21, it is sufficient to prove that, for any x belonging to the set CP_f of critical points of f, there exists a neighborhood W_x of x in M such that $f(W_x \cap CP_f)$ has measure zero. This will be proved using Sard's theorem in euclidean spaces.

Let $x \in CP_f$. Let also (ϕ, U, V) and (ξ, T, S) be a local parametrization of M around x and a chart of N at $f(x)$, respectively. Assume $f(V) \subseteq T$. Then, consider the local representation of f

$$g : U \to S, \quad u \mapsto \xi \circ f \circ \phi(u)$$

The set of critical values of g coincides with the set of critical values of $f_{|V} = f(CP_f \cap V)$, since ϕ and ξ are diffeomorphisms. By Theorem 17, the set of critical values of g has measure zero in $\mathbb{R}^n$ and then, by Definition 18 and Remark 20, $f(CP_f \cap V)$ has measure zero in N. ∎

Next result is a consequence of Sard's theorem and it is particularly important in applications (we will use it, in particular, in the chapter for the construction of the degree).

Proposition 24 *Let M and N be C^r manifolds of dimensions m, n and let $f : M \to N$ be a C^r function. If*

$$r > \max\{m - n, 0\}$$

then the set of regular values of f is dense in N.

Proof. If $C \subseteq N$ is not dense, then $N \backslash C$ contains an open set and hence it cannot have measure zero. ∎

We conclude this section with a proposition which points out some differences among the notions of open set, dense set and full measure set.

Proposition 25 *Let S be a subset of $\mathbb{R}^n$. The following results hold:*

1 *there are dense subsets of S of measure zero;*

2 *there are full measure subsets of S which are not open in S;*

Proof. 1. $\mathbb{Q}$ is a null set which is dense in $\mathbb{R}$. 2. $\mathbb{R}\backslash\mathbb{Q}$ is a full measure set and it is not open in $\mathbb{R}$. ∎

3. A transversality result

In this section the word transversality is not used as in Chapter 5, but with a different meaning, which is frequent in the literature on general equilibrium models, and which will be introduced in the second part of the book.

We are quite often interested in studying systems of nonlinear equations depending on parameters, that is, of the form $f(x, \omega) = 0$, where x is a vector of unknown variables and ω is a vector of parameters. In this case, we want to study the dependence of the solution set on the parameters.

The main result of this section is the next theorem.

Theorem 26 *Let M, Ω and N be C^r manifolds of dimensions m, p and n, respectively. Let $f : M \times \Omega \to N$ be a C^r function. Assume*

$$r > \max\{m - n, 0\}$$

If y is a regular value for f, then there exists a full measure subset Ω^* of $\subseteq \Omega$ such that for any $\omega \in \Omega^*$, y is a regular value for f_ω (where, as usual, $f_\omega : M \to N$, $x \mapsto f(x, \omega)$).

Proof. Since y is a regular value of f, then either $E \equiv f^{-1}(y)$ is empty or it is a C^r $(m+p-n)$-dimensional submanifold of $M \times \Omega$. If $E = \emptyset$, the conclusion is immediate. Otherwise, consider the projection

$$\pi : E \to \Omega, \quad (x, \omega) \mapsto \omega$$

Since E and Ω are C^r manifolds, π is C^r, with $r > \max\{m-n, 0\}$, and since $\dim E - \dim \Omega = m - n$, from Sard's theorem it follows that the set Ω^* of regular values of π is a full measure subset of Ω.

We want to prove now that, for any $\omega \in \Omega^*$, y is a regular value for f_ω. Let $\omega \in \Omega^*$. If $f_\omega^{-1}(y)$ is empty, there is nothing to prove. Suppose otherwise that there is $x \in M$ such that $f_\omega(x) = y$. We check that

$$d(f_\omega)_x : T_x M \to T_y N$$

is onto. As (x, ω) is a regular point for f,

$$df_{(x,\omega)} : T_{(x,\omega)}(M \times \Omega) \to T_y N$$

is onto. Hence, given any $w \in T_y N$, there exists $v \in T_{(x,\omega)}(M \times \Omega)$ such that

$$df_{(x,\omega)}(v) = w$$

The vector v can be uniquely represented as a pair (v_1, v_2), where $v_1 \in T_x M$ and $v_2 \in T_\omega \Omega$. If $v_2 = 0$, then, by Proposition 40,

$$w = df_{(x,\omega)}(v_1, 0) = d(f_\omega)_x(v_1)$$

that is, $d(f_\omega)_x^{-1}(w)$ is nonempty. If $v_2 \neq 0$, consider

$$d\pi_{(x,\omega)} : T_{(x,\omega)}E \to T_\omega \Omega$$

which is onto; then

$$d\pi_{(x,\omega)}(\widetilde{v}_1, v_2) = v_2$$

for a suitable $\widetilde{v}_1 \in T_x M$. Recalling Theorem 13

$$\ker df_{(x,\omega)} = T_{(x,\omega)}E$$

and hence

$$df_{(x,\omega)}(\widetilde{v}_1, v_2) = 0$$

Finally

$$d(f_\omega)_x(v_1 - \widetilde{v}_1) = df_{(x,\omega)}(v_1 - \widetilde{v}_1, 0) = df_{(x,\omega)}\big((v_1, v_2) - (\widetilde{v}_1, v_2)\big) = w$$

The previous formula shows that $v_1 - \widetilde{v}_1 \in (d(f_\omega)_x)^{-1}(w)$. ■

Corollary 27 *Under the assumptions of Theorem 26, let Ω^* be a full measure subset of Ω such that, for any $\omega \in \Omega^*$, y is a regular value of f_ω. Then*

1. *if* $\dim M < \dim N$, *then* $f_\omega^{-1}(y) = \emptyset$ *for any* $\omega \in \Omega^*$,
2. *if* $\dim M \geq \dim N$, *then, for any* $\omega \in \Omega^*$, $f_\omega^{-1}(y)$ *is an* $(m-n)$*-dimensional submanifold of* M.
3. *if the projection* $\pi : f^{-1}(y) \to \Omega$ *is proper, then* Ω^* *is open in* Ω.

Proof. 1. If $\dim M < \dim N$, for each $\omega \in \Omega$, f_ω has no regular points. Then, since y is a regular value for f_ω, for any $\omega \in \Omega^*$, $f_\omega^{-1}(y) = \emptyset$.

2. If $\dim M \leq \dim N$, the result follows from Theorem 49.

3. The statement is a direct consequence of Lemma 8. ■

4. Full and zero measure in convex sets

The purpose of this section is to show how the two concepts of null set and full measure set (Definitions 4 and 19) can be extended to encompass ordinary canonical simplices and their relative interiors (the former are not always manifolds), in order to allow for a meaningful discussion of the robustness of some results obtained in Part II. At the same time, we stick to the approach outlined in the introduction to Section 2 thereby trying to touch upon several fundamental aspects of Lebesgue measure theory as marginally as possible. On the other hand, we cannot avoid introducing a few basic concepts of affine geometry, although in a very elementary way.

Definition 28 Let $\{v_i\}_{i=1}^r \subseteq \mathbb{R}^n$. An **affine combination** of the set $\{v_i\}_{i=1}^r$ is a linear combination of $\{v_i\}_{i=1}^r$ with coefficients adding up to 1. In other words, $v \in \mathbb{R}^n$ is an affine combination of $\{v_i\}_{i=1}^r$ if there exists $\alpha \in \mathbb{R}^r$ such that

$$\sum_{i=1}^{r} \alpha_i v_i = v \qquad \text{and} \qquad \sum_{i=1}^{r} \alpha_i = 1$$

In other words, an affine combination is like a convex combination where the nonnegativity assumption on the coefficients is dropped.

Definition 29 A subset A of $\mathbb{R}^n$ is an **affine subspace** of $\mathbb{R}^n$ if it is closed with respect to affine combination, that is, if the following holds:

$$\left(\{v_i\}_{i=1}^r \subseteq A \quad \text{and} \quad \sum_{i=1}^{r} \alpha_i = 1 \right) \Rightarrow \sum_{i=1}^{r} \alpha_i v_i \in A$$

Definition 30 For every subset B of $\mathbb{R}^n$, the **affine hull** of B, or **affine subspace generated** by B, denoted $Aff(B)$, is the intersection of all affine subspaces of $\mathbb{R}^n$ which contain B.

It can be shown that $Aff(B)$ is the set of all affine combinations of finite subsets of B.

The proof of the following proposition is left as an exercise.

Proposition 31 *A subset A of $\mathbb{R}^n$ is an affine subspace of $\mathbb{R}^n$ if and only if there exists a uniquely determined vector subspace D of $\mathbb{R}^n$ such that*

$$\forall a \in A, \qquad A - \{a\} = D$$

Definition 32 Let A be an affine subspace of $\mathbb{R}^n$, and let D the vector subspace of $\mathbb{R}^n$ such that $A - \{a\} = D$ for each $a \in A$. D is called the **direction space** of A.

Definition 33 Let A be an affine subspace of $\mathbb{R}^n$. The **dimension** of A is the dimension of its direction space.

Remark 34 There is only one affine subspace of $\mathbb{R}^n$of dimension n, namely, $\mathbb{R}^n$it self. An affine subspace of $\mathbb{R}^n$of dimension 0 is a singleton. An affine subspace of $\mathbb{R}^n$of dimension 1 is a line.

Definition 35 An **affine**, or linear affine, **function** of $\mathbb{R}^n$ into $\mathbb{R}^m$ is a function $L : \mathbb{R}^n \to \mathbb{R}^m$ such that, for each $r \in \mathbb{N}$ and each $\alpha \in \mathbb{R}^r$,

$$\sum_{i=1}^{r} \alpha_i = 1 \Rightarrow L\left(\sum_{i=1}^{r} \alpha_i x_i\right) = \sum_{i=1}^{r} \alpha_i L(x_i)$$

L is an **affine isomorphism** if, in addition, L is bijective **.**

In other words, L is affine if it maps affine subspaces into affine subspaces. It can be shown that any affine function is the composite of a linear function with a translation.

Definition 36 Let C be a convex subset of $\mathbb{R}^n$. The **dimension** of C is the dimension of $Aff(C)$.

Proposition 37 *Let A be an affine subspace of $\mathbb{R}^n$. A is null if and only if* $\dim A < n$.

Proof. Let $k \equiv \dim A$ be strictly less than n, let D be the direction subspace of A, and let $\{v_i\}_{i=1}^n$ be a basis of $\mathbb{R}^n$ such that $\{v_i\}_{i=n-k+1}^n$ is a basis of D ([78], Corollary 2, p. 54). The function

$$L : \mathbb{R}^n \to \mathbb{R}^n, \quad x \mapsto \sum_{i=1}^{r} x_i v_i$$

is a linear isomorphism. If S is the k-dimensional subspace of $\mathbb{R}^n$ of all vectors having the first $n-k$ coordinates equal to zero, $L(S) = D$. Let now a be any element of A, and consider the translation

$$\tau_a : \mathbb{R}^n \to \mathbb{R}^n, \quad x \mapsto x + a$$

By Proposition 31, $\tau_a(D) = A$, and hence $(\tau_a \circ L)(S) = A$. By Remark 12, S is null; by Lemma 15, A is null. The reverse implication is trivial, by Remark 34. ■

Proposition 38 *Let C be convex subset of $\mathbb{R}^n$. C is null if and only if* $\dim C < n$.

Proof. 1. It is immediate from the previous proposition, Corollary 6, and Definition 36, that when $\dim C < n$, C is null.

2. If $\dim C = n$, we claim that C has nonempty interior, which by Corollary 7 implies that C is not null. To this purpose, we can assume that $0 \in C$ without loss of generality (for each $c \in C$, $C - \{c\}$ has nonempty interior if and only if C has nonempty interior). In such a case, $Aff(C) = span\, C$, since any linear combination $\sum_{i=1}^r \alpha_i c_i$ of the set $\{c_i\}_{i=1}^r \subseteq C$ can be seen as an affine combination $\sum_{i=0}^r \alpha_i c_i$ of the set $\{c_i\}_{i=0}^r \subseteq C$ with $c_0 = 0$ and $\alpha_0 = 1 - \sum_{i=1}^r \alpha_i$. Since $\dim span\, C = n$, C contains a linearly independent set $\{c_i\}_{i=1}^r$ of n vectors. Hence C contains the set of all convex combinations of the set $\{c_i\}_{i=0}^r$ where $c_0 = 0$, that is, the set

$$T \equiv \left\{ \sum_{i=1}^n \alpha_i c_i : \sum_{i=1}^r \alpha_i \leq 1 \text{ and } \alpha \geq 0 \right\}$$

Consider the coordinate function

$$f : \mathbb{R}^n \to \mathbb{R}^n, \quad x \mapsto \alpha \text{ such that } x = \sum_{i=1}^n \alpha_i c_i$$

which is a linear isomorphism ([78], p. 48), and let $H^- \equiv \{\alpha \in \mathbb{R}^n : \sum_{i=1}^r \alpha_i < 1\}$, an open set (in fact, an open hypersemispace). Since f is continuous,

$$\widetilde{T} \equiv f^{-1}\left(\mathbb{R}^n_{++} \cap H^-\right) = \left\{ \sum_{i=1}^n \alpha_i c_i : \sum_{i=1}^r \alpha_i < 1 \text{ and } \alpha \gg 0 \right\}$$

is an open subset of $\mathbb{R}^n$ which is contained in T and hence in C. ■

Definition 39 Let K be a convex subset of $\mathbb{R}^n$, let k be the dimension of K, and let L be an affine function of $\mathbb{R}^n$ into $\mathbb{R}^k$ such that

$$L(Aff(K)) = \mathbb{R}^k$$

A subset H of K is **null** in K if $L(H)$ is null in $L(K)$. A subset H of K is of **full measure** in K if $L(K\backslash H)$ is null in $L(K)$.

Remark 40 By Proposition 38, if K is a convex subset of $\mathbb{R}^n$ which is not null, the above definition turns out to have exactly the same form of Definition 19. Indeed, in such a case $Aff(K) = \mathbb{R}^n$, and L can be taken to be the identity; thus any subset H of K is a **full measure set** of K if $K\backslash H$ is not null in $\mathbb{R}^n$.

Proposition 41 *Let*

$$S = \left\{ s \in \mathbb{R}^n : \sum_{k=1}^{n} s_k = 1 \text{ and } s \geq 0 \right\}$$

be the canonical $(n-1)$-dimensional simplex of $\mathbb{R}^n$, and let

$$\widetilde{S} = \left\{ s \in \mathbb{R}^n : \sum_{k=1}^{n} s_k = 1 \text{ and } s \gg 0 \right\}$$

Then $\widetilde{S}$ is of full measure in S.

Proof. The affine hull $Aff(S)$ of S is the affine hyperplane of equation

$$\sum_{i=1}^{r} x_i = 1$$

and the direction subspace $D(S)$ of $Aff(S)$ has equation

$$\sum_{i=1}^{r} x_i = 0$$

Thus indeed $\dim S = n-1$. Let

$$L \equiv \mathbb{R}^n \to \mathbb{R}^{n-1}, \ (x_1, \ldots, x_n) \mapsto (x_1, \ldots, x_{n-1})$$

Then L is affine, because it is linear, and maps $Aff(S)$ onto $\mathbb{R}^{n-1}$, because $\left(x_1, \ldots, x_{n-1}, 1 - \sum_{i=1}^{n-1} x_i\right) \in L^{-1}(x_1, \ldots, x_{n-1})$ for each $(x_1, \ldots, x_{n-1})$ in $\mathbb{R}^{n-1}$.

We have

$$S - \widetilde{S} = \bigcup_{i=1}^{n} F_i$$

where, for each i, F_i is the face of S belonging to the i-th coordinate hyperplane

$$F_i \equiv \{s \in S : s_i = 0\}$$

Let now

$$T \equiv L(S) \qquad \widetilde{T} \equiv L\left(\widetilde{S}\right) \qquad G_i \equiv L(F_i)$$

Then

$$\begin{aligned} T &= \left\{t \in \mathbb{R}^{n-1} : \sum_{k=1}^{n-1} t_k \leq 1 \text{ and } t \geq 0\right\} \\ \widetilde{T} &= \left\{t \in \mathbb{R}^{n-1} : \sum_{k=1}^{n-1} t_k < 1 \text{ and } t \gg 0\right\} \end{aligned}$$

and, for each $i \in \{1, \ldots, n-1\}$

$$G_i = \left\{t \in \mathbb{R}^{n-1} : \sum_{k=1}^{n-1} t_k \leq 1\,,\, t \geq 0 \text{ and } t_i = 0\right\}$$

whereas

$$G_n = \left\{t \in \mathbb{R}^{n-1} : \sum_{k=1}^{n-1} t_k = 1 \text{ and } t \geq 0\right\}$$

Then for each $i \in \{1, \ldots, n\}$, G_i is a (subset of a) null set of $\mathbb{R}^{n-1}$ by Proposition 11. Moreover,

$$T \backslash \widetilde{T} = \bigcup_{i=1}^{n} G_i$$

■

Appendix

We conclude the chapter with the following result.

Proposition 42 *If D is a full measure subset of $\mathbb{R}^{m+n}$, then The projections $\pi_1(D)$ and $\pi_2(D)$ are full measure subsets of $\mathbb{R}^m$ and $\mathbb{R}^n$, respectively.*

We have to clarify that we put the above proposition in this appendix, that is, in an isolated position with respect to the other similar results of the chapter, because the proof requires some tools of the theory of Lebesgue measure.

The proof makes use of the following lemma.

Lemma 43 *Let μ, $\mu_1$and μ_2 denote the Lebesgue measures defined on $\mathbb{R}^{m+n}$, $\mathbb{R}^m$, and $\mathbb{R}^n$ respectively. Then μ is the product measure of μ_1 and μ_2, that is, for any measurable sets $A \subseteq \mathbb{R}^m$ and $B \subseteq \mathbb{R}^n$, $\mu(A \times B) = \mu_1(A)\,\mu_2(B)$.*

The interested reader can find a detailed discussion of the above result in, e.g., Sections 12 and 18 of [15] (in particular, see Theorem 12.5, pp. 177-180,

Theorem 18.2, p. 236, and p. 241). In Section 3, p. 41, of the same book it is shown that a subset of $\mathbb{R}$ is null if and only if it has Lebesgue measure zero, and the statement holds true for any subset of $\mathbb{R}^k$ and any $k \in \mathbb{N}$.

Proof. Let E be the complement of D in $\mathbb{R}^{m+n}$, a null set. For each $k \in \mathbb{N}$, let

$$C_k \equiv [-k, k]^{m+n} \qquad C_{1,k} \equiv \pi_1 (C_k) = [-k, k]^m \qquad C_{2,k} \equiv \pi_2 (C_k) = [-k, k]^n$$

be the hypercubes of side $2k$ centered at 0 in $\mathbb{R}^{m+n}$, $\mathbb{R}^m$, and $\mathbb{R}^n$ respectively, and let

$$D_k \equiv D \cap C_k \qquad E_k \equiv E \cap C_k$$

It is clear that

$$D = \bigcup_{k \in \mathbb{N}} D_k \qquad E = \bigcup_{k \in \mathbb{N}} E_k$$

$$\mathbb{R}^m \backslash \pi_1 (D) = \bigcup_{k \in \mathbb{N}} C_{1,k} \backslash \pi_1 (D) \qquad \mathbb{R}^n \backslash \pi_2 (D) = \bigcup_{k \in \mathbb{N}} C_{2,k} \backslash \pi_2 (D_k)$$

where each E_k is null (by Corollary 6). By Proposition 10, it is enough to show that $C_{1,k} \backslash \pi_1 (D)$ and $C_{2,k} \backslash \pi_2 (D_k)$ are null sets of $\mathbb{R}^m$ and $\mathbb{R}^n$, respectively, for each $k \in \mathbb{N}$. Given $k \in \mathbb{N}$, we shall discuss $C_{1,k} \backslash \pi_1 (D)$ alone, since the argument for $C_{2,k} \backslash \pi_2 (D_k)$ is entirely symmetrical. Since the set

$$F_k \equiv (C_{1,k} \backslash \pi_1 (D)) \times C_{2,k}$$

is contained in E_k, it is a null set (again by Corollary 6). By the lemma above,

$$\mu_1 (C_{1,k} \backslash \pi_1 (D)) \, \mu_2 (C_{2,k}) = 0$$

Since $\mu_2 (C_{2,k}) = (2k)^n > 0$, $\mu_1 (C_{1,k} \backslash \pi_1 (D)) = 0$ and $C_{1,k} \backslash \pi_1 (D)$ is null.
■

Chapter 7

HOMOTOPY AND DEGREE THEORY

1. Introduction

For almost a century, degree theory has been a very important tool in analysis of existence and multiplicity results for solutions to nonlinear equations in euclidean spaces and manifolds. For example, the study of ordinary and partial differential equations has been considerably improved by degree theory. From the point of view taken in the present chapter, we give a simplified account of the matter by saying that degree theory consists in giving an estimate of the number of solutions to the equation $f(x) = y$ for a function $f : M \to N$, where M and N are C^2 boundaryless manifolds and f is continuous.

The version of degree theory we deal with is rather standard and simple with respect to other approaches. In the particular case in which $f : M \to N$ is C^1, $y \in N$ is a regular value, $f^{-1}(y)$ is finite and the equation $f(x) = y$ is considered, our method only distinguishes between an odd number of solutions (hence at least one) and an even number of solutions (hence possibly none); it is called modulo 2 degree theory or degree theory mod 2. However, the approach we follow in presenting its main concepts is not completely standard, and hopefully more suitable for an intuitive grasping of the material.

For the sake of completeness, we mention that a different construction of the theory of degree can be developed, by resort to the concept of orientation for manifolds. It is the so called Brouwer degree, for which the reader is referred for example to [89] or [71].

Section 2 introduces the notion of homotopy and discusses the main properties which are going to be used in the remaining sections of the chapter and throughout the rest of the book. Section 3 is dedicated to degree theory, and proceeds (as commonly done) first defining the degree for C^2 functions at a regular value, then for C^2 functions in general, and finally for functions which are only C^0. In Section 4 we present a simple way of computing the degree for C^1 functions at regular values and Section 5 points out some applications of the degree to the investigation of the existence of solutions to nonlinear equations which are particularly important in the economics problems in the second part of the book. The final appendix contains the computational details of some examples which are used in various sections of the chapter in order to illustrate the definitions provided.

2. Homotopies

Definition 1 Consider an element r of $\mathbb{N}_0^*$ (that is, a nonnegative integer or ∞), two boundaryless manifolds M and N, and two C^r functions $f, g : M \to N$. f is said to be C^r **homotopic** to g, denoted by $f \simeq_r g$, if there exists a function $H : M \times [0,1] \to N$ such that

1 H is of class C^r;

2 for any $x \in M$, $H(x,0) = f(x)$ and $H(x,1) = g(x)$.

H is called a C^r **homotopy** between f and g; in addition f and g are said to be **homotopic** by H. If $r = 0$, f is frequently said to be **continuously homotopic** to g, and H is called a **continuous homotopy** between f and g.

Example 2 (The "straight-line" homotopy) Consider a manifold M and two arbitrary C^r functions $f, g : M \to \mathbb{R}^n$. Then f and g are C^r homotopic by

$$H : M \times [0,1] \to \mathbb{R}^n, \quad (x,t) \mapsto (1-t) f(x) + tg(x)$$

H is called a straight-line homotopy, because, for each x and as t varies, "the point $f(x,t)$ moves from $f(x)$ to $g(x)$" along the straight-line segment joining $f(x)$ to $g(x)$.

Example 3 (A homotopy between a constant function and $id_{\mathbb{R}^n}$) As a particular case of Example 2, consider the identity of $\mathbb{R}^n$ and let g be a constant function, $g : \mathbb{R}^n \to \mathbb{R}^n$, $x \mapsto c$. A C^∞ homotopy between these functions is

$$H : \mathbb{R}^n \times [0,1] \to \mathbb{R}^n, \quad (x,t) \mapsto tx + (1-t)c$$

Example 4 (The identity of $\mathbb{R}^n \setminus \{0\}$ and its radial projection on the sphere S^{n-1}) As another particular case of Example 2, let $\dot{\mathbb{R}}^n$ be the set $\mathbb{R}^n \setminus \{0\}$, endowed with the restrictions of the usual euclidean norm and metric, and with the induced topology. The function

$$H : \dot{\mathbb{R}}^n \times [0,1] \to \dot{\mathbb{R}}^n, \quad (x,t) \mapsto (1-t)x + t\frac{x}{\|x\|}$$

is a C^∞ homotopy between the functions

$$id_{\dot{\mathbb{R}}^n} \qquad \text{and} \qquad g : \dot{\mathbb{R}}^n \to \dot{\mathbb{R}}^n, \ x \mapsto \frac{x}{\|x\|}$$

Example 5 (A homotopy between $id_{\mathbb{R}^2}$ and $-id_{\mathbb{R}^2}$) The function

$$H : \mathbb{R}^2 \times [0,1] \to \mathbb{R}^2, \ (x, y, t) \mapsto \begin{pmatrix} x & y \end{pmatrix} \begin{pmatrix} \cos \pi t & \sin \pi t \\ -\sin \pi t & \cos \pi t \end{pmatrix}$$

is C^∞, with $H(x, y, 0) = (x, y)$ and $H(x, y, 1) = -(x, y)$; the function $H_t : (x, y) \mapsto H(x, y, t)$ is a counterclockwise rotation of angle πt.

Proposition 6 *Let M and N be two manifolds. The relation $\simeq_r$ is an equivalence relation on the set of all C^r functions from M to N.*

Proof. ($f \simeq_r f$) The function having all its t-partial functions equal to f

$$H : M \times [0,1] \to N, \ (x, t) \mapsto f(x)$$

is a homotopy between f and itself; H is C^r whenever f is so.

($f \simeq_r g \Rightarrow g \simeq_r f$) The function $\widetilde{H}$ obtained by complementing to 1 the variable t in the formula defining any given function $H : M \times [0,1] \to N$

$$\widetilde{H} : M \times [0,1] \to N, \ (x, t) \mapsto H(x, 1-t)$$

is a homotopy between g and f whenever H is a homotopy between f and g; $\widetilde{H}$ is C^r if H is so.

($f \simeq_r g$ and $g \simeq_r h \Rightarrow f \simeq_r h$) The continuous homotopy case is dealt with easily: the function which glues together at $t = \frac{1}{2}$ the composition of F with id_M times the "linear blow" of $\left[0, \frac{1}{2}\right]$ to $[0,1]$ and the composition of G with id_M times the linear blow of $\left[\frac{1}{2}, 1\right]$ to $[0,1]$

$$H : M \times [0,1] \to N, \ (x, t) \mapsto \begin{cases} F(x, 2t) & t \in \left[0, \frac{1}{2}\right] \\ G(x, 2t-1) & t \in \left[\frac{1}{2}, 1\right] \end{cases}$$

is a continuous homotopy between f and h whenever F and G are continuous homotopies between f and g and between g and h, respectively (see also the pasting lemma, [90] p. 108).

Even if F and G are C^r homotopies, the homotopy H of the previous case is generally not C^r, if $r > 0$; more precisely, H can be not differentiable at $t = \frac{1}{2}$. In order to build a homotopy which is C^r if F and G are C^r, let us define the following functions:

$$\psi : \mathbb{R} \to [0,1], \ t \mapsto \begin{cases} e^{\frac{1}{t-\frac{3}{4}} - \frac{1}{t-\frac{1}{4}}} & \text{if } t \in \left(\frac{1}{4}, \frac{3}{4}\right) \\ 0 & \text{otherwise} \end{cases}$$

$$\Psi : \mathbb{R} \to [0,1], \; t \mapsto \int_{1/4}^{t} \psi(\tau) d\tau \qquad \Phi : \mathbb{R} \to [0,1], \, t \mapsto \frac{\Psi(t)}{\Psi(3/4)}$$

Notice that Φ is C^∞ and that

$$\begin{aligned} \Phi(t) = 0 & \quad \text{if } t \leq \tfrac{1}{4} \\ 0 < \Phi(t) < 1 & \quad \text{if } t \in \left(\tfrac{1}{4}, \tfrac{3}{4}\right) \\ \Phi(t) = 1 & \quad \text{if } t \geq \tfrac{3}{4} \end{aligned}$$

The functions

$$\begin{aligned} \widetilde{F} : M \times [0,1] \to N, & \quad (x,t) \mapsto F(x, \Phi(t)) \\ \widetilde{G} : M \times [0,1] \to N, & \quad (x,t) \mapsto G(x, \Phi(t)) \end{aligned}$$

are C^r homotopies between f and g and between g and h, respectively, if F and G are C^r; moreover, $\widetilde{F}(x,t) = g(x) \; \forall t \in \left[\frac{3}{4}, 1\right]$ and $\widetilde{G}(x,t) = g(x)$, $\forall t \in \left[0, \frac{1}{4}\right]$. Thus the function

$$\widetilde{H} : M \times [0,1] \to N, \quad (x,t) \mapsto \begin{cases} \widetilde{F}(x, 2t) & \text{if } t \in \left[0, \frac{1}{2}\right] \\ \widetilde{G}(x, 2t-1) & \text{if } t \in \left[\frac{1}{2}, 1\right] \end{cases}$$

is a C^r homotopy between f and h. ■

Remark 7 If N is a convex manifold, then the proof of transitivity is trivial. It is sufficient to consider the following function:

$$H : M \times [0,1], \quad (x,t) \mapsto (1-t) F(x,t) + t G(t,x)$$

which is a homotopy between f and h, as the reader can check as an exercise.

Given a function $F : M \times [0,1] \to N$, for each $t \in [0,1]$, consider the t-partial function of F

$$F_t : M \to N, \quad x \mapsto F(x,t)$$

Lemma 8 *Given a continuous function $F : M \times [0,1] \to N$ and consider $y \in N$. If $F^{-1}(y)$ is compact, the same is true for every t-section $\left[F^{-1}(y)\right]_t = F_t^{-1}(y)$.*

Proof. Let $t \in [0,1]$. Then, $x \in F_t^{-1}(y)$ if and only if $(x,t) \in F^{-1}(y)$. This implies that $x \in \pi_1\left[F^{-1}(y)\right]$; thus $F_t^{-1}(y)$ is a closed set because F_t is continuous, and it is a subset of $\pi_1\left[F^{-1}(y)\right]$. Since $F^{-1}(y)$ is compact and π_1 is continuous, then $\pi_M[F^{-1}(y)]$ is compact; finally the claim follows by continuity of π_1. ■

Example 9 The converse of Lemma 8 is generally not true, that is, compactness of $F_t^{-1}(y)$ for each $t \in [0,1]$ does not imply compactness of $F^{-1}(y)$. For example, consider the functions $f : \mathbb{R} \to \mathbb{R}$, $x \mapsto e^x$ and $g : \mathbb{R} \to \mathbb{R}$, $x \mapsto e^x - 1$, and the homotopy

$$F : \mathbb{R} \times [0,1] \to \mathbb{R}, \quad (x,t) \mapsto (1-t)\, e^x + t\,(e^x - 1)$$

We have

$$F_t^{-1}(0) = \begin{cases} \emptyset & t = 0 \\ \{\ln t\} & t \in (0,1] \end{cases}$$

Therefore, $F_t^{-1}(0)$ is compact for each t, whereas $F^{-1}(0) = \{(\ln t, t)\}_{t \in (0,1]}$ is not compact.

The following proposition is a straightforward consequence of the Corollary 53 of Chapter 5.

Proposition 10 *Let $r \geq 1$, M and N be two C^r boundaryless manifolds of the same dimension. Let also $f, g : M \to N$ and $y \in N$ satisfy the following assumptions:*

1. *f and g are C^r and y is a regular value for f and g;*
2. *there exists a C^r homotopy H between f and g, such that $H^{-1}(y)$ is compact and y is a regular value for H and for ∂H.*

Then $H^{-1}(y)$ is a compact C^r manifold with boundary of dimension one and

$$\partial\left(H^{-1}(y)\right) = \left(f^{-1}(y) \times \{0\}\right) \cup \left(g^{-1}(y) \times \{1\}\right)$$

Moreover, $\partial\left(H^{-1}(y)\right)$ is a compact C^r boundaryless manifold of dimension zero.

3. Degree theory mod 2

This section is devoted to the construction of a modulo 2 degree theory for continuous functions between manifolds of the same dimension.

To briefly sketch the outline of the matter, consider a continuous function $f : M \to N$, an element $y \in N$ and an open subset U of M such that $f^{-1}(y) \cap U$ is compact. We associate to the triple (f, U, y) an element of $\{0, 1\}$, which we call the degree of (f, U, y). In other words, denoted by $\mathcal{T}$ the set of all triples (f, U, y), where f, U and y are as above, the degree can be seen a function $\mathcal{D} : \mathcal{T} \to \{0, 1\}$.

As said in the introduction, a degree theory is mainly an important tool to study equations between manifolds. To this purpose, a meaningful concept of degree is required to verify some crucial properties (see Theorem 23). Among such properties we state here the following three ones:

1 (normalization) $\deg(id_M, M, y) = 1$, where M is any manifold, id_M is the identity and y is any element of M;

2 (additivity) $\deg(f, M, y) = \deg(f, U_1, y) + \deg(f, U_2, y)$, where U_1 and U_2 are disjoint subsets of M and $f^{-1}(y) \subseteq U_1 \cup U_2$;

3 (homotopy invariance) $\deg(f, M, y) = \deg(g, M, y)$, where f and g are homotopic by H and $H^{-1}(y)$ is compact.

The importance of the above properties lies also in the fact that there exists exactly one possible degree with such properties, that is, there exists exactly one function $\mathcal{D} : \mathcal{T} \to \{0, 1\}$ verifying 1, 2 and 3 above. The so-called axiomatic approach to the degree defines the unique possible degree starting from the previous properties (see [80] or [46]).

The degree of any triple (f, U, y) naturally depends on each element f, U and y. On the other hand, in the particular case when M is compact and N is connected (or, more generally, if f is proper, N being still connected), the degree of a triple (f, U, y) is independent of y, allowing us to consider it as a global attribute of the function f. This result is particularly important not only for the study of nonlinear equations, but also to investigate topological properties of compact manifolds (as spheres or closed balls) and continuous functions defined on compact manifolds. One of the classical and perhaps the most famous topological application is the Brouwer fixed point theorem, which states that a continuous function f of a closed ball of $\mathbb{R}^n$ into itself admits a fixed point, that a point x such that $f(x) = x$.

Topological applications are not considered in this book; the reader can see for example [89], [71] or [80].

The construction of the degree will take three steps. In the first step, we will take into account only admissible triples (f, U, y) such that f is C^2 and y is a regular value for f. In the second one, we will remove the regularity assumption on y, thus defining the degree for admissible triples (f, U, y) such that f is still C^2. In the last step, we will extend the definition of degree to arbitrary admissible triples.

Assumption on M and N In what follows and unless otherwise stated, M and N will be two C^2 boundaryless manifolds of the same dimension n.

The assumption C^1 for manifolds will be sufficient for the construction of partition of unity and tubular neighborhood (see the following Subsection 3.3). These are the unique results in the chapter for which C^2 manifolds are not required.

Definition 11 Let M and N be two boundaryless C^2 manifolds of the same dimension and let $f : M \to N$ be a continuous function. An **admissible triple** is a triple (f, U, y) such that:

1 U is an open subset of M;

2 y is an element of N;

3 $f^{-1}(y) \cap U$ is compact.

Remark 12 Observe that if $f^{-1}(y)$ is compact, we have immediately that (f, M, y) is an admissible triple.

Remark 13 Let (f, M, y) be an admissible triple and U be an open subset of M. If $f^{-1}(y) \cap \delta U = \emptyset$, then (f, U, y) and $\left(f, M \backslash \overline{U}, y\right)$ are admissible triples. Indeed, $f^{-1}(y) \cap U$ and $f^{-1}(y) \cap M \backslash \overline{U}$ are both closed in $f^{-1}(y)$ and then compact. In particular, (f, U, y) and $\left(f, M \backslash \overline{U}, y\right)$ are admissible triples if $f^{-1}(y) \subset U$.

Example 14 Let

$$f : \mathbb{R} \to \mathbb{R}, \quad x \mapsto \begin{cases} e^{-\frac{1}{x-1}} & \text{if } x > 1 \\ 0 & \text{if } -1 \leq x \leq 1 \\ e^{\frac{1}{x+1}} & \text{if } x < -1 \end{cases}$$

Then f is C^∞, $f^{-1}(0) = [-1, 1]$ is compact and $(f, \mathbb{R}, 0)$ is admissible. If $U \equiv (0, 2)$, then $f^{-1}(0) \cap \delta U = \{0\} \neq \emptyset$ and $f^{-1}(0) \cap U = (0, 1]$, which is closed in U but not compact; thus $(f, U, 0)$ is not an admissible triple. This example shows that the condition $f^{-1}(y) \cap \delta U = \emptyset$ cannot be dispensed with in Remark 13.

3.1 Definition of degree for C^2 functions at a regular value

3.1.1 Definition and examples

Our starting definition of degree concerns the special case in which f is C^2 and y is a regular value for f.

Definition 15 Let (f, M, y) be an admissible triple, such that $f : M \to N$ is a C^2 function and $y \in N$ is a regular value for f. The **degree** of (f, M, y) is given by the following formula:

$$\deg(f, M, y) \equiv \left[\# f^{-1}(y)\right]_2 \tag{7.1}$$

In other words

$$\deg(f, M, y) = \begin{cases} 0 & \text{if } \# f^{-1}(y) \text{ is even} \\ 1 & \text{if } \# f^{-1}(y) \text{ is odd} \end{cases} \tag{7.2}$$

The definition is meaningful, since, by Corollary 16 of Chapter 4, $f^{-1}(y)$ is a finite set.

Remark 16 In formula (7.2) we have shorten the notation writing $0, 1$ for $[0]_2, [1]_2$. In the sequel the operations regarding the degree will be performed according to the arithmetic of $\mathbb{Z}_2$[1].

Example 17 (The degree of the real power functions) Let $\{f_n\}_{n\in\mathbb{N}}$ be the family of real power functions

$$f_n : \mathbb{R} \to \mathbb{R}, \quad x \mapsto x^n$$

If $n = 1$, $f_1 = id_{\mathbb{R}}$, and we have just seen that its degree is equal to 1 at any y. If $n > 1$, y is a regular value for f_n if and only if $y \neq 0$, in which case

$$\begin{array}{ll} \deg(f_n, \mathbb{R}, y) = 1 & \text{if } n \text{ is odd} \\ \deg(f_n, \mathbb{R}, y) = 0 & \text{if } n \text{ is even} \end{array}$$

$\deg(f_n, \mathbb{R}, 0)$ is undefined as yet; it is going to be defined in next section, making $\deg(f_n, \mathbb{R}, y)$ a constant function of y on all $\mathbb{R}$.

Example 18 Let $f : S^1 \to S^1$ be the function defined in Example 7 of Chapter 4. For each $X \in S^1$, $\# f^{-1}(X) = 4$; $P = (0, 1)$ is the only critical value of f. Then $\deg(f, S^1, X)$ is defined for each $X \neq P$, and it is equal to 0.

Example 19 Consider the function[2]

$$F : \mathbb{R}^2 \to \mathbb{R}^2, \quad (x, y) \mapsto (x^2 - y^2, 2xy)$$

and notice that

$$\det DF(x, y) = \det \begin{pmatrix} 2x & -2y \\ 2y & 2x \end{pmatrix} = 4(x^2 + y^2)$$

so that dF is onto except at $(0, 0)$, and (u, v) is a regular value for F if and only if $(u, v) \neq (0, 0)$. Consider now

$$f : S^1 \to S^1, \quad (x, y) \mapsto F(x, y)$$

[1]Recall that in $\mathbb{Z}_2$, $1 + 1 = 0$ and $0 + 1 = 1 + 0 = 1$.

[2]The readers who are familiar with the theory of functions of complex variables will recognize in this example the square function $\mathbb{C} \to \mathbb{C}$, $z \mapsto z^2$, up to the standard identification of $\mathbb{C}$ with $\mathbb{R}^2$.

Observe that $df_{(x,y)}$ is an isomorphism because $dF_{(x,y)}$ is so, and every $(u,v) \in S^1$ is a regular value for f. Moreover, $\left(f, S^1, (u,v)\right)$ is an admissible triple for every $(u,v) \in S^1$, since f is C^∞, $f^{-1}(u,v)$ is closed, and S^1 is compact. In order to compute $\deg\left(f, S^1, (u,v)\right)$ we look for solutions to the following system:

$$\begin{cases} x^2 - y^2 = u & (1) \\ 2xy = v & (2) \end{cases} \tag{7.3}$$

By a simple computation (see Appendix), system (7.3) is seen to have two solutions for all $(u,v) \in S^1$, namely

$$(x_1, y_1) = \left(\sqrt{\frac{1+u}{2}}, \operatorname{sign} v \sqrt{\frac{1-u}{2}}\right)$$

$$(x_2, y_2) = \left(-\sqrt{\frac{1+u}{2}}, -\operatorname{sign} v \sqrt{\frac{1-u}{2}}\right)$$

Therefore,

$$\deg\left(f, S^1, (u,v)\right) = 0$$

for all $(u,v) \in S^1$.

Example 20 Consider the function[3]

$$G : \mathbb{R}^2 \to \mathbb{R}^2, \quad (x,y) \mapsto (e^x \cos y, e^x \sin y)$$

and notice that

$$\det DG(x,y) = \det \begin{pmatrix} e^x \cos y & -e^x \sin y \\ e^x \sin y & e^x \cos y \end{pmatrix} = e^{2x}$$

so that $dG_{(x,y)}$ is a linear isomorphism for every $(x,y) \in \mathbb{R}^2$, and every $(u,v) \in \mathbb{R}^2$ is a regular value for G. Let now, for any $n \in \mathbb{N}$,

$$M_n \equiv \mathbb{R} \times (-\pi, (2n-1)\pi)$$

The triple $(G, M_n, (u,v))$ is admissible for every $(u,v) \in \mathbb{R}^2$, since g_n is C^∞ and $G^{-1}(u,v) \cap M_n$ is finite for any n, as we are going to see. To this end, consider the system

$$\begin{cases} e^x \cos y = u & (1) \\ e^x \sin y = v & (2) \end{cases} \tag{7.4}$$

[3] The readers who are familiar with the theory of functions of complex variables will recognize in this example the exponential function $\mathbb{C} \to \mathbb{C}$, $z \mapsto \exp z$, up to the standard identification of $\mathbb{C}$ with $\mathbb{R}^2$.

If $(u, v) = (0, 0)$, system (7.4) has no solutions; otherwise (see Appendix), it has exactly one solution in any subset of $\mathbb{R}^2$ of the form $\mathbb{R} \times I$, where I is a halfopen interval of length 2π. In $\mathbb{R} \times (-\pi, \pi]$, the solution can be written as follows:

$$x = \ln \sqrt{u^2 + v^2}$$
$$y = \operatorname{sign} v \arccos \frac{u}{\sqrt{u^2 + v^2}}$$

Thus

$$\deg(G, M_1, (u, v)) = 1 \qquad \text{if } v \neq 0 \text{ or } v = 0 \text{ and } u > 0$$
$$\deg(G, M_1, (u, v)) = 0 \qquad \text{if } v = 0 \text{ and } u \leq 0$$

Similarly, in $\mathbb{R} \times (-\pi, (2n - 1)\pi]$ system (7.4) has exactly n solutions, namely:

$$x_k = \ln \sqrt{u^2 + v^2}$$
$$y_k = \operatorname{sign} v \arccos \frac{u}{\sqrt{u^2 + v^2}} + 2(k - 1)\pi$$

with $k = 1, ..., n$. Thus

$$\deg(G, M_2, (u, v)) = 0 \qquad \text{if } v \neq 0 \text{ or } v = 0 \text{ and } u \geq 0$$
$$\deg(G, M_2, (u, v)) = 1 \qquad \text{if } v = 0 \text{ and } u < 0$$

and

$$\deg(G, M_n, (u, v)) = \deg(G, M_1, (u, v)) \qquad \text{if } n \text{ is odd}$$
$$\deg(G, M_n, (u, v)) = \deg(G, M_2, (u, v)) \qquad \text{if } n \text{ is even}$$

Example 21 (The degree of a planar bilinear function) Consider the bilinear (affine) function

$$h : \mathbb{R}^2 \to \mathbb{R}^2, \quad (x, y) \mapsto \begin{pmatrix} u(x, y) \\ v(x, y) \end{pmatrix} = \begin{pmatrix} (x + 1)(y + 1) \\ (x - 1)(y - 1) \end{pmatrix}$$

The determinant of the jacobian matrix of h is equal to $2(x - y)$; it is null on the straight line of equation $x = y$, the set CP_h of critical points of h. Then the set C_h of critical values for h can be parametrically described as follows:

$$u = (x + 1)^2 \tag{7.5}$$
$$v = (x - 1)^2 \tag{7.6}$$

(7.5)-(7.6) imply that C_h is the parabola γ of equation

$$(u - v)^2 - 8(u + v) + 16 = 0 \tag{7.7}$$

It is clear that h is a symmetric function, that is,

$$\forall (x, y) \in \mathbb{R}^2, \qquad h(x, y) = h(y, x)$$

and hence that the solutions to the equation system

$$h(x, y) = (u, v) \tag{7.8}$$

if existing, are even in number unless $u = v$; that is, unless $h(x, y)$ is a critical value for h. In fact, in the Appendix it is shown that there are two solutions to (7.8) when the lefthand side of (7.7) is positive, and none when it is negative. Therefore, for each $(u, v) \notin \gamma$,

$$\deg\left(h, \mathbb{R}^2, (u, v)\right) = 0$$

If $(u, v) \in \gamma$, $h^{-1}(u, v)$ is a singleton, but $\deg\left(h, \mathbb{R}^2, (u, v)\right)$ is as yet undefined. It will be defined in next section.

Exercise 22 Consider a generic (affine) bilinear function in the plane

$$k : \mathbb{R}^2 \to \mathbb{R}^2, \ (x, y) \mapsto \begin{pmatrix} u(x, y) \\ v(x, y) \end{pmatrix} = \begin{pmatrix} axy + bx + cy + d \\ pxy + qx + ry + s \end{pmatrix}$$

under the assumptions: $ap \neq 0$ and $(aq - bp)(ar - cp) \neq 0$. Show that conclusions similar to the above hold for k; (hint: there exist χ, ψ, affine isomorphisms of $\mathbb{R}^2$, such that $k = \psi^{-1} \circ h \circ \chi$).

3.1.2 Properties of the degree

As in the introduction of this section, denote by $\mathcal{T}$ the set of all admissible triples. Up to now the degree is not defined for all triples in $\mathcal{T}$, but just for these (f, M, y) such that f is C^2 and y is a regular value. The general definition of degree for all admissible triples will be given in Section 3.3.

In Theorem 23 below, we present a set of properties concerning the general definition of degree, applied to all triples in $\mathcal{T}$. We will prove them step by step, starting from the notion of degree which is now available, that is for C^2 functions and regular values. The proof of Theorem 23 will be completed in Section 3.3.

Theorem 23 (Properties of the degree) *The following properties of the degree hold:*

1 (normalization) given any manifold M,

$$\deg(id_M, M, y) = 1 \tag{7.9}$$

where y is any element of M and, as usual, id_M is the identity of M;

2 *(existence) for any admissible triple* (f, M, y),

$$\text{if } \deg(f, M, y) = 1 \quad \text{then } f^{-1}(y) \neq \emptyset \tag{7.10}$$

3 *(additivity) given any admissible triple* (f, M, y), *if* U_1 *and* U_2 *are open, disjoint subsets of* M *such that* $f^{-1}(y)$ *is contained in* $U_1 \cup U_2$, *then*

$$\deg(f, M, y) = \deg(f, U_1, y) + \deg(f, U_2, y) \tag{7.11}$$

4 *(excision) given any admissible triple* (f, M, y), *if* U *is an open neighborhood of* $f^{-1}(y)$ *in* M, *then*

$$\deg(f, M, y) = \deg(f, U, y) \tag{7.12}$$

Let us now prove that the above properties hold in the subset of $\mathcal{T}$ of triples (f, M, y), with f of class C^2 and y regular value for f, that is for the degree introduced in Definition 15.

Theorem 24 *The properties of the degree of Theorem 23 hold for the subset of* $\mathcal{T}$ *of triples* (f, M, y), *where* f *is* C^2 *and* y *is a regular value for* f.

Proof. Properties 1-3 are more or less obvious, and the reader is asked to prove them as an exercise. Property 4 is a consequence of property 3 and the definition of degree. Indeed, as $M \backslash \overline{U}$ is open in M and it is disjoint from $f^{-1}(y)$, then

$$\deg\left(f, M \backslash \overline{U}, y\right) = 0$$

Now observe that $f^{-1}(y)$ is contained in $U \cup M \backslash \overline{U}$ (actually in U) and compact; then property 3 holds and thus

$$\deg(f, M, y) = \deg(f, U, y) + \deg\left(f, M \backslash \overline{U}, y\right) = \deg(f, U, y)$$

■

Another important property is given in the following lemma; it establishes a sort of local constance of the degree as a function of y.

Lemma 25 *Let* (f, M, y) *be an admissible triple where* f *is* C^2 *and* y *is a regular value for* f. *Let also* U *be an open neighborhood of* $f^{-1}(y)$ *in* M *with compact closure. Then, there exists an open neighborhood* V *of* y *in* N *such that, for each* $y' \in V$,

$$\deg\left(f, U, y'\right) = \deg(f, U, y)$$

Proof. If $f^{-1}(y) = \emptyset$, then

$$\deg(f, U, y) = 0$$

As $\overline{U}$ is compact, then $f(\overline{U})$ is compact as well and thus $V \equiv N \backslash f\left(\overline{U}\right)$ is an open neighborhood of y. Then for each $y' \in V$

$$\deg\left(f, U, y'\right) = 0$$

Let, otherwise, $f^{-1}(y) \neq \emptyset$. From Corollary 37 of Chapter 4 it follows that there exists an open neighborhood V of y in N such that

1 $\#\left(f^{-1}(y') \cap U\right) = \#\left(f^{-1}(y) \cap U\right)$, for each $y' \in V$,

2 each $y' \in V$ is a regular value for $f_{|U}$.

Then the claim follows by Definition 15. ■

In the first three examples of previous subsection the lemma holds trivially, since the degree is constant. In general, however, not only $\deg(f, M, y)$ depends on y, but it can happen that, for some $y \in N$, every neighborhood of y contains elements y' such that $\deg(f, M, y') \neq \deg(f, M, y)$. Hence it is important to notice that the lemma asserts that $\deg(f, U, y') = \deg(f, U, y)$ for all y' in some neighborhood V of y, *provided* U contains $f^{-1}(y)$ and $\overline{U}$ is compact (as a matter of fact, properness of f on $\overline{U}$ is sufficient, even if $\overline{U}$ is not compact). This is illustrated in the following example.

Example 26 (The function $g : \mathbb{R} \to \mathbb{R},\ x \mapsto e^x$) For each $y \in \mathbb{R}$ the triple $(g, \mathbb{R}, y)$ is admissible, because g is injective and y is regular for g. On the other hand, it is clear that

$$\begin{array}{ll} \deg(g, \mathbb{R}, y) = 1 & \text{for } y > 0 \\ \deg(g, \mathbb{R}, y) = 0 & \text{for } y < 0 \end{array}$$

so that every neighborhood of 0 contains (positive) elements y such that

$$\deg(g, \mathbb{R}, y) = 1 \neq \deg(g, \mathbb{R}, 0)$$

Indeed, $\mathbb{R}$ is not compact, nor is g proper, since, for example, $[0, 1]$ is compact and $g^{-1}([0, 1]) = (-\infty, 0]$ is not. However, if $U = (a, b)$, then $\overline{U}$ is compact, and $\deg(g, U, y) = 0$ for every $y < e^a$. Thus, for each $c < 0$, (c, e^a) is a neighborhood of 0 completely consisting of regular values at which $\deg(g, U, y) = \deg(g, U, 0)$.

The situation is completely analogous in Example 20. Let n be any positive integer number and let g_n be the corresponding function defined there. Fix $\overline{u} < 0$, and consider the (regular) value $(\overline{u}, 0)$. Every open neighborhood V of $(\overline{u}, 0)$ contains elements $(u', 0)$ with $u' < 0$, as well as elements (u'', v'') with $u'' < 0$ and $v'' \neq 0$; yet we have proved that

$$\deg\left(g_n, M_n, (u', 0)\right) = \deg\left(g_n, M_n, (\overline{u}, 0)\right) \neq \deg\left(g_n, M_n, (u'', v'')\right)$$

However, if U is any open neighborhood of

$$g_n^{-1}(\overline{u},0) = \{(\ln|\overline{u}|,(2k-1)\pi)\}_{k=1,\dots,n-1}$$

with compact closure contained in M_n, then

$$\underline{y} \equiv \inf_{(x,y)\in U} y = \min_{(x,y)\in \overline{U}} y > -\pi$$

$$\overline{y} \equiv \sup_{(x,y)\in U} y = \max_{(x,y)\in \overline{U}} y < (2n-1)\pi$$

and, with the positions

$$\alpha \equiv \min\left\{\cos\underline{y},\cos\overline{y}\right\} > -1$$

$$W \equiv \left\{(u,v) : \frac{u}{\sqrt{u^2+v^2}} < \alpha\right\}$$

W is an open neighborhood of $(\overline{u},0)$, the same is true for $V \equiv W \cap f(U)$, and

$$\deg(g_n,U,(u,v)) = \deg(g_n,U,(\overline{u},0))$$

for each (u,v) in V.

Exercise 27 Provide a similar argument about $(0,0)$, a regular value for the function g_n.

Next theorem establishes another important property of the degree, namely, the homotopy invariance.

Theorem 28 (Homotopy lemma) *Let $f : M \to N$, $g : M \to N$ be C^2 and $y \in N$. Suppose the following conditions hold:*

1. *(f,M,y) and (g,M,y) are admissible;*
2. *y is a regular value for f and g;*
3. *there exists a C^2 homotopy $H : M \times [0,1] \to N$ between f and g such that $H^{-1}(y)$ is compact.*

Then

$$\deg(f,M,y) = \deg(g,M,y)$$

Proof. If $f^{-1}(y)$ and $g^{-1}(y)$ are both empty, the claim is trivially true. Let then either $f^{-1}(y)$ or $g^{-1}(y)$ be not empty, and suppose first that y is a regular value for H as well. Recalling Proposition 10, $H^{-1}(y)$ is a compact, one-dimensional manifold, with the following boundary:

$$\partial H^{-1}(y) = \left(f^{-1}(y) \times \{0\}\right) \cup \left(g^{-1}(y) \times \{1\}\right)$$

Thus

$$\#\partial H^{-1}(y) = \#\left(f^{-1}(y) \times \{0\}\right) + \#\left(g^{-1}(y) \times \{1\}\right) = \#f^{-1}(y) + \#g^{-1}(y)$$

Since every compact one-dimensional manifold with boundary has an even number of points in its boundary (Theorem 10, Chapter 5), it follows immediately that

$$\left[\#f^{-1}(y)\right]_2 = \left[\#g^{-1}(y)\right]_2$$

Consequently

$$\deg(f, M, y) = \deg(g, M, y)$$

Remove now the assumption that y is a regular value for H. For each (x, t) in $H^{-1}(y)$, let U_x be an open neighborhood of x in M with compact closure. Clearly $U_x \times [0, 1]$ is an open neighborhood of (x, t) in $M \times [0, 1]$. Then

$$H^{-1}(y) \subseteq \left(\bigcup\nolimits_{(x,t)\in H^{-1}(y)} U_x\right) \times [0, 1]$$

Since $H^{-1}(y)$ is compact, there exists a finite subset K of $H^{-1}(y)$ such that

$$H^{-1}(y) \subseteq U \times [0, 1]$$

where

$$U \equiv \bigcup_{(x,t)\in K} U_x$$

In addition, $\overline{U} = \cup_{(x,t)\in K} \overline{U}_x$ ([90], p. 100). Then $\overline{U}$ is compact, as it is a finite union of compact sets. By the excision property proved in Theorem 24,

$$\deg(f, U, y) = \deg(f, M, y) \tag{7.13}$$

$$\deg(g, U, y) = \deg(g, M, y) \tag{7.14}$$

Let us prove that $\deg(f, U, y) = \deg(g, U, y)$. Since $\overline{U}$ is compact, from Lemma 25, there is a neighborhood V of y in N such that every z in V is a regular value for the restriction of f and g to U, and the following holds:

$$\deg(f, U, y) = \deg(f, U, z) \tag{7.15}$$

$$\deg(g, U, y) = \deg(g, U, z) \tag{7.16}$$

Consider a regular value $\overline{z}$ for $H_{|U\times[0,1]}$, with $\overline{z}\in V$ (the existence of such a $\overline{z}$ follows from Sard's theorem[4]).

In addition, observe that, since $\delta U\times[0,1]$ is compact, then $H(\delta U\times[0,1])$ is compact as well. Since this latter set does not contain y, we can choose the above $\overline{z}$ close enough to y in such a way that $H^{-1}(\overline{z})$, which is compact in $\overline{U}\times[0,1]$, is actually contained in $U\times[0,1]$. Applying the first part of the proof to the restriction of H to $U\times[0,1]$, we obtain

$$\deg(f,U,\overline{z})=\deg(g,U,\overline{z})$$

Finally, from (7.13)-(7.14) and from (7.15)-(7.16), it follows

$$\deg(f,M,y)=\deg(g,M,y)$$

■

We would like to stress here the importance of a homotopy invariance-type property for a correct construction of a degree theory. Such a property is particularly useful in applications, when it is often the case that one needs to compute the degree of a triple (f,M,y), and a such computation is quite hard. Suppose that f is homotopic to g by a homotopy H, with $H^{-1}(y)$ compact, and suppose that the degree has been defined in such a way that a homotopy invariance property holds. If the computation of $\deg(g,M,y)$ is possible, then $\deg(f,M,y)$ is obtained as $\deg(f,M,y)=\deg(g,M,y)$.

Remark 29 As the reader can notice, in the proof of the above theorem we apply Sard's theorem to H, which is defined from an $(m+1)$-dimensional manifold ($U\times[0,1]$) to an m-dimensional manifold N. That is the reason why we need to assume that H (and consequently f and g) is C^2 and not C^1.

Example 9 at the end of last section shows that we cannot weaken the compactness assumption on $H^{-1}(y)$ by assuming only that $H_t^{-1}(y)$ is compact for each t. For the homotopic functions $f:\mathbb{R}\to\mathbb{R}$, $x\mapsto e^x$ and $g:\mathbb{R}\to\mathbb{R}$, $x\mapsto e^x-1$, $f^{-1}(0)=\emptyset$ and $g^{-1}(0)=\{0\}$ are compact sets and 0 is a regular value. However, $\deg(f,\mathbb{R},0)=0$ and $\deg(g,\mathbb{R},0)=1$.

Let us see an application of the properties of the degree, in particular the homotopy lemma, to the problem of solving a nonlinear equation.

Example 30 Consider the following system of two nonlinear equations:

$$\begin{cases} x+2y+\cos x=0\\ 3x+y+\ln(x+2)=0 \end{cases} \tag{7.17}$$

[4]The reader can verify as an exercise that in this case the assumptions of Sard's theorem hold

The set of solutions to this system coincides with $f^{-1}(0,0)$, where f is the C^∞ function

$$f : (-2,+\infty) \times \mathbb{R} \to \mathbb{R}^2, \quad (x,y) \mapsto (x + 2y + \cos x \,,\, 3x + y + \ln(x+2))$$

If we are able to find an open subset U of $(-2,+\infty)\times\mathbb{R}$ such that $\deg(f, U, (0,0))$ is well defined and equal to one, then from the existence property (Theorem 23, property 2) we can conclude that (7.17) has at least one solution in U.

To solve this problem, observe first that the determinant of the jacobian matrix of f at (x,y)

$$\det Df(x,y) = \det \begin{pmatrix} 1 - \sin x & 2 \\ 3 + \frac{1}{x+2} & 1 \end{pmatrix}$$

is equal to $-5 - \sin x - \frac{2}{x+2}$, which is strictly negative in all the domain. Hence f has no critical points, and $(0,0)$ is a regular value for f (as any other element of $\mathbb{R}^2$).

Moreover, $f^{-1}(0,0)$ is compact and to see this we prove that it is bounded and closed in $\mathbb{R}^2$ (and not simply closed in $(-2,+\infty)\times\mathbb{R}$). First, observe that $x + 2y + \cos x = 0$ implies

$$\frac{x}{2} + y \in \left[-\frac{1}{2}, \frac{1}{2}\right] \tag{7.18}$$

which, together with $3x + y + \ln(x+2) = 0$, yields

$$\frac{5}{2}x + \ln(x+2) \in \left[-\frac{1}{2}, \frac{1}{2}\right] \tag{7.19}$$

The function $x \mapsto \frac{5}{2}x + \ln(x+2)$ is increasing everywhere, and takes value $-\frac{5}{2} < -\frac{1}{2}$ for $x = -1$ and value $\frac{5}{2} + \ln 3 > \frac{1}{2}$ for $x = 1$; from (7.19), x certainly belongs to $[-1,1]$, and then, from (7.18), y belongs to $[-1,1]$ as well. Thus $f^{-1}(0,0) \subseteq U \equiv [-1,1] \times [-1,1]$ and the triple $(f, U, (0,0))$ is admissible; $(0,0)$ is a regular value for $f_{|U}$.

Our aim is now to prove that $\deg(f, U, (0,0)) = 1$. Consider the homotopy

$$H : (-2,+\infty) \times \mathbb{R} \times [0,1] \to \mathbb{R}^2$$

$$(x,y,t) \mapsto (x + 2y + t\cos x \,,\, 3x + y + t\ln(x+2))$$

H is clearly C^∞; in addition, replicating for every section $H_t^{-1}(0,0)$ the arguments put forth for $f^{-1}(0,0) = H_1^{-1}(0,0)$, it is possible to conclude that $H_t^{-1}(0,0) \subseteq U$ for each $t \in [0,1]$. Hence the inverse image

$$H^{-1}(0,0) = \bigcup_{t\in[0,1]} H_t^{-1}(0,0) \times \{t\}$$

is contained in $U \times [0,1]$, and closed in $\mathbb{R}^2$; then, it is compact.

Let

$$g \equiv H_0 : (-2, +\infty) \times \mathbb{R} \to \mathbb{R}^2, \quad (x, y) \mapsto (x + 2y, 3x + y)$$

It is immediate to see that $(0,0)$ is a regular value for g and that $g^{-1}(0,0) = \{(0,0)\}$. Now we can apply the homotopy lemma, obtaining

$$\deg(f, U, (0,0)) = \deg(g, U, (0,0)) = 1$$

Thus we have proved that (7.17) has at least one solution in U.

Example 31 (Example 21, continued) Consider now the following bilinear (affine) function in the plane,

$$\overline{h} : \mathbb{R}^2 \to \mathbb{R}^2, \quad (x, y) \mapsto \begin{pmatrix} u(x,y) \\ v(x,y) \end{pmatrix} = \begin{pmatrix} (x+1)\,y \\ (x-1)\,y \end{pmatrix}$$

The set of critical points of $\overline{h}$ is the x-axis, which is entirely mapped into the origin. Thus $(0,0)$ is the only critical value for $\overline{h}$, and $\overline{h}^{-1}(0,0)$ is a noncompact set, so $\left(\overline{h}, \mathbb{R}^2, (0,0)\right)$ is not admissible and the degree will remain undefined for this triple. Every pair $(u, v) \neq (0,0)$ is a regular value for $\overline{h}$. If $u = v$, $\overline{h}^{-1}(u, v)$ is empty, whereas if $u \neq v$, $\overline{h}^{-1}(u, v)$ contains the single element $\left(\frac{u+v}{u-v}, \frac{u-v}{2}\right)$ (see Appendix). Thus

$$\begin{aligned} &\deg\left(\overline{h}, \mathbb{R}^2, (u,v)\right) = 0 && \text{if } u = v \neq 0 \\ &\deg\left(\overline{h}, \mathbb{R}^2, (u,v)\right) = 1 && \text{if } u \neq v \end{aligned}$$

It is immediate to see that in any neighborhood of an element (u, v), with $u = v$, the degree is not constant, that is, local constance of the degree does not hold. However, fix an open set U in $\mathbb{R}^2$ with compact closure. If V is an open neighborhood having empty intersection with $f\left(\overline{U}\right)$, we obtain $\deg\left(\overline{h}, U, (u,v)\right) = 0$ for each $(u, v) \in V$; by this restriction of the domain, the local constance of the degree trivially holds.

Exercise 32 Consider a generic bilinear (affine) function k in the plane as in Exercise 22,

$$k : \mathbb{R}^2 \to \mathbb{R}^2, \quad (x, y) \mapsto \begin{pmatrix} u(x,y) \\ v(x,y) \end{pmatrix} = \begin{pmatrix} axy + bx + cy + d \\ pxy + qx + ry + s \end{pmatrix}$$

removing this time the assumption $(aq - bp)(ar - cp) \neq 0$ (but keeping the assumption $ap \neq 0$). Suppose first that $aq - bp = 0$ and $ar - cp \neq 0$, or $aq - bp \neq 0$ and $ar - cp = 0$. Show that conclusions similar to the above hold for k; (hint: see Exercise 22). Suppose then that both $aq - bp$ and $ar - cp$ are equal to 0. Show that k can be seen to be affinely diffeomorphic to the following function:

$$u = xy$$
$$v = xy$$

and draw your conclusions concerning $\deg\left(k, \mathbb{R}^2, (u,v)\right)$ whenever (u,v) is a regular value for k.

3.2 Definition of degree for C^2 functions at an arbitrary value

Let $f : M \to N$ be a C^2 function and (f, M, y) be an admissible triple, where y is any element of N, that is, y is not necessarily regular for f. Our aim here is to define the degree of the triple (f, M, y). To do it by a direct application of Definition 15 when y is a critical value would have several shortcomings. We shall mention two among them. First, $f^{-1}(y)$ may be an infinite set, even when it is compact, and thus formula (7.1) is meaningless. Second, even if $f^{-1}(y)$ is finite, so that formula (7.1) could work, a homotopy invariance-type property would not hold. The following two examples illustrate these shortcomings.

Example 33 Consider the simple homotopy

$$H : \mathbb{R} \times [0,1] \to \mathbb{R}, \quad (x,t) \mapsto x^2 - t$$

Applying Definition 15 to all triples $(H_t, \mathbb{R}, 0)$, including $(H_0, \mathbb{R}, 0)$ even if 0 is a critical value for H_0, we have

$$\deg(H_t, \mathbb{R}, 0) = \begin{cases} 0 & \text{if } t > 0 \\ 1 & \text{if } t = 0 \end{cases}$$

The example above suggests that any sound definition of $\deg(f, M, y)$ at a critical value y cannot be based on the cardinality of the inverse image $f^{-1}(y)$. Hence a different approach appears to be necessary.

The construction which is going to be explained in details below consists in defining $\deg(f, M, y)$ in terms of the degree of f at regular values sufficiently close to y. By Sard's theorem, we know that every neighborhood of y in N contains regular values for f (this actually holds even if f is just C^1). If z is a regular value for f, and if $f^{-1}(z)$ is compact, then $\deg(f, M, z)$ is well defined. Of course $\deg(f, M, z)$ depends, in general, on z, so that a definition of $\deg(f, M, y)$ as $\deg(f, M, z)$, for some regular value z of f close to y with $f^{-1}(z)$ compact, could be meaningless. Nevertheless, one may hope to find a suitable neighborhood V of y in N, such that $\deg(f, M, z)$ does not depend on

z, provided z is a regular value for f belonging to V and $f^{-1}(z)$ is compact. In other words, a possible definition of $\deg(f, M, y)$ as the degree of f at regular values sufficiently close to y could be a consequence of the following assertion:

- *given an admissible triple (f, M, y), where f is C^2 and y is a critical value for f, there exists a neighborhood V of y in N such that* $\deg(f, M, z)$ *does not depend on z, provided z is a regular value for f belonging to V and $f^{-1}(z)$ is compact.*

However, the above assertion is false even if y is a regular value for f, as it has been discussed in previous subsection (Examples 26 and 31), and as it happens in last example. At any rate, it is not difficult to slightly modify Example 26 in order to obtain a function f having a critical value y, in every neighborhood of which there are regular values z_1, z_2 for f such that $\deg(f, M, z_1) \neq \deg(f, M, z_2)$.

Example 34 Consider the function $f : \mathbb{R} \to \mathbb{R}$, $x \mapsto x^2 e^x$. The triple $(f, \mathbb{R}, 0)$ is admissible and 0 is a critical value for f. The only other critical value for f is $\max f = 4e^{-2}$. According to Definition 15, we have

$$\begin{cases} \deg(f, \mathbb{R}, z) = 0 & \text{if } z < 0 \\ \deg(f, \mathbb{R}, z) = 1 & \text{if } 0 < z < 4e^{-2} \text{ or } z > 4e^{-2} \end{cases}$$

In other words, given any neighborhood V of 0 there exist $z_1, z_2 \in V$ such that $\deg(f, \mathbb{R}, z_1) \neq \deg(f, \mathbb{R}, z_2)$.

However, arguing as in Example 26, if we take an open neighborhood U of $\{0\} = f^{-1}(0)$, with compact closure, and a connected neighborhood V of 0 such that $V \cap f(\delta U) = \emptyset$, then $\deg(f, U, z) = 0$ for each $z \in V$. Therefore, a possible definition of $\deg(f, \mathbb{R}, 0)$ would be

$$\deg(f, \mathbb{R}, 0) \equiv \deg(f, U, z)$$

with U any open neighborhood of $\{0\} = f^{-1}(0)$ with compact closure and z is any element belonging to a connected neighborhood V of 0 such that $V \cap f(\delta U) = \emptyset$.

This procedure can be generalized to any C^2 function $f : M \to N$, yielding a definition of degree for any admissible triple (f, M, y). To this end, let us first prove two lemmas.

Lemma 35 *Let $f : M \to N$ be a C^2 function and (f, M, y) be an admissible triple. Consider a nonempty open neighborhood U of $f^{-1}(y)$ in M having compact closure, and an open neighborhood V of y in N such that*

1 V is C^2 diffeomorphic to $\mathbb{R}^n$,

2 $V \cap f(\delta U) = \emptyset$.

For any z_1, z_2 belonging to V and regular values for $f_{|U}$,

$$\deg(f, U, z_1) = \deg(f, U, z_2)$$

Before starting the proof of the lemma, observe that the existence of such a V verifying 1 follows from the fact that any open ball in $\mathbb{R}^n$ is diffeomorphic to $\mathbb{R}^n$. Moreover, conditions 2 holds because the restriction of f to $\overline{U}$ is proper and, therefore, $f(\delta U)$ is closed and does not contain y.

Proof. If $f^{-1}(V) \cap U = \emptyset$, $\deg(f, U, z) = 0$ for any $z \in V$ and the claim is true.

Suppose instead that $W \equiv f^{-1}(V) \cap U$ is not empty. W is an open subset of U, containing $f^{-1}(y)$. It is important to observe that $g : W \to V$, $g(x) \equiv f(x)$, is a proper function. Indeed, let K be any compact subset of V. Clearly, $f^{-1}(K) \cap \overline{U}$ is compact. Moreover, since $V \cap f(\delta U) = \emptyset$, $f^{-1}(K) \cap \overline{U} = f^{-1}(K) \cap U = g^{-1}(K)$, which turns out to be compact.

Let now $\psi : V \to \mathbb{R}^n$ be a C^2 diffeomorphism, and consider the function $h \equiv \psi \circ g : W \to \mathbb{R}^n$. Let $z_1, z_2 \in V$ be regular values for $f_{|U}$. Hence, they are regular values for g, and thus $r_1 \equiv \psi(z_1)$ and $r_2 \equiv \psi(z_2)$ are regular values for h. Consider the homotopy

$$H : W \times [0,1] \to \mathbb{R}^n, \quad (x,t) \mapsto h(x) - (tr_1 + (1-t) r_2)$$

This homotopy is of course C^2. In addition, $H^{-1}(0)$ is a compact subset of $W \times [0,1]$. To see this, observe that, given any $(x,t) \in W \times [0,1]$, $H(x,t) = 0$ if and only if $h(x) = tr_1 + (1-t) r_2$, that is, if and only if $x \in h^{-1}(tr_1 + (1-t) r_2)$.

Therefore, denoted, as usual, by π_1 the projection of $W \times [0,1]$ onto W, $\pi_1\left(H^{-1}(0)\right) = h^{-1}(\mathcal{L}) = g^{-1}\left(\psi^{-1}(\mathcal{L})\right)$, where $\mathcal{L}$ is the line segment in $\mathbb{R}^n$ connecting r_1 to r_2. Since $\mathcal{L}$ is compact, $\pi_1\left(H^{-1}(0)\right)$ is compact because h is proper, as composition of proper functions (see Proposition 26 of Chapter 4). By continuity of H, $H^{-1}(0)$ is closed in $W \times [0,1]$, hence in $U \times [0,1]$ and in $\overline{U} \times [0,1]$, because $V \cap f(\delta U) = \emptyset$. Finally, $H^{-1}(0)$ is compact since it is a closed subset of a compact set.

In addition, since r_1 and r_2 are regular values for h, 0 is a regular value for H_0 and H_1. Therefore, we can apply the homotopy lemma (Theorem 28) and obtain

$$\deg(H_0, W, 0) = \deg(H_1, W, 0)$$

Now it is immediate to see that

$$\#H_0^{-1}(0) = \#\left(f^{-1}(z_2) \cap U\right)$$
$$\#H_1^{-1}(0) = \#\left(f^{-1}(z_1) \cap U\right)$$

and hence that

$$\deg(f, U, z_1) = \deg(f, U, z_2)$$

■

Lemma 36 *Let $f: M \to N$ be a C^2 function and (f, M, y) be an admissible triple. Consider two nonempty open neighborhoods U_1 and U_2 of $f^{-1}(y)$ in M, both having compact closure, and an open neighborhood V of y in N such that*

1 V is C^2 diffeomorphic to $\mathbb{R}^n$,

2 $V \cap f(\delta U_1 \cup \delta U_2) = \emptyset$.

Then, for any $z \in V$ which is a regular value for $f_{|U_1}$ and $f_{|U_2}$,

$$\deg(f, U_1, z) = \deg(f, U_2, z)$$

Proof. By Lemma 35, given any two elements z and $\overline{z}$ of V, which are regular values for $f_{|U_1}$ and $f_{|U_2}$,

$$\deg(f, U_1, z) = \deg(f, U_1, \overline{z})$$
$$\deg(f, U_2, z) = \deg(f, U_2, \overline{z})$$

Therefore, to prove the claim it is sufficient to show the existence of some $\overline{z} \in V$, which is a regular value for $f_{|U_1}$ and $f_{|U_2}$, such that $(f, U_1, \overline{z})$ and $(f, U_2, \overline{z})$ are admissible and

$$\deg(f, U_1, \overline{z}) = \deg(f, U_2, \overline{z})$$

Suppose the above equality never holds in V. Then, consider a sequence $(z_n)_{n \in \mathbb{N}}$ of regular values for $f_{|U_1}$ and $f_{|U_2}$, convergent to y, and such that, for each n,

$$\deg(f, U_1, z_n) \neq \deg(f, U_2, z_n) \tag{7.20}$$

This implies that, for each n, $\# f_{|U_1}^{-1}(z_n) \neq \# f_{|U_2}^{-1}(z_n)$, and, therefore, the inverse image of z_n in $(U_1 \cup U_2) \setminus (U_1 \cap U_2)$ is nonempty. Let then x_n be an element of $(U_1 \cup U_2) \setminus (U_1 \cap U_2)$ such that $f(x_n) = z_n$. Since the set $\{x_n\}_{n \in \mathbb{N}}$ is contained in $(\overline{U}_1 \cup \overline{U}_2) \setminus (U_1 \cap U_2)$, which is compact, there exists a subsequence $(x_{n_k})_{k \in \mathbb{N}}$ of (x_n) converging to some $\overline{x} \in (\overline{U}_1 \cup \overline{U}_2) \setminus (U_1 \cap U_2)$. By continuity of f, $f(\overline{x}) = y$. This is impossible, because $\left(f_{|U_1 \cup U_2}\right)^{-1}(y)$ is contained in $U_1 \cap U_2$. ■

The above two lemmas justify the following definition of degree for admissible triples (f, M, y) with f of class C^2 and y arbitrary. This is an extension of Definition 15 since y is not assumed to be regular.

Definition 37 Let $f: M \to N$ be C^2 and (f, M, y) be admissible. Consider

1 an open neighborhood U of $f^{-1}(y)$ with compact closure,

2 an open neighborhood V of y in N, C^2 diffeomorphic to $\mathbb{R}^n$, and not intersecting $f(\delta U)$,

3 an element z of V, regular value for $f_{|U}$.

The **degree** of (f, M, y) is given by the following formula:

$$\deg(f, M, y) \equiv \deg(f, U, z)$$

We would stress here that this definition is well given since the degree of the triple (f, U, y) remains the same for any choice of U, V, and $z \in V$, verifying conditions 1-3 above, and this is ensured by the two lemmas just proved.

Proceeding as in previous section, we prove the four properties stated in Theorem 23 of this extension of the notion of degree.

Theorem 38 *The properties of the degree of Theorem 23 hold for the subset of T of admissible triples (f, M, y), where f is C^2.*

Proof. 1. (normalization) There is nothing to prove, since any element y of a manifold M is a regular value for $id : M \to M$.
2. (existence) Take an open subset U of M, containing $f^{-1}(y)$ and having compact closure. Let $(z_n)_{n\in N}$ be a sequence of regular values of $f_{|U}$ converging to y, such that, for all n,

$$\deg(f, U, z_n) = \deg(f, M, y) \neq 0$$

By the existence property of the degree at regular values, $f^{-1}(z_n) \cap U \neq \emptyset$ for all n. Let $(x_n)_{n\in N}$ be a sequence of elements of U, such that $f(x_n) = z_n$ for each n. Since the sequence is contained in $\overline{U}$, which is compact, without loss of generality we can assume that (x_n) converges to some $\overline{x} \in \overline{U}$. Since f is continuous and z_n converges to y, it follows that $f(\overline{x}) = y$ and thus $\overline{x} \in U$.
3. (additivity) Let U_1 and U_2 be two open disjoint subsets of M such that $f^{-1}(y) \subseteq U_1 \cup U_2$. According to Definition 37, there exists a neighborhood V of y in N such that for each $z \in V$, if z is a regular value of $f_{|U_1 \cup U_2}$, the following holds:

$$\deg(f, U_1 \cup U_2, y) = \deg(f, U_1 \cup U_2, z)$$
$$\deg(f, U_1, y) = \deg(f, U_1, z)$$
$$\deg(f, U_2, y) = \deg(f, U_2, z)$$

By the additivity property of the degree at regular values, we have

$$\deg(f, U_1 \cup U_2, z) = \deg(f, U_1, z) + \deg(f, U_2, z)$$

and hence

$$\deg(f, U_1 \cup U_2, y) = \deg(f, U_1, y) + \deg(f, U_2, y)$$

4. (excision) Analogously to the proof of Theorem 24, this property is a consequence of the above existence and additivity properties. ■

Going back to our examples, observe what follows:

1 for the real power functions (Example 17), $\deg(f_n, \mathbb{R}, 0)$ is equal to 1 if n is odd, and it is equal to 0 if n is even. Thus, for each n, $\deg(f_n, \mathbb{R}, y)$ is constant for $y \in \mathbb{R}$.

2 Similarly, for the function f of Example 18, $\deg\left(f, S^1, P\right) = 0$, making $\deg\left(f, S^1, X\right)$ constant for $X \in S^1$.

3 Again, the bilinear function h of Example 21 has constant degree, since now $\deg\left(h, \mathbb{R}^2, (u, v)\right) = 0$ for each $(u, v) \in \gamma$.

4 Finally, the 0-partial function of the homotopy of Example 33 is just the ordinary square function, which has been already discussed at point 1. Thus $\deg(H_0, \mathbb{R}, 0) = 0$, making $\deg(H_t, \mathbb{R}, 0)$ constant across the whole homotopy.

Let us now prove the homotopy invariance property for the current definition of degree.

Theorem 39 (Homotopy invariance) *Let $H : M \times [0, 1] \to N$ be a C^2 function and y be an element of N such that $H^{-1}(y)$ is compact. Then*

$$\deg(H_0, M, y) = \deg(H_1, M, y)$$

Proof. Consider any open subset U of M, such that $\overline{U}$ is compact and $U \times [0, 1]$ contains $H^{-1}(y)$[5]. By the excision property, it is sufficient to prove that

$$\deg(H_0, U, y) = \deg(H_1, U, y)$$

Since $\overline{U} \times [0, 1]$ is compact, for all z in N, $H^{-1}(z) \cap \left(\overline{U} \times [0, 1]\right)$ is compact as well. In addition, $H(\delta U \times [0, 1])$ is compact and does not contain y. Therefore, there exists a neighborhood V of y in N, such that, for any $z \in V$,

$$H^{-1}(z) \cap (\delta U \times [0, 1]) = \emptyset$$

[5]For the existence of such U see the proof of the homotopy lemma (Theorem 28).

In other words

$$H^{-1}(z) \cap (U \times [0,1]) = H^{-1}(z) \cap (\overline{U} \times [0,1])$$

which implies that $H^{-1}(z) \cap (U \times [0,1])$ is compact, and hence $\deg(H_t, U, z)$ is well defined for any t.

Let now z be an element of V and a regular value for $(H_0)_{|U}$ and $(H_1)_{|U}$ (the existence of z follows from Sard's Theorem). The homotopy lemma can be applied, and

$$\deg(H_0, U, z) = \deg(H_1, U, z)$$

Since z can be chosen in such a way that

$$\deg(H_0, U, y) = \deg(H_0, U, z)$$
$$\deg(H_1, U, y) = \deg(H_1, U, z)$$

we conclude that

$$\deg(H_0, U, y) = \deg(H_1, U, y)$$

■

Example 40 We can apply the homotopy invariance (Theorem 39) to the function

$$f : \mathbb{R}^2 \to \mathbb{R}^2, \quad (x, y) \mapsto \left(\frac{e^x}{1+x^2} - 6y, x + \cos(xy)\right)$$

in order to show that $f^{-1}(0) \neq \emptyset$. Define

$$g : \mathbb{R}^2 \to \mathbb{R}^2, \quad (x, y) \mapsto (-6y, x)$$

The functions f and g are homotopic by

$$H : \mathbb{R}^2 \times [0,1] \to \mathbb{R}^2, \quad (x, y, t) \mapsto \left(t\frac{e^x}{1+x^2} - 6y, x + t\cos(xy)\right)$$

$H^{-1}(0)$ is compact. To see this, take $((x_n, y_n, t_n))_{n \in \mathbb{N}}$ a sequence in $H^{-1}(0)$. Up to a subsequence, (t_n) converges to some $\overline{t} \in [0,1]$. Now, since, for each n,

$$x_n + t_n \cos(x_n y_n) = 0$$

and $-1 \leq \cos(x_n y_n) \leq 1$, it follows, again up to a subsequence, that (x_n) converges to some $\overline{x} \in [-1,1]$. Finally, since for each n,

$$t_n \frac{e^{x_n}}{1+x_n^2} - 6y_n = 0$$

we also have that $\{y_n\}$ converges to some $\overline{y}$. By continuity of H, it also follows that $(\overline{x}, \overline{y}, \overline{t}) \in H^{-1}(0)$. Then, by the above homotopy invariance,

$$\deg(f, \mathbb{R}^2, 0) = \deg(g, \mathbb{R}^2, 0)$$

g is an isomorphism, $0 \in \mathbb{R}^2$ is a regular value and

$$\deg\left(g, \mathbb{R}^2, 0\right) = 1$$

then

$$\deg\left(f, \mathbb{R}^2, 0\right) = 1$$

Finally, by the existence property proved in Theorem 38, $f^{-1}(0)$ is nonempty. The reader can notice that comparing this computation to that of Example 30, here it is not necessary to prove that 0 is a regular value for both f and g.

3.3 Extension to continuous functions

In this final step we shall extend the definition of degree to continuous functions. The extension is based on an approximation method, which is summarized as follows. Consider a continuous function $f : M \to N$ and an element y of N such that $f^{-1}(y)$ is compact. In order to assign a degree to the triple (f, M, y), we will consider a C^2 function g which is, in a sense we will specify later, sufficiently close to f in a suitable neighborhood U of $f^{-1}(y)$. Then, we will define

$$\deg(f, M, y) \equiv \deg(g, U, y)$$

We will clearly need to prove that the value $\deg(f, M, y)$ is independent of the choice of g and U. Finally, we will give the complete proof of Theorem 23 and the general homotopy invariance property.

As a first tool in our approximation method we need the following two lemmas.

Lemma 41 (Bump functions) *Let $x \in \mathbb{R}^n$ and $r \in \mathbb{N}$. For any two real numbers δ and ε with $0 \leq \varepsilon < \delta$, there exists a C^r function $g : \mathbb{R}^n \to [0,1]$, which is strictly positive only on the open ball $B_\delta(x)$ and equal to one only in the closed ball $\overline{B}_\varepsilon(x)$.*

Proof. Consider the function

$$h_{\varepsilon,\delta} : \mathbb{R} \to [0,1], \quad t \mapsto \begin{cases} 1 & \text{if } |t| \leq \varepsilon \\ \left[1 - \left(\dfrac{|t| - \varepsilon}{\delta - \varepsilon}\right)^{r+1}\right]^{r+1} & \text{if } \varepsilon < |t| < \delta \\ 0 & \text{if } |t| \geq \delta \end{cases}$$

The reader can check as an exercise that $h_{\varepsilon,\delta}$ is C^r (hint: the functions $t \mapsto t^{r+1}$, $t \mapsto (1-t)^{r+1}$ have the first r derivatives equal to zero at zero and at one, respectively).

The function

$$g : \mathbb{R}^n \to [0,1], \quad z \mapsto h_{\varepsilon,\delta}(\|z - x\|)$$

satisfies the requirements. ■

Remark 42 The reason why the functions referred to in Lemma 41 are called "bump functions" has to do with the shape of their graph. In what follows, we will only need "thin" bump functions, that is, bump functions for which $\varepsilon = 0$.

Remark 43 It is possible to prove the existence of a C^∞ bump function, that is, a bump function which is C^∞ and not only C^r. The proof is slightly more technical and makes use of the C^∞ function

$$h : \mathbb{R} \to \mathbb{R}, \quad t \mapsto \begin{cases} e^{-1/t} & t > 0 \\ 0 & t \leq 0 \end{cases}$$

For our purposes, however, a C^r bump function will be sufficient.

Lemma 41 is a classical result in Calculus which is used, in this context, to prove the following lemma.

Lemma 44 (Finite partition of unity) *Let M be a C^r manifold, $r \geq 1$, and assume that M admits a finite open covering $\{U_i\}_{i=1,\dots,s}$ such that each U_i is C^r diffeomorphic to an open subset of $\mathbb{R}^n$. Then there exists a family of C^r functions $\{\alpha_i\}_{i=1,\dots,s}$, with $\alpha_i : M \to [0,1]$, satisfying the following properties:*

1 *for each i, $\operatorname{supp}(\alpha_i)$, the support of α_i, is contained in U_i;*

2 *for all $x \in M$, $\sum_{i=1}^{s} \alpha_i(x) = 1$.*

The family $\{\alpha_i\}_{i=1,\dots,s}$ is called a ***partition of unity*** *of M associated with $\{U_i\}_{i=1,\dots,s}$.*

Proof. We recall that $\operatorname{supp}(\alpha_i)$ is the closure of the set in which α_i does not vanish. For each $i = 1, \dots, s$, let $\phi_i : U_i \to \phi_i(U_i)$ denote a diffeomorphism between U_i and $\phi_i(U_i)$, open in $\mathbb{R}^n$. Since a manifold is a paracompact topological space (Theorem 39 of Chapter 2), there exists an open covering $\{V_\lambda\}_{\lambda\in\Lambda}$ of M such that

1) for every $\lambda \in \Lambda$, there exists i such that $\overline{V_\lambda} \subseteq U_i$,

2) for every $\lambda \in \Lambda$, $\overline{V_\lambda}$ is compact,

3) $\{V_\lambda\}_{\lambda\in\Lambda}$ is locally finite, that is, for every $x \in M$, there exists a neighborhood M_x of x in M (which we may suppose open) such that M_x intersects only a finite number of V_λ's.

Moreover, let $\{W_\lambda\}_{\lambda\in\Lambda}$ a "shrinking" of $\{V_\lambda\}_{\lambda\in\Lambda}$, that is, an open covering of M such that, for every $\lambda \in \Lambda$, $\overline{W}_\lambda$ is contained in V_λ (in particular, $\overline{W}_\lambda$ is compact). For the existence of a shrinking of an open covering, the reader can see [90], p. 224.

Consider a function $f : \Lambda \to \{1, ..., s\}$ such that, for each $\lambda \in \Lambda$, $\overline{V}_\lambda \subseteq U_{f(\lambda)}$. For each given $\lambda \in \Lambda$, let $Z_\lambda \equiv \phi_{f(\lambda)}(W_\lambda)$. $\overline{W}_\lambda$ is compact and consequently $\overline{Z}_\lambda = \phi_{f(\lambda)}(\overline{W}_\lambda)$ is compact. This implies that there exists a finite set of open balls of $\mathbb{R}^n$, $\{B_j\}_{j=1,...,k_\lambda}$, with each $\overline{B}_j$ contained in $\phi_{f(\lambda)}(V_\lambda)$, such that

$$\overline{Z}_\lambda \subseteq \bigcup_{j=1}^{k_\lambda} B_j \tag{7.21}$$

By Lemma 41, there exist k_λ C^r bump functions $g_{\lambda,j} : \mathbb{R}^n \to [0, 1]$, such that $g_{\lambda,j}(z) > 0$ if and only if $x \in B_j$. Define

$$g_\lambda : \mathbb{R}^n \to \mathbb{R}_+, \quad z \mapsto \sum_{j=1}^{k_\lambda} g_{\lambda,j}(z)$$

Of course g_λ is C^r, and

$$\begin{array}{ll} g_\lambda(z) > 0 & \text{if } z \in \overline{Z}_\lambda \\ g_\lambda(z) = 0 & \text{if } z \in \mathbb{R}^n \backslash \bigcup_{j=1}^{k_\lambda} B_j \end{array}$$

Then the function

$$h_\lambda : M \to \mathbb{R}_+, \quad x \mapsto \begin{cases} g_\lambda\left(\phi_{f(\lambda)}(x)\right) & x \in \phi_{f(\lambda)}^{-1}\left(\bigcup_{j=1}^{k_\lambda} B_j\right) \\ 0 & x \in M \backslash \phi_{f(\lambda)}^{-1}\left(\bigcup_{j=1}^{k_\lambda} B_j\right) \end{cases}$$

is well defined and C^r.

Thus we have obtained a family $\{h_\lambda\}_{\lambda\in\Lambda}$ of C^r functions from M to $\mathbb{R}_+$, which can be associated with the locally finite open covering $\{V_\lambda\}_{\lambda\in\Lambda}$, in the following sense:

a) for every $\lambda \in \Lambda$ and for every $x \in \overline{W}_\lambda$, $h_\lambda(x) > 0$ (observe that from (7.21), $\phi_{f(\lambda)}^{-1}\left(\bigcup_{j=1}^{k_\lambda} B_j\right) \supseteq \phi_{f(\lambda)}^{-1}\left(\overline{Z}_\lambda\right) = \overline{W}_\lambda$),

b) hence, for every $x \in M$, there is at least an index λ such that $h_\lambda(x) > 0$,

c) for every $\lambda \in \Lambda$, h_λ is identically zero in $M \backslash V_\lambda \subseteq M \backslash \phi^{-1}_{f(\lambda)}\left(\bigcup_{j=1}^{k_\lambda} B_j\right)$,

d) hence, for every $x \in M$, there exists a neighborhood of x in which h_λ is not identically zero only for a finite subset of indices in Λ.

Let us now "normalize" the family $\{h_\lambda\}_{\lambda\in\Lambda}$, by defining a new family $\{\widetilde{\alpha}_\lambda\}_{\lambda\in\Lambda}$ as follows:

$$\widetilde{\alpha}_\lambda : M \to [0,1], \quad x \mapsto \frac{h_\lambda(x)}{\sum_{c\in\Lambda} h_c(x)}$$

The above definition makes sense, because

- from b), for every $x \in M$, $\sum_{c\in\Lambda} h_c(x)$ is strictly greater than zero,
- from d), for every $x \in M$, on M_x the sum $\sum_{c\in\Lambda} h_c$ is actually a finite sum.

In addition, $\widetilde{\alpha}_\lambda$ is C^r and it is identically zero on $M \backslash V_\lambda$ which is closed in M. In particular,

$$\text{supp}\,(\widetilde{\alpha}_\lambda) \subseteq \overline{V}_\lambda \subseteq U_{f(\lambda)} \tag{7.22}$$

Finally, for $i = 1, ..., s$, let $\Lambda_i \equiv f^{-1}(i)$, and define

$$\alpha_i : M \to [0,1], \quad x \mapsto \sum\nolimits_{\lambda\in\Lambda_i} \widetilde{\alpha}_\lambda(x) \tag{7.23}$$

Since, for every $x \in M$, $\widetilde{\alpha}_\lambda$ is not identically zero on a suitable neighborhood of x only for a finite subset of indices in Λ, it follows that the finite family $\{\alpha_i\}_{i=1,...,s}$ is well defined, that each α_i is C^r, and that x belongs to supp (α_i) if and only if it belongs to supp $(\widetilde{\alpha}_\lambda)$ for some λ. From (7.22),

$$\text{supp}\,(\alpha_i) \subseteq \bigcup\nolimits_{\lambda\in\Lambda_i} \overline{V}_\lambda \subseteq U_i$$

and it is easy to see that the α_i's sum up to one everywhere on M. ■

Let us now prove an approximation result for continuous functions with values in euclidean spaces.

Proposition 45 *Let M be a C^2 manifold, $f : M \to \mathbb{R}^l$ be continuous, and C be a compact subset of M. For any $\varepsilon > 0$, there exist an open subset U of M, containing C, and a C^2 function $g : U \to \mathbb{R}^l$, such that $\|f(x) - g(x)\| < \varepsilon$ for every $x \in U$.*

Proof. Assume first that C is nonempty. Fix a positive ε. For any $x \in C$, consider

$$W_x \equiv \{y \in M : \|f(y) - f(x)\| < \varepsilon\}$$

which is an open subset of M. Moreover, let V_x be an open neighborhood of x in M, which is C^2 diffeomorphic to an open subset of $\mathbb{R}^n$, and let $U_x \equiv W_x \cap V_x$. C is contained in $\bigcup_{x \in C} U_x$ and, since C is compact, there is a finite set $\{x_i\}_{i=1,\ldots,s}$ such that $U \equiv \bigcup_{i=1,\ldots,s} U_{x_i}$ contains C. By Lemma 44, there exists a finite partition of unity of U associated with $\{U_{x_i}\}_{i=1,\ldots,s}$, that is, a family of C^2 functions, $\{\alpha_i\}_{i=1,\ldots,s}$, $\alpha_i : U \to [0, 1]$, such that

i) for all i, supp $(\alpha_i) \subseteq U_{x_i}$,

ii) for all $x \in U$, $\sum_{i=1}^{s} \alpha_i(x) = 1$.

Consider the function

$$g : U \to \mathbb{R}^l, \quad x \mapsto \sum_{i=1}^{s} \alpha_i(x) f(x_i)$$

Clearly, g is C^2 since so are the α_i's, and we have

$$\begin{aligned}\|g(x) - f(x)\| &= \left\| \sum_{i=1}^{s} \alpha_i(x) f(x_i) - \sum_{i=1}^{s} \alpha_i(x) f(x) \right\| \\ &\leq \sum_{i=1}^{s} \alpha_i(x) \|f(x_i) - f(x)\|\end{aligned}$$

Given any $x \in U$, for each i, either $x \in U_{x_i} \subseteq W_{x_i}$, and then $\|f(x_i) - f(x)\| < \varepsilon$, or $\alpha_i(x) = 0$. Therefore, for all $x \in U$,

$$\sum_{i=1}^{s} \alpha_i(x) \|f(x_i) - f(x)\| < \varepsilon$$

and the result is proved when C is nonempty.

If C is empty, given a positive ε, take any $\overline{x} \in M$ and define

$$U \equiv \{y \in M : \|f(y) - f(\overline{x})\| < \varepsilon\}$$

Of course, U is an open subset of M containing C. Then, the constant function $g : U \to \mathbb{R}^l$, $x \mapsto f(\overline{x})$ is the required approximation of f. ■

Proposition 45 is a preliminary result for a more general approximation theorem, which deals with continuous functions between manifolds. What we need in order to define the degree of a triple (f, M, y), for a continuous $f : M \to N$, is a C^2 function g such that

1 g is defined in M or in a neighborhood U of $f^{-1}(y)$,

2 g takes values in N (not simply in $\mathbb{R}^l$),

3 (g, U, y) is an admissible triple,

4 g is an approximation of f on U,

5 if $h : U \to N$ is another C^2 function as close as g to f, or more, and if (h, U, y) is admissible, then $\deg(h, U, y) = \deg(g, U, y)$.

If those conditions do hold, we are able to define $\deg(f, M, y)$ as $\deg(g, U, y)$. Proposition 45 guarantees the existence of an approximating function $\widetilde{f} : U \to \mathbb{R}^l$, that is a function which, in general, does not take values in N. This implies that the degree of the triple $(\widetilde{f}, U, y)$ is meaningless if $l > \dim N$ (in other words if N is not open in $\mathbb{R}^l$).

However, in order to show that a function g verifying conditions 1-5 exists, we use Proposition 45 and we introduce the concept of tubular neighborhood, which play a crucial role if $l > \dim N$. Given a manifold N of $\mathbb{R}^l$, a tubular neighborhood of N in $\mathbb{R}^l$ is, roughly speaking, a neighborhood $\mathcal{N}$ of N in $\mathbb{R}^l$ together with a submersion $\rho : \mathcal{N} \to N$ such that, for each $z \in \mathcal{N}$, $\rho(z)$ is the unique closest point to z in N. From a different point of view, ρ is a projection on N such that $z - \rho(z)$ is normal to N at $\rho(z)$.

Now, starting from Proposition 45, if there exists a tubular neighborhood $\mathcal{N}$ of N (or, as we will see, a tubular neighborhood of a suitable open subset of N), and if $\widetilde{f} : U \to \mathbb{R}^l$ is a C^2 function, sufficiently close to f in order to take values in $\mathcal{N}$, then $g \equiv \rho \circ \widetilde{f}$ is a well defined function satisfying the previous conditions 1 and 2, and we are able to prove that it satisfies conditions 3-5 as well.

Definition 46 Let N be a C^r n-dimensional manifold of $\mathbb{R}^l$, with $r \geq 1$. A **tubular neighborhood** of N in $\mathbb{R}^l$ is a pair $(\mathcal{N}, \rho)$ such that

1 $\mathcal{N}$ is an open subset of $\mathbb{R}^l$ containing N;

2 $\rho : \mathcal{N} \to N$ is a C^r submersion;

3 for any $z \in \mathcal{N}$, $\rho(z)$ is the unique closest point to z in N.

Theorem 47 (Existence of tubular neighborhoods) *Consider an n-dimensional C^r manifold N of $\mathbb{R}^l$, with $r \geq 1$. Given any compact subset C of N, there exist an open neighborhood V of C in N and a tubular neighborhood $(\mathcal{V}, \rho)$ of V in $\mathbb{R}^l$.*

Proof. If $n = l$ and recalling that N is assumed to be boundaryless, then N is open in $\mathbb{R}^l$. Therefore, given any compact subset C of N, an open neighborhood

V of C and a tubular neighborhood of V can be chosen respectively as N and (N, id_N).

Suppose otherwise that $n < l$. Let Y be an open neighborhood of C in N and let NY be the normal bundle of Y in $\mathbb{R}^l$ (see Chapter 3 for the definition and properties of normal bundle). Consider the function

$$h : NY \to \mathbb{R}^l, \quad (y, n) \mapsto y + n$$

Of course h is C^r since it can be viewed as the restriction to NY of the linear function

$$L : \mathbb{R}^l \times \mathbb{R}^l \to \mathbb{R}^l, \quad (y, n) \mapsto y + n$$

Let us show that every element (y, n) of NY is a regular point for h. The tangent space to NY at (y, n) is $T_yY \times N_yY$, where N_yY is the normal space to N at y. Observe that

$$dh_{(y,n)} = \left(dL_{(y,n)}\right)_{|T_yY \times N_yY} = L_{|T_yY \times N_yY}$$

and hence

$$\text{Im } dh_{(y,n)} = L\left(T_yY \times N_yY\right) = T_yY + N_yY = \mathbb{R}^l$$

so that $dh_{(y,n)}$ is onto and therefore an isomorphism. Apply the inverse function theorem to the submanifold $Y \times \{0\}$ of NY. Precisely, for each $(y, 0) \in Y \times \{0\}$ there exist open neighborhoods T of $(y, 0)$ in NY and U of $y = h(y, 0)$ in $\mathbb{R}^l$ such that $h_{|T}$ is a C^r diffeomorphism between T and U. Still using the Inverse Function Theorem we prove that

- there exists an open neighborhood of $C \times \{0\}$ in NY diffeomorphic, by h, to an open neighborhood of C in $\mathbb{R}^l$.

To see this we have only to prove that there exists an open neighborhood W' of $C \times \{0\}$ in NY such that h is one-to-one on W'. If, otherwise, we suppose that h is not one-to-one on any neighborhood of $C \times \{0\}$, this is equivalent to assume that, there exist two sequences $(x_n)_{n \in \mathbb{N}}$ and $(z_n)_{n \in \mathbb{N}}$ in NY, respectively converging to $\overline{x}$ and $\overline{z}$, both belonging to $C \times \{0\}$, such that $x_n \neq z_n$ and $h(x_n) = h(z_n)$ for all n.

By continuity of h, $h(\overline{x}) = h(\overline{z})$ and, since h is one-to-one on $C \times \{0\}$, $\overline{x} = \overline{z}$. This implies that every neighborhood of $\overline{x} = \overline{z}$ contains a pair of (different) points x_n and z_n such that $h(x_n) = h(z_n)$; this is not possible by the inverse function theorem.

Therefore let W' be an open neighborhood of $C \times \{0\}$ in NY diffeomorphic to an open neighborhood of C in $\mathbb{R}^l$. Of course, $W'_0 \equiv W' \cap (Y \times \{0\})$ is an open subset of $Y \times \{0\}$ containing $C \times \{0\}$. Let Y' be an open subset of Y, with compact closure and with $C \subseteq Y'$ such that $\overline{Y}' \times \{0\} \subseteq W'_0$.

Now, consider the distance function from points of $\overline{Y}' \times \{0\}$ to $NY \setminus W'$

$$d : \overline{Y}' \times \{0\} \to \mathbb{R}, \quad (y,0) \mapsto \inf_{(z,n) \in NY \setminus W'} \|(y,0) - (z,n)\|_{\mathbb{R}^l \times \mathbb{R}^l}$$

Observe that $d(y,0) > 0$, for each $(y,0) \in \overline{Y}' \times \{0\}$; in addition, since the distance function is continuous and $\overline{Y}' \times \{0\}$ is compact, d has a positive minimum δ. Consequently $W \equiv Y' \times B_{\delta/2}(0)$ is an open neighborhood of $C \times \{0\}$ contained in W'. In addition, $\overline{W}$ is compact and contained in W'.

Now, observe that the construction of W is given in order to define the following function

$$\pi : \overline{W} \to \overline{Y}' \times \{0\}, \quad (y,n) \mapsto (y,0)$$

which is a C^r submersion. Moreover, let $V \equiv h(Y' \times \{0\})$ and $\mathcal{V} \equiv h(W)$. The function

$$\rho : \overline{\mathcal{V}} \to \overline{V}, \quad x \mapsto h(\pi(h^{-1}(x)))$$

is a C^r submersion since so is π and h is a diffeomorphism. Our aim is to prove that $(\mathcal{V}, \rho_{|\mathcal{V}})$ is a tubular neighborhood of V. Of course $\mathcal{V}$ is a neighborhood of V in $\mathbb{R}^l$. It remains to prove that, given $x \in \mathcal{V}$, $\rho(x)$ is the unique closest point to x in V. Consider the distance function from x

$$v \mapsto ||v - x||$$

and let v_x a (not necessarily unique) minimizer on $\overline{V}$. Then consider any C^1 curve γ in Y such that $\gamma(0) = v_x$. The C^1 function

$$||\gamma(t) - x||^2$$

has a minimum at $t = 0$. Then

$$\frac{d}{dt}(||\gamma(t) - x||^2)_{t=0} = 2\gamma'(0)\,(v_x - x) = 0$$

Then $\gamma'(0)$ and $x - v_x$ are orthogonal in $\mathbb{R}^l$ and, by the generality of such a curve γ, it follows that $x - v_x$ is orthogonal to Y at v_x, that is, $x - v_x \in N_{v_x}Y$. Moreover, let $\widetilde{v}_x \equiv \rho(x)$, which belong to V, since $x \in \mathcal{V}$. Since h is one-to-one on W and $x = h(v_x, x - v_x) = h(\widetilde{v}_x, x - \widetilde{v}_x)$, then $\widetilde{v}_x = v_x$. Therefore v_x belongs to V (not just to $\overline{V}$) and it is the unique closest point to x in V. ■

Theorem 48 *Let M and N be two C^2 manifolds and assume N contained in $\mathbb{R}^l$. Let $f : M \to N$ be a continuous function and C be a compact subset of*

M. Then, for any $\varepsilon > 0$, there exist an open neighborhood U of C in M, and a C^2 function $g : U \to N$ such that $\|f(x) - g(x)\| < \varepsilon$ for every $x \in U$.

Moreover, g is homotopic to f by means of a continuous homotopy $H : U \times [0,1] \to N$ such that $\|H(x,t) - f(x)\| < \varepsilon$, for all $(x,t) \in U \times [0,1]$.

Proof. Fix a positive ε. Assume first that C is nonempty. Let V be an open neighborhood of $f(C)$ in N and $(\mathcal{V}, \rho)$ be a tubular neighborhood of V in $\mathbb{R}^l$. By compactness of $f(C)$ there exists a positive $\bar{\varepsilon}$ such that, for each $y \in f(C)$, the open ball $B_{\bar{\varepsilon}}(y)$ in $\mathbb{R}^l$ is contained in $\mathcal{V}$. Consider a positive ε' such that $\varepsilon' < \frac{1}{2}\min\{\varepsilon, \bar{\varepsilon}\}$. The compactness of C implies the existence of a finite number of elements $x_1, x_2, \ldots, x_s$ of C such that

$$C \subseteq U \equiv \bigcup_{i=1}^{s} U_i, \quad \text{with } U_i \equiv \{x \in M : \|f(x) - f(x_i)\| < \varepsilon'\}, \quad i = 1, \ldots, s$$

Observe that U_i's are open in M. By the proof of Proposition 45, we can define a C^2 function $\widetilde{f} : U \to \mathbb{R}^l$ such that $\|f(x) - \widetilde{f}(x)\| < \varepsilon'$ for every $x \in U$. In addition, for any $x \in U$, $\widetilde{f}(x) \in \bigcup_{y \in f(C)} B_{\bar{\varepsilon}}(y)$. To see this, given $x \in U$, let i be such that $x \in U_i$. Then

$$\|\widetilde{f}(x) - f(x_i)\| \leq \|\widetilde{f}(x) - f(x)\| + \|f(x) - f(x_i)\| < 2\varepsilon' < \bar{\varepsilon}$$

Now consider the function $g : U \to N$, $g \equiv \rho \circ \widetilde{f}$. Recalling the definition of tubular neighborhood (in particular the definition of ρ), it follows that, for any $x \in U$,

$$\|g(x) - \widetilde{f}(x)\| \leq \|f(x) - \widetilde{f}(x)\| < \varepsilon'$$

Therefore

$$\|g(x) - f(x)\| \leq \|g(x) - \widetilde{f}(x)\| + \|f(x) - \widetilde{f}(x)\| < 2\varepsilon' < \varepsilon$$

Thus the first part of the theorem is proved, that is we have obtained the existence of the approximating function g. Now, define $H : U \times [0,1] \to N$ as

$$H(x,t) = \rho((1-t)f(x) + tg(x))$$

Roughly speaking, H can be regarded as a sort of projection on N, by means of ρ, of the homotopy $(x,t) \mapsto (1-t)f(x) + tg(x)$, which takes values in $\mathbb{R}^l$ (precisely in $\mathcal{V}$). Observe that, given x, when t varies, $H(x,t)$ is the point in the segment between $f(x)$ and $g(x)$. Thus, by the definition of ρ and since $\|g(x) - f(x)\| < \varepsilon$, we have

$$\|H(x,t) - f(x)\| < \varepsilon$$

for all $(x,t) \in U \times [0,1]$ and the proposition is proved in the case when C is nonempty.

If $C = \emptyset$, the function produced in the first part of the proof for a suitable nonempty open set C' works for C as well. ■

Proposition 49 *Let $f : M \to N$ be continuous and (f, M, y) be admissible. Let N be contained in $\mathbb{R}^l$. Let U_1, U_2 be two open subsets of M, both containing $f^{-1}(y)$ and such that $\overline{U}_1$ and $\overline{U}_2$ are compact.*

Consider an open neighborhood V of y in N, containing $f(\overline{U}_1) \cup f(\overline{U}_2)$ and a tubular neighborhood $(\mathcal{V}, \rho)$ of V in $\mathbb{R}^l$. Moreover, let ε_1 and ε_2 be two positive real number such that, for $i = 1, 2$, we have

1. *$\varepsilon_i < \|f(x) - y\|$, for all $x \in \delta U_i$,*
2. *for every $z \in f(\overline{U}_i)$ the open ball $B_{\varepsilon_i}(z)$, of $\mathbb{R}^l$, is contained in $\mathcal{V}$.*

If $g_1 : \overline{U}_1 \to N$ and $g_2 : \overline{U}_2 \to N$ are two C^2 functions such that

$$\|g_i(x) - f(x)\| < \varepsilon_i$$

for all $x \in \overline{U}_i$, $i = 1, 2$, then the triples (g_1, U_1, y), (g_2, U_2, y) are admissible and

$$\deg(g_1, U_1, y) = \deg(g_2, U_2, y)$$

Exercise 50 Prove that an open neighborhood V of y in N, containing $f(\overline{U}_1) \cup f(\overline{U}_2)$ and having a tubular neighborhood in $\mathbb{R}^l$ actually exists.

Exercise 51 Prove that ε_1 and ε_2 as in the statement exist.

Proof. Let U be an open subset of M, containing $f^{-1}(y)$, and such that its closure, $\overline{U}$, is compact and contained in $U_1 \cap U_2$. Let $\overline{\varepsilon}$ be a sufficiently small real number such that the following conditions hold:

1. $0 < \overline{\varepsilon} < \min\{\varepsilon_1, \varepsilon_2\}$,
2. $\overline{\varepsilon} < \|f(x) - y\|$, for all $x \in (\overline{U}_1 \cup \overline{U}_2) \backslash U$,
3. for every $z \in f(\overline{U})$, $B_{\overline{\varepsilon}}(z) \subseteq \mathcal{V}$.

The existence of such $\overline{\varepsilon}$ is seen as in exercise 51. By Theorem 48 it follows that there exists a C^2 function $\overline{g} : \overline{U}_1 \cup \overline{U}_2 \to N$ such that $\|\overline{g}(x) - f(x)\| < \overline{\varepsilon}$ for any $x \in \overline{U}_1 \cup \overline{U}_2$. In addition, $\overline{g}(x) \neq y$ for all $x \in (\overline{U}_1 \cup \overline{U}_2) \backslash U$. Indeed,

$$\overline{\varepsilon} < \|f(x) - y\| \leq \|f(x) - \overline{g}(x)\| + \|\overline{g}(x) - y\| < \overline{\varepsilon} + \|\overline{g}(x) - y\|$$

Therefore, $\overline{g}^{-1}(y)$, which is compact, is contained in U_1 and U_2. Hence the triples $(\overline{g}, U_1, y)$, $(\overline{g}, U_2, y)$ and $(\overline{g}, U, y)$ are admissible and, by the excision property of the degree for C^2 functions (Theorem 38), it follows

$$\deg(\overline{g}, U_1, y) = \deg(\overline{g}, U, y) = \deg(\overline{g}, U_2, y) \qquad (7.24)$$

Let us now show that

$$\deg(g_1, U_1, y) = \deg(\overline{g}, U_1, y) \tag{7.25}$$

and

$$\deg(g_2, U_2, y) = \deg(\overline{g}, U_2, y) \tag{7.26}$$

Define the homotopy $G(x,t) \equiv tg_1(x) + (1-t)\overline{g}(x)$, with $(x,t) \in \overline{U}_1 \times [0,1]$. G takes value in $\mathbb{R}^l$ and

$$\|G(x,t) - f(x)\| < \varepsilon_1$$

for any $(x,t) \in \overline{U}_1 \times [0,1]$. Consequently, Im G is contained in $\mathcal{V}$, and the homotopy

$$H : \overline{U}_1 \times [0,1] \to N, \quad (x,t) \mapsto \rho \circ G(x,t)$$

is well defined. In addition, $H^{-1}(y)$ is compact in $U_1 \times [0,1]$. Then from Theorem 39, (7.25) is proved. Analogously we obtain (7.26). By (7.24), (7.25) and (7.26) we have

$$\deg(g_1, U_1, y) = \deg(g_2, U_2, y)$$

■

Now we are in the position to give a definition of degree for general admissible triples.

Definition 52 (General definition of degree) Let M and N be two C^2 manifolds and assume N contained in $\mathbb{R}^l$. Let $f : M \to N$ be a continuous function and (f, M, y) be admissible. Consider:

1 an open subset U of M, with compact closure, and such that $f^{-1}(y) \subseteq U$;

2 an open neighborhood V of $f(\overline{U})$ in N and a tubular neighborhood $(\mathcal{V}, \rho)$ of V;

3 a positive real number $\overline{\varepsilon}$ such that

  (a) $\overline{\varepsilon} < \|f(x) - y\|$, for all $x \in \delta U$,

  (b) for every $z \in f(\overline{U})$ the open ball $B_{\overline{\varepsilon}}(z)$ in $\mathbb{R}^l$ is contained in $\mathcal{V}$;

4. a C^2 function $g : \overline{U} \to N$ such that $\|g(x) - f(x)\| < \overline{\varepsilon}$, for all $x \in \overline{U}$.

The **degree** of (f, M, y) is defined by the following formula

$$\deg(f, M, y) \equiv \deg(g, U, y) \tag{7.27}$$

Observe that (g, U, y) is admissible, by Proposition 49

An interesting and straightforward property of the above definition is the local constance of the degree with respect to y. The following proposition extends Lemma 25.

Proposition 53 *Let $f : M \to N$ be continuous and let (f, M, y) be admissible. Let also U be an open neighborhood of $f^{-1}(y)$ in M with compact closure. Then, there exists an open neighborhood V of y in N such that, for each $y' \in V$,*

$$\deg\left(f, U, y'\right) = \deg\left(f, U, y\right)$$

Proof. Let V be an open neighborhood of y with compact closure and such that $f(\delta U) \cap \overline{V} = \emptyset$. As $f(\delta U)$ and $\overline{V}$ are both compact their distance, i.e.,

$$d \equiv \min_{s \in f(\delta U),\, t \in \overline{V}} ||s - t||$$

is positive. Let $0 < \varepsilon < d$. Now, apply Definition 52 and consider a C^2 function

$$g : \overline{U} \to N$$

such that $||g(x) - f(x)|| < \overline{\varepsilon}$, for all $x \in \overline{U}$. Observe that formula (7.27) holds for every $y' \in V$. Then

$$\deg(f, U, y') = \deg(g, U, y') \tag{7.28}$$

By Lemma 35,

$$\deg(g, U, y') = \deg(g, U, y'') \tag{7.29}$$

if y', y'' are regular value for g belonging to V. On the other hand, (7.29) is true for all $y', y'' \in V$ (not only for regular values), by Definition 37. Finally, the claim follows by (7.28). ■

The following corollary is a straightforward consequence of the above proposition. The proof is left as an exercise.

Corollary 54 *Let $f : M \to N$ be proper and N connected. Then* $\deg(f, M, y)$ *is constant with respect to y.*

We are now able to conclude the proof of Theorem 23, partially given proving Theorems 24 and 38.

Proof of Theorem 23. 1. (normalization) Nothing needs to be proved.

2. (existence) Let $f : M \to N$ be continuous and (f, M, y) be an admissible triple. Assume that $f^{-1}(y) = \emptyset$. Let U be an open subset of M with compact closure $\overline{U}$, and let $g : \overline{U} \to N$ be a C^2 function, close enough to f, such that (g, U, y) is admissible and

$$\deg(f, M, y) = \deg(g, U, y)$$

The continuous function $\|f(x) - y\|$ has a positive minimum ε in $\overline{U}$ as it is compact. Without loss of generality, by Theorem 48 the function g can be chosen in such a way that $\|f(x) - g(x)\| < \frac{1}{2}\varepsilon$, $x \in \overline{U}$, then it follows immediately that $g^{-1}(y) \cap \overline{U}$ is empty. Analogously, we can choose a regular value z for g, such that $\|z - y\| < \frac{1}{2}\varepsilon$ and also

$$\deg(g, U, y) = \deg(g, U, z)$$

By means of this choice of z, we have that $g^{-1}(z) \cap \overline{U}$ is empty. Therefore, by the excision property (Theorem 38) it follows that $\deg(g, U, z) = 0$ and finally $\deg(f, M, y) = 0$, which contradicts the assumption.
3. (additivity) Using an analogous strategy as the above point 2, let U_1 and U_2 be open subsets of M, with compact closure, such that

- $\overline{U}_i \subseteq U, i = 1, 2,$
- $f^{-1}(y) \subseteq (U_1 \cup U_2).$

Then, let $g : \overline{U}_1 \cup \overline{U}_2 \to N$ be a C^2 function, close to f in such a way that

$$\begin{aligned} \deg(f, M, y) &= \deg(g, U_1 \cup U_2, y) \\ \deg(f, U_1, y) &= \deg(g, U_1, y) \\ \deg(f, U_2, y) &= \deg(g, U_2, y) \end{aligned} \tag{7.30}$$

In addition, let $z \in N$ be a regular value for g, sufficiently close to y in such a way that

$$\begin{aligned} \deg(g, U_1 \cup U_2, y) &= \deg(g, U_1 \cup U_2, z) \\ \deg(g, U_1, y) &= \deg(g, U_1, z) \\ \deg(g, U_2, y) &= \deg(g, U_2, z) \end{aligned} \tag{7.31}$$

By the additivity property proved in Theorem 24, we obtain the additivity property for this general case.
4. (excision) Once more, nothing needs to be proved. ∎

Theorem 55 (General homotopy invariance) *Let $H : M \times [0,1] \to N$ be continuous and $y \in N$ be such that $H^{-1}(y)$ is compact. Then*

$$\deg(H_0, M, y) = \deg(H_1, M, y)$$

Proof. Let U be an open subset of M with compact closure such that $H^{-1}(y) \subseteq U \times [0, 1]$. It is possible to prove that, given any positive ε, there exists a C^2 function, $G : \overline{U} \times [0, 1] \to N$, such that

$$\|H(x, t) - G(x, t)\| < \varepsilon$$

for all $(x,t) \in \overline{U} \times [0,1]$. The existence of such a G does not strictly follow from the approximation result in Theorem 48, since this deals with functions defined on boundaryless manifolds, whereas $M \times [0,1]$ has clearly nonempty boundary. However, let $\delta > 0$ and consider

$$\widetilde{H} : M \times (-\delta, 1+\delta) \to N, \quad (x,t) \mapsto \begin{cases} H(x,t) & (x,t) \in M \times [0,1] \\ H(x,0) & (x,t) \in M \times (-\delta, 0) \\ H(x,1) & (x,t) \in M \times (1, 1+\delta) \end{cases}$$

Applying now Theorem 48 to $\widetilde{H}$ (which is trivially continuous), for any $\varepsilon > 0$, there exists a C^2 function, $G : \overline{U} \times [0,1] \to N$, such that

$$\|H(x,t) - G(x,t)\| < \varepsilon$$

for all $(x,t) \in \overline{U} \times [0,1]$.

If ε is sufficiently small, the approximating function G is such that $G^{-1}(y)$ (which is compact) is contained in $U \times [0,1]$. Then, by the homotopy invariance property for C^2 functions (Theorem 39),

$$\deg(G_0, U, y) = \deg(G_1, U, y) \tag{7.32}$$

Moreover, recalling the definition of degree for continuous functions, Definition 52, G can be chosen close enough to H in such a way that

$$\begin{aligned} \deg(H_0, U, y) &= \deg(G_0, U, y) \\ \deg(H_1, U, y) &= \deg(G_1, U, y) \end{aligned} \tag{7.33}$$

Therefore,

$$\deg(H_0, U, y) = \deg(H_1, U, y)$$

and, finally, the claim follows by the excision property. ■

4. Computation of degree for C^1 functions

The general definition of degree (Definition 52) covers all possible triples we consider admissible (Definition 11) and it is actually the final step in the construction of degree. On the other hand, such a definition does not lead us to an easy method to compute the degree of a triple (f, M, y). To have a simple procedure, we need that f is C^2 and $y \in N$ is a regular value. In this case $\deg(f, M, y)$ can be obtained simply computing $\# f^{-1}(y)$.

In the particular case when f is just C^1 (y remaining regular value) a question arises: does the formula

$$\deg(f, M, y) \equiv [\# f^{-1}(y)]_2$$

remains valid? Observe that the right term of the above formula is meaningful for C^1 functions, since $f^{-1}(y)$ is finite. The answer to the above question is luckily positive and aim of this section is to prove the above equality.

Proposition 56 *Let M and N be two C^2 manifolds. Let $f : M \to N$ be a C^1 function and $y \in N$ be a regular value for f. Assume $f^{-1}(y)$ is compact. Then, the degree of (f, M, y) is given by the following formula:*

$$\deg(f, M, y) \equiv \left[\# f^{-1}(y)\right]_2 \tag{7.34}$$

Proof. If $f^{-1}(y) = \emptyset$, the claim is trivially true. Otherwise, let $f^{-1}(y) = \{x_1, ..., x_r\}$ and let $U_1, ..., U_r$ be open pairwise disjoint subsets of M, such that $f^{-1}(y) \cap U_i = \{x_i\}$, for each i. By the additivity property (Theorem 23) it is sufficient to prove that, for each i,

$$\deg(f, U_i, y) = 1 \tag{7.35}$$

Let i be given and denote x_i by $\overline{x}$. Since $df_{\overline{x}}$ is a linear isomorphism between $T_{\overline{x}}M$ and T_yN, we can apply the inverse function theorem (Theorem 48 in Chapter 3) and conclude, without loss of generality, that f is a C^1 diffeomorphism between U_i and $V_i \equiv f(U_i)$. Still without loss of generality, we can suppose that U_i and V_i are C^2 diffeomorphic[6] to $\mathbb{R}^n$ (where n is the dimension of M and N). Let

$$\phi : \mathbb{R}^n \to U_i \text{ and } \psi : V_i \to \mathbb{R}^n$$

be two C^2 diffeomorphisms, and consider

$$g : \mathbb{R}^n \to \mathbb{R}^n, \quad z \mapsto \psi \circ f \circ \phi(z)$$

The function g is clearly a C^1 diffeomorphism, so that its differential at $\overline{z} \equiv \phi^{-1}(\overline{x})$ is a linear isomorphism and, in particular, a C^∞ function. Consider

$$G : \mathbb{R}^n \times \mathbb{R} \to \mathbb{R}^n, \quad (z, t) \mapsto t\left(g(\overline{z}) + dg_{\overline{z}}(z - \overline{z})\right) + (1 - t)g(z)$$

Notice that the restriction of G to $\mathbb{R} \times [0, 1]$ is the straight line homotopy between g and its first order approximation at $\overline{z}$.

The differential of G along z at $(\overline{z}, 0)$, $D_zG_{(\overline{z},0)}$, is an isomorphism from $\mathbb{R}^n$ to itself since it coincides with $dg_{\overline{z}} = d\psi_y \circ df_{\overline{x}} \circ d\phi_{\overline{z}}$, which is an isomorphism as well. Then, we can apply the implicit function theorem to G to say that there exists an open neighborhood $W \times (a, b)$ of $(\overline{z}, 0)$ in $\mathbb{R}^n \times \mathbb{R}$ such that, for each $t \in (a, b)$ there exists a unique function $\zeta : (a, b) \to W$ such that $G(\zeta(t), t)$ is constant, coinciding with $\psi(y)$ for each t. Since $G(\overline{z}, t) = \psi(y)$ for all $t \in \mathbb{R}$, then $\zeta(t) = \overline{z}$, for any $t \in (a, b)$.

Now, consider some c, with $0 < c < b$, and the restriction of G to $W \times [0, c]$. $G^{-1}(\psi(y))$ is equal to $\overline{z} \times [0, c]$ and then it is compact. Let $W' \equiv \phi(W)$ which is an open neighborhood of $\overline{x}$, contained in U_i and define

$$H : W' \times [0, 1] \to V, \quad (x, t) \mapsto \psi^{-1}\left(G\left(\phi^{-1}(x), ct\right)\right)$$

[6] Recall that manifolds are assumed to be C^2.

To better understand the properties of H observe that

i) H is a C^1 homotopy taking values in N and such that $H_0 = f_{W'}$; the factor c is only needed to define the homotopy in $W' \times [0,1]$ instead of $W' \times [0,c]$;

ii) $H^{-1}(y)$ coincides with $\{\overline{x}\} \times [0,1]$ and then it is compact;

iii) H_1 is such that $x \mapsto \psi^{-1}\left(g(\overline{z}) + dg_{\overline{z}}\left(\phi^{-1}(x) - \overline{z}\right)\right)$; precisely, H_1 is C^2, y is a regular value for this function and $H_1^{-1}(y) = \overline{x}$.

Condition iii) implies that

$$\deg(H_1, W', y) = 1$$

Moreover, applying the general homotopy invariance property to H, that is Theorem 55,

$$\deg(f, W', y) = \deg(H_1, W', y) = 1$$

and, finally, by the excision property applied to f with respect to W', we have

$$\deg(f, U_i, y) = 1$$

■

5. Existence results for nonlinear equations

In this section we present two different results giving sufficient conditions to prove existence of solutions for a nonlinear equation $f(x) = y$ defined on manifolds. In the first subsection f is continuous and we give a direct application of the general homotopy invariance property described in Theorem 55 obtaining a theorem which is going to be used several times in the second part of the book.

In the second subsection we study one particular case in which f is C^1 and M and N are abstract manifolds. This is due to the fact that in the second part of the book we tackle a particular problem in a Grassmannian manifold.

5.1 Case A: manifolds in euclidean spaces

Theorem 57 *Let M and N C^2 be two boundaryless manifolds of the same dimension, and let $f, g : M \to N$ be such that*

1 *f is C^0 and g is C^1;*

2 *y is a regular value for g;*

3 *$\#g^{-1}(y)$ is odd;*

4 *there exists a continuous homotopy H from f to g such that $H^{-1}(y)$ is compact.*

Then $f^{-1}(y) \neq \emptyset$.

Proof. Observe that, since $H^{-1}(y)$ is compact, then (f, M, y) and (g, M, y) are admissible. By Proposition 56,

$$\deg(g, M, y) = 1$$

Then, applying the general homotopy invariance, we have

$$\deg(f, M, y) = 1$$

Finally the claim follows by the existence property of the degree. ■

5.2 Case B: abstract manifolds

In this subsection we assume that M and N are C^2 abstract manifolds of the same dimension, say n. In addition M and N are assumed Hausdorff and paracompact (see [90] Section 4 in Chapter 6) in order to apply Whitney's theorem. We present here an existence result for nonlinear equations which will be used in the second part of the book.

Definition 58 Let M and N be two C^r manifolds, $r \geq 1$ and $f : M \to N$ be C^r. An element $y \in N$ is said to be a **regular value** for f if

1 for any $x \in f^{-1}(y)$,

2 for any local parametrization (ϕ, U, V) of M around x,

3 for any chart (ξ, T, S) of N at y such that $f(V) \subseteq T$,

the differential at $\phi^{-1}(x)$ of the local representation of f, $\xi \circ f \circ \phi$ is surjective.

The reader can easily observe that this definition of regular value is equivalent to that given in Chapter 4 if the above M and N are euclidean manifolds.

Theorem 59 *Let M and N be two C^2 boundaryless abstract manifolds of the same dimension n. Suppose that M and N are Hausdorff and paracompact. Let $g : M \to N$ be a function and assume that*

1 g is C^1 and proper

2 $y \in N$ is a regular value for g;

3 $\#g^{-1}(y)$ is odd;

4 N is connected;

Then, for any $z \in N$, $g^{-1}(z) \neq \emptyset$.

Proof. Recall that the paracompactness and Hausdorff assumption on M and N guarantees that such manifolds are C^2 diffeomorphic to $\widetilde{M}$ and $\widetilde{N}$, which are C^2 manifolds of $\mathbb{R}^{2n}$. Let

$$\phi : \widetilde{M} \to M, \quad \text{and} \quad \psi : N \to \widetilde{N}$$

be C^2 diffeomorphisms and consider

$$h : \widetilde{M} \to \widetilde{N}, \quad x \mapsto \psi \circ g \circ \phi(x)$$

Clearly h is C^1 and proper as composition of C^1 proper functions. Moreover denote $\widetilde{y} \equiv \psi(y)$. It is immediate to see that

a) $h^{-1}(\widetilde{y})$ is finite and odd;

b) $\widetilde{y}$ is a regular value for h.

Then $\deg\left(h, \widetilde{M}, \widetilde{y}\right) = 1$. Let us now check that $\deg\left(h, \widetilde{M}, \widetilde{z}\right) = 1$ for every $\widetilde{z} \in \widetilde{N}$. To this aim fix any $\widetilde{z} \in \widetilde{N}$. By Corollary 54, there exists a neighborhood $\widetilde{V}$ of $\widetilde{z}$ such that

$$\deg\left(h, \widetilde{M}, \widetilde{z}\right) = \deg\left(h, \widetilde{M}, \widetilde{z}'\right)$$

for any $\widetilde{z}' \in \widetilde{V}$. Then, the degree depends continuously on points in $\widetilde{N}$. Since $\widetilde{N}$ is connected, it follows that $\deg\left(h, \widetilde{M}, \widetilde{z}\right)$ is constant and, consequently, equal to 1 for each $\widetilde{z} \in \widetilde{N}$. Therefore, $h^{-1}(\widetilde{z})$ is nonempty for each $\widetilde{z}$ and, hence, $g^{-1}(z)$ is nonempty for each $z \in N$. ■

6. Appendix

Ad example 19

Taking into account that $x^2 + y^2 = u^2 + v^2 = 1$, adding 1 to both sides of the first equation of system (7.3) and squaring both sides of the second one yields the system

$$\begin{cases} 2x^2 = 1 + u \\ 4x^2y^2 = 1 - u^2 \end{cases}$$

or, equivalently,

$$\begin{cases} x^2 = \dfrac{1+u}{2} \\ y^2 = \dfrac{1-u}{2} \end{cases}$$

Each of the above equations has two opposite solutions, but for a solution (x, y) of the original system it must be taken into account that (2) of (7.3) prescribes that x and y be concordant or discordant according to the sign of v.

Ad example 20

By squaring both sides of (7.4), summing, and substituting the result in (1) of (7.4), we obtain the system

$$\begin{cases} e^x = \sqrt{u^2 + v^2} & (1) \\ \cos y = \dfrac{u}{\sqrt{u^2 + v^2}} & (2) \end{cases} \tag{7.36}$$

The above system is not equivalent to (7.4), since (2) of (7.4) requires

$$\operatorname{sign} \sin y = \operatorname{sign} v \tag{7.37}$$

Thus we shall consider the system composed by the three equations (1) and (2) of (7.36) and (7.37). Equation (1) of (7.36) has a unique solution for $x \in \mathbb{R}$, namely

$$x = \ln \sqrt{u^2 + v^2}$$

On the other hand, equations (2) of (7.36) and (7.37) have in $\mathbb{R}$ infinitely many solutions; if $\overline{y}$ is one of them, then the solution set is

$$\{\overline{y} + 2k\pi\}_{k \in \mathbb{Z}}$$

If instead y is restricted to any halfopen interval of length $2n\pi$, the solutions are exactly n; if $\overline{y}$ is the smallest one, the solution set is

$$\{\overline{y} + 2\,(k-1)\,\pi\}_{k=1,\ldots,n}$$

Remembering that the version of the sign function which we actually use takes value 1 when its argument is 0, for the unique solution to the system of (2) of (7.36) and (7.37) belonging to $(-\pi, \pi]$ a particularly convenient form is available, namely

$$\overline{y} = \operatorname{sign} v \arccos \frac{u}{\sqrt{u^2 + v^2}}$$

Ad example 21

Consider the equation system

$$\begin{cases} (x+1)\,(y+1) = u \\ (x-1)\,(y-1) = v \end{cases} \tag{7.38}$$

which is symmetric in the two variables, and hence is solved by the couple (y, x) whenever it is solved by the couple (x, y). In order to obtain an explicit form for its solutions, it is enough to add and subtract both equations of (7.38), yielding the equivalent system

$$\begin{cases} xy = \dfrac{u+v}{2} - 1 \\ x + y = \dfrac{u-v}{2} \end{cases} \tag{7.39}$$

By (7.39) it is clear that there are solutions to system (7.38) if and only if there are solutions to the second degree equation

$$2z^2 - (u-v)\,z + u + v - 2 = 0 \tag{7.40}$$

Thus we have three possible cases:

1 $(u-v)^2 - 8\,(u+v) + 16 < 0$. Equation (7.40) and system (7.38) have no solutions. The inverse image $h^{-1}\,(u,v)$ is empty; (u,v) is a regular value for h.

2 $(u-v)^2 - 8\,(u+v) + 16 = 0$. Equation (7.40) has the unique solution

$$z = \frac{u-v}{4}$$

and system (7.38) has the unique solution

$$\left(\frac{u-v}{4}, \frac{u-v}{4}\right)$$

The inverse image $h^{-1}\,(u,v)$ contains a single element (x,y) for which $x=y$, a critical point; (u,v) is a critical value for h.

3 $(u-v)^2 - 8\,(u+v) + 16 > 0$. Equation (7.40) has the two distinct solutions

$$z_1 = \frac{u-v-\sqrt{(u-v)^2 - 8\,(u+v) + 16}}{4}$$
$$z_2 = \frac{u-v+\sqrt{(u-v)^2 - 8\,(u+v) + 16}}{4}$$

and system (7.38) has the two symmetric solutions

$$(z_1, z_2) \qquad \text{and} \qquad (z_2, z_1)$$

The inverse image $h^{-1}\,(u,v)$ contains two elements having distinct components; (u,v) is a regular value for h.

Ad example 31

Recall the definition of $\overline{h}$

$$u = (x+1)\,y \tag{7.41}$$
$$v = (x-1)\,y \tag{7.42}$$

and notice that $\overline{h}$ maps vertical lines into positively sloping 45° degree lines, and horizontal lines (the x-axis excluded) into lines through the origin. The determinant of $D\overline{h}(x,y)$ is equal to $-2y$; thus $CP_{\overline{h}}$ is the x-axis and $C_{\overline{h}}$ is a single point, namely, the origin. Every other couple (u,v) is a regular value for $\overline{h}$; however, two cases must be distinguished. If $u = v \neq 0$, $\overline{h}^{-1}(u,v)$ is empty, since $x+1 = x-1$ is impossible. Otherwise, subtracting (7.42) from (7.41), and substituting back,

$$x = \frac{u+v}{u-v}$$
$$y = \frac{u-v}{2}$$

that is, $\overline{h}^{-1}(u,v) = \left\{\left(\frac{u+v}{u-v}, \frac{u-v}{2}\right)\right\}$, a singleton. Hence

$$\mathrm{Im}\,\overline{h} = \left(\mathbb{R}^2 \backslash \ell\right) \cup \{(0,0)\}$$

where ℓ denotes the straight line of equation $u = v$.

II

Chapter 8

EXCHANGE ECONOMIES

The main goal of this chapter is to present a description of a private ownership competitive exchange economy and to show properties of equilibria in that framework.

In Section 1, we describe the main aspects characterizing an exchange economy and the definition of equilibrium. The qualification "exchange" stresses the fact that any consideration concerning production is left aside[1].

In Section 2, we describe the main properties of the households' demand function, a parametric representation of the solution to the households' maximization problem.

Sections 3, 4 and 5 are devoted to the characterization of equilibria and to a proof of their existence in terms of what we call the extended system, i.e., the system of first order conditions of the households' problem and the market clearing conditions. A crucial role in the proof is played by the concept of Pareto optimal allocation and by the homotopy technique. The former is defined and analyzed in Section 4, the latter is developed in Section 5, which contains a few other important lemmas. Observe that our approach to existence does not require an independent analysis of the demand map: in fact, if the reader is interested just in the existence result, she can skip Section 2[2].

In Section 6, uniqueness and C^1 dependence of equilibria on exogenous variables [3] are analyzed using the concept of regular economy.

[1] Some of the assumptions of this chapter, or comments to them, have been phrased with reference to production as well as to exchange, in order not to repeate them anew in Chapter 9.

[2] Demand functions are referred to in the proof of existence of equilibria only when discussing the rank of the partial jacobian (with respect to choice variables and multiplier) of the functions appearing on the lefthand side of the first order conditions.

[3] An exogenous element describes anything pre-determined, given, not explained by the model. An endogenous element is any non-exogenous element, and it has to be determined, explained in the model.

In Section 7, our approach is compared with some of the existing ones in the literature.

The appendices contain technical material about the household's problem.

Since part of the material presented in the chapter is standard, we have omitted some of the proofs, giving appropriate references.

1. The model

The object of this chapter is the analysis of a group of individuals, called households consumers, who own bundles of commodities. Households can engage in transactions concerning commodities in order to improve their well being. We have used two fundamentals concepts in the above sentences: commodities and households. In this section, we discuss in some detail those features of the model.

Sometimes we will make assumptions which are more restrictive than needed in order to obtain the results we are about to state. Throughout the exposition we will try to keep a clear distinction between assumptions which can be removed without essentially changing the conclusions reached, and more essential assumptions which can hardly be dispensed with.

1.1 The commodity space

"Commodities can be divided into goods and services. Each commodity is completely specified by its physical characteristics, its location and date at which it is available."[4]

Assumption c1 There is a finite number C of completely homogeneous commodities .

A commodity is denoted by an element $c \in \{1, ..., C\} \equiv \mathcal{C}$.

Assumption c2 Each commodity is perfectly divisible.

By perfect divisibility we mean here that in principle any real number of a commodity can be owned, consumed if the commodity is a consumption good, used in production if it is a production factor, and produced if it is a producible good.

Assumption c3 There is only one period of time in which exchange takes place.

There is no need to stress that this assumption is extremely strong. Introduction of time and uncertainty is the content of Chapters 10-15.

[4][12], p. 382.

Consistently with the above assumption, the third characteristic of a commodity (the date of availability) is for the time being ignored.

Assumption c4 Commodities are only available in nonnegative quantities.

This is a rather natural and widespread assumption.

Definition 1 A **commodity bundle** is a C-dimensional vector, an ordered list of quantities of the existing commodities.

Definition 2 The **commodity space** is $\mathbb{R}_+^C$. Commodity bundles are usually denoted with the letter x. Thus $x \equiv (x^c)_{c \in \mathcal{C}}$ is the commodity bundle containing, for each $c \in \mathcal{C}$, the quantity x^c of commodity c.

1.2 Households

Assumption h1 There is a finite number H of individuals, called households or consumers[5].

A household is denoted by an element $h \in \{1, ..., H\} \equiv \mathcal{H}$.

The basic goal of each household is to choose a commodity vector. Her choices are based on her **preferences** over a subset of the commodity space, called her **consumption set**. Her choices are limited by the economic environment to a subset of the consumption set called her **budget set**. First of all, that set depends upon the households property rights. In all the book, we consider a private ownership economy: households own a commodity vector which constitutes their **endowments**. In Chapter 9, we will assume that households may also own shares of firms. Moreover, commodities can be exchanged with other commodities at some exchange ratios described by **prices**. Endowments and prices will in fact be used to define budget sets.

In what follows, we describe the basic concepts mentioned above: consumption sets; preferences and the related concept of utility functions; budget sets and their key ingredients, endowments and prices.

1.2.1 Consumption sets

The consumption set of household $h \in \mathcal{H}$ is a subset X_h of the commodity space whose elements are the vector of commodities the individual can conceivably consume, given the physical or institutional constraints imposed by the environment. An element $x_h \equiv (x_h^c)_{c=1}^C \in X_h$ is called consumption vector and can be interpreted as the vector of commodities consumed by household h.

[5]We give the same meaning to both terms.

Common assumptions on X_h are that it is convex, bounded from below ([78] p. 369) and unbounded. To easily use differential techniques, we make the following stronger assumption[6].

Assumption h2 $\forall h \in \mathcal{H}$, $X_h = \mathbb{R}^C_{++} \equiv \{x \in \mathbb{R}^C : x \gg 0\}$.

1.2.2 Preferences

Fundamental characteristics of a household are her "tastes" or "preferences". When considering two vectors in her consumption set, a household is usually able to make one of the following statements: "I prefer x to x'","I prefer x' to x" or "I am indifferent between x and x'". We formalize the above considerations as follows.

Definition 3 A **binary relation** on an arbitrary set S is a subset R of $S \times S$.

If $(x, y) \in R$, we also write $x\mathcal{R}y$, which we read "x is in relation $\mathcal{R}$ with y";

if $(x, y) \notin R$, we also write $\neg (x\mathcal{R}y)$, which we read "x is not in relation $\mathcal{R}$ with y".

Definition 4 A binary relation on the consumption set is called a **preference relation**. We denote it by $\succeq$, which is read as "at least as good as" or "weakly preferred to".

Definition 5 Consider $x, y \in \mathbb{R}^C_{++}$.

$$[x \succ y] \equiv [x \succeq y \text{ and } \neg (y \succeq x)]$$

$x \succ y$ is read "x is preferred to y";

$$[x \sim y] \equiv [x \succeq y \text{ and } y \succeq x]$$

$x \sim y$ is read "x is indifferent to y".

Definition 6 **Indifference set** of $x \equiv \{x' \in \mathbb{R}^C_{++} : x' \sim x\} \equiv I(x)$;

upper contour set of $x \equiv \{x' \in \mathbb{R}^C_{++} : x' \succeq x\} \equiv UC(x)$;

lower contour set of $x \equiv \{x' \in \mathbb{R}^C_{++} : x \succeq x'\} \equiv LC(x)$.

We now present some properties we impose on $\succeq$.

Pr1 $\succeq$ is complete, i.e., $\forall x, y \in \mathbb{R}^C_{++}$, $x \succeq y$ or $y \succeq x$.

Pr2 $\succeq$ is transitive, i.e., $\forall x, y, z \in \mathbb{R}^C_{++}$, if $x \succeq y$ and $y \succeq z$, then $x \succeq z$.

[6] For the possibility of considering $\mathbb{R}^C_{+}$ instead of $\mathbb{R}^C_{++}$ as the consumption set, see the comment after Assumption u4 below.

Pr3 $\succeq$ is continuous, i.e., $\forall x \in \mathbb{R}^C_{++}$, $LC(x)$ and $UC(x)$ are closed sets.

Assumption[7] Pr1 says that households are able to "compare" each possible pair of commodity bundles, while "introspection quickly reveals how hard it is to evaluate alternatives that are far from the realism of common experience"[8]. Completeness also implies that individuals "know" completely the characteristics of each element of the choice set. This is quite a formidable assumptions and a lot of recent contributions, even in general equilibrium theory, have been focused on (lack of) information problems.

Assumption Pr 2 "implies that it is impossible to face the decision maker with a sequence of pairwise choices in which her preferences appear to cycle (...). Like the completeness property, the transitivity assumption can be hard to satisfy when evaluating alternatives far from common experience. As compared to the completeness property, however, it is also more fundamental in the sense that substantial portions of economic theory would not survive if economic agents could not be assumed to have transitive preferences"[9].

Assumption Pr 3 "says that the consumer's preferences cannot exhibit "jumps", with, for example, the consumer preferring each element in sequence $(x^n)_{n\in\mathbb{N}}$ to the corresponding element in sequence $(y^n)_{n\in\mathbb{N}}$ but suddenly reversing her preferences at the limiting points of these sequences x and y"[10].

A classical example of preferences not satisfying the continuity assumption is given by the lexicographic preferences. The two-dimensional version of them, denoted by $\mathcal{L}$, is defined as follows: $\forall x \in \mathbb{R}^2_{++}$, $\forall y \in \mathbb{R}^2_{++}$

$$x\mathcal{L}y \text{ if and only if } [x_1 > y_1] \text{ or } [(x_1 = y_1) \text{ and } (x_2 > y_2)]$$

Thus a household with lexicographic preferences gives some value to the second commodity, but it gives infinitely more value to the first - see for example [87], pp. 46-47, for details.

1.2.3 Utility functions

"For analytical purposes, it is very helpful if we can summarize the consumer's preferences by means of a utility function because mathematical programming techniques can then be used to solve the consumer's problem."[11]

[7] In some works, the assumption of reflexivity (defined as $\forall x \in X, x \succeq x$) is added to the above assumption. In fact, reflexivity is implied by completeness.

[8] [87], p. 6.

[9] [87], p. 7.

[10] [87], p. 46.

[11] [87], p. 46.

Definition 7 A function $u : \mathbb{R}^C_{++} \to \mathbb{R}$ is a utility function representing the preference relation $\succeq$ if

$$\forall x, y \in \mathbb{R}^C_{++},\ x \succeq y \Leftrightarrow u(x) \geq u(y)$$

Theorem 8 *Assume that $\succeq$ defined on an arbitrary subset of $\mathbb{R}^C$ is complete, transitive, and continuous. Then, there exists a continuous utility function u representing $\succeq$.*

Proof. See, for example, [70], Theorem A.1, p. 52. ■

We are also going to make the following assumptions directly on the utility function.

Assumption u1 For $h \in \mathcal{H}$, $u_h : \mathbb{R}^C_{++} \to \mathbb{R}$ is a C^2 function.

Assumption u1 "is introduced particularly for reasons of mathematical convenience. We use it when (...) we employ the analytic calculus in our reasoning".[12]

It is possible to give conditions in terms of preferences which imply that the utility function is C^2 - see [42]. "Intuitively, what is required is that indifference sets be smooth surfaces that fit together nicely so that the rates at which commodities substitute for each other depend differentiably on the consumption levels[13]."

Assumption u2 For $h \in \mathcal{H}$, u_h is differentiably strictly increasing, i.e., for every $x_h \in \mathbb{R}^C_{++}$,

$$Du_h(x_h) \gg 0$$

Assumption u2 indeed implies (by standard calculus arguments) that households' utility functions are strictly increasing as mappings between $\mathbb{R}^C_{++}$ and $\mathbb{R}$, i.e.[14], for any $x'_h, x_h \in \mathbb{R}^C_{++}$,

$$x'_h > x_h \Rightarrow u_h(x'_h) > u_h(x_h)$$

Therefore, Assumption u2 requires that no matter what the household is consuming, she prefers a bundle with slightly more of anything. In other words, all commodities are "goods", not "bads", such as garbage.

Some weaker assumptions are often presented in the literature. One among those which still allows to get some basic results of our model[15] is called "local non-satiation", which requires that u_h has no local maxima.

[12] [84], p. 24.

[13] [87], p. 49. The word "smooth" is used with an intuitive meaning.

[14] In fact, the reader should be aware that different authors give different definitions of monotonicity of functions $f : \mathbb{R}^n \to \mathbb{R}$.

[15] See Remarks 34, 45, and 50 below.

Assumption u2' u_h is locally non satiated, i.e., $\forall\ x \in \mathbb{R}^C_{++}$ and $\forall \varepsilon > 0$ there exists $x' \in \mathbb{R}^C_{++}$ such that $\|x - x'\| < \varepsilon$ and $u_h(x') > u_h(x)$.

Assumption u3 For $h \in \mathcal{H}$, u_h is differentiably strictly quasiconcave, i.e.,

$$Du_h(x_h)\, v = 0 \Rightarrow vD^2 u_h(x_h)\, v < 0 \quad \forall x_h \in \mathbb{R}^C_{++}, \forall v \in \mathbb{R}^C \setminus \{0\}$$

Assumption u3 implies that for each x_h, the upper contour set of x_h is a strictly convex set [16]. Therefore, the intuitive notion involved is that households prefer bundles in which commodities are fairly distributed to those which are concentrated on a few goods.

The assumption allows us to get uniqueness of the solution to the household choice problem - see problem $(Ph1)$ and Lemma 25 - and to characterize that solution by means of the so called first order conditions - see Theorem 27.

Assumption u4 For $h \in \mathcal{H}$, and for any $\underline{u} \in \mathbb{R}$,

$$Cl_{\mathbb{R}^C}\left\{x \in \mathbb{R}^C_{++} : u_h(x) \geq \underline{u}\right\} \subseteq \mathbb{R}^C_{++}$$

where $Cl_{\mathbb{R}^C}$ denotes the closure in the standard topology of $\mathbb{R}^C$.

Assumption u4 implies that if the amount consumed of a given good becomes very small, then the amount of some other good has to become very large to leave the household utility level unchanged. The assumption allows to get "interior solutions" to the household problem. In the case $C = 2$, indifference curves of utility functions satisfying u4 do not touch the axes. The assumption can be removed, enlarging at the same time the consumption set from $\mathbb{R}^C_{++}$ to $\mathbb{R}^C_{+}$, without changing the nature of all the results of this chapter, as shown in [16], [102], [120].

A different version of Assumption u4 commonly made in the literature is the following one.

Assumption u4' For any $\underline{u} \in \mathbb{R}$,

$$Cl_{\mathbb{R}^C}\left\{x \in \mathbb{R}^C_{++} : u_h(x) = \underline{u}\right\} \subseteq \mathbb{R}^C_{++}$$

Clearly, $u4 \Rightarrow u4'$. In Lemma 71 in the Appendix, we show that in fact $\langle u4' \textit{ and } u2 \rangle \Rightarrow u4$.

Definition 9 The set of all utility functions satisfying Assumptions u1 to u4 is denoted by $\widehat{\mathcal{U}}$. Moreover $\mathcal{U} \equiv \widehat{\mathcal{U}}^H$ with generic element $u \equiv (u_h)_{h=1}^H$.

[16] $S \subseteq \mathbb{R}^C$ is a strictly convex set if

$$\forall x, x' \in \mathbb{R}^C,\ \forall \lambda \in (0,1), \qquad (1-\lambda)x + \lambda x' \in IntS$$

For further properties of differentiably striclty quasi-concavity see also Section 3.1 in Chapter 1.

1.2.4 Initial endowments

Definition 10 The **endowment set** is the subset of the commodity space consisting of all commodity bundles which in principle a household may own.

Assumption h3 Every household initially owns a positive quantity of all commodities.

It is reasonable to say, and it is generally assumed, that all households own at least something of some commodity, let it be just labor. As a matter of fact, a perspicuous description of the position occupied by a household in a society based on organized exchange comes from consideration of the commodities owned and *not owned* by that household before entering the system of admissible transactions. However, we are here completely disregarding the issue, in order to put ourselves in a position which is much more simple from a technical point of view.

Definition 11 The **endowment set** of each household is $\mathbb{R}^C_{++}$. For each $h \in \mathcal{H}$ the commodity bundle owned by household h is denoted $e_h \equiv (e_h^c)_{c \in C}$. Moreover, $e \equiv (e_h)_{h \in \mathcal{H}}$.

An economy is completely characterized by an element $(e, u) \in \mathbb{R}^{CH}_{++} \times \mathcal{U}$. In this chapter, we take the utility function vector u as given. Therefore, we can give the following definition.

Definition 12 An **exchange economy** is an element $e \in \mathbb{R}^{CH}_{++}$.

1.2.5 Prices

All commodities are traded in markets at publicly observable prices, expressed in monetary unit terms. A price for commodity $c \in C$ is a real number $\widetilde{p}^c$ which indicates the amount of units of money which has to be paid to get one unit of commodity c. The price vector for all commodities is denoted by $\widetilde{p} \equiv (\widetilde{p}^c)_{c \in C} \in \mathbb{R}^C$.[17]

"No theory of money is offered here, and it is assumed that the economy works without the help of a good serving as a medium of exchange."[18] Therefore, money is only a unit of measure.

Prices are obviously used to describe the exchange ratio between goods. Assuming that $\widetilde{p}^{c'} \neq 0$, $\frac{\widetilde{p}^c}{\widetilde{p}^{c'}}$ gives the amount of good c' that may be exchanged for one unit of good c.

[17] In Definition 19, we are going to define p in terms of $\widetilde{p}$. Of course, p will be the normally used symbol for price vectors.

[18] [40], p. 28.

In some occasions throughout the book, it will be useful to choose a commodity, say commodity C, as the unit of measure of the exchanges. That can be done considering a price vector $\frac{\tilde{p}}{p^C} \equiv p$. In that case the price of commodity C is identically equal to 1, and, for $c \neq C$, p^c indicates the (real) number of units of commodity C that can be exchanged for one unit of commodity c.

"Observe that there is nothing that logically requires price to be positive. A negative price simply means that a "buyer" is actually paid to consume the commodity (which is not illogical for commodities that are "bads", such as pollution")[19]. On the other hand, consistently with assumption u2, we assume that all commodities are goods (in a strict sense) and therefore all prices are strictly positive. The *price set* is then $\mathbb{R}^C_{++}$.

1.2.6 Competitive budget set

The budget set of household h is the subset of the consumption set which contains elements the household can afford to buy given the *economic constraints* defining the economy under analysis. The budget set is the set of solutions to a system of inequalities called budget constraints . In this chapter, we study the case of an "exchange economy" "competitive" "complete market" budget set, whose characteristics are a consequence of the Assumptions h2, and h3 done above, and of Assumption h4 and h5 below. The formal definition of budget set is given in Definition 15.

Definition 13 An economy is **competitive** if individuals are "price takers", i.e., they behave as if they could not influence prices with their behavior.

Assumption h4 Households live in a competitive economy.

Loosely speaking, the above assumptions is likely to be valid when individuals are "small" and it is clearly violated in many situations. It is anyway one of the fundamental assumptions of general equilibrium models.

We have now to make an assumption on how households can exchange goods.

Definition 14 Markets are **complete** if each household h can trade any good with any other good.

Assumption h5 Markets are complete.

Observe that markets are incomplete if there exist distinct subsets of the set $\mathcal{C}$ of goods, and goods in each subset can be exchanged only with other goods in the same subset. There are extremely good reasons to think that incompleteness of markets due to geographical or even legal restrictions is quite an important

[19][87], p. 20.

case to be studied in the economic environment described by our simplifying assumptions of the present chapter. On the other hand, the most studied case of incomplete markets is that one which follows from the removal of Assumption c3: it is hard to think that any good today can be exchanged with any other good in any possible situation in a year from now. As already mentioned, Chapters 10-14, analyze the far reaching consequences of the removal of Assumption c3 and h5.

A way of formalizing the assumption of complete or incomplete markets is to characterize the budget set in terms of a *unique* budget constraint or many budget constraints, respectively.

Keeping in mind households' property rights which, in this chapter, are summarized in the ownership of the endowment vector of commodities - see Assumption h2 - we can finally use Assumption h3, h4 and h5 to say that a consumption vector is affordable if its value $\widetilde{p}x_h$ is smaller than household's wealth $\widetilde{p}e_h$. Therefore Assumptions h2, h3, h4 and h5 are summarized in the concept of exchange economy, competitive, complete market budget set presented below.

Definition 15 In an exchange economy, for each h, given a price vector $\widetilde{p} \in \mathbb{R}^C_{++}$ and an endowment vector $e_h \in \mathbb{R}^C_{++}$, household h's **exchange economy, competitive, complete market budget set** is

$$B(\widetilde{p}, e_h) \equiv \left\{x_h \in \mathbb{R}^C_{++} : \widetilde{p}x_h \leq \widetilde{p}e_h\right\}$$

Remark 16 Saying that households can choose any element of their budget set implies that they are able to buy or sell any amount of commodity they wish under their wealth constraint. Some authors make that condition explicit part of the definition of price taking behavior.

Remark 17 Household h can decide to consume a quantity x_h^c of good c which can be in any relationship with e_h^c. In fact,

i. if $x_h^c \geq e_h^c$, she has to buy $(x_h^c - e_h^c)$ units of good c, and in that case we say that she has an *excess demand* of that good;

ii. if $x_h^c < e_h^c$, she has to sell $(e_h^c - x_h^c)$ units of good c, and in that case we say that she has an *excess supply* of that good.

In most of the economics literature, the amount x_h^c is called the "demand" of good c (by household h). We too are going to use that terminology, considering the term "demand" interchangeable with the term "consumption". That use is different from the somehow common meaning of "demand" of good c as the amount the household wants to buy on the market, which in our terminology is called "excess demand". Similar consideration applies to the use of the expressions "supply" and "excess supply".

We then make the following assumption.

Assumption h6. Each household chooses the consumption vector which gives her the highest level of utility, given the above described physical and institutional constraint ($x_h \in \mathbb{R}^C_{++}$) and economic constraints ($\widetilde{p}x_h \leq \widetilde{p}e_h$).

We can then define household h's maximization problem as follows. For given $\widetilde{p} \in \mathbb{R}^C_{++}$, $e_h \in \mathbb{R}^C_{++}$,

$$(Ph0) \qquad \max_{x_h \in \mathbb{R}^C_{++}} \; u_h(x_h) \qquad s.t. \qquad \widetilde{p}x_h \leq \widetilde{p}e_h$$

Remark 18 For any $\alpha > 0$,

$$\left\{x_h \in \mathbb{R}^C_{++} : \widetilde{p}x_h \leq \widetilde{p}e_h\right\} = \left\{x_h \in \mathbb{R}^C_{++} : \alpha\widetilde{p}x_h \leq \alpha\widetilde{p}e_h\right\}$$

Therefore, there is no loss of generality in "normalizing" prices using commodity C price (i.e., choosing the above α equal to $1/\widetilde{p}^C$). In this way, commodity C becomes the unit of measure of exchanges. Commodity C is called the numeraire good .

Definition 19 $p^c \equiv \frac{\widetilde{p}^c}{\widetilde{p}^C}$, $p^\backslash \equiv \left(p^1, ..., p^{C-1}\right) \in \mathbb{R}^{C-1}_{++}$ and $p \equiv \left(p^\backslash, 1\right)$.

We can then rewrite problem $(Ph0)$, as follows. For given $p^\backslash \in \mathbb{R}^{C-1}_{++}$, $e_h \in \mathbb{R}^C_{++}$,

$$(Ph1) \qquad \max_{x_h \in \mathbb{R}^C_{++}} \; u_h(x_h) \qquad s.t. \qquad px_h \leq pe_h$$

2. Definition of equilibrium

Summarizing, we have described the following basic concepts of our model:

1 The commodity space $\mathbb{R}^C_+$ (Assumptions c1,...,c4) ;

2 A finite set $\mathcal{H}$ of households (Assumptions h1). Each of them described by

   (a) a consumption set $\mathbb{R}^C_{++}$ (Assumptions h2);

   (b) a utility function $u_h : \mathbb{R}^C_{++} \to \mathbb{R}$ (Assumptions u1,...,u4);

   (c) an initial endowment $e_h \in \mathbb{R}^C_{++}$ (Assumptions h3);

3 prices $p \in \mathbb{R}^C_{++}$.

Remark 20 Due to the assumptions on the utility functions, the economy we described is sometimes called a "smooth" (exchange) economy. In fact, in some of the literature, utility function are assumed to be C^∞ or smooth.

We denote the vector of all households consumption vectors by $x \equiv (x_h)_{h=1}^H$, which is called consumption allocation or simply allocation.

We call equilibrium a situation in which prices are such that the total consumption of each good is equal to the total resources of that good, i.e., the sum of each household's endowments. Formally, we have the following definition of (Walrasian) equilibrium.

Definition 21 A vector $(x, p^{\backslash}) \in \mathbb{R}^{CH}_{++} \times \mathbb{R}^{C-1}_{++}$ is an *equilibrium* allocation-price vector for an economy $e \in \mathbb{R}^{CH}_{++}$ if

1 households maximize, i.e., for $h = 1, ..., H$, x_h solves problem $(Ph1)$ at $p \in \mathbb{R}^{C}_{++}$, $e_h \in \mathbb{R}^{C}_{++}$;

2 x satisfies the following market clearing conditions

$$\sum_{h=1}^{H} x_h = \sum_{h=1}^{H} e_h \tag{8.1}$$

Definition 22 For every $e \in \mathbb{R}^{CH}_{++}$, let $X(e) \subseteq \mathbb{R}^{CH}_{++}$ be the set of equilibrium allocations associated with an economy e.

The "Edgeworth box" is a very convenient device which completely describes an exchange economy with *two goods and two households.*

By means of the Edgeworth box we can give the following geometric representation of the economy. Assume that households' endowments are $e_1 = (e_1^1, e_1^1)$ and $e_2 = (e_2^1, e_2^2)$.

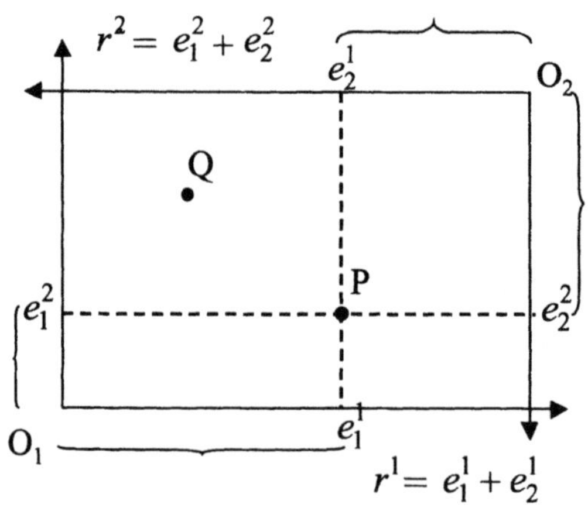

Figure 8.1. Edgeworth box 1

The length and the height of the box are total resources $r^1 = e_1^1 + e_2^1$ and $r^2 = e_1^2 + e_2^2$, respectively; the lower left origin O_1 is referred to household 1, while the upper right O_2 is referred to household 2. The point P corresponds to the initial endowments. Each point in the Edgeworth box represents a non-wasteful allocation or redistribution of resources caused by an exchange of goods between households (see for example point Q).

In the following figure, we add households' tastes representation in terms of indifference curves to the Edgeworth box. The shape of the indifference curves is consistent with Assumptions $u1$- $u4$.

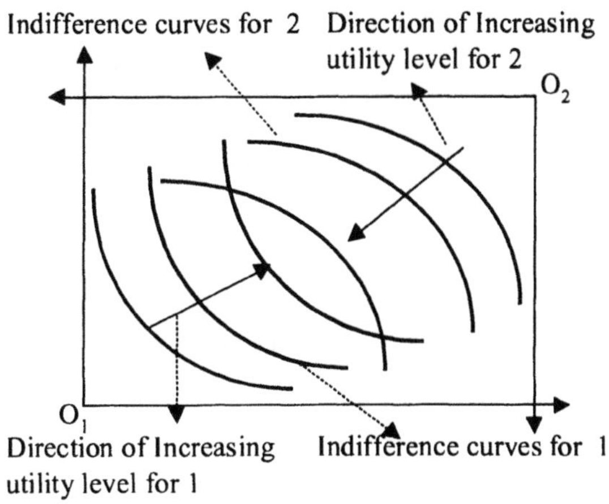

Figure 8.2. Edgeworth box 2

Moreover, increasing satisfaction levels for household 1 (2) correspond to higher (lower) indifference curves.

Finally we introduce prices in the Edgeworth box; since any point in the Edgeworth box represents the endowments of both households (viewed in mirror image), they have a common budget line that is $\{x \in \mathbb{R}^2_{++} : px = pe_h\}$ (even if their wealths are different). Given a price vector $p = (p^1, p^2)$, each household chooses the consumption bundle $\bar{x}_h$ which solves her maximization problem. Therefore we can draw the following figures.

Given the initial endowment e, $(\bar{x}, p)$ in figure 8.3 verifies conditions in Definition 21 On the other hand, $(\bar{x}, p)$ in figure 8.4 is not an equilibrium allocation-price vector. The amount $(x_1^2 + x_2^2)$ of good 2 that households want to consume is smaller than the available quantity r^2 of that good. On the other hand, the chosen amount $(x_1^1 + x_2^1)$ of aggregate consumption of good 1 is bigger than the available resources r^1.

Given the initial endowment e, (x, p) in figure 8.4 is not an equilibrium allocation-price vector: there is an excess demand of good 1 and an excess supply of good 2: there is an excess demand of good 1 and an excess supply of good 2.

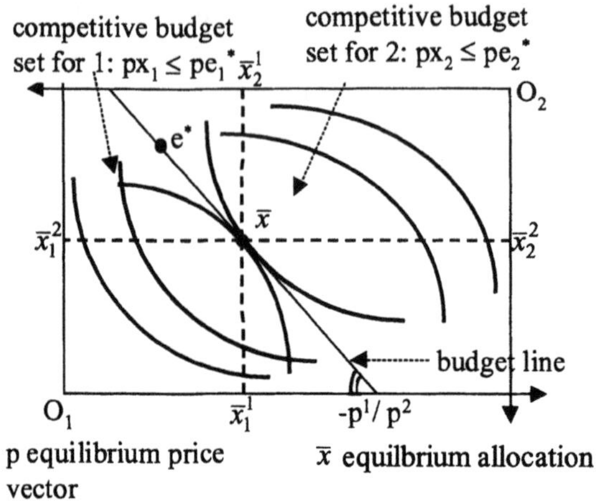

Figure 8.3. Equilibrium allocation

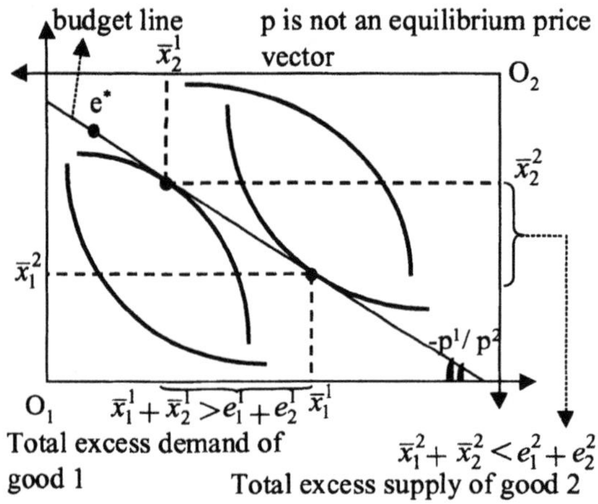

Figure 8.4. Non-equilibrium allocations

3. The demand function

From Assumption $u2$, households do not want to throw away goods. Therefore, a solution to $(Ph1)$ has to satisfy the budget constraint as an equality, as formalized below.

Definition 23 Household h's maximization problem is defined as follows. For given $p^\backslash \in \mathbb{R}^{C-1}_{++}$, $e_h \in \mathbb{R}^C_{++}$,

$$(Ph) \qquad \max_{x_h \in \mathbb{R}^C_{++}} \quad u_h(x_h) \qquad s.t. \qquad -p(x_h - e_h) = 0$$

Remark 24 We write the constraint of problem (Ph) in a somehow controintuitive manner to be able to easily apply Lagrange Theorem, in fact, to get a (strictly) positive value of the Lagrange multiplier associated with that constraint - see Theorem 27.

Lemma 74 in the Appendix shows that the solution sets to problems $(Ph1)$ and (Ph) coincide if the solution set of $(Ph1)$ is not empty.

Lemma 25 *For each* $p^{\backslash} \in \mathbb{R}^{C-1}_{++}$, $e_h \in \mathbb{R}^{C}_{++}$, *a solution to problem* (Ph) *exists and is unique.*

Proof. From Lemma 77 in the Appendix, we know that for any $(p^{\backslash}, e_h) \in \mathbb{R}^{C-1}_{++} \times \mathbb{R}^{C}_{++}$, there exists a solution to (Ph). Therefore we are left with showing that such solution is unique. For ease of notation, drop the subscript h.

Suppose that there exist x' and x'', $x' \neq x''$, which solve (Ph) at (p, e). Defining $\widetilde{x} \equiv \frac{1}{2}x' + \frac{1}{2}\, x''$, it follows that

1 $p\widetilde{x} \equiv \frac{1}{2}px' + \frac{1}{2}px'' = pe$, i.e., $\widetilde{x}$ belong to the budget set of problem (Ph).

2 From Proposition 26 and 28 in Chapter 1, we know that Assumption u2 implies that if $x' \neq x''$ and $\lambda \in (0, 1)$, then

$$u\left((1-\lambda)\, x' + \lambda x''\right) > \min\left\{u\left(x'\right), u\left(x''\right)\right\}$$

Therefore, we have $u\left(\widetilde{x}\right) > u\left(x'\right) = u\left(x''\right)$.

1. and 2. contradict the fact that x' and x'' are solutions to (Ph). ■

Due to the above lemma, we can give the following definition.

Definition 26 The demand function is the solution function to (Ph), i.e.,

$$x_h : \mathbb{R}^{C-1}_{++} \times \mathbb{R}^{C}_{++} \to \mathbb{R}^{C}_{++}, \quad \left(p^{\backslash}, e_h\right) \mapsto \arg\max\,(Ph)$$

Theorem 27 *The following statements hold.*

1 $x_h = x_h\left(p^{\backslash}, e_h\right) \Leftrightarrow$ *there exists* $\mu_h \in \mathbb{R}_{++}$ *such that* (x_h, μ_h) *is the unique solution to system*

$$\left\{\begin{array}{lcll} Du_h\left(x_h\right) - \mu_h p & = & 0 & (1) \\ -p\left(x_h - e_h\right) & = & 0 & (2) \end{array}\right. \tag{8.2}$$

in the exogenous variables $\left(p^{\backslash}, e_h\right)$*;*

2 $x_h\left(p^{\backslash}, e_h\right)$ *is a* C^1 *function.*

Proof. 1. The result follows showing that (Ph) satisfies necessary and sufficient conditions for Lagrange theorem (see Subsection 3.2.1 in Chapter 1). The lagrangean function for problem (Ph) is

$$\mathcal{L}\left(x_h, \mu_h, p^\backslash, e_h\right) = u_h\left(x_h\right) + \mu_h\left(-p\left(x_h - e_h\right)\right)$$

and the associated first order conditions are just system (8.2).

[⇒] Necessary conditions are satisfied because $D_{x_h}\left[p\left(x_h - e_h\right)\right] = p \neq 0$. Therefore, if $x_h = x_h\left(p^\backslash, e_h\right)$, then there exists $\mu_h \in \mathbb{R}$ such that $\left(x_h, \mu_h\right)$ is the unique solution to system (8.2). The fact that $\mu_h > 0$ follows from equation (1) in that system, Assumption u2 and the fact that $p \in \mathbb{R}^C_{++}$.

[⇐] Sufficient conditions for Lagrange theorem are satisfied because, from Assumptions u3, u_h is differentiably strictly quasiconcave, the constraint is linear and, from what said above, a solution $\left(x_h, \mu_h\right)$ to system (8.2) is such that μ_h is strictly positive.

2. Define

$$F_h : \mathbb{R}^C_{++} \times \mathbb{R}_{++} \times \mathbb{R}^{C-1}_{++} \times \mathbb{R}^C_{++} \to \mathbb{R}^C \times \mathbb{R}$$

$$F_h : \left(x_h, \mu_h, p^\backslash, e_h\right) \mapsto \begin{pmatrix} Du_h\left(x_h\right) - \mu_h p \\ -p\left(x_h - e_h\right) \end{pmatrix}$$

The result follows as an application of the implicit function theorem; more precisely it is enough to show that

$$D_{(x_h, \mu_h)} F_h\left(x_h, \mu_h, p^\backslash, e_h\right) \tag{8.3}$$

computed at the solution of system (foch), has full row rank $(C+1)$.

Suppose $D_{(x_h, \mu_h)} F_h$ does not have full rank; then there would exist $\Delta x \in \mathbb{R}^C$ and $\Delta\mu \in \mathbb{R}$ such that $\Delta \equiv (\Delta x, \Delta\mu) \neq 0$ and

$$D_{(x_h, \mu_h)} F_h\left(x_h, \mu_h, p^\backslash, e_h\right)(\Delta x, \Delta\mu) = 0$$

or

$$\begin{array}{lll} D^2 u_h\left(x_h\right)\Delta x - p\Delta\mu & = 0 & (a) \\ -p\Delta x & = 0 & (b) \end{array} \tag{8.4}$$

The idea of the proof is to contradict Assumption u3.

Claim 1. $\Delta x \neq 0$.

Suppose otherwise, i.e., $\Delta x = 0$. But then, from equation (a) in system (8.4), we would have $p\Delta\mu = 0$ and, since $p \neq 0$, $\Delta\mu = 0$, contradicting the assumption that $\Delta \equiv (\Delta x, \Delta\mu) \neq 0$.

Claim 2. $Du_h(x_h)\Delta x = 0$.

From (1) in (foch), we have

$$Du_h(x_h)\Delta x - \mu_h p \Delta x = 0$$

using (b) in (8.4), the desired result follows.

Claim 3. $\Delta x D^2 u_h \Delta x = 0$.

Premultiplying (a) by Δx, we get

$$\Delta x D^2 u_h (x_h) \Delta x - \Delta x p \Delta \mu = 0$$

Using (b), the result follows.

Claims 1, 2 and 3 contradict Assumption u3. ■

4. The extended approach

Following the seminal work by [108], we use first order conditions and market clearing conditions (8.1) to characterize equilibria. The main advantage of this approach is that all the information about the model is contained in the extended equilibrium system. A discussion of the relationship between extended approach and a "reduced" approaches based on the aggregate excess demand function is presented in Subsection 8.2.

We now introduce the system of first order conditions to problems (Ph) for $h \in \mathcal{H}$, and a *subset*[20] of conditions (8.1) which characterizes equilibria.

# eq.	matching var.		eq.	
C	x_h	$(h1)_{h\in\mathcal{H}}$	$Du_h(x_h) - \mu_h p$	$= 0$
1	μ_h	$(h2)_{h\in\mathcal{H}}$	$-p(x_h - e_h)$	$= 0$
$C-1$	$p^\backslash$	(M)	$-\sum_{h=1}^H x_h^\backslash + \sum_{h=1}^H e_h^\backslash$	$= 0$

(8.5)

where eq. and var. stands for equations and variables respectively, $x_h^\backslash \equiv (x_h^c)_{c\neq C}$ and $e_h^\backslash \equiv (e_h^c)_{c\neq C}$.

Next to each "row" in system (8.5), we write the number of corresponding equations and the vectors of endogenous variables with the same number of components (the "matching" variables). Observe that the number of equations is equal to the number of endogenous variables.

Definition 28 Define

$$\xi \equiv \left(x, \mu, p^\backslash\right) \in \Xi$$

where $\Xi \equiv \mathbb{R}^{CH}_{++} \times \mathbb{R}^{H}_{++} \times \mathbb{R}^{C-1}_{++}$, and

$$F : \Xi \times \mathbb{R}^{CH}_{++} \to \mathbb{R}^{\dim \Xi}, \quad (\xi, e) \mapsto \text{left hand side of (8.5)}$$

System (8.5) and the above function are called extended equilibrium system and function, respectively.

[20] Proposition 29 explains why just a *subset* of equilibrium conditions is enough for our purposes.

Proposition 29 *(Walras law)* $\left(x, p^{\backslash}\right)$ *is an equilibrium allocation-price vector associated with an economy e if and only if there exists μ such that* $F\left(x, \mu, p^{\backslash}, e\right) = 0$.

Proof. As shown in Theorem 27, for $h = 1, ..., H$, x_h solves problem (Ph) at $p^{\backslash} \in \mathbb{R}_{++}^{C-1}$, $e_h \in \mathbb{R}_{++}^{C}$ if and only if there exists $\mu_h \in \mathbb{R}_{++}$ such that (x_h, μ_h) solves the system of first order conditions, i.e., equations $(h1)$ and $(h2)$ in system (8.5).

To show that if $F\left(x, \mu, p^{\backslash}, e\right) = 0$ then $\sum_{h=1}^{H}\left(x_h^C - e_h^C\right) = 0$, observe that summing up consumers' budget constraints, we get

$$-p\left(\sum_{h=1}^{H} x_h - \sum_{h=1}^{H} e_h\right) = 0 \tag{8.6}$$

Using the fact that $p \gg 0$ and the expression (8.6), we get the desired result. ■

The above proposition shows that the market clearing condition for good C is "redundant", a fact called Walras law. That result follows from the *equality* characterizing the budget constraint and it is a consequence of the monotonicity assumption on utility functions (Assumption u2).

Consistently with our extended approach, and because of Proposition 29, we then define an equilibrium vector as the vector of endogenous variables of system 8.5.

Definition 30 (Equilibrium) For each economy $e \in \mathbb{R}_{++}^{CH}$, if $F(\xi, e) = 0$, we call $\xi \equiv \left(x, \mu, p^{\backslash}\right)$ an equilibrium vector for e.

The set $F_e^{-1}(0)$ of equilibria associated with e is denoted by $\Xi(e)$.

We want to show that for each economy e there exists a solution to system $F(\xi, e) = 0$. To get that result, we are going to use Theorem 57 in Chapter 7. According to it, we identify

f	with	F_e
y	with	0
M	with	Ξ
N	with	$\mathbb{R}^{\dim \Xi}$

To verify the assumptions of the theorem, in Section 5, we construct a so-called test economy, which is a key step to define the function G - which plays the role of function g in Theorem 57 in Chapter 7. Then in Section 6, we introduce the needed homotopy H_e.

It turns out that the test economy is nothing but a Pareto optimal allocation, whose definition is presented and discussed in some detail in Section 5. Then, we show the other needed results: there exists $\xi^* \in \Xi$ such that $\{\xi^*\} = G^{-1}(0)$ - Lemma 46; $DG(\xi^*)$ has full rank - Lemma 47; $H_e^{-1}(0)$ is compact - Lemma 48.

5. Pareto optimal allocations

One of the key issue of economic theory is *efficiency* of resource allocations among households. We say that x is an *efficient allocation* or a *Pareto optimal allocation* if there is no other feasible allocation x', i.e., such that $\sum_{h=1}^{H} x'_h = \sum_{h=1}^{H} x_h$, at which every household is at least as well off and someone is strictly better off. In the Edgeworth box of figure 8.5, $\overline{x}$ is not Pareto optimal:

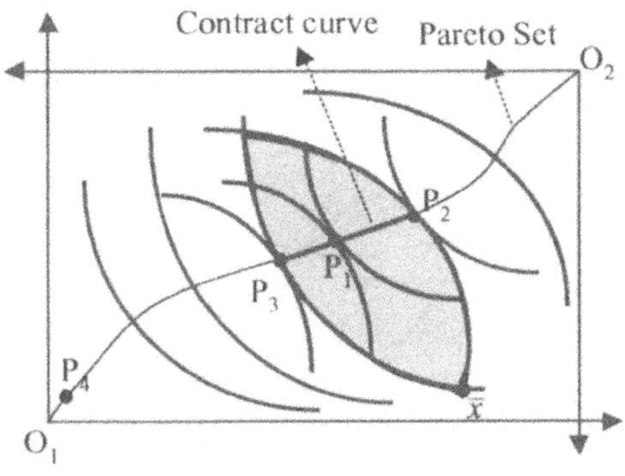

Figure 8.5. Pareto optimal allocations

every allocation in the shaded region corresponds to a higher utility level of at least one consumer and a not lower level of the other one. We say that every point in the shaded areas constitutes a *Pareto improvement* upon x.

On the other hand, all the points on the curve from O_1 to O_2 - without the end points [21]- are Pareto optimal allocations. Moreover, the curve from P_3 to P_2, called *contract curve*, is the part of Pareto set where both consumers do at least as well as their initial endowments.

It should be clear that Pareto optimality is hard to be considered a fairness criterion. In fact, any point on the Pareto set arbitrarily close to O_1 is Pareto optimal, even if in O_1 consumer 1 is "very rich" and consumer 2 "very poor".

A simple question arises now. What can we say about the relationship between Pareto optimal and equilibrium allocations?

We first present the so called *First Fundamental Theorem of Welfare Economics*[22], which simply shows that every equilibrium allocation is Pareto optimal. That theorem can be considered a formal expression of Adam Smith's "invisible hand". In a perfectly competitive environment, the only possible justification for intervention in the economy is some redistributive objective. The *Second Fundamental Theorem of Welfare Economics* offers a partial converse

[21] Since we assumed that $\mathbb{R}^2_{++}$ is the consumption set, O_1 and O_2 are not admissible consumption vectors.
[22] Probably a better terminology would be "First Fundamental Theorem of *Efficiency* Economics".

result. Roughly speaking, it says that, under some convexity assumptions, a planner can achieve any Pareto optimal allocations, first taxing and subsidizing households and then letting the market work. As figure 8.7 illustrates, a transfer of good 1 that moves the endowment vector to e'' will have the price $p = (p^1, p^2)$ and an allocation x as an equilibrium. A transfer of good 2 changes endowments to e' does so as well. In fact, if all commodities can be easily transferred, then we could equally well move the endowment vector to the allocation x[23].

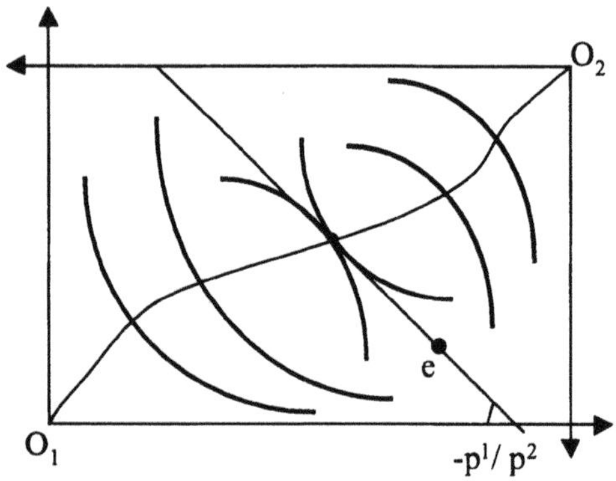

Figure 8.6. First Fundamental Theorem of Welfare Economics

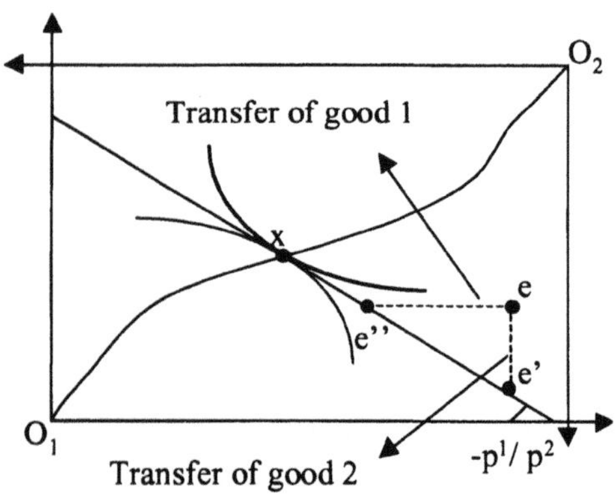

Figure 8.7. Second Fundamental Theorem of Welfare Economics

From the above Edgeworth boxes, two important properties of an economy coinciding with a Pareto optimal allocation are apparent. First, the equilibrium allocation associated with a Pareto optimal allocation is that allocation itself: indifference curves of both households are tangent to each other and to the

[23]Figure 8.7 and its description are taken from [87] p.525

budget line in that allocation. Moreover, given our *strict* convexity assumptions (see Assumption u3), there is no other price (and allocation) which is an equilibrium for the starting Pareto optimal allocation. The above facts suggest why Pareto optimal allocations play a fundament role in the way we choose to show existence of equilibria.

In fact, the homotopy used in the application of Theorem 57 in Chapter 7 is just a way of linking an arbitrary economy with an economy coinciding with a Pareto optimal allocation.

In the remaining part of this section, we give a formal definition of Pareto optimality and a standard (and simple) proof of the First Fundamental Theorem of Welfare Economics. Then, we introduce a maximization problem whose unique solution, characterized by a system of equations, is a Pareto optimal allocation. Those results are used in this section to show the Second Fundamental Theorem of Welfare Economics, and in next section to construct the above mentioned homotopy.

Definition 31 $\mathcal{E} \equiv \left\{x \in \mathbb{R}^{CH}_{++} : \exists \left(\mu, p, e\right) \text{ for which } F\left(x, \mu, p^{\backslash}, e\right) = 0\right\}$ is called the set of equilibrium allocations associated with e[24].

Definition 32 An allocation $x^* \in \mathbb{R}^{CH}_{++}$ is Pareto optimal if there does not exists $x' \in \mathbb{R}^{CH}_{++}$ such that

1 $\sum_{h=1}^{H} x'_h = \sum_{h=1}^{H} x^*_h$, and

2 $\left(u_h\left(x'_h\right)\right)_{h=1}^{H} > \left(u_h\left(x^*_h\right)\right)_{h=1}^{H}$.

The set of Pareto optimal allocations is denoted by $\mathcal{PO}$.

Theorems 33 and 44 show that $\mathcal{E} = \mathcal{PO}$.

Theorem 33 ***(First Fundamental Theorem of Welfare Economics)*** *$\mathcal{E} \subseteq \mathcal{PO}$, i.e., any equilibrium allocation is Pareto optimal.*

Proof. Suppose otherwise, then there exists $x \in \mathcal{E}$, i.e., $F\left(x, \mu, p^{\backslash}, e\right) = 0$ for some $\left(\mu, p^{\backslash}, e\right)$, and $x \notin \mathcal{PO}$, i.e., there exists $x' \in \mathbb{R}^{CH}_{++}$ such that

$$i.\ \sum_{h=1}^{H} x'_h = \sum_{h=1}^{H} x_h, \qquad ii.\ \left(u_h\left(x'_h\right)\right)_{h=1}^{H} > \left(u_h\left(x_h\right)\right)_{h=1}^{H}$$

From ii., we have that

a. for every h, $u_h\left(x'_h\right) \geq u_h\left(x_h\right)$, and

[24]Recall that the function F is introduced in Definition 28.

b. there exist h' such that $u_{h'}\left(x'_{h'}\right) > u_{h'}\left(x_{h'}\right)$.

Claim 1. a implies that $px'_h \geq px_h = pe_h$.

If $px'_h < px_h$, household h could choose a "better" consumption equal to $x_h^+ \equiv x'_h + (0, ..., 0, px_h - px'_h) > x'_h$.

In fact, x_h^+ belongs to her budget set:

$$p\left(x'_h + \left(0, ..., 0, px_h - pe'_h\right)\right) = px_h = pe_h$$

Moreover, it gives a higher utility than x_h :

$$u_h\left(x_h^+\right) > u_h\left(x'_h\right) \geq u_h\left(x_h\right)$$

where the first strict inequality comes from Assumption u2, contradicting the fact that x_h solves problem (Ph).

Claim 2. b implies that $px'_{h'} > px_{h'}$.

If $px'_{h'} \leq px_{h'}$, household h' could choose her consumption *equal to* $x'_{h'}$, which belongs to her budget set and gives a higher utility than $x_{h'}$, again contradicting the fact that $x_{h'}$ solves problem (Ph').

Summing up the inequalities obtained in the above Claims with respect to h we get

$$p\sum_{h=1}^{H} x'_h > p\sum_{h=1}^{H} x_h$$

which contradicts the assumption that $\sum_{h=1}^{H} x'_h = \sum_{h=1}^{H} x_h$. ■

Remark 34 In the above theorem, the only needed assumption about utility functions is Assumption $u2'$ - local non-satiation.

To prove the converse of Theorem 33, that is $\mathcal{PO} \subseteq \mathcal{E}$, we define r as the total resources associated with e and the set $\underline{U}^r$ of utility level vectors attainable with resources r. Given r and $\underline{U}^r$, we characterize the set of Pareto Optima as the solution set to a maximization problem (8.7) or (8.8) - see Lemma 38; then in Lemma 39 we prove that for any $(r, \underline{U}^r)$ there exists a unique solution to problem (8.8). This result allows us to show that for any economy which is already a Pareto optimal allocation, there exists a unique equilibrium vector $\xi^* = (x^*, \mu^*, p^{*\backslash})$ such that x^* coincides with the Pareto optimal allocation itself (Lemma 42). As a consequence of the above results, $\mathcal{PO} \subseteq \mathcal{E}$ is proved.

Definition 35 Given an economy $e \in \mathbb{R}_{++}^{CH}$, the total resources associated with e are $r \equiv \sum_{h=1}^{H} e_h \in \mathbb{R}_{++}^{C}$.

Definition 36 Define $\underline{U}^r$ as

$$\underline{U}^r \equiv \{(\underline{u}_h)_{h=1}^H \in \mathbb{R}^H : \exists\, x \in \mathbb{R}_{++}^{CH} \text{ such that } \sum_{h=1}^H x_h = r, \text{ and } \forall h,\ u_h(x_h) - \underline{u}_h = 0\}$$

i.e., $\underline{U}^r$ is the set of utility level vectors attainable with resources r. Define also

$$\underline{U}^r_{\setminus 1} \equiv \left\{(\underline{u}_h)_{h\neq 1} \in \mathbb{R}^{H-1} : \exists\ \underline{u}_1 \in \mathbb{R} \text{ such that } \left(\underline{u}_1, (\underline{u}_h)_{h\neq 1}\right) \in \underline{U}^r\right\}$$

Given our smooth and strictly quasiconcave assumptions on utility functions, we can characterize Pareto optimal allocations in terms of the solutions to the following maximization problem.

For given $\left(r, \underline{u}_{\setminus 1}\right) \in \mathbb{R}_{++}^C \times \underline{U}^r_{\setminus 1}$,

$$\begin{array}{lll} \max\limits_{x\in\mathbb{R}_{++}^{CH}} & u_1(x_1) & \\ s.t. & (1)\quad u_h(x_h) \geq \underline{u}_h & for\ h \neq 1 \\ & (2)\quad \sum_{h=1}^H x_h \leq r & \end{array} \tag{8.7}$$

In problem (8.7), we want to maximize the attainable utility level of consumer 1 for given utility level of the other consumers and given resources.

Remark 37 Since utility functions are strictly increasing (from Assumption u2), the solution set to problem (8.7) coincides with the solution set to the following problem

$$\begin{array}{llll} \max\limits_{x\in\mathbb{R}_{++}^{CH}} & u_1(x_1) & & \\ s.t. & u_h(x_h) - \underline{u}_h = 0 & for\ h \neq 1 & (1) \\ & -\sum_{h=1}^H x_h + r = 0 & & (2) \end{array} \tag{8.8}$$

Lemma 38 *The following statements hold.*

1 x^* *solves problem (8.8) at* $\left(r, \underline{u}_{\setminus 1}\right) \Rightarrow x^*$ *is Pareto optimal;*

2 x^* *is Pareto optimal* $\Rightarrow$ x^* *solves problem (8.8) for* $r = \sum_{h=1}^h x_h^*$ *and* $(\underline{u}_h)_{h\neq 1} = (u_h(x_h^*))_{h\neq 1}$.

Proof. 1. Suppose otherwise. Then, there exists $\widehat{x}$ which satisfies constraint (2) in problem (8.7) - which has the same set of solutions as problem (8.8) - and such that $(u_h(\widehat{x}_h))_{h=1}^H > (u_h(x_h^*))_{h=1}^H$.

If $u_1(\widehat{x}_1) > u_1(x_1^*)$, we contradict the definition of x^* as a solution to problem (8.7).

If there exists $h' \neq 1$ such that $u_{h'}(\widehat{x}_{h'}) > u_{h'}\left(x_{h'}^*\right)$, by continuity of the utility function, there exists $\varepsilon > 0$ such that $u_{h'}(\widehat{x}_{h'} - \varepsilon \mathbf{1}) > u_{h'}\left(x_{h'}^*\right)$, with $\mathbf{1} \equiv (1, ..., 1) \in \mathbb{R}^C$, while $u_1(\widehat{x}_1 + \varepsilon \mathbf{1}) > u_1(x_1^*)$, again contradicting the definition of x^*.

2. Suppose otherwise. Then, there exists $\widehat{x}$ such that

$$u_1(\widehat{x}_1) > u_1(x_1^*) \tag{8.9}$$

$$u_h(\widehat{x}_h) - u_h(x_h^*) = 0 \qquad for\ h \neq 1 \tag{8.10}$$

$$-\sum_{h=1}^{H} \widehat{x}_h + \sum_{h=1}^{H} x_h^* = 0 \tag{8.11}$$

(8.9), (8.10) and (8.11) contradict the fact that x^* is a Pareto optimal allocation. ■

We now write the lagrangean function associated with problem (8.8).

$$\mathcal{L}\left(x, \theta_{\setminus 1}, \gamma, r, \underline{u}_{\setminus 1}\right) = u_1(x_1) + \sum_{h=2}^{H} \theta_h \left(u_h(x_h) - \underline{u}_h\right) + \gamma \left(-\sum_{h=1}^{H} x_h + r\right)$$

where $(\theta_h)_{h=2}^H \in \mathbb{R}^{H-1}$ and $\gamma \in \mathbb{R}^C$ are the vectors of Lagrange multipliers.

For given $\left(r, \underline{u}_{\setminus 1}\right) \in \mathbb{R}_{++}^C \times \underline{U}_{\setminus 1}^r$, the system of first order conditions in the unknowns $\left(x, \theta_{\setminus 1}, \gamma\right) \in \mathbb{R}_{++}^{CH} \times \mathbb{R}^{H-1} \times \mathbb{R}^C$ is

$$\begin{cases} Du_1(x_1) - \gamma & = 0 \quad (1) \\ \theta_h Du_h(x_h) - \gamma & = 0 \quad (2) \\ (u_h(x_h) - \underline{u}_h)_{h \neq 1} & = 0 \quad (3) \\ -\sum_{h=1}^{H} x_h + r & = 0 \quad (4) \end{cases} \tag{8.12}$$

Define $(\theta_h)_{h \neq 1} \equiv \theta_{\setminus 1}$ and $\theta_1 = 1$.

Lemma 39 *The following statements hold.*

1 There exists a unique solution x^* to problem (8.8) at $\left(r, \underline{u}_{\setminus 1}\right) \in \mathbb{R}_{++}^C \times \underline{U}_{\setminus 1}^r$.

2 There exists $\left(\theta_{\setminus 1}^*, \gamma^*\right)$ such that $\left(x^*, \theta_{\setminus 1}^*, \gamma^*\right)$ is the unique solution to system (8.12) at $\left(r, \underline{u}_{\setminus 1}\right)$.

Proof. 1.

Existence. Since $\underline{u}_{\backslash 1} \in \underline{U}^r_{\backslash 1}$ by Definition 36, the constraint set is not empty. From the extreme value theorem ([90] Theorem 6.4, p.175) we are left to show that the constraint set K is compact. Take an arbitrary $x^0 \equiv \left(x_h^0\right)_{h=1}^H$ belonging to that set. Add the constraint $u_1(x_1) \geq u_1\left(x_1^0\right)$. The (possibly empty) solution set of this modified maximization problem is the same as problem (8.8). From constraint (1) in problem (8.8) and from the added constraint, Assumption u4 allows to conclude that for all h, K is contained in a closed subset of $\mathbb{R}^C$. Moreover, from the fact that the consumption set is $\mathbb{R}^C_{++}$, K is bounded from below by zero. From constraint (2) in problem (8.8), K is bounded from above. Hence K is compact.

Uniqueness. Since the constraint functions are quasiconcave , the constraint set is convex. The objective function is strictly quasiconcave .

2.

It follows from Kuhn-Tucker Theorems (see Subsection 3.2.2 in Chapter 1), whose assumptions are verified below.

Necessity of Kuhn-Tucker conditions follows from the fact that the jacobian matrix of the constraint function has full row rank. The computation of that matrix is described below.

$$
\begin{array}{r} \\ u_2(x_2) - \underline{u}_2 \\ \ldots \\ u_H(x_H) - \underline{u}_H \\ -\sum_{h=1}^H x_h + r \end{array}
\begin{array}{c} \begin{array}{ccc} x_2 & \ldots \quad x_H & x_1 \end{array} \\ \left(\begin{array}{cccc} Du_2 & & & \\ & & & \\ & & Du_H & \\ -I & & -I & -I \end{array}\right) \end{array}
$$

Sufficiency of Kuhn-Tucker conditions follows from the fact that the objective function is pseudoconcave (see Table 1.6 in Chapter 1) and the constraint functions are quasiconcave.

Observe that from equation (1) and (2) in system (8.12), γ and $\theta_{\backslash 1}$ are uniquely determined and strictly positive. ■

As the following lemma shows, $\left(r, \underline{u}_{\backslash 1}\right)$ can be substituted with

$$\left(\sum_{h=1}^H x_h^*, (u_h(x_h^*))_{h \neq 1}\right)$$

in the statement of the previous lemma.

Lemma 40 *The following statements hold.*

1 x^* *is the unique solution to problem (8.8) for* $r = \sum_{h=1}^H x_h^*$ *and* $\left(\underline{u}_h\right)_{h\neq 1} = (u_h(x_h^*))_{h \neq 1}$*;*

2 $\left(x^*, \theta^*_{\setminus 1}, \gamma^*\right)$ *is the unique solution to system (8.12) for* $r = \sum_{h=1}^{H} x^*_h$ *and* $\left(\underline{u}_h\right)_{h \neq 1} = \left(u_h\left(x^*_h\right)\right)_{h \neq 1}$. *In fact,*

$$\text{for each } h \text{ and } c,\ \theta^*_h = \frac{D_{x_1^c} u_1\left(x_1^*\right)}{D_{x_h^c} u_h\left(x_h^*\right)} > 0 \quad \text{and} \quad \gamma^* = Du_1\left(x_1^*\right) \gg 0$$

Proof. From part 1 of Lemma 39, we know that there exists a unique solution x^{**} to problem (8.8) at $\left(\sum_{h=1}^{H} x^*_h, \left(u_h\left(x^*_h\right)\right)_{h \neq 1}\right)$. From part 2 of that lemma we know that there exists a unique solution $\left(x^{**}, \theta^{**}_{\setminus 1}, \gamma^{**}\right)$ to system (8.12) at $\left(\sum_{h=1}^{H} x^*_h, \left(u_h\left(x^*_h\right)\right)_{h \neq 1}\right)$. It is then enough to show that $\left(x^*, \theta^*_{\setminus 1}, \gamma^*\right)$ is a solution to system (8.12) at $\left(\sum_{h=1}^{H} x^*_h, \left(u_h\left(x^*_h\right)\right)_{h \neq 1}\right)$.

That system is

$$\begin{cases} Du_1\left(x_1\right) - \gamma & = 0 \quad (1) \\ \theta_h Du_h\left(x_h\right) - \gamma & = 0 \quad (2) \\ \left(u_h\left(x_h\right) - u_h\left(x^*_h\right)\right)_{h \neq 1} & = 0 \quad (3) \\ -\sum_{h=1}^{H} x_h + \sum_{h=1}^{H} x^*_h & = 0 \quad (4) \end{cases} \tag{8.13}$$

The desired results follow from the definition of $(x^*, \theta^*, \gamma^*)$ as a solution to system (8.12) at $\left(r, \underline{u}_{\setminus 1}\right)$, i.e., from Lemma 39, and from the inspection of (8.13). ∎

Remark 41 Statement 2 in Lemma 40 is the result of this section we are going to use to show existence of equilibria.

The following theorem shows that given an economy which is a Pareto optimal allocation, there exists a unique equilibrium allocation associated with it, which is the starting Pareto optimal allocation itself. In other words, for every economy $x^* \in \mathcal{PO}$, there exists a unique associated equilibrium $\xi^* \equiv \left(x^*, \mu^*, p^{\setminus *}\right)$.

Lemma 42 *Let* $F_{x^*} : \left(x, \mu, p^{\setminus}\right) \mapsto F\left(x, \mu, p^{\setminus}, x^*\right)$.
Then $F_{x^*}^{-1}(0) = \left\{\left(x^*, \mu^*, p^{\setminus *}\right)\right\}$, *where*

$$\mu^*_h = \frac{\gamma^{*C}}{\theta^*_h} = D_{x_h^C} u_h\left(x^*_h\right) \text{ for every } h, \quad \text{and} \quad p^* = \frac{\gamma^*}{\gamma^{*C}} = \frac{Du_1\left(x_1^*\right)}{D_{x_1^C} u_1\left(x_1^*\right)}$$

Proof. From statement 2 in Lemma 40, we have that $\left(x^*, \theta^*_{\setminus 1}, \gamma^*\right)$ is the unique solution to system 8.13. Then

$$\begin{cases} Du_1\left(x_1^*\right) - \gamma^{*C}\frac{\gamma^*}{\gamma^{*C}} & = 0 & & (1) \\ Du_h\left(x_h^*\right) - \frac{\gamma^{*C}}{\theta_h^*}\frac{\gamma^*}{\gamma^{*C}} & = 0 & \text{for } h \neq 1 & (2) \\ \frac{\gamma^*}{\gamma^{*C}}\left(x_h^* - x_h^*\right) & = 0 & \text{for every } h & (3) \\ -\sum_{h=1}^H x_h^{*\setminus} + \sum_{h=1}^H x_h^{*\setminus} & = 0 & & (4) \end{cases}$$

which shows that $\left(x^*, \mu^*, p^{\setminus *}\right) \in F_{x^*}^{-1}(0)$. To show uniqueness, we are now left with showing that

$$F\left(\widehat{x}, \widehat{\mu}, \widehat{p}^{\setminus}, x^*\right) = 0 \Rightarrow \left(\widehat{x}, \widehat{\mu}, \widehat{p}^{\setminus}\right) = \left(x^*, \mu^*, p^{\setminus *}\right)$$

1. $\widehat{x} = x^*$.

Suppose otherwise, i.e., $\widehat{x} \neq x^*$. Consider $\widetilde{x} \equiv \frac{1}{2}\left(\widehat{x} + x^*\right)$.
Since $F\left(\widehat{x}, \widehat{\mu}, \widehat{p}^{\setminus}, x^*\right) = 0$, then $\sum_h \widehat{x}_h = \sum_h x_h^*$ and

$$\sum_{h=1}^H \widetilde{x}_h = \frac{1}{2}\left(\sum_{h=1}^H \widehat{x}_h + \sum_{h=1}^H x_h^*\right) = \sum_{h=1}^H x_h^* \tag{8.14}$$

From Assumption u3 (strict quasiconcavity of the utility functions), we then have

$$u_h\left(\widetilde{x}_h\right) > u_h\left(x_h^*\right) \text{ for every } h \tag{8.15}$$

But (8.14) and (8.15) contradict the Pareto optimality of x^*.

2. $\left(\widehat{\mu}, \widehat{p}\right) = \left(\mu^*, p^*\right)$.

$F\left(x^*, \widehat{\mu}, \widehat{p}, x^*\right) = 0 \Rightarrow$

$$Du_1\left(x_1^*\right) - \widehat{\mu}_1\widehat{p} = 0 \qquad \text{and} \qquad Du_h\left(x_h^*\right) - \widehat{\mu}_h\widehat{p} = 0$$

Therefore,

$$\widehat{\mu}_1 = D_{x_1^C}u_1\left(x_1^*\right) = \mu_1^*$$

where the last equality follows from the definition of μ_1^*. Moreover,

$$\widehat{p} = \frac{Du_1\left(x_1^*\right)}{\widehat{\mu}_1} = \frac{Du_1\left(x_1^*\right)}{D_{x_1^C}u_1\left(x_1^*\right)} = p^*$$

$$\widehat{\mu}_h = D_{x_h^C}u_h\left(x_h^*\right) = \mu_h^*$$

where, again,we used the definitions of p^* and μ^*. ■

Remark 43 Observe that from the above proof we get also

$$p^{C*} = 1 \text{ and for every } h, \quad \frac{Du_h(x_h^*)}{D_{x_h^C} u_h(x_h^*)} = \frac{\gamma^*}{\theta_h^*} \frac{\theta_h^*}{\gamma^{C*}} = p^*$$

Theorem 44 ***(Second Fundamental Theorem of Welfare Economics)*** *$\mathcal{PO} \subseteq \mathcal{E}$, i.e., any Pareto optimal allocation is an equilibrium allocation.*

Proof. The result is part of previous lemma. In fact, a Pareto optimal allocation is the unique equilibrium allocation associated with an economy which is the Pareto optimal allocation itself. ■

Remark 45 Local non-satiation and quasiconcavity - which are implied by Assumption u2 and u3, respectively - are the assumptions on utility functions needed to show the above theorem[25].

6. Existence of equilibria

In this section, given an economy e, we are going to construct a homotopy H from the equilibrium function F_e to a well chosen function G. Applying Theorem 57, in Chapter 7 we will be able to show existence of equilibria.

Consider an arbitrary economy $e \in \mathbb{R}^{CH}_{++}$, a Pareto optimal allocation x^* feasible at e, i.e., such that $\sum_{h=1}^{H} x_h^* = \sum_{h=1}^{H} e_h$. *For given e and* x^*, define

$$\begin{array}{lll} \dots & & \\ Du_h(x_h) - \mu_h p & = 0 & (h1) \\ -px_h + p\left[(1-\tau)e_h + \tau x_h^*\right] & = 0 & (h2) \\ \dots & & \\ -\sum_{h=1}^{H} x_h^{\backslash} + \left[(1-\tau)\sum_{h=1}^{H} e_h^{\backslash} + \tau \sum_{h=1}^{H} x_h^{*\backslash}\right] & = 0 & (M) \end{array} \qquad (8.16)$$

$$H_e : \Xi \times [0,1] \to \mathbb{R}^{\dim \Xi}, \quad (\xi, \tau) \mapsto \text{left hand side of system (8.16)}$$

and[26]

$$G : \Xi \to \mathbb{R}^{\dim \Xi}, \quad \xi \mapsto H_e(\xi, 1)$$

where Ξ is presented in Definition 28.

Observe that

$$H_e(\xi, 0) = F(\xi, e) = F_e(\xi)$$

[25] [87], p. 552.
[26] Of course, a more precise, but heavier, notation would be H_{e,x^*} instead of H_e.

and

$$H_e(\xi, \tau) = F(\xi, (1-\tau)e + \tau x^*)$$

We are now ready to verify the assumptions of Theorem 57 in Chapter 7: $\{\xi^*\} = G^{-1}(0)$ - Lemma 46; $DG(\xi^*)$ has full rank - Lemma 47; $H^{-1}(0)$ is compact - Lemma 48.

Lemma 46 $G^{-1}(0) = \{\xi^*\}$ *with* $\xi^* = (x^*, \mu^*, p^{\backslash *})$ *defined in Lemmas 40 and 42.*

Proof. The result follows from the fact that $G(\xi) \equiv F(\xi, x^*)$ and therefore $G^{-1}(0) = F_{x^*}^{-1}(0)$, and from Lemma 42. ■

Lemma 47 $\operatorname{rank} DG(\xi^*) = \dim \Xi$.

Proof. The computation of $DG(\xi^*) = D_{(x,\mu,p^{\backslash})}F(e^*, \mu^*, p^{\backslash *}, e^*)$ is described below.

$$\begin{array}{l} \\ Du_1(x_1) - \mu_1 p \\ p(x_1 - x_1^*) \\ \ldots \\ Du_h(x_h) - \mu_h p \\ p(x_h - x_h^*) \\ \\ Du_H(x_H) - \mu_H p \\ p(x_H - e_H^*) \\ \sum_{h=1}^{H}\left(x_h^{\backslash} - x_h^{*\backslash}\right) \end{array}
\begin{array}{c} \begin{array}{cccccccc} x_1 & \mu_1 & & x_h & \mu_h & \ldots & x_H & \mu_H & p^{\backslash} \end{array} \\
\left(\begin{array}{ccccccccc} D_1^2 & p^{*T} & & & & & & & \mu_1^* \widetilde{I} \\ p^* & 0 & & & & & & & x_1^* - x_1^* = 0 \\ & & \ldots & & & & & & \\ & & & D_h^2 & p^{*T} & & & & \mu_h^* \widetilde{I} \\ & & & p^* & 0 & & & & 0 \\ & & & & & \ldots & & & \\ & & & & & & D_H^2 & p^{*T} & \mu_H^* \widetilde{I} \\ & & & & & & p^* & 0 & 0 \\ \widetilde{I}^T & & \ldots & \widetilde{I}^T & & & \widetilde{I}^T & & \end{array}\right) \end{array}$$

where

$$\widetilde{I} \equiv \begin{bmatrix} I_{C-1} \\ 0 \end{bmatrix}_{C \times (C-1)}$$

and for each h, $D_h^2 \equiv D^2 u_h(x_h^*)$.

Take $\Delta \equiv (\Delta x_1, \Delta\mu_1, ..., \Delta x_H, \Delta\mu_H, \Delta p^{\backslash}) \in \mathbb{R}^{(C+1)H+C-1}$, with $\Delta x_h \in \mathbb{R}^C$ and $\Delta\mu_h \in \mathbb{R}$ for each h, and $\Delta p^{\backslash} \in \mathbb{R}^{C-1}$. We want to show that if $D_{(x,\mu,p)}F(x^*, \mu^*, p^*, x^*)\Delta = 0$, then $\Delta = 0$. The system is written in detail below.

$$
\begin{array}{lll}
D_1^2\Delta x_1 - p^*\Delta\mu_1 + \mu_1^*\widetilde{I}\Delta p^\backslash & = 0 & (1.1) \\
-p^*\Delta x_1 & = 0 & (1.2) \\
\ldots & & \\
D_h^2\Delta x_h - p^*\Delta\mu_h + \mu_h^*\widetilde{I}\Delta p^\backslash & = 0 & (h.1) \\
-p^*\Delta x_h & = 0 & (h.2) \\
\ldots & & \\
D_H^2\Delta x_H - p^*\Delta\mu_H + \mu_H^*\widetilde{I}\Delta p^\backslash & = 0 & (H.1) \\
-p^*\Delta x_H & = 0 & (H.2) \\
\widetilde{I}^T\Delta x_1 + \ldots + \widetilde{I}^T\Delta x_h + \ldots + \widetilde{I}^T\Delta x_H & = 0 & (M)
\end{array}
\tag{8.17}
$$

We distinguish two cases: Case 1. For each h, $\Delta x_h = 0$ and Case 2. There exists h' such that $\Delta x_{h'} \neq 0$.

Case 1.

In this case $(h.1)$, $h = 1, ..., H$, in (8.17) becomes

$$(h.1*) \qquad -p^*\Delta\mu_h + \mu_h^*\widetilde{I}\Delta p^\backslash = 0$$

whose last equation implies that $\Delta\mu_h = 0$. Then $(h.1*)$ becomes $\mu_h^*\Delta p^\backslash = 0$. Since $\mu_h^* > 0$, it follows that $\Delta p^\backslash = 0$. Then, $\Delta = 0$, as desired.

Case 2.

We are going to show that this case cannot hold, since Claim 2 and 3 below are inconsistent.

Claim 1. $Du_h\left(x_h^*\right)\Delta x_h = 0,$ for $h = 1, ..., H$.

Postmultiplying $(h.1)$ in (8.16) computed at ξ^*, i.e., $Du_h\left(x_h^*\right) - \mu_h^* p^* = 0$, by Δx_h, we get

$$Du_h\left(x_h^*\right)\Delta x_h - \mu_h^* p^*\Delta x_h = Du_h\left(x_h^*\right)\cdot\Delta x_h = 0$$

where the last equality follows from $(h.2)$ in (8.17).

Claim 2. $\sum_{h=1}^{H}\left(\Delta x_h \frac{D_h^2}{\mu_h^*}\Delta x_h\right) = 0.$

For $h = 1, ..., H$, define $\Delta x_h^\backslash$ as the vector Δx_h without its last component. Premultiplying $(h.1)$ in (8.17) by $\frac{\Delta x_h}{\mu_h^*}$, we get

$$
\begin{aligned}
0 &= \Delta x_h\frac{D_h^2}{\mu_h^*}\Delta x_h - \frac{\Delta x_h}{\mu_h^*}p^*\Delta\mu_h - \Delta x_h\begin{pmatrix}\Delta p^{\backslash T}\\ 0\end{pmatrix} \\
&= \Delta x_h\frac{D_h^2}{\mu_h^*}\Delta x_h - \Delta x_h\begin{pmatrix}\Delta p^{\backslash T}\\ 0\end{pmatrix}
\end{aligned}
$$

where the second equality follows from $(h.2)$ in (8.17). Taking the sum over h, we get

$$0 = \sum_{h=1}^{H} \left(\Delta x_h \frac{D_h^2}{\mu_h^*} \Delta x_h \right) - \left(\sum_{h=1}^{H} \Delta x_h^{\backslash} \right) \Delta p^{\backslash} = \sum_{h=1}^{H} \left(\Delta x_h \frac{D_h^2}{\mu_h^*} \Delta x_h \right)$$

where the second equality comes from (M) in (8.17).

Claim 3. $\sum_{h=1}^{H} \left(\Delta x_h \frac{D_h^2}{\mu_h^*} \Delta x_h \right) < 0$.

Partition $\{1, ..., H\}$ in the two sets H_0 and H_1 such that if $h \in H_0$, then $\Delta x_h = 0$ and if $h \in H_1$, then $\Delta x_h \neq 0$. From the assumption of Case 2, $H_1 \neq \emptyset$.

If $h \in H_0$, then $\Delta x_h \frac{D_h^2}{\mu_h^*} \Delta x_h = 0$.

If $h \in H_1$, from Claim 1, the fact that $\mu_h^* > 0$ and Assumption u3, we get $\Delta x_h^T \frac{D_h^2}{\mu_h^*} \Delta x_h < 0$ and finally

$$\sum_{h=1}^{H} \left(\Delta x_h \frac{D_h^2}{\mu_h^*} \Delta x_h \right) = \sum_{h \in H_0} \left(\Delta x_h \frac{D_h^2}{\mu_h^*} \Delta x_h \right) + \sum_{h \in H_1} \left(\Delta x_h \frac{D_h^2}{\mu_h^*} \Delta x_h \right) < 0$$

■

We are now left with the last step in the proof of existence.

Lemma 48 *For each* $e \in \mathbb{R}_{++}^{CH}$, $H_e^{-1}(0)$ *is compact.*

Proof. We are going to show that $H_e^{-1}(0)$ is sequentially compact, i.e., that any sequence $(\xi^v, \tau^v)_{v \in \mathbb{N}}$ in $H_e^{-1}(0)$ admits a subsequence which converges to an element of $H_e^{-1}(0)$.

First of all observe that, up to a subsequence, since $\{\tau^v : v \in \mathbb{N}\} \subseteq [0, 1]$, $(\tau^v)_{v \in \mathbb{N}}$ converges and so do $(\overline{e}_h^v)_{v \in \mathbb{N}}$, $(\overline{e}^v)_{v \in \mathbb{N}}$, $(\overline{r}^v)_{v \in \mathbb{N}}$ where

$$\begin{aligned} \overline{e}_h^v &\equiv (1 - \tau^v) e_h + \tau^v x_h^* \\ \overline{e}^v &\equiv (1 - \tau^v) e + \tau^v x^* \\ \overline{r}^v &\equiv (1 - \tau^v) \textstyle\sum_{h=1}^{H} e_h^{\backslash} + \tau^v \sum_{h=1}^{H} x_h^{*\backslash} \end{aligned}$$

say to $\overline{e}_h$, $\overline{e}$ and $\overline{r}$, respectively.

Claim 1. $(x^v)_{v \in \mathbb{N}}$ admits a subsequence converging to $\overline{x} \in \mathbb{R}_{++}^{CH}$.

It suffices to show that $\{x^v : v \in \mathbb{N}\}$ is contained in a compact subset of $\mathbb{R}_{++}^{CH}$.

α. Boundedness from below. Since $\{x^v : v \in \mathbb{N}\} \subseteq \mathbb{R}_{++}^{CH}$ is bounded from below by zero.

β. Boundedness from above. From the fact that

$$F(x^v, \mu^v, p^v, (1 - \tau^v) e + \tau^v x^{*v}) = 0$$

we have that for any $h' = 1, ..., H$, $x^v_{h'} = \sum_{h=1}^{H} \overline{e}^v_h - \sum_{h \neq h'} x^v_h$. Therefore, since $(\overline{e}^v_h)_{v \in \mathbb{N}}$ converges and $\{x^v : v \in \mathbb{N}\}$ is bounded from below , the result follows.

γ. Closedness. By definition of x^v, for any $v \in \mathbb{N}$ and for any h, we have that

$$u_h(x^v_h) \geq u_h(\overline{e}^v_h) \geq \min_{e_h \in \{\overline{e}^v_h : v \in N\} \cup \{\overline{e}_h\}} u_h(e_h) \equiv \underline{u}_h$$

where the min is well defined because $\{\overline{e}^v_h : v \in \mathbb{N}\} \cup \{\overline{e}_h\}$ is a compact set (which can be proved using an argument similar to [90], Example 2, p.164).

Therefore, $\{x^v_h : v \in \mathbb{N}\} \subseteq Cl_{\mathbb{R}^C} \{x_h \in \mathbb{R}^C : u_h(x_h) \geq \underline{u}_h\}$, which is a closed set (in the topology of $\mathbb{R}^C$) contained in $\mathbb{R}^C_{++}$ from Assumption u4. Therefore, $\left\{x^v = (x^v_h)_{h=1}^H : v \in \mathbb{N}\right\}$ is contained in a closed set, too.

Claim 2. $(\mu^v)_{v \in \mathbb{N}}$ admits a subsequence converging to $\overline{\mu} \in \mathbb{R}^H_{++}$.

Since $p^C = 1$, for $h = 1, ..., H$, from $(h.1)$ in (8.16) ,we have

$$D_{x^C_h} u_h(x^v_h) = \mu^v_h$$

Taking $\lim_{v \to +\infty}$ of both sides, we get

$$\lim_{v \to +\infty} \mu^v_h = \lim_{v \to +\infty} D_{x^C_h} u_h(x^v_h) = D_{x^C_h} u_h(\overline{x}_h) \equiv \overline{\mu}_h > 0$$

where last strict inequality comes from Assumption u2.

Claim 3. $\left(p^{\setminus v}\right)_{v \in \mathbb{N}}$ admits a subsequence converging to $\overline{p}^{\setminus} \in \mathbb{R}^{C-1}_{++}$.

Again from $(h.1)$ in (8.16), $D_{x^{\setminus}_h} u_h(x^v_h) - \mu^v_h p^{\setminus} = 0$. Therefore, we get

$$\lim_{v \to +\infty} p^{\setminus v} = \lim_{v \to +\infty} \frac{D_{x^{\setminus}_h} u_h(x^v_h)}{\mu^v_h} = \frac{D_{x^{\setminus}_h} u_h(x_h)}{\mu_h} \equiv \overline{p}^{\setminus} > 0$$

where again the last strict inequality comes from Assumption u2.

Finally, since $\{(\xi^v, \tau^v) : v \in \mathbb{N}\} \subseteq H_e^{-1}(0)$ means that for every v

$$F(\xi^v, (1 - \tau^v) e + \tau^v e^*) = 0$$

taking limits of both sides and using the continuity of F, we get

$$F\left(\overline{\xi}, (1 - \overline{\tau}) e + \overline{\tau} e^*\right) = 0$$

i.e., $\left(\overline{\xi}, \overline{\tau}\right) \in H_e^{-1}(0)$, as desired. ■

From the previous lemmas and from Theorem 57 in Chapter 7, we finally have the following existence result.

Theorem 49 ***(Existence of Equilibria)*** *For every $u \in \mathcal{U}$ and every economy $e \in \mathbb{R}^{CH}_{++}$, there exists $\xi \in \Xi$ such that $F(\xi, e) = 0$.*

Remark 50 Existence of equilibria can be obtained under conditions which are considerably weaker than those assumed in this chapter. In fact, see for example [87], it suffices that for each $h \in \mathcal{H}$,

a. the consumption set X_h is closed and convex;

b. the preference relation $\succeq_h$ on X_h is continuous, locally non-satiated and convex[27];

c. there exists $\widehat{x}_h \in X_h$ such that $e_h \geq \widehat{x}_h$.

7. Regular economies

Having[28] solved the existence problem, we now discuss the structure of the set of equilibria for an arbitrary economy e in terms of its cardinality and dependence on e. It is well known[29] that *uniqueness* of equilibria can only be obtained by imposing stronger assumptions on the structure of the model. On the other hand some weaker results can be obtained.

Consider an economy e and an associated equilibrium ξ. Assume that in any neighborhood of ξ, there are infinitely many other equilibria for the economy e. In this situation the economic explanation of equilibrium is essentially indeterminate [30]. It is then interesting to study how common is to have an economy e for which the set of equilibria is *finite,* and therefore such that for every equilibrium ξ in $\Xi(e)$, there is a neighborhood of ξ in which ξ is the unique equilibrium allocation for e.

Even under our relatively strong assumptions, $\Xi(e)$ may be *not even discrete*. In the simplest case of an exchange economy with two consumers and two commodities, one can easily find in the associated Edgeworth box a set $\Xi(e)$ made up of a continuum of points ([44] and [87] p. 591).

A way to analyze the above problem is to define a concept of *regular* economy such that (a) the set of nonregular economies is a negligible subset of the endowment space, and (b) every regular economy has a finite set of equilibria. In fact, it would be interesting to show still another property of regular economies. Specifically, if e' is "close to" e, then one would like the set of equilibria $\Xi(e')$ to be "close to" $\Xi(e)$. Otherwise an arbitrary small error in the determination of the characteristics of e would yield an entirely different set of equilibria, thus depriving the theory of much of its explanatory power. Therefore it is also desirable for the definition of regularity to be such that (c) in a neighborhood of a regular economy, the set of equilibria depends in a *continuously differentiable* manner on the economy. We call singular or critical an economy which is not regular. Consider the following figure which illustrates in a suggestive manner

[27] $\succeq_h$ is convex if the Upper Contour set $UC(x)$ is convex.

[28] In this brief introduction to the concept of regular economies,we follow closely [44].

[29] See, for example, [87].

[30] For a discussion of equilibria indeterminacy see also also subsection 4 in Chapter 12.

the structure of the equilibrium manifold. Three critical economies e^1, e^2 and

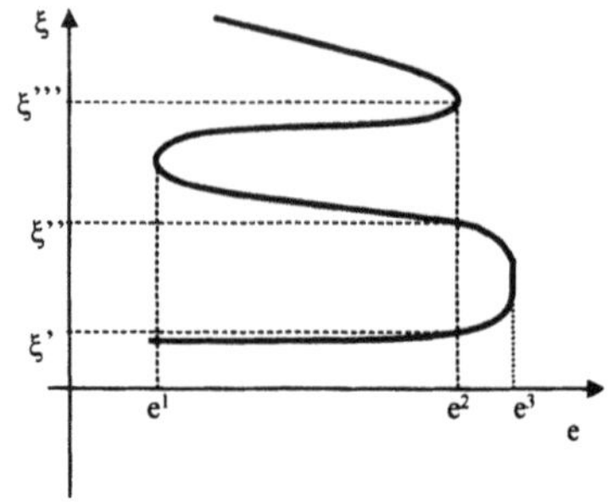

Figure 8.8. Regular economies

e^3 are displayed in figure 8.8. As said in (a), the set of critical economies is small. The "pathologies" associated with critical economies are made clear by the figure. For instance, the economy e^1 has a discrete set of (two) equilibria, but a continuous displacement of the economy in a neighborhood of e^1 produces a sudden change in the set of equilibria. In particular, consider e on the right of e^1 and the associated equilibrium ξ on the upper branch of the equilibrium manifold. As e crosses e^1, the equilibrium state of the system jumps to the lower branch of M. The economy e^3 has a continuum of equilibria. Although the economy e^2 is critical, it has a discrete set of (three) equilibria, and in a neighborhood of e^2, the set of equilibria around ξ' and ξ'' (but not ξ''') depends in a *continuously differentiable* manner on the economy.

We now introduce a definition of regular economies and show that they have the nice properties discussed above (Theorems 55 and 56) and constitute a "very large" subset of the set of all economies (Theorem 57).

Definition 51 $e \in \mathbb{R}^{CH}_{++}$ is a regular economy if 0 is a regular value for F_e. Let $\mathcal{R}$ be the set of regular economies.

Lemma 52 0 *is a regular value for* F.

Proof. We have to prove that for every $(\xi, e) \in F^{-1}(0)$, $D_{(\xi,e)}F(\xi, e)$ has full row rank $(C+1)H + C - 1$. It suffices to show that for *every* $(\xi, e) \in F^{-1}(0)$ and for every $h \in \{1, ..., H\}$, $D_{(x,\mu,e_h)}F(\xi, e)$ has full rank.

The computation of that partial jacobian matrix is described below.

$$\begin{array}{ll} & \\ (1.1) & Du_1(x_h) - \mu_1 p \\ (1.2) & p(x_1 - e_1) \\ \ldots & \\ (h.1) & Du_h(x_h) - \mu_h p \\ (h.2) & p(x_h - e_h) \\ \ldots & \\ (H.1) & Du_H(x_H) - \mu_H p \\ (H.2) & p(x_H - e_H) \\ (M) & \sum_h \left(x_h^{\backslash} - e_h^{\backslash}\right) \end{array} \begin{array}{c} \begin{array}{cccccccccc} x_1 & \mu_1 & & x_h & \mu_h & \ldots & x_H & \mu_H & e_h^{\backslash} & e_h^C \end{array} \\ \left(\begin{array}{cccccccccc} D_1^2 & p^T & & & & & & & & \\ p & 0 & & & & & & & & \\ & & .. & & & & & & & \\ & & & D_h^2 & p^T & & & & & \\ & & & p & 0 & & & & -p^{\backslash} & -1 \\ & & & & & .. & & & & \\ & & & & & & D_H^2 & p^T & & \\ & & & & & & p & 0 & & \\ \widetilde{I}^T & & .. & \widetilde{I}^T & & .. & \widetilde{I}^T & & -I_{C-1} & \end{array}\right) \end{array} \tag{8.18}$$

where $\widetilde{I} \equiv \begin{bmatrix} I_{C-1} \\ 0 \end{bmatrix}$. Observe that $D_{(x,\mu)}\left(F_h(.)\right)_{h=1}^{H}$ is a block diagonal matrix and each block

$$D_{(x_h,\mu_h)}F_h(x_h, \mu_h, p, e_h)$$

has full rank $C+1$ from the proof of Theorem 27. Then, the upper left block of the matrix corresponding to the first order conditions of each household's maximization problem and to the variables $(x_h, \mu_h)_{h\in\mathcal{H}}$ has full rank. Therefore, we are left with "perturbing" (see Section 4 in Chapter 1) for a description of this method) the last $C-1$ rows of the jacobian matrix 8.18, which is done below

$$(M) \quad \leftarrow e_h^{\backslash} \quad \rightsquigarrow \quad (h.2) \quad \leftarrow e_h^C$$

■

Proposition 53 *$F^{-1}(0)$ is a C^1 submanifold of $\Xi \times \mathbb{R}_{++}^{CH}$ of dimension CH.*

Proof. From Theorem 49, $F^{-1}(0) \neq \emptyset$. From the previous lemma and from Theorem 9 in Chapter 4, the desired result follows. ■

Lemma 54

$$pr: F^{-1}(0) \to \mathbb{R}_{++}^{CH}, \quad (\xi, e) \mapsto e \tag{8.19}$$

is proper.

Proof. Take a sequence $(e^v)_{v\in\mathbb{N}}$ in $\mathbb{R}_{++}^{CH}$. By assumption, it admits a convergent subsequence, say to $\overline{e} \in \mathbb{R}_{++}^{CH}$. We want to show that the sequence $(\xi^v, e^v)_{v\in\mathbb{N}}$ in $F^{-1}(0)$ does converge to an element of $F^{-1}(0)$. The proof follows the same steps as the proof of Lemma 48. ■

Theorem 55 *If* $e \in \mathcal{R}$, *then*

1 *there exists* $r \in \mathbb{N}$ *such that*

$$F_e^{-1}(0) = \{\xi^i\}_{i=1}^r$$

2 *there exists an open set* $Y \subseteq \mathbb{R}^{CH}_{++}$, *and for each* i, *an open set* $V_i \subseteq \mathbb{R}^{\dim \Xi}$ *and a unique function* $g_i : Y \to V_i$ *such that*

(a) $(\xi^i, e) \in V_i \times Y$,

(b) g_i *is* C^1,

(c) $g_i(e) = \xi^i$,

(d) *for every* $e' \in Y$, $F(g_i(e'), e') = 0$,

Proof. 1. From the Theorem 49, for every e, $F_e^{-1}(0) \neq \emptyset$. Moreover, from Lemma 48, $F_e^{-1}(0)$ is compact. Then the result follows from Corollary 16 in Chapter 4.

2. From the implicit function theorem ([3], Theorem 13.7, p. 374), we have that for each i, there exist an open sets $Y_i \subseteq \mathbb{R}^{CH}_{++}$, an open set $V_i \subseteq \mathbb{R}^{\dim \Xi}$ and a unique function $\widehat{g}_i : Y_i \to V_i$ such that $(\xi^i, e) \in V_i \times Y_i$ and statements 2.b, 2.c and 2.d hold. Then defined $Y \equiv \cap_{i=1}^r Y_i$, an open neighborhood of e, for each i, it is enough to take $g_i \equiv \widehat{g}_{i|Y}$. ■

Theorem 56 *Given* $e \in \mathcal{R}$. *Then*

1 *There exists* $r \in \mathbb{N}$ *such that* $pr^{-1}(e) = \{(\xi^i, e)\}_{i=1}^r$;

2 *there exist an open neighborhood* Y *of* e *in* $\mathbb{R}^{CH}_{++}$, *and for each* i *an open neighborhood* U_i *of* (ξ^i, e) *in* $F^{-1}(0)$, *such that:*

(a) $U_j \cap U_k = \emptyset$ *if* $j \neq k$;

(b) $pr^{-1}(Y) = \cup_{i=1}^r U_i$;

(c) $pr_{|U_i} : U_i \to Y$ *is a diffeomorphism.*

Proof. It follows from Corollary 44 in Chapter 4, identifying the function f and pr there with the functions F and pr here, and the variables x and z there with ξ and e here. In particular, Assumption 1 of that Corollary follows from Theorem 52; Assumption 2 from the assumption that $e \in \mathcal{R}$; Assumption 3 from Theorem 49; Assumption 4 from Lemma 54. ■

Theorem 57 *The set* $\mathcal{R}$ *has the following properties.*

1 $\mathcal{R}$ *is of full Lebesgue measure in* $\mathbb{R}^{CH}_{++}$,

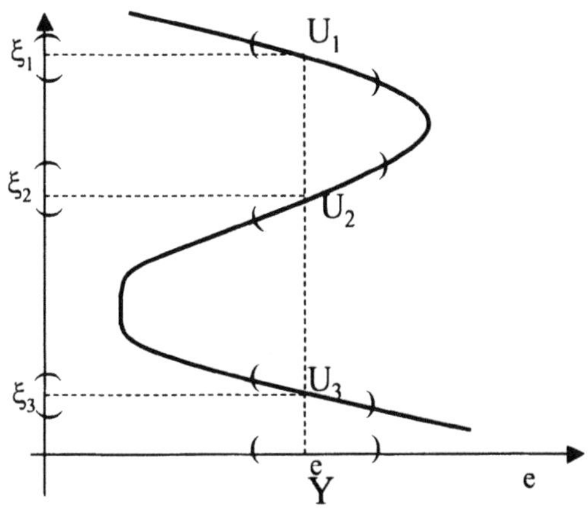

Figure 8.9. Regularity of e

2 *$\mathcal{R}$ is open in* $\mathbb{R}^{CH}_{++}$.

Proof. 1. From Theorem 52, we know that 0 is a regular value for F. We can now use the transversality theorem, see Theorem 26 in Chapter 6 identifying

f	with	F
M	with	Ξ
Ω	with	$\mathbb{R}^{CH}_{++}$
N	with	$\mathbb{R}^{\dim \Xi}$

to conclude the desired result. Observe that, since utility functions are C^2, the inequality condition contained in the set of assumptions of that theorem is verified.

2. The result follows from Lemma 54 and part 3 in Corollary 27 in Chapter 6. ■

8. Reduced approaches

In this section we study the relationship between our approach and what we can call a "reduced" approach. In the latter framework, equilibria are defined in terms of fewer equations than those used in this chapter. A crucial role is played by the aggregate excess demand function, which is defined below.

Definition 58 For $c = 1, ..., C$, the aggregate excess demand function for commodity c is

$$z^c : \mathbb{R}^{C-1}_{++} \times \mathbb{R}^{CH}_{++} \to \mathbb{R}$$
$$z^c : (p^{\backslash}, e) \mapsto \sum_{h=1}^{H} \left(x_h^c \left(p^{\backslash}, e_h \right) - e_h^c \right)$$

where $x\left(p^{\setminus}, e_h\right) \equiv \left(x^c\left(p^{\setminus}, e_h\right)\right)_{c=1}^{C}$ is the demand by household h introduced in Definition 26.

Under the maintained assumptions of this chapter, the above function is C^1 (see Theorem 27). Define also

$$z^{\setminus} : \mathbb{R}_{++}^{C-1} \times \mathbb{R}_{++}^{CH} \to \mathbb{R}^{C-1}$$
$$z^{\setminus} : \left(p^{\setminus}, e\right) \mapsto \left(z^c\left(p^{\setminus}, e\right)\right)_{c=1}^{C-1}$$

Using Walras' law - see Proposition 29 - we can use $z^{\setminus}$ to define equilibrium prices.

Definition 59 The price vector $p^{\setminus} \in \mathbb{R}_{++}^{C-1}$ is an equilibrium price associated with the economy e if $z^{\setminus}\left(p^{\setminus}, e\right) = 0$.

The allocation-price equilibrium set is

$$E \equiv \left\{\left(p^{\setminus}, e\right) \in \mathbb{R}_{++}^{C-1} \times \mathbb{R}_{++}^{CH} : z^{\setminus}\left(p^{\setminus}, e\right) = 0\right\}$$

We can also define a "reduced" projection as follows

$$pr^r : E \to \mathbb{R}_{++}^{CH}, \quad \left(p^{\setminus}, e\right) \mapsto e$$

Existence of equilibrium prices for each economy e is equivalent to one of the following two statements:

$$\forall e \in \mathbb{R}_{++}^{CH}, \quad \left(z_e^{\setminus}\right)^{-1}(0) \neq \emptyset \tag{8.20}$$

or

$$\forall e \in \mathbb{R}_{++}^{CH} \quad \left(pr^r\right)^{-1}(e) \neq \emptyset \tag{8.21}$$

A reduced approach has been followed by most authors in the literature on smooth economies. The most important contributions are still those by [41], [51] and [9]. In what follows, we are going to present two reduced approaches whose goal is to show (8.20) and (8.21), respectively. The latter approach has also been used extensively in the literature on incomplete markets, to which Chapters 10-15 are devoted. In fact, we are going to use the "reduced projection" in the proof of existence of equilibria in the model with real assets presented in Chapter 13.

In the remaining part of this section, we show how the main steps in proving existence and regularity for both reduced approaches follow easily from the results obtained in the extended approach, in the case described in this chapter. More specifically, we first analyze the relationship between existence of equilibria and regular economies in the extended approach and in the proposed reduced approaches; then we prove the desired results in both of them; finally, we analyze the main mathematical differences between the extended and the reduced approach.

8.1 Existence of equilibria and regular economies in the extended and the reduced approaches

The relationship among the concept of equilibria in the different approaches is described in the following lemma.

Lemma 60

$$\begin{array}{lr} pr^r\left(p^\backslash, e\right) = e & \Leftrightarrow \\ z^\backslash\left(p^\backslash, e\right) = 0 & \Leftrightarrow \\ \exists\,(x, \mu) \in \mathbb{R}^{CH}_{++} \times \mathbb{R}^{H}_{++} \textit{ such that } F\left(x, \mu, p^\backslash, e\right) = 0 & \end{array}$$

Proof. The first equivalence follows from the definition of *pr* (recall that its domain is $\left(z^\backslash\right)^{-1}(0)$). The second equivalence follows from the definition of $z^\backslash$ and part 1 in Theorem 27. ■

An obvious consequence of the above lemma is that our equilibrium existence result in terms of the function F implies existence results in terms of $z^\backslash$ and pr^r. The natural question about the relationships among regular economies in the extended and the reduced approaches still remains open. In fact, in the latter case we still do not even have an appropriate definition of those economies. We then proceed as follows. First, we present definitions of regular economies in the reduced approaches and we show that they are equivalent to the one presented above (Proposition 65). Then, to further appreciate the links between the different approaches, *without using our existence result*, we show how the "basic ingredients" of the extended approach , i.e., Lemmas 46, 47, 48, can be used to directly show existence of equilibria in the two proposed reduced approaches: the equivalence results contained in Lemma 60 and Proposition 65 suffices to the goal.

Definition 61 e is a Debreu-regular economy [31] if 0 is a regular value of

$$z_e^\backslash : \mathbb{R}^{C-1}_{++} \to \mathbb{R}^{C-1}, \quad p^\backslash \mapsto z^\backslash(p, e)$$

We denote by $\mathcal{R}^D$ the set of Debreu-regular economy

Definition 62 e is a Balasko-regular economy if e is a regular value of pr^r. We denote the set of Balasko-regular economy by $\mathcal{R}^B$ ([9], p. 87).

We are going to show that the above notions of regularity are equivalent to that one proposed in this chapter. For that purpose, we need some definitions

[31] This definition of regularity has been introduced by Debreu in [41]. See also [51], [108] and [110].

and Lemma 63 below. Define

$$F_h : \Xi \times \mathbb{R}^{CH}_{++} \to \mathbb{R}^{C+1}$$
$$F_h : (\xi, e) \mapsto \begin{pmatrix} Du_h(x_h) - \mu_h p \\ -p(x_h - e_h) \end{pmatrix}$$

$$F_{\mathcal{H}} \equiv (F_h)_{h=1}^{H}$$

and

$$Z^{\backslash} : \Xi \times \mathbb{R}^{CH}_{++} \to \mathbb{R}^{C-1}, \quad (\xi, e) \mapsto \sum_{h=1}^{H} \left(x_h^{\backslash} - e_h^{\backslash} \right)$$

Lemma 63 *0 is a regular value for* $z^{\backslash}$.

Proof. If $\left(z^{\backslash}\right)^{-1}(0) = \emptyset$, we are done. Assume otherwise, and take $\left(\widehat{p}^{\backslash}, \widehat{e}\right) \in \left(z^{\backslash}\right)^{-1}(0)$. We are going to use Theorem 40 in Chapter 4 to show the desired result. Identify the notation of that theorem with the notation presented in the table.

Theorem 40 in Chapter 4	
X	$\mathbb{R}^{CH}_{++} \times \mathbb{R}^{H}_{++}$
Z	$\mathbb{R}^{CH}_{++}$
x	(x, μ)
z	e
f	$\overline{F}_{\mathcal{H}} \equiv (F_{\mathcal{H}})_{\widehat{p}^{\backslash}}$
g	$\overline{Z}^{\backslash} \equiv \left(Z^{\backslash}\right)_{\widehat{p}^{\backslash}}$
m	CH
l	$C-1$
y	0

From Lemma 60, there exists $(\widehat{x}, \widehat{\mu})$ such that $(\widehat{x}, \widehat{\mu}) \in \overline{F}_{\mathcal{H}}^{-1}(0)$. Let's check that the three assumptions of Theorem 40 in Chapter 4 are verified.

1 $\overline{F}_{\mathcal{H}}$ and $\overline{Z}^{\backslash}$ are clearly C^1.

2 $\overline{F}_{\mathcal{H}}(\widehat{x}, \widehat{\mu}, \widehat{e}) = 0$ from what said above.

3 $\operatorname{rank} D_{(x,\mu)} \overline{F}_{\mathcal{H}}(\widehat{x}, \widehat{\mu}, \widehat{e})$ is full from the proof of part 2 of Theorem 27.

Observe that, as stated in Conclusion 1 of Theorem 40 in Chapter 4, 3 implies that there exist an open set $U \subseteq \mathbb{R}^{CH}_{++}$ containing $\widehat{e}$ and a unique C^1 function $(\widehat{x}, \widehat{\mu}) : U \to \mathbb{R}^{\dim \Xi}$ such that $(\widehat{x}, \widehat{\mu})(\widehat{e}) \equiv (\widehat{x}(\widehat{e}), \widehat{\mu}(\widehat{e})) = (\widehat{x}, \widehat{\mu})$, and

$$\text{for each } e \in U, \ \overline{F}_{\mathcal{H}}((\widehat{x}, \widehat{\mu})(e), e) = 0$$

Therefore, $\forall e \in U$, $\widehat{x}(e) = x\left(\widehat{p}^{\backslash}, e\right)$, i.e., the demand function vector for all households for given $\widehat{p}^{\backslash}$, and

$$\overline{Z}^{\backslash}\left(\left(\widehat{x},\widehat{\mu}\right)(e), e\right) = \sum_{h=1}^{H}\left(x_h^{\backslash}\left(\widehat{p}^{\backslash}, e_h\right) - e_h^{\backslash}\right) = z^{\backslash}\left(\widehat{p}^{\backslash}, e\right) \qquad (8.22)$$

Using (8.22), from Conclusion 2 of Theorem 40 in Chapter 4, if

$$\text{rank}\left[D_{(x,\mu,e)}\left(\overline{F}_{\mathcal{H}}, \overline{Z}^{\backslash}\right)\left(\widehat{x},\widehat{\mu},\widehat{e}\right)\right] \text{ is full} \qquad (8.23)$$

then

$$\text{rank}\, D_e z^{\backslash}\left(\widehat{p}^{\backslash}, \widehat{e}\right) \text{ is full, i.e., } e \text{ is Debreu-regular}$$

To conclude the proof, it is enough to observe that (8.23) is the content of the proof of Theorem 52. ■

Lemma 63 and Theorem 9 in Chapter 4 clearly imply the following result.

Proposition 64 $\left(z^{\backslash}\right)^{-1}(0)$ *is a* C^1 CH*-dimensional submanifold of* $\mathbb{R}^{C-1}_{++} \times \mathbb{R}^{CH}_{++}$.

Proposition 65 *e is a regular economy* ⇔ *e is a Debreu-regular economy* ⇔ *e is a Balasko-regular economy* .

Proof. We use Theorem 40 in Chapter 4 to show the first equivalence and Corollary 43 to show that if e is a Debreu-regular economy, then e is a Balasko-regular economy. Theorem 4.3.6 in [9] proves the remaining implication.

1. e is a regular economy ⇔ e is a Debreu-regular economy.

From Lemma 60, $F_e^{-1}(0) = \emptyset$ if and only if $\left(z_e^{\backslash}\right)^{-1}(0) = \emptyset$. Otherwise, there exists $(\widehat{x},\widehat{\mu})$ such that $\left(\widehat{x},\widehat{\mu},\, \widehat{p}^{\backslash}\right) \in F_e^{-1}(0)$ if and only if $\widehat{p}^{\backslash} \in \left(z_e^{\backslash}\right)^{-1}(0)$.

Identify the notation of Theorem 40 in Chapter 4 with the notation presented in the table.

Theorem 40 in Chapter 4	
X	$\mathbb{R}^{CH}_{++} \times \mathbb{R}^{H}_{++}$
Z	$\mathbb{R}^{C-1}_{++}$
x	(x,μ)
z	$p^{\backslash}$
f	$(F_{\mathcal{H}})_e \equiv \left(\widetilde{F}_{\mathcal{H}}\right)_e$
g	$\left(Z^{\backslash}\right)_e \equiv \left(\widetilde{Z}^{\backslash}\right)_e$
m	$C-1$
l	$C-1$
y	0

Let's check that the three assumptions of Theorem 40 in Chapter 4 are verified.

1 $\widetilde{F}_{\mathcal{H}}$ and $Z^{\backslash}$ are clearly C^1.

2 $\widetilde{F}_{\mathcal{H}}\left(\widehat{x},\widehat{\mu},\,\widehat{p}^{\backslash}\right)=0$ by what said above.

3 $\operatorname{rank} D_{(x,\mu)}\widetilde{F}_{\mathcal{H}}\left(\widehat{x},\widehat{\mu},\,\widehat{p}^{\backslash}\right)$ is full from the proof of part 2 of Theorem 27.

Observe that, as stated in Conclusion 1 of Theorem 40 in Chapter 4, 3 implies that there exist an open set $U\subseteq\mathbb{R}^{C-1}_{++}$ containing $\widehat{p}^{\backslash}$ and a unique C^1 function $(\widehat{x},\widehat{\mu}):U\rightarrow\mathbb{R}^{HC}_{++}\times\mathbb{R}^{H}_{++}$ such that $(\widehat{x},\widehat{\mu})\left(\widehat{p}^{\backslash}\right)=(\widehat{x},\widehat{\mu})$, and

$$\text{for every } p^{\backslash}\in U,\ \widetilde{F}_{\mathcal{H}}\left((\widehat{x},\widehat{\mu})\left(p^{\backslash}\right),p^{\backslash}\right)=0$$

Therefore, $\forall p^{\backslash}\in U$, $\widehat{x}\left(p^{\backslash}\right)=x\left(p^{\backslash},e\right)$, i.e., the demand function vector for all households for given e, and

$$\widetilde{Z}^{\backslash}\left((\widehat{x},\widehat{\mu})\left(p^{\backslash}\right),p^{\backslash}\right)=\sum_{h=1}^{H}\left(x_h^{\backslash}\left(p^{\backslash},e_h\right)-e_h^{\backslash}\right)=z^{\backslash}\left(p^{\backslash},e\right)\qquad(8.24)$$

Using 8.24, from Conclusion 2 of Theorem 40 in Chapter 4,

$$\operatorname{rank}\left[D_{(x,\mu,p^{\backslash})}\left(\widetilde{F}_{\mathcal{H}},\widetilde{Z}^{\backslash}\right)\left(\widehat{x},\widehat{\mu},\,\widehat{p}^{\backslash}\right)\right]\ \text{is full, i.e., } e \text{ is regular}$$

if and only if

$$\operatorname{rank} D_{p^{\backslash}}z^{\backslash}\left(\widehat{p}^{\backslash},e\right)\ \text{is full, i.e., } e \text{ is Debreu-regular.}$$

2. e is a Debreu-regular economy $\Rightarrow$ e is a Balasko-regular economy.

Identify the notation of Corollary 43 in Chapter 4 with the notation presented in the table.

Corollary 43 in Chapter 4	
X	$\mathbb{R}^{C-1}_{++}$
Z	$\mathbb{R}^{CH}_{++}$
x	$p^{\backslash}$
z	e
f	$z^{\backslash}$
pr	pr^{r}
y	0

Observe that the assumptions of the corollary are satisfied:

1. 0 is a regular value for $z^{\backslash}$ from Lemma 63;

2. 0 is a regular value for $z_e^{\backslash}$ from the definition of e a Debreu-regular economy.

Then from the conclusion of Corollary 43, e is a regular value for pr^{r}, i.e., the desired result. ■

8.1.1 Equilibria in terms of aggregate excess demand function

In this case we want to apply the same existence theorem we used in the extended approach. We can identify the notation of Theorem 57 in Chapter 7 with that one used in the present reduced approach as described in the following table.

Theorem 57 in Chapter 7	reduced approach
M	$\mathbb{R}^{C-1}_{++}$
N	$\mathbb{R}^{C-1}$
f	$z^{\backslash}_{e}$
g	$z^{\backslash}_{e^*}$
H	h

where $e \in \mathbb{R}^{CH}_{++}$ is an arbitrary endowment and x^* is a Pareto optimal allocation feasible at e, and

$$h : \mathbb{R}^{C-1}_{++} \times [0,1] \to \mathbb{R}^{C}, \quad \left(p^{\backslash}, \tau\right) \mapsto z^{\backslash}_{(1-\tau)e+\tau x^*}\left(p^{\backslash}\right)$$

where we recall that $z^{\backslash}_{(1-\tau)e+\tau x^*}\left(p^{\backslash}\right) \equiv z^{\backslash}\left(p^{\backslash}, (1-\tau)\, e + \tau x^*\right)$. While Assumptions 1 and 2 of Theorem 57 in Chapter 7 clearly hold, the lemma below proves the remaining assumptions.

Lemma 66 *The following statements hold.*

1 $\left(z^{\backslash}_{x^*}\right)^{-1}(0) = \left\{p^{\backslash *}\right\}$, *where* $p^* \equiv \dfrac{Du_1(x_1^*)}{D_{x_1^C} u_1(x_1^*)}$;

2 *0 is a regular value for* $z^{\backslash}_{x^*}$;

3 $h^{-1}(0)$ *is compact.*

Proof. 1 follows from Lemma 60 and Lemma 46.
2 follows from Proposition 65 and Lemma 47.
3 can be proved using the same argument used in Lemma 48. ∎

Theorem 67 *For each* $e \in \mathbb{R}^{CH}_{++}$, $\exists p^{\backslash} \in \mathbb{R}^{C-1}_{++}$ *such that* $\left(z^{\backslash}_{e}\right)^{-1}(0) \neq \emptyset$.

Proof. It follows from Theorem 57 in Chapter 7 and the above lemma. ∎

Theorem 68 *If* $e \in \mathcal{R}^{D}$, *then the following statements hold.*

1 there exists $r \in \mathbb{N}$ *such that*

$$z_e^{-1}(0) = \left\{p^{\backslash i}\right\}_{i=1}^{r}$$

2 *for each i, there exists an open set $Y \subseteq \mathbb{R}^{CH}_{++}$, an open set $U_i \subseteq \mathbb{R}^{C-1}_{++}$ and a unique function $g_i : Y \rightarrow V_i$ such that*

(a) $\left(p^{\backslash i}, e\right) \in V_i \times Y$,

(b) g_i *is* C^1,

(c) $g_i\,(e) = p^{\backslash i}$,

(d) *for every* $e' \in Y$, $z^{\backslash}\left(g_i\,(e'), e'\right) = 0$.

Proof. The proof is strictly analogous to that one of Theorem 55. ∎

8.1.2 Equilibria in terms of projection

To show existence in the reduced approach framework, Theorem 59 in Chapter 7 has been used in the literature. We can identify the notation of that theorem with that one used in our reduced approach as described in the following table

Theorem 59	reduced approach
M	E
N	$\mathbb{R}^{CH}_{++}$
g	pr^{r}
y	$e^* \in \mathcal{PO}$

(8.25)

Theorem 59 requires the domain E of the function pr^r to be a C^2 manifold. Since, from Definition 59, $E = \left(z^{\backslash}\right)^{-1}(0)$, we need $z^{\backslash}$ to be a C^2 function. Therefore, to apply Theorem 59, the way we stated and proved it, we have to assume that, for each h, u_h is C^r with $r \geq 3$. In fact, that theorem could be weakened assuming that both domain and value set of the function g are C^1 manifold - see [86], Section J.2, pp.46-47. Moreover, $\mathbb{R}^{CH}_{++}$ is clearly a connected manifold.

Assumptions 1 to 4 of Theorem 59 are clearly satisfied.

The lemma below proves the remaining assumptions.

Lemma 69 *The following statements hold.*

1 $pr^{-1}\left(x^*\right) = \left\{p^{\backslash *}, x^*\right\}$, *where* $p^* \equiv \dfrac{Du_1\left(x_1^*\right)}{D_{x_1^C} u_1\left(x_1^*\right)}$;

2 x^* *is a regular value for* pr;

3 pr *is proper*.

Proof. The proof is word by word the proof of Lemma 66. ∎

Theorem 70 *For each* $e \in \mathbb{R}^{CH}_{++}$, $\exists p^{\backslash} \in \mathbb{R}^{C-1}_{++}$ *such that* $pr^{-1}\,(e) \neq \emptyset$

Proof. It follows from Theorem 59 in Chapter 7 and the above lemma. ∎

The regularity theorem for the projection is nothing but Theorem 56.

8.2 On the relationship between reduced and extended approach

As we tried to explain in the previous subsection, a reduced approach to general equilibrium model describes equilibria in terms of fewer equations than first order conditions to individuals maximization problems and market clearing conditions. As relevant examples, we discussed the cases in which equilibria are characterized by the aggregate excess demand function or the projection function - usually called the *natural* projection - from the equilibrium set to the economy space.

In those frameworks, the first step is to show that the individual maximization problem has a unique solution and the resulting demand function is C^1.

To show that the household's problem does admit a solution, the standard strategy requires to show that

a) the constraint set is compact.

Then, uniqueness follows from strict quasiconcavity of the objective function. Showing that the demand function is C^1, as an application of the implicit function theorem, requires to show that

b) the partial jacobian matrix of
(the left hand side of the first order conditions of the household's problem)
with respect to (choice variables and multipliers) has full row rank.

To be able to apply differential techniques, since the price-endowment equilibrium set is the domain of the natural projection, it is indispensable to show that

c) the equilibrium set is a manifold.

Moreover, to talk of regular values of the natural projection it is necessary to show that

d) the differential of the natural projection,
a function *defined on a manifold*, is onto.

In the extended approach, the proof of a) is not needed. Existence of a solution to the consumer problem is provided showing existence of equilibria. b) is implicitly used in the proof of "regularity" of an economy coinciding with a Pareto optimal allocation - see Lemma 47. c) is not needed to show existence of equilibria. The rank computation involved to show c) is used only to show properties of regular economies.

Since the domain of the "extended" function F defining equilibria is an open subset of an Euclidean space, computations related to regularity are done in terms of standard derivatives, without involving the more complicated computations mentioned in d). As a result, in the extended approach rank computations are easier than those done in a reduced approach, as the reader may probably see comparing our proofs with those, say, in [9] and in [10].

In the extended approach, the natural "extended" projection - see (8.19) - is a useful tool for its *topological properties* (properness), not for its *differential* nature.

Moreover, in the reduced approach, if a global diffeomorphism approach is not followed, a quite frequent use of first order conditions is needed, implying anyway a use of the extended approach in a hidden way.

Another difference between the extended and the reduced approaches lies in the different existence theorems used by the different approaches. For example, [9] uses a degree theorem based on the concept of *degree of a function.* In the extended approach, we use Theorem 57 in Chapter 7, which does not mention explicitly the word degree, and in fact requires the simpler concept of *degree of a function at a point.* Observe that, since the extended equilibrium function F is not proper, we cannot apply Theorem 59 in Chapter 7[32].

Moreover, a simpler aspect of Theorem 57 in Chapter 7 with respect to Theorem 59 is that to apply the latter we do not need to show that the value set of the analyzed function is connected.

9. Appendices

9.1 Relationship between Assumption u4 and u4'

Lemma 71 *If Assumption u2 and u4' hold, then Assumption u4 holds.*

Proof. Define

$$Cl_{\mathbb{R}^C}\left\{x \in \mathbb{R}^C_{++} : u_h\left(x\right) = \underline{u}\right\} \equiv S^0$$

and

$$Cl_{\mathbb{R}^C}\left\{x \in \mathbb{R}^C_{++} : u_h\left(x\right) \geq \underline{u}\right\} \equiv S^+$$

Suppose otherwise, i.e., there exists a sequence $(x^n)_{n\in\mathbb{N}}$ in S^+ such that $x^n \to \overline{x} \in \delta\left(\mathbb{R}^C_{++}\right)$. Up to reordering of the components, define $x^n \equiv (x^n_1, x^n_2) \in \mathbb{R}^{C_1} \times \mathbb{R}^{C_2} = \mathbb{R}^C$, such that $x^n_1 \to 0$. Take $e \equiv (e_1, e_2) \in S^0 \subseteq S^+$. From Assumption e1, $e \gg 0$, therefore there exists $\overline{n}$ such that if $n > \overline{n}$, we have $x^n_1 \ll e_1$.

For any $n > \overline{n}$, we want to show that

$$\text{there exists } x^{*n}_2 \text{ such that } 0 \leq x^{*n}_2 \leq x^n_2 \text{ for which } u\left(x^n_1, x^{*n}_2\right) = u\left(e_1, e_2\right)$$

Suppose there is no such x^{*n}_2. Then it must be the case that for any x_2 such that $0 \leq x_2 \leq x^n_2$, we have

$$u\left(x^n_1, x_2\right) > u\left(e_1, e_2\right) \tag{8.26}$$

[32]Observe that the above theorem deals with abstract manifolds, but applies to our case as well, since the concept of abstract manifold is more general than that one of manifolds in euclidean spaces - see Chapter 2 and especially Appendix 2.

(It cannot be $u\left(x_1^n, x_2\right) < u\left(e_1, e_2\right)$, because $u\left(x_1^n, x_2^n\right) \geq u\left(e_1, e_2\right)$). But then, $\lim_{m \to +\infty}\left(x_1^n, \frac{x_2}{m}\right) = \left(x_1^n, 0\right) \in \delta\left(\mathbb{R}_{++}^C\right)$, with $x_1^n \ll e_1$, and for sufficiently large m, $\frac{x_2^n}{m} \ll e_2$. Therefore, for sufficiently large m,

$$\left(x_1^n, \frac{x_2^n}{m}\right) \ll e \equiv \left(e_1, e_2\right) \tag{8.27}$$

and (8.26) and (8.27) contradict the monotonicity of $u(x)$, which is implied by Assumption u2.

Since $x_2^n \to \overline{x}_2$, for any $\varepsilon > 0$, for sufficiently large n, $0 \leq x_2^{*n} \leq x_2^n \leq \overline{x}_2 + \varepsilon \mathbf{1}$, with $\mathbf{1} \in \mathbb{R}^{C_2}$. Therefore, up to a subsequence, $x_2^{*n} \to \overline{\overline{x}}_2$. Therefore, for any n,

$$u\left(x_1^n, x_2^{*n}\right) = u\left(e_1, e_2\right) \tag{8.28}$$

$$\left(x_1^n, x_2^{*n}\right) \to \left(0, \overline{\overline{x}}_2\right) \in \delta\left(\mathbb{R}_{++}^C\right) \tag{8.29}$$

and (8.28) and (8.29) contradict Assumption u4. ■

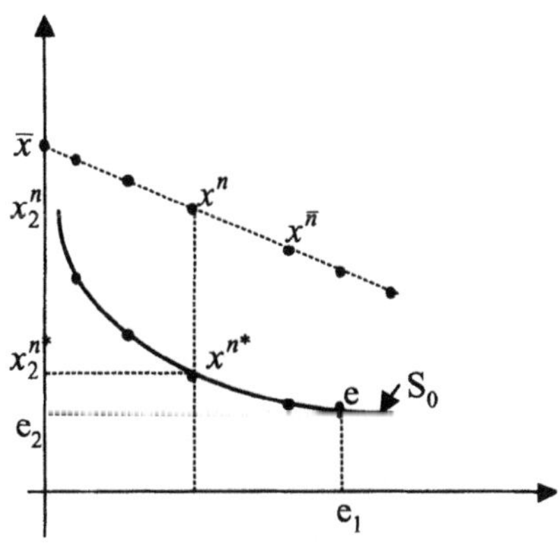

Figure 8.10. u2 and u4 imply u4'

9.2 Equivalent maximization problems

We split the proof into some lemmas.

Lemma 72 *Consider two subsets B and B_1 of $\mathbb{R}^C$ and the problems*

$$(P) \qquad \max_{x \in \mathbb{R}^C} \quad u(x) \quad s.t. \quad x \in B$$

and

$$(P1) \qquad \max_{x \in \mathbb{R}^C} \quad u(x) \quad s.t. \quad x \in B_1$$

Define $X^* \equiv \arg\max(P)$ *and* $X_1^* \equiv \arg\max(P1)$. *If* $B \subseteq B_1$, $X_1^* \subseteq B$, $X_1^* \neq \emptyset$, *then* $X_1^* = X^*$.

Proof. 1. $X_1^* \subseteq X^*$. Take $x_1^* \in X_1^* \neq \emptyset$. Then, by assumption, $x_1^* \in B$. For any $x \in B \subseteq B_1, u(x_1^*) \geq u(x)$, i.e., $x_1^* \in X^*$.

2. $X^* \subseteq X_1^*$.

From part 1 above and since, by Assumption $X_1^* \neq \emptyset$, we can take $x^* \in X^*$. Then $x^* \in B \subseteq B_1$. Since $X_1^* \subseteq B$, for any $x_1^* \in X_1^*$, $u(x^*) \geq u(x_1^*)$. By definition of X_1^*, for any $x_1 \in B_1$, $u(x_1^*) \geq u(x_1)$. From the two above inequalities, for any $x_1 \in B_1$, $u(x^*) \geq u(x_1)$, i.e., $x^* \in X_1^*$. ■

Remark 73 Observe that we can have $X_1^* = \emptyset$ and $X^* \neq \emptyset$, as shown in Figure 8.11

Define B_h and B_{h1} as the constraint sets for problems (Ph) and $(Ph1)$, respectively. Define X_h^* and X_{h1}^* as the solution sets for problems (Ph) and $(Ph1)$, respectively.

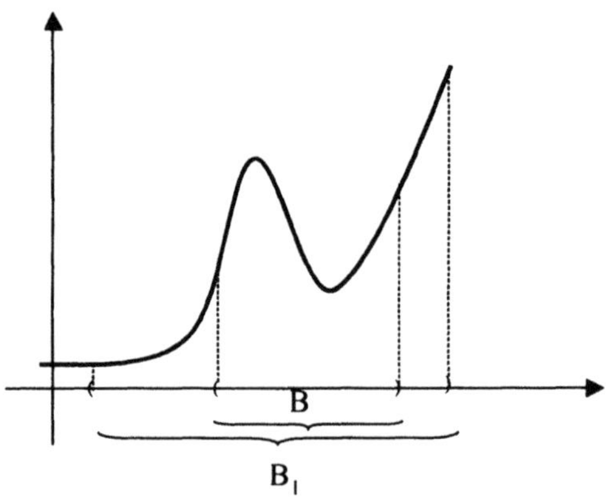

Figure 8.11. $X_1^* = \emptyset$ and $X^* \neq \emptyset$

Lemma 74 *If* $X_{h1}^* \neq \emptyset$, *then* $X_{h1}^* = X_h^*$.

Proof. To apply Lemma 72, we have to show that $B_h \subseteq B_{h1}$, which is obvious, and $X_{h1}^* \subseteq B_h$. Suppose otherwise, i.e., there exists $x_h \in X_{h1}^*$, and therefore $x_h \in B_{h1}$, but $x_h \notin B_h$ so that $px_h < pe_h$. Take $\varepsilon = pe_h - px_h > 0$, and $x_h^* = x_h + (0, ..., 0, \varepsilon)$. (Recall that we set $p^C = 1$.) Then,
i. $px_h^* \leq pe_h$ (in fact, $px_h^* = pe_h$) and therefore, $x_h^* \in B_{h1}$.
ii. $u_h(x_h^*) > u_h(x_h)$, because of Assumption u2.
But i. and ii. contradict the fact that $x_h \in X_{h1}$. ■

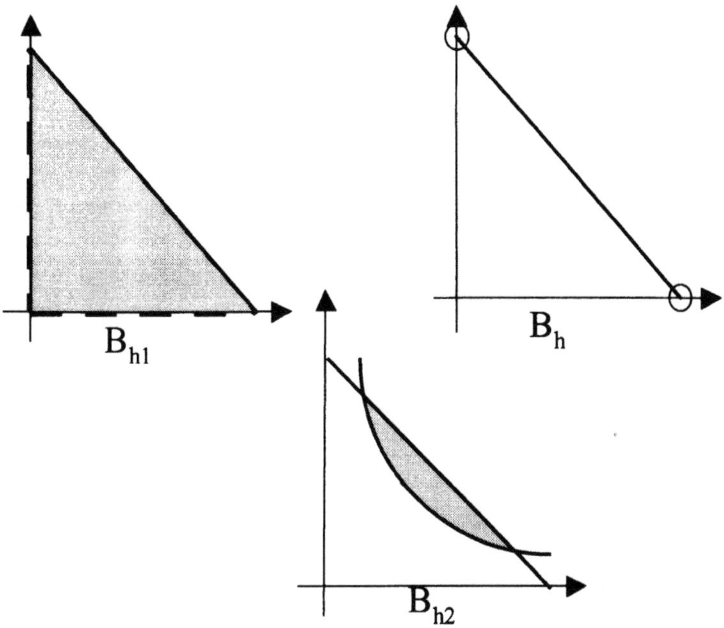

Figure 8.12. Household's budget sets

Remark 75 Clearly the constraint sets of both problems above are not compact. To be able to apply the extreme value theorem ([90] Theorem 6.4, p.175), we need to show that they have the same solution as another problem whose feasible set is in fact compact.

Consider

$$(Ph2) \quad \max_{x_h \in \mathbb{R}^C_{++}} u_h(x_h)$$
$$s.t. \quad p(x_h - e_h) \leq 0$$
$$u_h(x_h) \geq u_h(e_h)$$

Define B_{h2} as the constraint set for $(Ph2)$ and $X^*_{h2} \equiv \arg\max(Ph2)$.

To study the relationship among X^*_h, X^*_{h1} and X^*_{h2}, we need the following lemmas.

Lemma 76 *If $M, R \subseteq \mathbb{R}^C$, M is closed in R and $Cl_{\mathbb{R}^C} M \subseteq R$, then M is closed in $\mathbb{R}^C$.*

Proof. It follows from [90], Theorem 6.4 p. 95. ■

Lemma 77 *The following statements hold.*

1 $X^*_{h2} \neq \emptyset$;

2 $X_{h2}^{*} = X_{h1}^{*}$;

3 $X_{h}^{*} \neq \emptyset$.

Proof. For ease of notation, drop the subscript h.

1. Since u is continuous, as an application of the extreme value theorem ([90] Theorem 6.4, p.175), we have to show that B_2 is compact. B_2 is the intersection of the two sets
$\{x \in \mathbb{R}^C : \widetilde{p}x \leq \widetilde{p}e\}$, which is clearly closed in $\mathbb{R}^C$, and
$\{x \in \mathbb{R}^C_{++} : u(x) \geq u(e)\} \equiv M_e$.
Since $u(x)$ is a continuous function, M_e is closed in $\mathbb{R}^C_{++}$, and from Assumption $u4$, $Cl\, M_e \subseteq \mathbb{R}^C_{++}$. Therefore, from Lemma 76, M_e is closed in $\mathbb{R}^C$. Therefore, B_2 is the intersection of the two closed sets in $\mathbb{R}^C$ and therefore it is closed in $\mathbb{R}^C$.
We are now left with showing that B_2 is bounded.
It is bounded from below by 0. It is bounded from above because of the following argument. Take an arbitrary $x \in B_2$, so that $px \leq pe$. Then, since $p \in \mathbb{R}^C_{++}$, for $c = 1, ..., C$, $x^c \leq \left(pe - \sum_{c' \neq c} p^{c'} x^{c'}\right) / p^c$. Since e is given and x is bounded from below, x^c is bounded from above.

2. To show this results, we proceed showing that a. $X_{h2}^{*} \subseteq X_1^{*}$; b. $X_{h2}^{*} \supseteq X_1^{*}$.

a. $X_2^{*} \subseteq X_1^{*}$.
Take $x_2^{*} \in X_2^{*} \subseteq B_2$. Observe that for any $x \in B_2$ and $x' \in B_1 \backslash B_2$, it is the case that

$$u(x) > u(x') \tag{8.30}$$

Therefore
i. $x_2^{*} \in B_2 \subseteq B_1$, and
ii. for any $x \in B_1$,
if $x \in B_2$, $\quad u(x_2^{*}) > u(x)$, from the fact that $x_2^{*} \in X_2^{*}$,
if $x \in B_1 \backslash B_2$, $\quad u(x_2^{*}) > u(x)$, from (8.30), and the fact that $x_2^{*} \in B_2$

b. $X_2^{*} \supseteq X_1^{*}$.
Take $x_1^{*} \in X_1^{*}$.
i. Suppose that $x_1^{*} \notin B_2$, i.e., $x_1^{*} \in B_1 \backslash B_2$, but then $u(x_1^{*}) < u(e)$, and $e \in B_1$, contradicting the fact that $x_1^{*} \in X_1^{*}$.
ii. since $x_1^{*} \in X_1^{*}$, for any $x \in B_2 \subseteq B_1$, $u(x_1^{*}) > u(x)$.

3. follows immediately from Lemma 74 and Statements 1 and 2. ∎

Chapter 9

PRODUCTION ECONOMIES

In this chapter, we introduce the possibility of production in the general equilibrium framework: another class of agents, firms [1], enters the model. We first briefly discuss the concept of firm and we introduce the definition of equilibrium in a production economy. Then, following exactly the same strategy used in the previous chapter, we show existence of equilibria and properties of regular economies.

1. Set up of the model

Unless otherwise stated, we maintain all the Assumptions made in Chapter 8.

Assumption f1. There is a finite number F of firms.

Firms are denoted by subscript f; each of them uses goods, called inputs, to produce other goods, called outputs.

Definition 1 Any possible production activity of firm f is described by a vector $y_f \in \mathbb{R}^C$, called production vector . Input components of the vector y_f are negative and output components are positive. Moreover, $y \equiv (y_f)_{f=1}^F$.

For example a vector $y_f = (-4, 0, 0, -2, 3)$ indicates that firm f needs 4 units of good 1 and 2 units of good 4 to produce 3 units of good 5. Given the above sign convention, we can introduce the following definition.

Definition 2 For $f = 1, ..., F$, firm f's profit is py_f.

[1] We use firms and producers as having the same meaning.

Definition 3 The set of all feasible production vectors for firm f is called the **production set** $Y_f \subseteq \mathbb{R}^C$. If $y_f \in Y_f$, then y_f can be obtained as a result of the production process by firm f; if $y_f \notin Y_f$, that is not the case. Moreover, $Y \equiv \times_{f=1}^{F} Y_f$.

In our differentiable framework, it is very often convenient to describe the production set using the function presented below.

Definition 4 The function

$$t_f : \mathbb{R}^C \to \mathbb{R}, \quad t_f : y_f \mapsto t_f(y_f)$$

such that

$$Y_f = \{y_f \in \mathbb{R}^C : t_f(y_f) \geq 0\}$$

is called the transformation function of firm f. Moreover, $t \equiv (t_f)_{f=1}^{F}$ is the vector of transformation function of all firms. We make the following

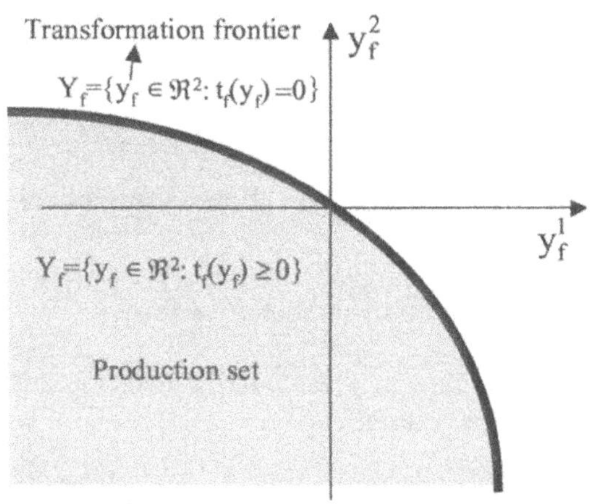

Figure 9.1. Production set

assumptions on the transformation function.

Assumption f2. For $f = 1, ..., F$, t_f

1 is C^2;

2 allows inactivity, i.e., $t_f(0) \geq 0$;

3 is differentiably strictly decreasing, i.e., for all $y_f \in \mathbb{R}^C$, $Dt_f(y_f) \ll 0$;

4 is differentiably strictly concave, i.e., for all $y_f \in \mathbb{R}^C$, $D^2 t_f(y_f)$ is negative definite ;

5 satisfies the no free lunch property, i.e., if $t_f(y_f) \geq 0$ and $y_f \geq 0$, then $y_f = 0$.

Definition 5 The set of all transformation functions satisfying Assumptions f2 is denoted by $\widehat{\mathcal{T}}$. Moreover $\mathcal{T} \equiv \widehat{\mathcal{T}}^F$ with generic element $t \equiv (t_f)_{f=1}^F$.

Assumption f2.1 allows the use of differential techniques and implies that $Y_f = t_f^{-1}([0, +\infty))$ is a closed set. Assumption f2.2 simply means that a firm can be inactive. Assumption f2.3 implies that t_f is strictly decreasing and therefore that the so called "free disposal" property holds, i.e.,

$$y'_f \leq y_f \text{ and } t_f(y_f) \geq 0 \Rightarrow t_f(y'_f) \geq 0$$

The above condition can then be interpreted as the possibility of disposing of goods at "no costs". For example, since $t_f(0) \geq 0$, then $t_f(0, ..., 0, -1) \geq 0$, i.e., it is possible to destroy one unit of good C without using any unit of other goods. Assumption f2.4 implies concavity of t_f and therefore convexity of Y_f (see Proposition 21 and Table 1.6 in Chapter 1), which basically amounts to Nonincreasing returns to scale, as shown in Proposition 6 below. Finally, Assumption f2.5 simply says that firms cannot produce something from nothing.

Proposition 6 *Given Assumption f2, the following properties hold:*

1 (Nonincreasing Returns to Scale) $\forall y_f \in Y_f$ *and* $\forall \alpha \in [0, 1]$, $\alpha y_f \in Y_f$;

2 $t_f(0) = 0$;

3 (Irreversibility) if $y_f \in Y_f$ *and* $-y_f \in Y_f$, *then* $y_f = 0$.

Proof. 1. From Assumption f2.2, $0 \in Y_f$; then, from Assumption f2.4, $\forall \alpha \in [0, 1]$, $(1 - \alpha) 0 + \alpha y_f = \alpha y_f \in Y_f$.

2. If $t_f(0) > 0$, then from the continuity of t_f, Assumption f2.5 (no free lunch) would be violated.

3. Suppose otherwise. Then, by the strict concavity of t_f, we have

$$0 = t_f(0) = t_f\left(\frac{1}{2} y_f + \frac{1}{2}(-y_f)\right) > \frac{1}{2} t_f(y_f) + \frac{1}{2} t_f(-y_f) \geq 0$$

a contradiction. ■

To show existence of equilibria, we also need to impose some properties on the set $\sum_{f=1}^F Y_f$, which will be used in the crucial Lemma 11.

Definition 7 $\sum_{f=1}^F Y_f$ is called aggregate production set.

It describes the production vectors which are feasible if all production set are used together. We denote by w a generic element of $\sum_{f=1}^F Y_f$.

Unfortunately, assumptions on each Y_f are not necessarily shared by the aggregate production set, as the following proposition shows.

Proposition 8 *1. $\sum_f Y_f$ satisfies no free lunch $\overset{\not\Leftarrow}{\Rightarrow}$ for each f, Y_f satisfies no free lunch .*

2. $\sum_f Y_f$ satisfies irreversibility $\overset{\not\Leftarrow}{\Rightarrow}$ for each f, Y_f satisfies irreversibility.

Proof. 1.
[$\Rightarrow$] For each f, take $y_{f'} = 0$ for $f' \neq f$.
[$\not\Leftarrow$] Consider the case of $F = C = 2$, and

$$t_1\left(y_1^1, y_1^2\right) = 2 - e^{6y_1^1} - e^{y_1^2}, \qquad t_2\left(y_2^1, y_2^2\right) = 2 - e^{y_2^1} - e^{6y_2^2}$$

Observe that the above transformation functions satisfy Assumptions f2. Moreover,

$$t_1\left(-\frac{1}{6}, \ln\left(2 - e^{-1}\right)\right) = 0, \quad t_2\left(\ln\left(2 - e^{-1}\right), -\frac{1}{6}\right) \text{ and } -\frac{1}{6} + \ln\left(2 - e^{-1}\right) > 0$$

2. If follows from the same argument given for 1 above. ∎

Because of the above negative results, we assume the following properties on $\sum_{f=1}^F y_f$, which are going to be used in Lemma 11 and Proposition 10 below, crucial ingredients of the existence theorem - see Lemmas 26 and 34.

Assumption f3.

1 (no free lunch for aggregate production set) if $w \in \sum_{f=1}^F Y_f$ and $w \geq 0$, then $w = 0$;

2 (irreversibility for aggregate production set) if $w \in \sum_{f=1}^F Y_f$ and $-w \in \sum_{f=1}^F Y_f$, then $w = 0$.

Definition 9 The set of feasible consumption-production allocations for given total resources $r \in \mathbb{R}_{++}^C$ is

$$A_r \equiv \left\{ (x, y) \in \mathbb{R}_{++}^{CH} \times Y : \sum_{h=1}^H x_h - \sum_{f=1}^F y_f \leq r \right\}$$

Proposition 10 *For each $r \in \mathbb{R}_{++}^C$, the set A_r of feasible allocations is bounded.*

Proof. Since $\sum_{h=1}^H x_h \gg 0$, if $(x, y) \in A_r$, then

$$y \in \widehat{Y}^r \equiv \left\{ y' \in Y : \sum_{f=1}^F y_f + r \gg 0 \right\}$$

Moreover, for each $h' \in \mathcal{H}$, if $(x, y) \in A_r$, then

$$0 \ll x_{h'} \ll \sum_{h=1}^{H} x_h \leq \sum_{f=1}^{F} y_f + r$$

Therefore, to show the desired result it is enough to prove that $\widehat{Y}_r$ is bounded, which is done in the lemma below. ■

Lemma 11 *$\widehat{Y}_r$ is bounded.*

Proof. Below we denote $\widehat{Y}_r$ simply by $\widehat{Y}$. Suppose $\widehat{Y}$ is not bounded. Then, we can find a sequence $(y^n)_{n\in\mathbb{N}} \equiv \left(\left(y_f^n\right)_{f=1}^{F}\right)_{n\in\mathbb{N}}$ in $\mathbb{R}^{CF}$ such that $\|y^n\| \to +\infty$. We can partition $\{1, ..., F\}$ into two subset $\mathcal{F}^\infty \neq \emptyset$ and $\mathcal{F}^0$ such that, up to a subsequence, if $f \in \mathcal{F}^\infty$, then $\left\|y_f^n\right\| \to +\infty$, and if $f \in \mathcal{F}^0$, then $\left\|y_f^n\right\| \to k_f \in \mathbb{R}_+$. Then up to a further subsequence for each $f \in \mathcal{F}^\infty$,

$$\frac{y_f^n}{\left\|y_f^n\right\|} \to \overline{y}_f$$

Since $0 \in Y_f$, and Y_f is convex, for sufficiently large n and each $f \in \mathcal{F}^\infty$,

$$\left(1 - \frac{1}{\left\|y_f^n\right\|}\right) 0 + \frac{1}{\left\|y_f^n\right\|} y_f^n = \frac{y_f^n}{\left\|y_f^n\right\|} \in Y_f$$

Since, from Assumption f2.1, Y_f is closed,

$$\overline{y}_f \in Y_f \text{ and } \left\|\overline{y}_f\right\| = 1 \tag{9.1}$$

From the definition of $\widehat{Y}$,

$$\sum_{f' \in \mathcal{F}^\infty} y_{f'}^n + \sum_{f'' \in \mathcal{F}^0} y_{f''}^n + r \gg 0$$

$$\sum_{f' \in \mathcal{F}^\infty} \frac{y_{f'}^n}{\sum_{f \in \mathcal{F}^\infty} \left\|y_f^n\right\|} + \sum_{f'' \in \mathcal{F}^0} \frac{y_{f''}^n}{\sum_{f \in \mathcal{F}^\infty} \left\|y_f^n\right\|} + \frac{r}{\sum_{f \in \mathcal{F}^\infty} \left\|y_f^n\right\|} \gg 0$$

and

$$\sum_{f' \in \mathcal{F}^\infty} \frac{\left\|y_{f'}^n\right\|}{\sum_{f \in \mathcal{F}^\infty} \left\|y_f^n\right\|} \frac{y_{f'}^n}{\left\|y_{f'}^n\right\|} + \sum_{f'' \in \mathcal{F}^0} \frac{y_{f''}^n}{\sum_{f \in \mathcal{F}^\infty} \left\|y_f^n\right\|} + \frac{r}{\sum_{f \in \mathcal{F}^\infty} \left\|y_f^n\right\|} \gg 0 \tag{9.2}$$

Then, since for each n and each $f' \in \mathcal{F}^{\infty}$,

$$0 \leq \frac{\left\| y_{f'}^{n} \right\|}{\sum_{f \in \mathcal{F}^{\infty}} \left\| y_{f}^{n} \right\|} \leq 1$$

up to a subsequence,

$$\lambda_{f'}^{n} \equiv \frac{\left\| y_{f'}^{n} \right\|}{\sum_{f \in \mathcal{F}^{\infty}} \left\| y_{f}^{n} \right\|} \rightarrow \lambda_{f'} \in [0,1] \tag{9.3}$$

Moreover,

$$\sum_{f' \in \mathcal{F}^{\infty}} \lambda_{f'}^{n} = \sum_{f' \in \mathcal{F}^{\infty}} \frac{\left\| y_{f'}^{n} \right\|}{\sum_{f \in \mathcal{F}^{\infty}} \left\| y_{f}^{n} \right\|} = 1$$

and

$$\sum_{f' \in \mathcal{F}^{\infty}} \lambda_{f'}^{n} \rightarrow \sum_{f' \in \mathcal{F}^{\infty}} \lambda_{f'} = 1 \tag{9.4}$$

Therefore, $\exists\, f^{*} \in \mathcal{F}^{\infty}$ such that

$$\lambda_{f^{*}} > 0 \tag{9.5}$$

Taking limits of both sides of (9.2), we get

$$\sum_{f \in \mathcal{F}^{\infty}} \lambda_{f} \overline{y}_{f} \geq 0$$

Now, we can distinguish two cases. Case 1. $\sum_{f \in \mathcal{F}^{\infty}} \lambda_{f} \overline{y}_{f} = 0$, and Case 2. $\sum_{f \in \mathcal{F}^{\infty}} \lambda_{f} \overline{y}_{f} > 0$. In both cases, we get a contradiction.
Case 1. If $\sum_{f \in \mathcal{F}^{\infty}} \lambda_{f} \overline{y}_{f} = 0$, from (9.1) we have

$$\lambda_{f^{*}} \overline{y}_{f^{*}} = - \sum_{f \in \mathcal{F}^{\infty} \setminus \{f^{*}\}} \lambda_{f} \overline{y}_{f} \in \sum_{f=1}^{F} Y_{f}$$

But also

$$\sum_{f \in \mathcal{F}^{\infty} \setminus \{f^{*}\}} \lambda_{f} \overline{y}_{f} \in \sum_{f=1}^{F} Y_{f}$$

In both statements above, we used the facts that $0 \in Y_f$ and $\lambda_f \overline{y}_f \in Y_f$.
But then, from Assumption f3.1 (irreversibility)

$$\lambda_{f^{*}} \overline{y}_{f^{*}} = - \sum_{f \in \mathcal{F}^{\infty} \setminus \{f^{*}\}} \lambda_{f} \overline{y}_{f} = 0$$

contradicting the fact that $\lambda_{f*} > 0$ and $\left\|\overline{y}_{f*}\right\| = 1$ - see (9.5) and (9.1).
Case 2. From (9.1) and (9.3), the convexity of Y_f and the fact that $0 \in Y_f$ for each f,

$$\sum_{f \in \mathcal{F}^{\infty}} \lambda_f \overline{y}_f \in \sum_{f=1}^{F} Y_f \tag{9.6}$$

Assumption of Case 2 and (9.6) contradict Assumption f3.2 (no free lunch for $\sum_{f=1}^{F} Y_f$).

Firms are owned by households, as formalized in the following definition.

■

Definition 12 s_{fh} is the share of firm f owned by household h. $s_f \equiv (s_{fh})_{h=1}^{H} \in \mathbb{R}^H$ and $s \equiv (s_f)_{f=1}^{f} \in \mathbb{R}^{FH}$.

The set of all shares of each firm f is

$$S \equiv \left\{ s_f \in [0,1]^H : \sum_{h=1}^{H} s_{fh} = 1 \right\}$$

The set of shares $s_h \equiv (s_{fh})_{f=1}^{F}$ of household h is $[0,1]^F$.

$s_{fh} \in [0,1]$ denotes the proportion of firm f 's profit owned by household h. The definition of S simply requires each firm to be completely owned by some households.

As for the households case, we assume that firms take prices.

Assumption f4. Firms are price takers.

On the basis of the above assumption, since firms are owned by households and higher profits increase households' wealth, we make the following assumption[2].

Assumption f5. Firms maximize profits.

As we will discuss in Remark 23 in Chapter 10, the above assumption becomes quite meaningless if we remove Assumption c3 and profits become uncertain.

On the base of the Assumptions f1-f5, each firm f chooses a *feasible* production vector y_f, i.e., y_f such that $t_f(y_f) \geq 0$, to maximize its profits py_f.

Definition 13 Firm f's profit maximization problem is defined as follows. For given $p \in \mathbb{R}^C_{++}$,

$$\max_{y_f \in \mathbb{R}^C} \; py_f \qquad s.t. \qquad t_f(y_f) \geq 0$$

[2]For a discussion of Assunption f5, see also [87] Section 5G, pp. 152-154.

Summarizing, each firm f is completely characterized by its production technology, i.e., its transformation function t_f, and share vector $s_f \equiv (s_{fh})_{h=1}^{H}$. Therefore, an economy with production is completely characterized by an element $(e, s, u, t) \in \mathbb{R}_{++}^{CH} \times S^F \times \mathcal{U} \times \mathcal{T}$. In this chapter, we take u and t for given. Therefore, we can introduce the following definition.

Definition 14 An **economy with production** is an element $(e, s) \in \mathbb{R}_{++}^{CH} \times S^F$. We also define $\pi \equiv (e, s)$ and $\Pi \equiv \mathbb{R}_{++}^{CH} \times S^F$.

The introduction of firms in the model clearly changes households' maximization problems . Since, for every f, household h receives a proportion s_{fh} of the profit py_f, her wealth is given by the value of both her initial endowment and her share of firms' profit. Therefore, household h's problem is the following one. For given $p \in \mathbb{R}_{++}^{C}$, $e_h \in \mathbb{R}_{++}^{C}$, $s_h \in [0,1]^F$,

$$\max_{x_h \in \mathbb{R}_{++}^{C}} \quad u_h(x_h) \qquad s.t. \qquad px_h \leq pe_h + \textstyle\sum_{f=1}^{F} s_{fh}(py_f)$$

Remark 15 As in the chapter on exchange economies, we normalize the price of good C. Observe that both household and firm maximization problems are unaffected by that normalization . With little and innocuous abuse of notation, we still denote the normalized prices by p and we define $p^{\backslash} \equiv (p^c)_{c \neq C} \in \mathbb{R}_{++}^{C-1}$ and $p \equiv (p^{\backslash}, 1)$, as we did in Chapter 8.

Market clearing conditions require the aggregate consumption to be equal to the sum of the aggregate production and initial endowments. Formally

$$\sum_{h=1}^{H} x_h = \sum_{h=1}^{H} e_h + \sum_{f=1}^{F} y_f \tag{9.7}$$

Definition 16 A vector $(x, y, p^{\backslash}) \in \mathbb{R}_{++}^{CH} \times \mathbb{R}^{CF} \times \mathbb{R}_{++}^{C-1}$ is an allocation-price **equilibrium** for an economy $\pi \in \Pi$ if

1 households maximize, i.e., for $h = 1, ..., H$, x_h solves the following maximization problem

$$\max_{x_h \in \mathbb{R}_{++}^{C}} \quad u_h(x_h) \qquad s.t. \qquad px_h \leq pe_h + \textstyle\sum_{f=1}^{F} s_{fh}(py_f) \tag{9.8}$$

at $p^{\backslash} \in \mathbb{R}_{++}^{C-1}$, $e_h \in \mathbb{R}_{++}^{C}$, $s_h \in [0,1]^F$;

2 firms maximize, i.e., for $f = 1, ..., F$, y_f solves the following problem

$$\max_{y_f \in \mathbb{R}^{C}} \quad py_f \qquad s.t. \qquad t_f(y_f) \geq 0 \tag{9.9}$$

at $p^{\backslash} \in \mathbb{R}^{C-1}_{++}$;

3 markets clear, i.e., Condition (9.7) holds.

As in the previous chapter we want to characterize equilibria in terms of an "extended system". A crucial step in that direction is the following proposition.

Proposition 17 *1 If a solution to problem (9.9) exists, it is unique.*

2 y_f is the solution to problem (9.9) $\Leftrightarrow$ there exists $\alpha_f \in \mathbb{R}_{++}$ such that (y_f, α_f) is the solution to the following system (of first order conditions of problem (9.9))

$$\begin{cases} p + \alpha_f Dt_f(y_f) & = 0 \quad (1) \\ t_f(y_f) & = 0 \quad (2) \end{cases} \tag{9.10}$$

Proof. To simplify notation, drop the subscript f.

1. Suppose there exist y, y' which are solutions to problem (9.9) and such that $y \neq y'$. Consider $y^\lambda \equiv (1-\lambda) y + \lambda y'$. Since the transformation function t is strictly concave, it follows that

$$t\left(y^\lambda\right) > (1-\lambda) t(y) + \lambda t(y') \geq 0$$

Since t is continuous, in fact C^2, there exists $\varepsilon > 0$ such that

$$z \in B_\varepsilon\left(y^\lambda\right) \Rightarrow t(z) > 0$$

Take $z' = y^\lambda + \frac{\varepsilon \cdot 1}{\sqrt{2C}}$. Then,

$$d\left(y^\lambda, z'\right) \equiv \left(\sum_{c=1}^{C}\left(\frac{\varepsilon 1}{\sqrt{2C}}\right)^2\right)^{\frac{1}{2}} = \left(C\left(\frac{\varepsilon}{C}\right)^2\right)^{\frac{1}{2}} = \left(\frac{\varepsilon^2}{2C}\right)^{\frac{1}{2}} = \frac{\varepsilon}{\sqrt{2C}} < \varepsilon.$$

Therefore $z' \in B\left(y^\lambda, \varepsilon\right)$ and

$$t\left(z'\right) > 0 \tag{9.11}$$

But,

$$pz' = py^\lambda + p\frac{\varepsilon 1}{\sqrt{2C}} > py^\lambda = py' \tag{9.12}$$

and (9.11) and (9.12) contradict the fact that y' solves problem (9.9).

2. The result follows showing that problem (9.9) satisfies necessary and sufficient conditions for Lagrange theorem (see Subsection 3.2.1 in Chapter 1). The lagrangean function for problem (9.9) is

$$\mathcal{L}(y, \alpha, p) = py + \alpha t(y)$$

and the associated first order conditions are just system (9.10).

We can apply Kuhn-Tucker Theorem- see Theorems 38, 40 in Chapter 1.

[⇒] Necessary conditions are satisfied since $Dt(y) \ll 0$. Therefore, if y_f is the solution to problem (9.9), then there exists $\alpha \in \mathbb{R}$ such that (y, α) is the unique solution to system (9.10). The fact that $\alpha > 0$ follows from equation (1) in that system, Assumption f2.3 and the fact that $p \in \mathbb{R}^C_{++}$. In fact, $\alpha = -\frac{p^1}{D_{y^1}t(y)} > 0$.

[⇐] Sufficient conditions of Kuhn-Tucker theorem are satisfied because of the following facts: the objective function is linear and therefore pseudo-concave; t is differentiably strictly concave and therefore quasiconcave. Since the Lagrange multiplier α is strictly positive, the constraint is binding, i.e., it holds with an equality. ■

Remark 18 Even under our Assumptions f1-f5, problem (9.9) may very well have no solution, as figure 9.2 shows.

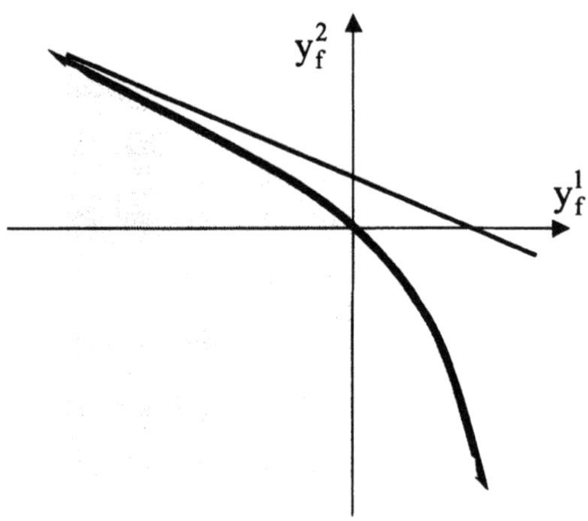

Figure 9.2. problem (9.9) has no solution

The graph of the boundary of the production set get closer to the straight line l as $y_1 \to -\infty$. If p is orthogonal to l, that line represents a level curve for the profit function and at those prices, the firm's problem has no solution.

We now introduce the system of first order conditions to both households' and firms' maximization problems and a *subset*[3] of conditions (9.7) which

[3] Proposition 20 explains why just a *subset* of equilibrium conditions is enough for our purposes.

characterizes equilibria .

# eq.	matching var.		eq.	
C	x_h	$(h.1)_{h\in\mathcal{H}}$	$Du_h(x_h) - \mu_h p$	$= 0$
1	μ_h	$(h.2)_{h\in\mathcal{H}}$	$-p\left(x_h - e_h - \sum_{f=1}^{F} s_{fh} y_f\right)$	$= 0$
C	y_f	$(f.1)_{f=1,\ldots,F}$	$p + \alpha_f Dt_f(y_f)$	$= 0$
1	α_f	$(f.2)_{f=1,\ldots,F}$	$t_f(y_f)$	$= 0$
$C-1$	$p^{\backslash}$	(M)	$-\sum_{h=1}^{H} x_h^{\backslash} + \sum_{h=1}^{H} e_h^{\backslash} + \sum_{f=1}^{F} y_f^{\backslash}$	$= 0$

(9.13)

where eq. and var stand for equations and variables respectively and we defined $y_f^{\backslash} \equiv \left(y_f^c\right)_{c\neq C}$ consistently with the definitions of $x_h^{\backslash}$ and $e_h^{\backslash}$, already presented in Chapter 8; we also wrote the "matching" variable, and their number, next to each equation.

Definition 19 Define

$$\xi \equiv \left(x, \mu, y, \alpha, p^{\backslash}\right) \in \Xi \equiv \mathbb{R}_{++}^{CH} \times \mathbb{R}_{++}^{H} \times \mathbb{R}^{CF} \times \mathbb{R}_{++}^{F} \times \mathbb{R}_{++}^{C-1}$$

$$F : \mathbb{R}^{\dim \Xi} \times \Pi \to \mathbb{R}^{\dim \Xi}, \quad F : (\xi, \pi) \mapsto \text{ left hand side of system (9.13)}$$

Function F is called the extended equilibrium function and ξ is the extended equilibrium vector. Observe that with innocuous abuse of notation, we are using the symbols ξ, Ξ and F with meaning which is different from that one of Chapter 8.

Proposition 20 $(x, y, p^{\backslash})$ *is an equilibrium allocation-price vector associated with an economy* π *if and only if there exists* (μ, α) *such that*

$$F\left(x, \mu, y, \alpha, p^{\backslash}, \pi\right) = 0$$

Proof. From Proposition 17, y_f is the solution to problem (9.9) if and only if there exists $\alpha_f \in \mathbb{R}_{++}$ solves equations $(f.1)$ and $(f.2)$ in (9.13).

With the same argument used in part 1 of Theorem 27 in Chapter 8, for $h \in \mathcal{H}$, x_h solves problem (9.8) at $p \in \mathbb{R}_{++}^{C}$, $e_h \in \mathbb{R}_{++}^{C}$, $s_h \in [0,1]^F$ if and only if there exists $\mu_h \in \mathbb{R}_{++}$ such that (x_h, μ_h) solves $(h.1)$ and $(h.2)$ in (9.13).

Finally, observe that summing up the consumers' budget constraints we get

$$p\left(\sum_{h=1}^{H} x_h - \sum_{h=1}^{H} e_h - \sum_{f=1}^{F} y_f\right) = 0$$

(i.e., Walras law). As in the case of exchange economies, using the fact that $p \gg 0$ and the above expression, we get the missing market clearing condition

$$\sum_{h=1}^{H} x_h^C - \sum_{h=1}^{H} e_h^C - \sum_{f=1}^{F} y_f^C = 0$$

Consistently with our extended approach, we then define an equilibrium vector as the vector of endogenous variables of system 9.13. ■

Definition 21 For each economy $\pi \in \Pi$, if $F(\xi, e) = 0$, we call $\xi \equiv (x, \mu, y, \alpha, p^{\backslash})$ an **equilibrium vector** for π.

2. Existence of equilibria

As in the previous chapter, we want to show that for each economy π there exists a solution to system $F(\xi, \pi) = 0$. To get that result, we use exactly the same strategy used in that chapter which is based on Theorem 57 in Chapter 7. In Subsection 2.1, we construct a so-called test economy, using a Pareto optimal allocation, whose definition is presented and discussed there. The test economy is a key step to construct the function G - which plays the role of function g in Theorem 57 in Chapter 7 - and the needed homotopy H_π in Subsection 2.2. Finally, we show the other needed results: $\{\xi^*\} = G^{-1}(0)$ - Lemmas 31 and 32; $DG(\xi^*)$ has full rank - Lemma 33; $H_\pi^{-1}(0)$ is compact - Lemma 34.

2.1 The test economy

Definition 22 An allocation in a production economy is an element $(x, y) \in \mathbb{R}_{++}^{CH} \times \mathbb{R}^{CF}$.

Definition 23 An allocation (x^*, y^*) is Pareto optimal in a production economy if there is no other allocation (x', y') such that

$$\begin{array}{l} \sum_{h=1}^{H} x_h' - \sum_{f=1}^{F} y_f' \leq \left(\sum_{h=1}^{H} x_h^* - \sum_{f=1}^{F} y_f^*\right) \quad \text{and} \\ \left(u_h\left(x_h'\right)\right)_{h=1}^{H} > \left(u_h\left(x_h^*\right)\right)_{h=1}^{H} \end{array}$$

Roughly speaking, (x^*, y^*) is Pareto optimal if it not possible to use the associated vector $\left(\sum_{h=1}^{H} x_h^* - \sum_{f=1}^{F} y_f^*\right)$ of "available" resources to obtain a new allocation which makes every households at least as well off and someone better off.

Given total resources $r \in \mathbb{R}_{++}^{C}$, define

$$\begin{array}{rl} \underline{U}^r \equiv & \{ \left(\underline{u}_h\right)_{h=1}^{H} \in \mathbb{R}^H : \text{there exists an allocation } (x, y) \text{ such that} \\ & \sum_{h=1}^{H} x_h - \sum_{f=1}^{F} y_f \leq r \text{ and } \left(u_h\left(x_h'\right)\right)_{h=1}^{H} = \left(u_h\left(x_h^*\right)\right)_{h=1}^{H} \} \end{array}$$

i.e., the set of utility level vectors attainable with resources r. Define also

$$\underline{U}^r_{\setminus 1} \equiv \left\{ (\underline{u}_h)_{h\neq 1} \in \mathbb{R}^{H-1} : \text{ there exists } \underline{u}_1 \in \mathbb{R} \text{ such that } \left(\underline{u}_1, (\underline{u}_h)_{h\neq 1}\right) \in \underline{U}^r \right\}$$

Given our smooth and strictly quasiconcave assumptions on utility functions, we can characterize Pareto optimal allocations in terms of the solutions to the following maximization problem. For given $\left(r, \underline{u}_{\setminus 1}\right) \in \mathbb{R}^C_{++} \times \underline{U}^r_{\setminus 1}$,

$$\begin{array}{llr} \max\limits_{(x,y)\in\mathbb{R}^{CH}_{++}\times\mathbb{R}^{CF}} & u_1(x_1) & \\ s.t. & u_h(x_h) - \underline{u}_h \geq 0 & \text{for } h \neq 1 \quad (1) \\ & -\sum_{h=1}^H x_h + r + \sum_{f=1}^F y_f \geq 0 & (2) \\ & t_f(y_f) \geq 0 & \text{for each } f \quad (3) \end{array} \tag{9.14}$$

Similarly to what done in the exchange economy case, in problem (9.14), we want to maximize the attainable utility level of consumer 1 for given utility level of the other consumers and given resources.

Remark 24 Since utility functions are strictly increasing (from Assumption u2), and transformation functions are strictly decreasing (from Assumption f2.3) the solution set to problem (9.14) coincides with the solution set to the following problem. For given $\left(r, \underline{u}_{\setminus 1}\right) \in \mathbb{R}^C_{++} \times \underline{U}^r_{\setminus 1}$,

$$\begin{array}{llr} \max\limits_{(x,y)\in\mathbb{R}^{CH}_{++}\times\mathbb{R}^{CF}} & u_1(x_1) & \\ s.t. & u_h(x_h) - \underline{u}_h = 0 & \text{for } h \neq 1 \quad (1) \\ & -\sum_{h=1}^H x_h + r + \sum_{f=1}^F y_f = 0 & (2) \\ & t_f(y_f) = 0 & \text{for each } f \quad (3) \end{array} \tag{9.15}$$

Lemma 25 *If (x^*, y^*) solves problem (9.14), then it is Pareto Optimal.*

Proof. The proof is strictly analogous to the proof of part 1 in Proposition 38 in Chapter 8. Suppose otherwise. Then, there exists $(\widehat{x}, \widehat{y})$ which satisfies constraints (2) and (3) in problem (9.14), and such that $(u_h(x'_h))_{h=1}^H > (u_h(x^*_h))_{h=1}^H$. If $u_1(\widehat{x}_1) > u_1(x^*_1)$, we contradict the definition of (x^*, y^*) as a solution to problem (9.14). If there exists $h \neq 1$ such that $u_h(\widehat{x}_h) > u_h(x^*_h)$, by continuity of the utility function, there exists $\varepsilon > 0$ such that $u_h(\widehat{x}_h - \varepsilon \mathbf{1}) > u_h(x^*_h)$, while $u_1(\widehat{x}_1 + \varepsilon \mathbf{1}) > u_1(x^*_1)$, again contradicting the definition of (x^*, y^*). ■

The lagrangean function associated with problem (9.14) is

$$\begin{aligned} \mathcal{L}\left(x, y, \theta, \gamma, \beta, r, \underline{u}_{\setminus 1}\right) = \; & u_1(x_1) + \sum_{h=2}^H \theta_h (u_h(x_h) - \underline{u}_h) + \\ & + \gamma \left(\sum_{h=1}^H (-x_h + e_h) + \sum_{f=1}^F y_f\right) + \sum_{f=1}^F \beta_f t_f(y_f) \end{aligned}$$

where $(\theta, \gamma, \beta) \in \mathbb{R}^{H-1} \times \mathbb{R}^{C} \times \mathbb{R}^{F}$ is the vector of Lagrange multipliers.

For given $\left(r, \underline{u}_{\setminus 1}\right) \in \mathbb{R}^{C}_{++} \times \mathbb{R}^{H-1}$, the system of first order conditions in the unknowns $(x, y, \theta, \gamma, \beta) \in \mathbb{R}^{CH}_{++} \times \mathbb{R}^{CF} \times \mathbb{R}^{H-1} \times \mathbb{R}^{C} \times \mathbb{R}^{F}$ is

$$\begin{cases} Du_1(x_1) - \gamma & = 0 \quad (1) \\ \theta_h Du_h(x_h) - \gamma & = 0 \quad (2) \\ \gamma + \beta_f Dt_f(y_f) & = 0 \quad (3) \\ (u_h(x_h) - \underline{u}_h)_{h \neq 1} & = 0 \quad (4) \\ -\sum_{h=1}^{H} x_h + \sum_{h=1}^{H} e_h + \sum_{f=1}^{F} y_f & = 0 \quad (5) \\ t_f(y_f) & = 0 \quad (6) \end{cases} \tag{9.16}$$

Define $(\theta_h)_{h \neq 1} \equiv \theta_{\setminus 1}$ and $\theta_1 = 1$.

Lemma 26 *The following statements hold.*

1 *There exists a unique solution (x^*, y^*) to problem (9.14) at $\left(r, \underline{u}_{\setminus 1}\right) \in \mathbb{R}^{C}_{++} \times \underline{U}^{r}_{\setminus 1}$.*

2 *There exists $\left(\theta^*_{\setminus 1}, \gamma^*, \beta^*\right)$ such that $(x^*, y^*, \theta^*, \gamma^*, \beta^*)$ is the unique solution to system (9.16) at $\left(r, \underline{u}_{\setminus 1}\right)$.*

Proof. 1.
From Definition 9, the constraint set of problem (9.14) can be rewritten as

$$C \equiv \{(x, y) \in A_r : \text{ for } h \neq 1,\ u_h(x_h) \geq \underline{u}_h\}$$

From Assumption f2.1, A_r is closed and from Proposition 10 is bounded. Then, from Assumption u4, and using Lemma 76 in Chapter 8 C is closed in $\mathbb{R}^{CH} \times \mathbb{R}^{CF}$ and therefore compact. Hence existence follows from the extreme value theorem ([90] Theorem 6.4, p.175).

Uniqueness. Since the utility functions and the transformation function are quasiconcave, the constraint set is convex - see Proposition 21 in Chapter 1. The objective function is strictly quasiconcave.

2. It follows from Kuhn-Tucker Theorems (see Subsection 3.2.2 in Chapter 1), whose assumptions are verified below.

Necessity of Kuhn-Tucker conditions follows from the fact that the jacobian matrix of the constraint function has full row rank. The computation of that matrix is described below.

$$
\begin{array}{l}
u_2(x_2) - \underline{u}_2 \\
u_H(x_h) - \underline{u}_H \\
\sum_{h=1}^{H}(-x_h + e_h) + \sum_{f=1}^{F} y_f \\
t_1(y_1) \\
t_F(y_F)
\end{array}
\begin{array}{c}
\begin{array}{ccccc} x_2 & x_H & x_1 & y_1 & y_F \end{array} \\
\begin{pmatrix}
Du_2(x_2) & & & & \\
& Du_H(x_H) & & & \\
-I & -I & -I & I & I \\
& & & Dt_f(y_f) & \\
& & & & Dt_F(y_F)
\end{pmatrix}
\end{array}
$$

The above matrix has clearly full rank.
Sufficiency of Kuhn-Tucker conditions follows from the facts that the objective function is pseudo-concave and the constraint functions are quasiconcave.

γ, $(\theta_h)_{h\neq 1}$ and β are uniquely determined and strictly positive from the equations (1), (2) and (3) in system (9.16) respectively. ■

Similarly to the exchange economy case, and as the following lemma shows, $\left(r, \underline{u}_{\setminus 1}\right)$ can be substituted with

$$\left(\sum_{h=1}^{H} x_h^* - \sum_{f=1}^{F} y_f^*, (u_h(x_h^*))_{h\neq 1}\right)$$

in the statement of the previous lemma.

Lemma 27 *The following statements hold.*

1 (x^*, y^*) *is the unique solution to problem (9.14) for* $r = \sum_{h=1}^{H} x_h^* - \sum_{f=1}^{F} y_f^*$, $\underline{u}_{\setminus 1} = (u_h(x_h^*))_{h\neq 1}$;

2 $(x^*, y^*, \theta^*, p^*, \gamma^*)$ *is the unique solution to system (9.16) for* $r = \sum_{h=1}^{H} x_h^*$ *and* $(\underline{u}_h)_{h\neq 1} = (u_h(x_h^*))_{h\neq 1}$.

Proof. The proof is very similar to the analogous lemma for the case of exchange economies - see Lemma 40 in Chapter 8. From part 1 of Lemma 26, we know that there exists a unique solution (x^*, y^*) to problem (9.14) at $\left(\sum_{h=1}^{H} x_h^*, (u_h(x_h^*))_{h\neq 1}\right)$. From part 2 of that lemma there exists a unique solution $(x^*, y^*, \theta^*, p^*, \gamma^*)$ to (9.16) at $\left(\sum_{h=1}^{H} x_h^* - \sum_{f=1}^{F} y_f^*, (u_h(x_h^*))_{h\neq 1}\right)$.

It is then enough to show that $(x^*, y^*, \theta^*, p^*, \gamma^*)$ is a solution to system (9.16) at $\left(\sum_{h=1}^{H} x_h^* - \sum_{f=1}^{F} y_f^*, (u_h(x_h^*))_{h\neq 1}\right)$. That system is

$$\begin{cases} Du_1(x_1) - \gamma & = 0 & (1) \\ \theta_h Du_h(x_h) - \gamma & = 0 & (2) \\ \gamma + \beta_f Dt_f(y_f) & = 0 & (3) \\ (u_h(x_h) - u_h(x_h^*))_{h \neq 1} & = 0 & (4) \\ -\sum_{h=1}^H x_h + \left(\sum_{h=1}^H x_h^* - \sum_{f=1}^F y_f^*\right) + \sum_{f=1}^F y_f & = 0 & (5) \\ t_f(y_f) & = 0 & (6) \end{cases} \tag{9.17}$$

The desired result follows from the definition of $(x^*, y^*, \theta^*, p^*, \gamma^*)$ as a solution to system (9.17) at $\left(r, \underline{u}_{\backslash 1}\right)$, i.e., from Lemma 26, and from the inspection of (9.17). ■

2.2 The homotopy

We now want to construct a homotopy to which we can apply Theorem 57 in Chapter 7. As in the case of the model of exchange economies, we are going to "link" an arbitrary economy $\pi = (e, s)$ with a well chosen economy $\pi^* = (e^*, s^*)$ which is constructed using (x^*, y^*), described in Lemma 27. The *basic idea* is to choose e^* equal to

$$\left(x_h^* - \sum_{f=1}^F s_{fh}^* y_f^*\right)_{h \in \mathcal{H}} \in \mathbb{R}_{++}^{CH} \tag{9.18}$$

and an arbitrary s^*at which condition 9.18 holds. Unfortunately, such s^* may not exist as the following example shows. Consider the allocations and endowments below

$$\begin{array}{llll} x_1^1 = 4, & x_2^1 = 6, & y^1 = 9, & e_1^1 = e_2^1 = \frac{1}{2} \\ x_1^2 = 1, & x_2^2 = 11, & y^2 = 9, & e_1^2 = e_2^2 = \frac{3}{2} \\ x_1^3 = 1 & x_2^3 = 1 & y^3 = -1 & e_1^3 = e_2^3 = \frac{3}{2} \end{array}$$

where we are assumed $H = 2$, $C = 3$, $F = 1$.[4]

Then, to satisfy condition (9.18), $s = (s_{11}, s_{12}) \equiv (s_1, s_2) = (s_1, 1 - s_1)$ has to be such that

$$\begin{cases} x_1 - s_1 y \gg 0 \\ x_2 - s_2 y \gg 0 \end{cases}$$

which implies

$$\begin{array}{lll} 1 - s_1 9 > 0, & \text{i.e.,} \quad s_1 < \frac{1}{9}, & \text{and} \\ 6 - (1 - s_1)\, 9 > 0 & \text{i.e.,} \quad s_1 > \frac{1}{3} & \end{array}$$

[4]We introduced good 3 to satisfy the "no-free-lunch" assumption on the technology.

which is impossible. Of course, it could be the case that there exists no Pareto optimal allocation (x^*, y^*) for which the conclusion of the above example holds true. Instead of studying the validity of the above statement, we proceed in the following different and, we believe, easier way.

The basic idea is to enlarge the set of shares to make them not only household and firm specific, but also *commodity* specific. In other words, we introduce shares of the type

$$\widehat{s}^c_{fh}$$

which can be interpreted as share of good c produced by firm f and owned by household h. The "true" shares can simply be identified with the subset of the enlarged set of shares for which

$$\widehat{s}^c_{fh} = \widehat{s}^1_{fh} \quad \text{for each} \quad c, f, h$$

Showing existence of equilibria for each element in the enlarged set of economies obviously implies existence for each "true" economy.

Definition 28 $\widehat{s}^c_{fh}$ is the share of good c produced by firm f owned by household h. $\widehat{s}_{fh} \equiv (s_{fh})_{c=1}^{C}$, $\widehat{s}_h \equiv (s_{fh})_{f=1}^{F} \in \mathbb{R}^{CF}$ and $s \equiv (s_f)_{h=1}^{H} \in \mathbb{R}^{CFH}$.

The set of all shares is

$$\widehat{S} \equiv \left\{ s \in [0,1]^{CFH} : \text{ for each } c \text{ and } f, \sum_{h=1}^{H} s^c_{fh} = 1 \right\}$$

Wealth of household h coming from her shares is then

$$\sum_{c=1}^{C} \sum_{f=1}^{F} p^c \widehat{s}^c_{fh} y^c_f$$

Therefore, we have to rewrite condition (9.18) as follows

$$\left(x^{c*}_h - \sum_{f=1}^{F} \widehat{s}^{c*}_{fh} y^{*c}_f \right)_{c,h} \in \mathbb{R}^{CH}_{++} \tag{9.19}$$

Lemma 29 *There exists $\widehat{s}^* \in \widehat{S}$ such that 9.19 is satisfied.*

Proof. We want to show that there exists $\widehat{s}^{c*}_{fh} \in \mathbb{R}^{CFH}$ such that

$$\begin{array}{ll} x^{c*}_{fh} - \sum_{f=1}^{F} \widehat{s}^{c*}_{fh} y^{c*}_f > 0 & \text{for each } c, h \\ \widehat{s}^{c*}_{fh} \geq 0 & \text{for each } c, f, h \\ \widehat{s}^{c*}_{fh} \leq 1 & \text{for each } c, f, h \\ \sum_{h=1}^{H} \widehat{s}^{c*}_f = 1 & \text{for each } c, f \end{array} \tag{9.20}$$

Take

$$\widehat{s}_{fh}^{c*} = \widehat{s}_{1h}^{*} \equiv \sigma_h^{c*} \quad \text{for each } c, f \tag{9.21}$$

Then we can rewrite (9.20) as follows

$$\begin{array}{ll} x_{fh}^{c*} - \sigma_h^{c*} \sum_{f=1}^{F} y_f^{c*} > 0 & \text{for each } c, h \\ \sigma_h^{c*} \geq 0 & \text{for each } c, h \\ \sigma_h^{c*} \leq 1 & \text{for each } c, h \\ \sum_{h=1}^{H} \sigma_h^{c*} = 1 & \text{for each } c \end{array} \tag{9.22}$$

Observe also that

1 If $\sum_{f=1}^{F} y_f^{c*} \leq 0$, the first condition in (9.22) is surely satisfied for that c;

2 Since $-\sum_h x_h + \sum_h e_h + \sum_f y_f = 0$, then $\sum_h x_h \gg \sum_f y_f$;

Then, the existence for each c, h of *per good* shares σ_h^{*c} which satisfy (9.22) is a consequence of the next lemma, where we drop the superscript c to simplify notation. ■

Lemma 30 *Given $H \geq 1$ strictly positive numbers $x_1, ..., x_H$ and a strictly positive number y such that*

$$\sum_{h=1}^{H} x_h > y$$

There exist H numbers $\sigma_1, ..., \sigma_H$ such that
1. for all h, $x_h - \sigma_h y > 0$,
2. for all h, $0 < \sigma_h \leq 1$, and
3. $\sum_{h=1}^{H} \sigma_h = 1$.

Proof. Let's proceed by induction on H.
1. The statement is true for $H = 1$. It is enough to choose $\sigma_1 = 1$.
2. If the desired statement is true for H, then it is true for $H + 1$.
Assume the statement is true for $\sigma^{old} \in \mathbb{R}^H$. Consider,

$$\sigma_h^{new} = \begin{cases} \sigma_h^{old} - \frac{1}{n} \frac{H}{H+1} & \text{if } h = 1, ..., H \\ \frac{1}{n} \frac{H}{H+1} & \text{if } h = H + 1, \end{cases}$$

with $n \in \mathbb{N}$ sufficiently large.

Let's verify that $(\sigma_h^{new})_{h=1}^{H}$ does verify the 3 conditions of the conclusions of the lemma.

Condition 1. For $h \neq H + 1$, it is true a fortiori. For $h = H + 1$, we must have $\frac{1}{n} \frac{H}{H+1} < \frac{x_{H+1}}{y}$, which is true for n sufficiently large.

Condition 2a. $\sigma_h > 0$. For $h \neq H + 1$, it is true for n sufficiently large. For $h = H + 1$, it is true for any n.

Condition 2b. $\sigma_h \leq 1$. It is true for any n.
Condition 3. $\sum_{h=1}^{H+1} \sigma_h^{new} = \left(\sum_{h=1}^{H} \sigma_h^{old}\right) - \frac{1}{n}\frac{H}{H+1} + \frac{1}{n}\frac{H}{H+1} = 1$. ∎
We can now consider the test economy

$$\left(\left(x_h^{c*} - \sum_{f=1}^{F} \widehat{s}_{fh}^{c*} y_f^{*c}\right)_{c,h}, \widehat{s}^*\right) \tag{9.23}$$

where (x^*, y^*) is described in Lemma 27, and $\widehat{s}^*$ satisfies the conditions described in the statement Lemma 29.

We can now redefine the extended equilibrium function F on the enlarged set of shares and then construct a homotopy which takes a convex combination of the economy corresponding to the "true economy" - i.e., such that $\widehat{s}_{fh}^{c} = \widehat{s}_{fh}^{1}$ for each c, f, h - and of the test economy (9.23) - see the homotopy $H_{\widehat{\pi}}$ defined below.

The former function is

$$\widehat{F} : \Xi \times \mathbb{R}_{++}^{CH} \times \widehat{S} \to \mathbb{R}^{\dim \Xi}$$

with $\widehat{F}(\xi, e, \widehat{s})$ equal the left hand side of system (9.13) with equation $(h.2)$ substituted by

$$-p\left(x_h - e_h\right) + \sum_{c=1}^{C} \sum_{f=1}^{F} p^c \widehat{s}_{fh}^{c} y_f^c$$

i.e.,

$$\begin{cases} \ldots & & & & \\ Du_h(x_h) - \mu_h p & = 0 & (h.1) & x_h & C \\ -p(x_h - e_h) + \sum_{c=1}^{C} \sum_{f=1}^{F} p^c \widehat{s}_{fh}^{c} y_f^c & = 0 & (h.2) & \mu_h & 1 \\ \ldots & & & & \\ p + \alpha_f Dt_f(y_f) & = 0 & (f.1) & y_f & C \\ t_f(y_f) & = 0 & (f.2) & \alpha_f & 1 \\ \ldots & & & & \\ -\sum_{h=1}^{H} x_h^{\backslash} + \sum_{h=1}^{H} e_h^{\backslash} + \sum_{f=1}^{F} y_f^{\backslash} & = 0 & (M) & p^{\backslash} & C-1 \end{cases}$$

Define $\widehat{\pi} \equiv (e, \widehat{s})$, $\widehat{\Pi} \equiv \mathbb{R}^{CH}_{++} \times \widehat{S}$ and $\widehat{\pi}^* \equiv \left(\left(x_h^{c*} - \sum_{f=1}^F \widehat{s}_{fh}^{c*} y_f^{*c} \right)_{c,h}, \widehat{s}^* \right)$.
For *given* $\widehat{\pi}$ *and* $\widehat{\pi}^*$ consider

$$\begin{cases} Du_h(x_h) - \mu_h p & = 0 & (h.1) \\ -px_h + p\left[(1-\tau)e_h + \tau\left(x_h^* - d\left(\widehat{s}_{fh}^*\right)\sum_{f=1}^F y_f^*\right)\right] + & & (h.2) \\ +p\left[+\sum_{f=1}^F \left[(1-\tau)d(\widehat{s}_{fh}) + \tau d\left(\widehat{s}_{fh}^*\right)\right] y_f\right] & = 0 & \\ \dots & & \\ p + \alpha_f Dt_f(y_f) & = 0 & (f.1) \\ t_f(y_f) & = 0 & (f.2) \\ \dots & & \\ -\sum_{h=1}^H x_h^{\backslash} + (1-\tau)\sum_{h=1}^H e_h^{\backslash} + & & (M) \\ +\tau\left(\sum_{h=1}^H x_h^{*\backslash} - \sum_{f=1}^F y_f^{*\backslash}\right) + \sum_{f=1}^F y_f^{\backslash} & = 0 & \end{cases} \tag{9.24}$$

$$H_{\widehat{\pi}} : \Xi \times [0,1] \to \mathbb{R}^n, \quad (\xi, \tau) \mapsto \text{left hand side of system (9.24)}$$

where for arbitrary $\widehat{s}_{fh}$,

$$d(\widehat{s}_{fh}) \equiv \begin{bmatrix} \widehat{s}_{fh}^1 & & \\ & \ddots & \\ & & \widehat{s}_{fh}^C \end{bmatrix} \in \mathcal{M}_{C,C}$$

and[5]

$$G : \Xi \to \mathbb{R}^{\dim \Xi}, \quad \xi \mapsto H_{\widehat{\pi}}(\xi, 1)$$

Observe that

$$H_{\pi}(\xi, 0) = \widehat{F}(\xi, \widehat{\pi}) = \widehat{F}_{\widehat{\pi}}(\xi)$$

and

$$H_{\widehat{\pi}}(\xi, \tau) = \widehat{F}(\xi, (1-\tau)\widehat{\pi} + \tau\widehat{\pi}^*)$$

We are now ready to present the other lemmas needed to show existence of equilibria.

Lemma 31 *There exists* ξ^* *such that* $\xi^* \in G_{\widehat{\pi}}^{-1}(0)$.

[5] Of course, a more precise, but heavier, notation would be $H_{\widehat{\pi}, x^*, y^*, \sigma^*}$ instead of $H_{\widehat{\pi}}$.

Proof. $G(\xi) = 0$ means

$$\begin{cases} Du_h(x_h) - \mu_h p & = 0 \quad (h.1) \\ -p\left(x_h - x_h^* - d\left(\widehat{s}_{fh}^*\right)\sum_{f=1}^F\left(y_{fh} - y_{fh}^*\right)\right) & = 0 \quad (h.2) \\ \dots & \\ p + \alpha_f Dt_f(y_f) & = 0 \quad (f.1) \\ t_f(y_f) & = 0 \quad (f.2) \\ \dots & \\ -\sum_{h=1}^H\left(x_h^\backslash - x_h^{*\backslash}\right) + \sum_{f=1}^F\left(y_f^\backslash - y_f^{*\backslash}\right) & = 0 \quad (M) \end{cases} \tag{9.25}$$

System (9.17) at $(x^*, y^*, \theta^*, \gamma^*, \beta^*)$ can be rewritten as

$$\begin{cases} Du_1(x_1^*) - \gamma^{*C}\frac{\gamma^*}{\gamma^{C*}} & = 0 \quad (1) \\ Du_h(x_h) - \frac{\gamma^{*C}}{\theta_h^*}\frac{\gamma^*}{\gamma^{C*}} & = 0 \quad (2) \\ \frac{\gamma^*}{\gamma^{C*}} + \frac{\beta_f^*}{\gamma^{C*}}Dt_f\left(y_f^*\right) & = 0 \quad (3) \\ \left(u_h(x_h^*) - u_h(x_h^*)\right)_{h\neq 1} & = 0 \quad (4) \\ -\sum_{h=1}^H x_h^* + \left(\sum_{h=1}^H x_h^* - \sum_{f=1}^F y_f^*\right) + \sum_{f=1}^F y_f^* & = 0 \quad (5) \\ t_f\left(y_f^*\right) & = 0 \quad (6) \end{cases} \tag{9.26}$$

Comparing system (9.26) and (9.25), it follows that the following vector is a solution to system (9.25):

$$\xi^* = \left((x_h^*)_{h=1}^H, \mu^* = \left(\frac{\gamma^{C*}}{\theta_h^*}\right)_{h=1}^H, y^*, \alpha^* = \left(\frac{\beta_f^*}{\gamma^{*C}}\right)_{f=1}^F, p^{*\backslash} = \left(\frac{\gamma^{c*}}{\gamma^{*C}}\right)_{c\neq C}\right)$$

where recall that $\theta_1^* \equiv 1$. ■

Lemma 32 $\{\xi^*\} = G^{-1}(0)$.

Proof. Suppose there exists $\xi' \equiv \left(x', \mu', y', \alpha'_f, p^{\backslash\prime}\right) \in G^{-1}(0)$, with $\xi' \neq \xi^{**}$. But then ξ' satisfies equations $(h.2)$ and (M) in system (9.25). Moreover, summing up equations $(h.2)$ with respect to h, we get

$$-p\left(\sum_{h=1}^H (x'_h - x_h^*) - \sum_{f=1}^F (y'_{fh} - y_{fh}^*)\right) = 0$$

and therefore

$$\sum_{h=1}^H (x'_h - x_h^*) - \sum_{f=1}^F (y'_{fh} - y_{fh}^*) = 0 \tag{9.27}$$

Consider now $(x'', y'') \equiv \frac{1}{2}(x', y') + \frac{1}{2}(x^*, y^*)$. Then (x'', y'') satisfies (9.27). Moreover, $t_f\left(y''_f\right) \geq 0$, since t_f is a concave function. Now, observe that, from equations $(h.1)$ and $(h.2)$ in system (9.25), for each h, x'_h solves

$$\max_{x_h \in \mathbb{R}^C_{++}} \quad u_h(x_h) \qquad s.t. \qquad p' x_h \leq p'\left(x^*_h + d\left(\widehat{s}^*_{fh}\right) \textstyle\sum_{f=1}^F \left(y'_{fh} - y^*_{fh}\right)\right)$$

Since $t_f\left(y^*_f\right) = 0$ and from equations $(f.1)$ and $(f.2)$ in system (9.25),

$$y'_f = \arg\max p' y_f \quad s.t. \quad t_f(y_f) \geq 0$$

we have that $\left(p'y'_f - p'y^*_f\right) \geq 0$. Therefore, for all h, $u_h(x'_h) \geq u_h(x^*_h)$. Finally, since utility functions are strictly quasiconcave,

$$\left(u_h\left(x''_h\right)\right)_{h=1}^H > \min\left\{\left(u_h\left(x'_h\right)\right)_{h=1}^H, \left(u_h\left(x^*_h\right)\right)_{h=1}^H\right\} \geq \left(u_h\left(x^*_h\right)\right)_{h=1}^H \tag{9.28}$$

Therefore, (x'', y'') satisfies (9.27) and (9.28) contradicting the fact that (x^*, y^*) is Pareto optimal. Therefore $(x', y') = (x^*, y^*)$ and the uniqueness of multipliers and prices follows from the same argument presented in 2. of the proof of Lemma 42 in Chapter 8. ■

Lemma 33 *$DG(\xi^*)$ has full row rank.*

Proof. The computation of $DG(\xi^*)$ is described below.

$$\begin{array}{l} \\ Du_h(x_h) - \mu_h p \\ -pz_h - pd\left(\widehat{s}^*_{fh}\right) \sum_{f=1}^F \left(y^*_f - y_f\right) \\ \cdots \\ p + \alpha_f Dt_f(y_f) \\ t_f(y_f) \\ \cdots \\ -\sum_{h=1}^H z^\setminus_h - \sum_{f=1}^F \left(y^{*\setminus}_f - y^\setminus_f\right) \end{array} \begin{array}{c} \begin{array}{ccccccc} x_h & \mu_h & \cdots & y_f & \alpha_f & \cdots & p^\setminus \end{array} \\ \left(\begin{array}{ccccccc} D^2 u_h & -p^{*T} & & & & & -\mu^*_h \widetilde{I} \\ -p^* & & & p^* d\left(\widehat{s}^*_{fh}\right) & & & \\ & & & & & & \\ & & & \alpha^*_f D^2 t_f\left(y^*_f\right) & Dt_f\left(y^*_f\right)^T & & \widetilde{I} \\ & & & Dt_f\left(y^*_f\right) & & & \\ & & & & & & \\ -\widetilde{I}^T & & & \widetilde{I}^T & & & \end{array}\right) \end{array}$$

where $z_h \equiv x_h - x^*_h$, $z^\setminus_h \equiv x^\setminus_h - x^{\setminus *}_h$, and

$$\widetilde{I} \equiv \begin{bmatrix} I_{C-1} \\ 0 \end{bmatrix}_{C \times (C-1)}$$

Take
$\Delta \equiv \left(\Delta x_1, \Delta\mu_1, ..., \Delta x_H, \Delta\mu_H, \Delta y_1, \Delta\alpha_1, ..., \Delta y_F, \Delta\alpha_F, \Delta p^\setminus\right) \in \mathbb{R}^{(C+1)H+(C+1)F+C-1}$, with $\Delta x_h \in \mathbb{R}^C$ and $\Delta\mu_h \in \mathbb{R}$ for each h, $\Delta y_f \in \mathbb{R}^C$ and $\Delta\alpha_f \in \mathbb{R}$ for each

f, and finally $\Delta p^{\setminus} \in \mathbb{R}^{C-1}$. We now show that $[DG(\xi^*)]\,\Delta = 0$ implies that $\Delta = 0$. The system is written below.

$$\begin{cases} D^2u_h\left(x_h^*\right)\Delta x_h - p\Delta\mu_h - \mu_h^*\widetilde{I}\Delta p^{\setminus} & = 0 \quad (h.1) \\ \left(\sum_{f=1}^{F} p^* d\left(\widehat{s}_{fh}^*\right)\Delta y_f\right) - p\Delta x_h & = 0 \quad (h.2) \\ \alpha_f^* D^2 t_f\left(y_f^*\right)\Delta y_f + Dt_f\left(y_f^*\right)\Delta\alpha_f + \widetilde{I}\Delta p^{\setminus} & = 0 \quad (f.1) \\ Dt_f\left(y_f^*\right)\Delta y_f & = 0 \quad (f.2) \\ \left(\sum_{f=1}^{F}\Delta y_f^{\setminus}\right) - \sum_{h=1}^{H}\left(\Delta x_h^{\setminus}\right) & = 0 \quad (M) \end{cases} \tag{9.29}$$

In the remaining part of the proof, we drop the superscript $*$.

Claim 1. There exists h' such that $\Delta x_{h'} \neq 0$.

Suppose not. From $(h.1)$ in (9.29), $\Delta\mu_h = 0$ and then also $\Delta p^{\setminus} = 0$. Finally, since, from Assumption f2.4 t_f is differentiably strict concavity,

$$\begin{bmatrix} \alpha_f D^2 t_f & Dt_f^T \\ Dt_f & 0 \end{bmatrix}$$

has full rank and then, form equations $(f.1)$ and $(f.2)$ in (9.29), $(\Delta y_f, \Delta\alpha_f) = 0$.

Claim 2. $\sum_h \Delta x_h D^2 u_h \Delta x_h + \sum_f \alpha_f \Delta y_f D^2 t_f \Delta y = 0$.

Premultiplying $(h.1)$ in (9.29) by $\frac{\Delta x_h}{\mu_h}$ and summing up with respect to h, we get

$$\sum_{h=1}^{H} \Delta x_h \frac{D^2 u_h}{\mu_h} \Delta x_h - \sum_{h=1}^{H} \frac{\Delta x_h}{\mu_h} p\Delta\mu_h - \sum_{h=1}^{H} \Delta x_h^{\setminus}\Delta p^{\setminus} = 0 \tag{9.30}$$

Observe that from equation $(f.1)$ in system (9.25), and equation $(f.2)$ in system (9.29), we get

$$-\alpha_f Dt_f \Delta y_f = p\Delta y_f$$

and

$$p\sum_f \Delta y_f = 0 \tag{9.31}$$

Now from the choice of $\widehat{s}^*$ and more precisely from condition (9.21), i.e., $\widehat{s}_{fh}^{c*} = \sigma_h^{c*}$ for each c, f, equation $(h.2)$ can be rewritten as

$$d\left(\widehat{s}_{fh}^*\right)\left(p\sum_{f=1}^{F}\Delta y_f\right) - p\Delta x_h = 0$$

from (9.31), we get

$$p\Delta x_h = 0 \tag{9.32}$$

Premultiplying $(f.1)$ by Δy, and using $(f.2)$, we get

$$0 = \alpha_f \Delta y D^2 t_f \Delta y + \Delta y_f D t_f \Delta \alpha_f + \Delta y_f^{\backslash} \Delta p^{\backslash}$$

and

$$-\alpha_f \Delta y D^2 t_f \Delta y = \Delta y_f^{\backslash} \Delta p^{\backslash} \tag{9.33}$$

Premultiplying (M) by $\Delta p^{\backslash}$, we get

$$\Delta p^{\backslash} \sum_f \Delta y_f^{\backslash} = \Delta p^{\backslash} \sum_h \Delta x_h^{\backslash} \tag{9.34}$$

From (9.33)and (9.34), we get

$$\Delta p^{\backslash} \sum_h \Delta x_h^{\backslash} = -\sum_f \alpha_f \Delta y_f D^2 t_f \Delta y_f \tag{9.35}$$

Substituting (9.32) and (9.35) in (9.30), we get the desired result.

But Claim 1 and 2 contradict the fact that utility functions are differentiably strictly quasiconcave and the transformation functions are strictly concave. ■

Lemma 34 *For each $\widehat{\pi} \in \widehat{\Pi}$, $H_{\widehat{\pi}}^{-1}(0)$ is compact.*

Proof. Take an arbitrary sequence $(\xi^v, \tau^v)_{v \in \mathbb{N}}$ in $H_{\widehat{\pi}}^{-1}(0)$.

First of all observe that, up to a subsequence, since $\{\tau^v : v \in \mathbb{N}\} \subseteq [0,1]$, $(\tau^v)_{v \in \mathbb{N}}$ converges and so does $(\overline{r}^n)_{n \in \mathbb{N}}$, where

$$\overline{r}^n \equiv (1 - \tau^n) \sum_{h=1}^{H} e_h^{\backslash n} + \tau^n \left(\sum_{h=1}^{H} x_h^{*\backslash} - \sum_{f=1}^{F} y_f^{*\backslash} \right)$$

say to $\overline{\overline{r}}$. Defined

$$\widetilde{r}^c \equiv \max_n \left(\{\overline{r}^{cn} : n \in \mathbb{N}\} \cup \{\overline{\overline{r}}^c\} \right) \quad \text{and} \quad \widetilde{r} \equiv (\widetilde{r}^c)_{c=1}^{C} \tag{9.36}$$

from Definition 9, we have

$\{(x^n, y^n) : n \in \mathbb{N}\} \subseteq A_{\widetilde{r}}$ where
$A_{\widetilde{r}} \equiv \left\{ (x, y) \in \mathbb{R}_{++}^{CH} \times \mathbb{R}^{CF} : \sum_{h=1}^{H} x_h - \sum_{f=1}^{F} y_f \leq \widetilde{r},\ t_f(y_f) \geq 0 \right\}$

simply because for each n, $r^n \leq \widetilde{r}$ and therefore $A_{r^n} \subseteq A_{\widetilde{r}}$. But from Proposition 10, $A_{\widetilde{r}}$ is bounded. Therefore, up to a subsequence, $y^n \to \bar{y}$ and $x^n \to \bar{x}$.

In order to show that $\bar{x} \in \mathbb{R}_{++}^{C}$, it is enough to prove that, for sufficiently large n,

$$\text{for each } h, \text{ there exists } \overline{\overline{e}}_h \text{ such that for each } n,\ u_h(x_h^n) \geq u_h(\overline{\overline{e}}_h) \tag{9.37}$$

If that is the case the desired result follows from Assumption u4. Below we show that (9.37) does hold.

Observe that from equations $(h.1)$ and $(h.2)$ in system (9.24), x_h^n solves the problem

$$\begin{array}{ll} \max\limits_{x_h \in \mathbb{R}^C_{++}} & u_h(x_h) \\ s.t. & \\ p^n x_h \leq & p^n\left((1-\tau^n)\, e_h + \tau^n \left(x_h^* - d\left(\widehat{s}^*_{fh}\right)\sum_{f=1}^F y_f^*\right)\right) \\ & +p^n\left(\sum_{f=1}^F \left((1-\tau^n)\, s_{fh} y_f^n + \tau^n d\left(\widehat{s}^*_{fh}\right) y_f^n\right)\right) \end{array} \tag{9.38}$$

The right hand side of the budget constraint can also be written as

$$p^n\left[(1-\tau^n)\, e_h + \tau^n x_h^*\right] + \left\{\sum_{f=1}^F (1-\tau^n)\, s_{fh} p^n y_f^n + \tau^n d\left(\widehat{s}^{c*}_{fh}\right) \sum_{f=1}^F \left(p^n y_f^n - p^n y_f^*\right)\right\} \tag{9.39}$$

We want to show that the term in curly brackets is non negative. From equations $(f.1)$ and $(f.2)$ in system (9.24),

$$p^n y_{fh}^n = \max_{y'_f} p^n y'_f \;\; s.t. \;\; t_f\left(y'_f\right) \geq 0$$

Since $t_f(0) = 0$ and $t_f\left(y_f^*\right) = 0$, it is then the case that

$$p^n y_f^n \geq p^n 0 = 0 \quad \text{and} \quad p^n y_f^n \geq p^n y_f^* \tag{9.40}$$

Therefore, household h solving problem (9.38) can afford to consume

$$\overline{e}_h^n \equiv (1-\tau^n)\, e_h + \tau^n x_h^*$$

Since $e_h, x_h^* \in \mathbb{R}^C_{++}$, the segment L_h joining e_h to x_h^*, i.e.,

$$L_h \equiv \left\{x_h \in \mathbb{R}^C_{++} : \exists \tau \in [0,1] \text{ such that } x_h = (1-\tau)\, e_h + \tau x_h^*\right\}$$

is a compact set contained in $\mathbb{R}^C_{++}$. Since $\{\overline{e}_h^n : n \in \mathbb{N}\} \subseteq L_h$, for each n

$$u_h\left(x_h^n\right) \geq u_h\left(\widetilde{\overline{e}}_h\right) \tag{9.41}$$

where

$$\widetilde{\overline{e}}_h \equiv \arg\min_{e_h \in L_h} u_h(e_h)$$

Taking limits of both sides of (9.41), we have

$$u_h\left(\overline{x}_h\right) \geq u_h\left(\widetilde{\overline{e}}_h\right)$$

and therefore, from Assumption u4, $\bar{x}_h \gg 0$, as desired. From $(h.1)$ in (9.24)

$$\mu_h^n \to \bar{\mu}_h > 0 \quad \text{and} \quad p^n \to \bar{p} \gg 0$$

From $(f.1)$

$$\alpha_f^n \to \overline{\alpha}_f \gg 0$$

Finally, since $\{(\xi^n, \tau^n) : n \in \mathbb{N}\} \subseteq H_{\widehat{\pi}}^{-1}(0)$ means that for every n

$$\widehat{F}\left(\xi^n, (1-\tau^n)\,\widehat{\pi} + \tau^n \widehat{\pi}^*\right) = 0$$

taking limits of both sides and using the continuity of $\widehat{F}$, we get

$$\widehat{F}\left(\overline{\xi}, (1-\overline{\tau})\,\widehat{\pi} + \overline{\tau}\widehat{\pi}^*\right) = 0$$

i.e., $\left(\overline{\xi}, \overline{\tau}\right) \in H_{\widehat{\pi}}^{-1}(0)$, as desired. ∎

From the previous lemmas and from Theorem 57 in Chapter 7, we finally have the following existence result.

Theorem 35 ***(Existence of equilibria)*** *For every $u \in \mathcal{U}$ and every $e \in \mathbb{R}_{++}^{CH}$, there exists $\xi \in \Xi$ such that $F(\xi, e) = 0$.*

3. Regularity

We now introduce the definition of regular economies in the case of economies with production and show that they have the same nice properties discussed for the case of exchange economies (see Theorems 42 and 43) and constitute a "very large" subset of the set of all economies (see Theorem 44).

Definition 36 $\pi \in \Pi$ is a regular economy if 0 is a regular value for F_π. Let $\mathcal{R}$ be the set of regular economies.

As we have already seen in Chapter 8, the nice properties of regular economies are proved by means of Theorem 9 in Chapter 4, Theorem 26 in Chapter 6 and some other related corollaries (see Section 7 in Chapter 8). In order to directly apply those results, the domain of extended equilibrium function F has to be a C^1 manifold. Unfortunately this is not the case here because S^F is not a C^1 manifold and therefore the domain $\Xi \times \mathbb{R}_{++}^{CH} \times S^F$ of F is not a C^1 manifold either. To overcome this problem, we consider a full measure subset $\widetilde{S}^F$ of S^F (see also Proposition 41 in Chapter 6) which is a C^∞ manifold, and we restrict the extended equilibrium function F to $\Xi \times \mathbb{R}_{++}^{CH} \times \widetilde{S}^F$.

Definition 37 $\widetilde{S} \equiv \left\{ s_f \in (0,1)^H : \sum_{h=1}^H s_{fh} = 1 \right\}$

Lemma 38 *$\widetilde{S}^F$ is a C^∞ manifold of dimension $F(H-1)$*

Proof. Consider the function $\Gamma : (0,1)^H \to \mathbb{R}$, $s_f \mapsto \sum_{h=1}^{H} s_{fh} - 1$. It is easy to check that 0 is a regular value for Γ and $\widetilde{S} = \Gamma^{-1}(0)$. Therefore from Theorem 9 in Chapter 4 $\widetilde{S}$ is a C^∞ manifold of dimension $(H-1)$. Then the result follows from Theorem 40 in Chapter 2. ■

Defined

$$\widetilde{F} : \Xi \times \mathbb{R}_{++}^{CH} \times \widetilde{S}^F \to \mathbb{R}^{\dim \Xi},\ \ (\xi, \pi) \mapsto F(\xi, \pi)$$

we can easily get the regularity results, following the same strategy we have seen in Section 7 in Chapter 8.

Lemma 39 *0 is a regular value for $\widetilde{F}$.*

Proof. We have to prove that for every $(\xi, \pi) \in \widetilde{F}^{-1}(0)$, $D\widetilde{F}(\xi, \pi)$ has full row rank $(C+1)H + (C+1)F + C - 1$. It suffices to show that for *every* $(\xi, e) \in \widetilde{F}^{-1}(0)$, $D_{(x,\mu,y,\alpha,e)}\widetilde{F}(\xi, e)$ has full rank. The computation of that jacobian matrix is described below.

$$\begin{array}{lll|cccccccc}
 & & & x_h & \mu_h & \ldots & y_f & \alpha_f & \ldots & e_h^{\backslash} & e_h^C \\
(h.1) & Du_h(x_h) - \mu_h p & & D^2 u_h & -p^T & & & & & & \\
(h.2) & -p\left(x_h - e_h - \sum_{f=1}^{F} s_{fh} y_f\right) & & -p & & & p\sum_{f=1}^{F} s_{fh} & & & p^{\backslash} & 1 \\
\ldots & & & & & & & & & & \\
(f.1) & p + \gamma_f Dt_f(y_f) & & & & & \alpha_f D^2 f & Df^T & & & \\
(f.2) & t_f(y_f) & & & & & Df & & & & \\
\ldots & & & & & & & & & & \\
(M) & \sum_{h=1}^{H}\left(x_h^{\backslash} \quad e_h^{\backslash}\right) + \sum_{f=1}^{F} y_f^{\backslash} & & -\widetilde{I}^T & & & \widetilde{I}^T & & & I &
\end{array}$$

The above matrix has clearly full row rank:

$$(h.1) \leftarrow x_h$$

$$(h.2) \leftarrow e_h^C$$

$$\begin{array}{lll}
(f.1) \text{ and } (f.2) & \leftarrow (y, \alpha_f) & \rightsquigarrow (h.2) \\
(h.2) & \leftarrow e_h^C &
\end{array}$$

$$\begin{array}{lll}
(M) & \leftarrow e_h^{\backslash} & \rightsquigarrow (h.2) \\
(h.2) & \leftarrow e_h^C &
\end{array}$$

■

Proposition 40 *$\widetilde{F}^{-1}(0)$ is a C^1 submanifold of $\Xi \times \mathbb{R}_{++}^{CH} \times \widetilde{S}^F$ of dimension $CH + F(H-1)$.*

Proof. From Theorem 49, $\widetilde{F}^{-1}(0) \neq \emptyset$. From the previous lemma and from Theorem 9 in Chapter 4, the desired result follows. ■

Lemma 41

$$pr : \widetilde{F}^{-1}(0) \to \mathbb{R}^{CH}_{++} \times \widetilde{S}^F, \quad (\xi, \pi) \mapsto \pi$$

is a proper function.

Proof. Take a sequence $(\pi^v)_{v \in \mathbb{N}}$ in $\mathbb{R}^{CH}_{++} \times \widetilde{S}^F$. By assumption, it admits a convergent subsequence, say to $\overline{\pi}, \in \mathbb{R}^{CH}_{++} \times \widetilde{S}^F$. We want to show that considered the sequence $(\xi^v)_{v \in \mathbb{N}}$ in $\widetilde{F}^{-1}(0)$ does converge to an element of $\widetilde{F}^{-1}(0)$. The proof follows exactly the same steps as the proof of Lemma 34. ■

The proofs of the following results is almost verbatim the same as those of the analogous results in Chapter 8 - see Theorems 55, 56 and 57- and uses Theorem 35, Lemmas 39 and 41.

Theorem 42 *If $\pi \in \mathcal{R}$, then*

1 *there exists $r \in \mathbb{N}$ such that*

$$F_{\pi}^{-1}(0) = \{\xi^i\}_{i=1}^{r}$$

2 *for each i, there exists an open set $Y \subseteq \mathbb{R}^{CH}_{++} \times \widetilde{S}^F$, an open set $V_i \subseteq \mathbb{R}^{\dim \Xi}$ and a unique function $g_i : Y \to V_i$ such that*

  (a) $(\xi^i, \pi) \in V_i \times Y$,

  (b) *g_i is C^1,*

  (c) $g_i(\pi) = \xi^i$,

  (d) *for every $\pi' \in Y$, $F(g_i(\pi'), \pi') = 0$.*

Theorem 43 *Given $\pi \in \mathcal{R}$. Then*

1 *There exists $r \in \mathbb{N}$ such that $pr^{-1}(\pi) = \{(\xi^i, \pi)\}_{i=1}^{r}$;*

2 *there exist an open neighborhood Y of π in $\mathbb{R}^{CH}_{++} \times \widetilde{S}^F$, and for each i an open neighborhood U_i of (ξ^i, π) in $F^{-1}(0)$, such that:*

  (a) $U_j \cap U_k = \emptyset$ *if* $j \neq k$;

  (b) $pr^{-1}(Y) = \cup_{i=1}^{r} U_i$;

  (c) *$pr_{|U_i} : U_i \to Y$ is a diffeomorphism.*

Theorem 44 *The set $\mathcal{R}$ has the following properties.*

1 *$\mathcal{R}$ is of full Lebesgue measure in $\mathbb{R}^{CH}_{++} \times S^F$,*

2 *$\mathcal{R}$ is open in* $\mathbb{R}^{CH}_{++} \times S^F$.

Proof. Following exactly the same argument in Theorem 57 Chapter 8, we get that $\mathcal{R}$ is an open and full measure set of $\mathbb{R}^{CH}_{++} \times \widetilde{S}^F$. From Proposition 41 in Chapter 6 $\mathbb{R}^{CH}_{++} \times \widetilde{S}^F$ is a full measure set in $\mathbb{R}^{CH}_{++} \times S^F$ and then we are done. ■

The following result is an example of how generic regularity can be used to show other generic properties. In fact, we show that firms are "typically active".

Theorem 45 *There exists an open and full measure subset Ω^* of $\mathbb{R}^{GH}_{++} \times S^F$ such that $\forall \pi \in \Omega^*$, $\forall \left(x, \mu, y, \alpha, p^{\backslash}\right) \in F_{\pi}^{-1}(0)$ and $\forall (c, f) \in \mathcal{C} \times \{1, ..., F\}$, we have that $y_f^c \neq 0$.*

Proof. We consider the equilibrium system (9.13) and we add the equation $y_f^c = 0$ for an arbitrary $(c, f) \in \mathcal{C} \times \{1, ..., F\}$. If in equilibrium firm f uses good c as an input or an output , then this system must have no solution. To fix ideas take $(c, f) = (1, 1)$. We want to prove that there exists an open and full measure subset of the endowment space for which this system has no solution. Openness follows from the properness of the projection from the equilibrium set to the endowment space. That proof goes along the same line of the proof of Lemma 34. To show the full measure result, as a consequence of the transversality theorem it is enough to prove that 0 is a regular value of the function

$$\widetilde{F}^f : \Xi \times \mathbb{R}^{GH}_{++} \times \widetilde{S}^F \to \mathbb{R}^{\dim \Xi + 1}, \quad (\xi, \pi) \mapsto (\text{ left hand side of } (9.13), y_1^1)$$

The computation of a partial jacobian matrix of $\widetilde{F}^f$ is described below, where for lighter notation we call 1 and F two generic firms and 1 and H two generic households.

		x_1	μ_1	x_H	μ_H	y_1	α_{f1}	y_F	α_{fF}	$e_1^{\backslash}$	e_1^C	$e_H^{\backslash}$	e_H^C	$p^{\backslash}$
$(h.11)$	$Du_1(x_1) - \mu_1 p$	D^2u_1	$-p^T$											$-\mu_1 \widetilde{I}$
$(h.12)$	$-p\left(x_1 - e_1 - \sum_f s_{1f} y_f\right)$	$-p$				$s_{11}p$		$s_{1F}p$		$p^{\backslash}$	1			$*$
$\cdots$														
$(h.H1)$	$Du_H(x_H) - \mu_H p$			D^2u_H	$-p^T$									$-\mu_H \widetilde{I}$
$(h.H2)$	$-p\left(x_H - e_H - \sum_f s_{Hf} y_f\right)$			$-p$		$s_{H1}p$		$s_{HF}p$				$p^{\backslash}$	1	$*$
$(f.11)$	$p + \alpha_1 Dt_1(y)$					$\alpha_1 D^2 t_1$	Dt_1^T							$\widetilde{I}$
$(f.12)$	$t_1(y)$					Dt_1								
$\cdots$														
$(f.F1)$	$p + \alpha_F Dt_F(y)$							$\alpha_F D^2 t_F$	Dt_F^T					$\widetilde{I}$
$(f.F2)$	$t_F(y)$							Dt_F						
(M)	$-\sum_h \left(x_h^{\backslash} - e_h^{\backslash}\right) + y^{\backslash}$	$-\widetilde{I}^T$		$-\widetilde{I}^T$		$\widetilde{I}^T$		$\widetilde{I}^T$		I		I		
(P)	y_1^1					10								

Using Theorem 39, it is enough to perturb the last equation, which is done below.

$$\begin{array}{ll}
(P) \leftarrow y_1^1 \rightsquigarrow & (f.11), (f.12), (h.12), (h.H2), (M) \\
(f.12) \leftarrow y_1^C \rightsquigarrow & (f.11), (h.12), (h.H2) \\
(f.11) \leftarrow \left(\alpha_1, p^{\setminus}\right) \rightsquigarrow & (h.11), (h.12), (h.H1), (h.H2), (f.F1) \\
(h.11) \leftarrow x_1 \rightsquigarrow & (h.12), (M) \\
(h.H1) \leftarrow x_H \rightsquigarrow & (h.H2), (M) \\
(f.F1), (f.F2) \leftarrow (y_F, \alpha_{F1}) \rightsquigarrow & (h.12), (h.H2), (M) \\
(M) \leftarrow e_1^{\setminus} \rightsquigarrow & (h.12), (h.H2) \\
(h.12) \leftarrow e_1^C & \\
(h.H2) \leftarrow e_H^C &
\end{array}$$

Therefore there exists a full measure and open set of Ω^* of $\mathbb{R}_{++}^{GH} \times \widetilde{S}^F$ such that $\forall \pi \in \Omega^*, \forall \left(x, \mu, y, \alpha, p^{\setminus}\right) \in F_\pi^{-1}(0)$ and $\forall (c, f) \in \mathcal{C} \times \{1, ..., F\}$, we have that $y_f^c \neq 0$. The result is then proved provided that $\mathbb{R}_{++}^{GH} \times \widetilde{S}^F$ is an open and full measure set of $\mathbb{R}_{++}^{GH} \times S^F$. ■

Chapter 10

TIME, UNCERTAINTY AND INCOMPLETE MARKETS

In this chapter, we discuss the model of an exchange economy presented in Chapter 8 *without Assumption c3*, i.e., we introduce time and uncertainty. In Section 1, we present that more general framework keeping Assumption h5 - market completeness, and we discuss the need for its removal and the introduction of a financial side of the economy. In Section 2, we formalize a way of introducing financial assets, define the notion of complete *financial* markets and study the relationship with complete markets. Finally, the definition of incomplete financial markets is given and the main fundamental results, which are going to be shown in the following chapters, are summarized.

1. A model with time, uncertainty and complete markets

As mentioned above, in this section we do introduce time and uncertainty, but, for the time being, we also make all the other assumptions presented in Chapter 8, adapted to the present framework. We therefore present an **exchange economy model with time, uncertainty and complete markets.**

We assume that there is a finite set $\mathcal{T} \equiv \{0, 1, ..., T\}$ of periods (today for $t = 0$, tomorrow for $t = 1, ...$) and, in each period, a finite set $\mathcal{S}_t \equiv \{1, ..., S_t\}$ of states of the world (tomorrow if it rains, tomorrow if it is sunny,...). In each period, households do not know which state of the world is going to occur in the following period.

A state of the world, or state of nature, or spot is "a complete description of a possible outcome of uncertainty, the description being sufficiently fine for any two distinct states of the world to be mutually exclusive"[1].

[1] [87], p. 688.

For simplicity, we assume that $\mathcal{T} = \{0,1\}$, i.e., there are 2 periods: today and tomorrow. The state of the world today is known to individuals, and it is called state 0. Tomorrow, S states of the world are possible. Therefore, the set of possible states of the world can be identified with the set $\mathcal{S}' \equiv \{0, 1, ..., S\}$ with generic element s, where $s = 0$ denotes the unique state in period 0 and $s \neq 0$ denotes state s in period 1 - see figure 10.1.

In this framework, a commodity may be defined in terms not only of its physical or chemical characteristics, but also in terms of the period or the state of nature in which it is available ([40]). A commodity is therefore described by an element (s, c) of the set $\mathcal{S}' \times \mathcal{C}$. Then examples of commodities are: commodity $(0, c')$ (apple today), (s, c'') (oranges tomorrow if it rains). The total number of goods available in the economy is $G \equiv (S+1)\,C$. If we accept the above definition of commodity, we can trivially introduce time and uncertainty in the original model. In fact, we get a model which is formally identical to the standard one, up to a renumbering of the commodities.

The time structure of the model can be described as follows. In state 0, households receive endowments of goods; exchange them and promise to deliver goods in other states in period 1; consume the goods they acquired. Uncertainty is resolved. In each state in period 1, households receive other endowments of goods; exchange goods and fulfill the obligations underwritten in state 0; consume the goods they acquired.

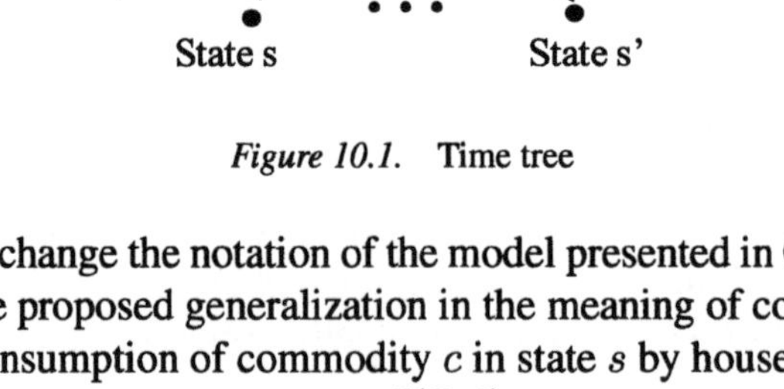

Figure 10.1. Time tree

We can then change the notation of the model presented in Chapter 9 to take into account the proposed generalization in the meaning of commodity.

$x_h^{s,c}$ is the consumption of commodity c in state s by household h.

$x_h^s \equiv (x_h^{s,c})_{c=1}^C$, $x_h \equiv (x_h^s)_{s=0}^S \in \mathbb{R}_{++}^{C(S+1)}$, $x \equiv (x_h)_{h=1}^H \in \mathbb{R}_{++}^{GH}$.

Similar notation is used for the endowments: $e_h^{s,c}$ is the endowment of commodity c in state s owned by household h.

$e_h^s \equiv (e_h^{s,c})_{c=1}^C$, $e_h \equiv (e_h^s)_{s=0}^S \in \mathbb{R}_{++}^G$, $e \equiv (e_h)_{h=1}^H \in \mathbb{R}_{++}^{GH}$.

$p^{s,c}$ is the price of commodity c in spot s denominated, or measured, in units of account.

$p^s \equiv (p^{sc})_{c=1}^C, p \equiv (p^s)_{s=0}^S \in \mathbb{R}^G$.

Household h's preferences are represented by an utility function

$$u_h : \mathbb{R}_{++}^G \to \mathbb{R}, \quad x_h \mapsto u_h(x_h) \tag{10.1}$$

defined on the "enlarged" consumption set $\mathbb{R}_{++}^G$.

Remark 1 The utility function in (10.1) is a generalization of the so-called "extended expected utility function"[2] which is defined as

$$u_h : \mathbb{R}_{++}^G \to \mathbb{R}, \quad x_h \mapsto \sum_{s=1}^S \pi^s v_h^s \left(x_h^0, x_h^s\right)$$

where π^s is the probability of occurrence of state $s \geq 1$, and $v_h^s : \mathbb{R}_{++}^{2C} \to \mathbb{R}$. If for each $s \geq 2$, $v_h^s = v_h^1$, the utility function in (10.1) is a standard expected utility function - see Chapter 6 in [87].

We assume that each household utility function satisfies the same assumptions presented in Chapter 8 (again with respect to the "enlarged" consumption set $\mathbb{R}_{++}^G$). The set of utility function satisfying those Assumptions is denoted by $\widehat{\mathcal{U}}$. Moreover, $\mathcal{U} \equiv \widehat{\mathcal{U}}^H$ with generic element u.

Definition 2 An economy in exchange economy model with time, uncertainty and complete markets is a pair $(e, u) \in \mathbb{R}_{++}^{GH} \times \mathcal{U}$.

Following Debreu, we can present a definition of equilibrium which uses the more comprehensive definition of commodity we introduced.

Definition 3 A vector $(x, p) \in \mathbb{R}_{++}^{GH} \times \mathbb{R}_{++}^G$ is an **equilibrium allocation-price vector in the exchange economy model with uncertainty and complete markets** for an economy $(e, u) \in \mathbb{R}_{++}^{GH} \times \mathcal{U}$ if

1 households maximize, i.e., for $h = 1, ..., H$, for given $p \in \mathbb{R}_{++}^G$ and $e_h \in \mathbb{R}_{++}^G$, x_h solves the problem

$$\max_{x_h \in \mathbb{R}_{++}^G} u_h(x_h) \qquad s.t. \qquad p(x_h - e_h) = 0 \tag{10.2}$$

2 x satisfies the following market clearing conditions

$$\sum_{h=1}^H (x_h - e_h) = 0$$

[2] [87], p. 202.

Remark 4 In problem (10.2), we assume that households know future prices, or they have expectations about prices which turn out to be true. This needless to say strong assumption is the core of the so-called "rational expectation" paradigm[3].

Removing Assumption c3 - absence of time - and retaining Assumption h5 - market completeness - is quite unrealistic, but it is equivalent to a more realistic assumption contained in a model introduced by [4].

In that models, goods can be directly exchanged with other goods *only* in the same spot. On the other hand, sufficiently numerous and diversified *financial* assets are available. In the remaining part of this section, we present an intuitive account of that result. In Section 2, we formally present it.

Consider a model with only two states of the world, α and β, in period 1. A financial asset is a contract described by a vector $\left(-q, r^{\alpha}, r^{\beta}\right)$, where q is the price households have to *pay* to obtain the right to *receive* r^{α} if state α occurs and r^{β} if state β occurs. r^{α} and r^{β} are called returns of the assets; price and returns in Arrow's model are denominated in units of account. Suppose that in our example the two following assets are available: $\left(-q^{\alpha}, 1, 0\right)$ and $\left(-q^{\beta}, 0, 1\right)$. Call them asset α and β, respectively. Then, it is clearly possible to exchange any good in any spot with any other good in any other spot using the above assets, and there is no difference between this model and the standard exchange economy model.

For example, to sell some units of good 1-today in order to buy some units of good 2-tomorrow-in-state-α, it is enough to sell the desired amount of good 1-today to buy some amount of asset $(-q^{\alpha}, 1, 0)$. If state α occurs, the return of that asset can be used to buy good 2.

Similarly, purchases of good 1-today using good 2-tomorrow-in-state-α can be financed by the sale of asset α: through that sale, households receive "money", or units of account, today and promise to deliver money tomorrow if α occurs.

To exchange good 1-in-state-α with good 2-in-state-β, it is enough to combine exchanges of assets α and asset β.

Remark 5 Observe that exchanges of goods between different states are made possible by the transfer of *wealth* between those states and transfers of wealth are obtained through exchanges of assets.

The following diagram summarizes how assets can be used to transfer wealth as indicated by arrows, while the associated exchanges of assets are explained

[3]For a discussion of rational expectations, see Chapter 9 in [87], and references quoted there. See also the brief discussion in the Introduction of the book.

below the diagram.

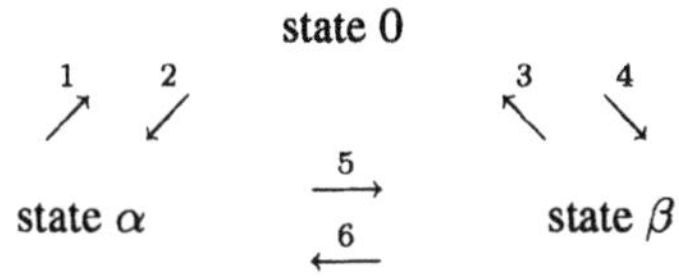

1 : sell asset α;
2 : buy asset α;
3 : sell asset β;
4 : buy asset β;
5 : combine 1 and 4 above, i.e., sell asset α and buy asset β;
6 : combine 3 and 2 above, i.e., sell asset β and buy asset α.

In the following section we present the topics described above in a rigorous manner.

2. A model with financial markets

In this section, we first describe a version of the model presented in the previous section with a financial sector. We then link *market completeness* with *financial market completeness* - a rank condition on the matrix which describes the financial structure of the model. We also verify that equilibrium allocations in the model with (uncertainty and) complete *financial* markets coincide with the equilibrium allocations in the exchange economy model with uncertainty and complete markets. Finally we define *incomplete* financial markets and the different types of assets we are going to analyze in the following chapters.

2.1 General set-up of the model

For readers convenience, and for future use, we repeat the main characteristics and the basic notation of the model introduced in the previous section.

There are 2 periods: today, called state 0, and tomorrow, when S states of the world are possible. Therefore, the set of possible states of the world can be identified with the set $\{0, 1, ..., S\} \equiv \mathcal{S}'$ with generic element s, where $s = 0$ denotes the unique state in period 0 and $s \neq 0$ denotes state s in period 1.

A commodity is an element (s, c) of the set $\mathcal{S}' \times \mathcal{C}$. The total number of goods available in the economy is $G \equiv (S + 1)\, C$.

$x_h^{s,c}$ is the consumption of commodity c in state s by household h.

$x_h^s \equiv (x_h^{s,c})_{c=1}^C$, $x_h \equiv (x_h^s)_{s=0}^S \in \mathbb{R}_{++}^{C(S+1)}$, $x \equiv (x_h)_{h=1}^H \in \mathbb{R}_{++}^{GH}$.

$e_h^{s,c}$ is the endowment of commodity c in state s owned by household h.

$e_h^s \equiv (e_h^{s,c})_{c=1}^C$, $e_h \equiv (e_h^s)_{s=0}^S \in \mathbb{R}_{++}^G$, $e \equiv (e_h)_{h=1}^H \in \mathbb{R}_{++}^{GH}$.

$p^{s,c}$ is the price of commodity c in spot s denominated in units of account.

$p^s \equiv (p^{sc})_{c=1}^C$, $p \equiv (p^s)_{s=0}^S \in \mathbb{R}_{++}^G$.

Household h's preferences are represented by an utility function

$$u_h : \mathbb{R}_{++}^G \to \mathbb{R}, \quad x_h \mapsto u_h(x_h)$$

We assume that utility functions satisfy the "standard smooth assumptions" introduced in Chapter 8 and summarized below.

Assumption u For $h \in \mathcal{H}$, $u_h : \mathbb{R}^G_{++} \to \mathbb{R}$ is

1 C^2,

2 differentiably strictly increasing,

3 differentiably strictly quasiconcave , and

4 for any $\underline{u} \in \mathbb{R}$, $Cl_{\mathbb{R}^G} \{x \in \mathbb{R}^G_{++} : u_h(x) \geq \underline{u}\} \subseteq \mathbb{R}^G_{++}$.

As in the previous section, the set of utility function satisfying Assumption u is denoted by $\widehat{\mathcal{U}}$, and $\mathcal{U} \equiv \widehat{\mathcal{U}}^H$ with generic element u.

The difference between the model we present here and that one of the previous section is that we substitute Assumption h5 with Assumption h5* and we introduce a financial side of the economy.

Assumption h5* Markets are complete *only inside each spot* $s = 0, 1, ..., S$.

There are I assets, or financial *instruments*, labelled by $i = 1, 2, ..., I$.

Definition 6 An **asset** i is an $(S + 1)$ dimensional vector

$$\left(-q^i, r^{1i}, ..., r^{si}, ..., r^{Si}\right)$$

where $q^i \in \mathbb{R}$ is the price of the asset i, measured in units of account, and $r^{si} \in \mathbb{R}$ is the return of asset i in state s, i.e., the amount of units of account that a household has the right to receive in state s if she purchased one unit of asset i in state 0.

As in most literature, we assume that returns are a function - possibly constant - of good prices, i.e., for each $s = 0, 1, ..., S$, and $i = 1, ..., I$, r^{si} is the function defined below.

$$r^{si} : \mathbb{R}^G_{++} \to \mathbb{R}, \quad p \mapsto r^{si}(p)$$

We also introduce the following definitions.

Definition 7 $q \equiv (q^i)_{i=1}^I$ is the **asset price** vector;

$$R : \mathbb{R}^G_{++} \to \mathcal{M}_{S,I}, \quad p \mapsto R(p) \equiv \begin{bmatrix} r^{11}(p) & \cdots & r^{1i}(p) & \cdots & r^{1I}(p) \\ \cdots & & & & \\ r^{s1}(p) & \cdots & r^{si}(p) & \cdots & r^{sI}(p) \\ \cdots & & & & \\ r^{S1}(p) & \cdots & r^{Si}(p) & \cdots & r^{SI}(p) \end{bmatrix}$$

is the **return matrix** function[4]; column i of $R(p)$ is denoted by ${}^{i}r(p)$ and describes the returns of asset i; row s is denoted by $r^{s}(p)$ and denotes the returns of each asset in state s; $\mathcal{R}_{S,I}$ is the set of all return matrix functions defined above; $R(p)$ is called return matrix at good prices p.

Definition 8 $b_h^i \in \mathbb{R}$ is the **demand of asset** i by household h, $b_h \equiv \left(b_h^i\right)_{i=1}^{I} \in \mathbb{R}^I$ is the asset demand vector or portfolio holding by household h; $b \equiv (b_h)_{h=1}^{H}$.

The time structure of the model is the following one.

In period zero, households exchange goods and assets and then consumption takes place.

In period one, one of S possible states of nature occurs. In each of them, assets pay their returns, households exchange goods and then consumption takes place.

Household h's maximization problem is as follows. For given $e_h \in \mathbb{R}_{++}^G$, $p \in \mathbb{R}_{++}^G$, $q \in \mathbb{R}^I$, $R \in \mathcal{R}_{S,I}$,

$$\begin{array}{lll} \max_{(x_h,b_h)\in\mathbb{R}_{++}^G\times\mathbb{R}^I} & u_h(x_h) & \\ s.t. & p^0x_h^0 + qb_h = p^0e_h^0 & \text{and} \\ & p^sx_h^s = p^se_h^s + {}^{s}r(p)\,b_h \quad \text{for } s \geq 1 & \end{array} \tag{10.3}$$

We assume there are no endowments of assets. Observe that in spot 0, each households finances purchases of goods and assets using her wealth in that spot. In each spot $s \geq 1$, purchases of goods are financed by the sale of endowments and asset returns. In fact, the term ${}^{s}r(p)\,b_h$ can take any sign since both ${}^{s}r(p)$ and b_h^i can be positive, negative or zero.

Defined[5]

$$\Phi : \mathbb{R}_{++}^G \to \mathcal{M}_{(S+1),G}, \quad p \mapsto \Phi(p) \equiv \begin{bmatrix} p^0 & & & \\ & p^1 & & \\ & & \ddots & \\ & & & p^S \end{bmatrix}$$

problem (10.3) can be rewritten as

$$\begin{array}{ll} \max_{(x_h,b_h)\in\mathbb{R}_{++}^G\times\mathbb{R}^I} & u_h(x_h) \\ s.t. & -\Phi(p)(x_h - e_h) + \begin{bmatrix} -q \\ R(p) \end{bmatrix} b_h = 0 \end{array} \tag{10.4}$$

[4]Recall that $\mathcal{M}_{S,I}$ is the set of all $S \times I$ dimensional matrices of real numbers.

[5]Recall our notation convention: a vector $x \in \mathbb{R}^n$ part of a partioned matrix is treated as a $1 \times n$ matrix, i.e., a "row" vector.

Definition 9 A **financial economy** is an element $(e, R, u) \in \mathbb{R}^{GH}_{++} \times \mathcal{R}_{S,I} \times \mathcal{U}$.

Definition 10 The vector $(x, b, p, q) \in \mathbb{R}^{GH}_{++} \times \mathbb{R}^{IH} \times \mathbb{R}^{G}_{++} \times \mathbb{R}^{I}$ is a **financial equilibrium demand-price vector** for the economy $(e, R, u) \in \mathbb{R}^{GH}_{++} \times \mathcal{R}_{S,I} \times \mathcal{U}$ if

1 for $h = 1, ..., H$, (x_h, b_h) solves problem (10.4) at (p, q, e, R, u), i.e., households maximize, and

2
$$\begin{array}{ll} \sum_{h=1}^{H} x_h = \sum_{h=1}^{H} e_h & \text{and} \\ \sum_{h=1}^{H} b_h = 0 & \end{array}$$
i.e., both commodity and asset markets clear.

In the pure exchange economy model presented in Chapter 8, since we assumed that utility functions are strictly increasing,without loss of generality, we restricted our analysis to strictly positive *commodity* prices. If that were not the case, there would be no equilibrium: each household would demand an infinite amount of free and useful commodities. A similar "no free lunch" or "no something-for-nothing" concept can be introduced for *asset prices* in relationship with asset returns.

Definition 11 $q \in \mathbb{R}^I$ is a **no-arbitrage asset price** vector for the return matrix $R(p)$, if there is no $a \in \mathbb{R}^I$ such that $\begin{bmatrix} -q \\ R(p) \end{bmatrix} a > 0$.

If q is a no-arbitrage price for $R(p)$, we write $q \in N(p)$. We also[6] let

$$N \equiv \left\{(p, q) \in \mathbb{R}^{G}_{++} \times \mathbb{R}^{I} : q \in N(p)\right\}$$

If q is a no-arbitrage price, then there is no portfolio holding which guarantees some strictly positive gain in some state and no loss in all other states.

Remark 12 If (x, b, p, q) is a financial equilibrium demand-price vector for the economy (e, R, u), it must be the case that q is a no-arbitrage price vector for $R(p)$. Otherwise, household would demand an infinite amount of the arbitrage portfolio a such that $\begin{bmatrix} -q \\ R(p) \end{bmatrix} a > 0$.

Because of the above remark, basically all our analysis is restricted to consider prices (p, q) in N.

[6] A more precise, but quite heavy notation would be $N\left(p, \mathcal{R}_{S,I}\right)$ for $N(p)$ and $N\left(\mathcal{R}_{S,I}\right)$ for N.

Definition 13 Available assets are **not redundant** at commodity price p, if rank $R(p)_{S\times I} = I$.

Assets are not redundant, if none of them can be obtained as a (linear) combination of other available assets. If available assets are redundant, some of them could be eliminated without changing the set of consumption vectors each household h can afford when solving problem (10.4).

We characterize Definition 11 in two equivalent ways.

Lemma 14 *The following statements are equivalent.*

1. *There is no* $a \in \mathbb{R}^I$ *such that* $\begin{bmatrix} -q \\ R(p) \end{bmatrix} a > 0$*;*
2. *There exists* $\lambda \in \mathbb{R}^{S+1}_{++}$ *such that* $\lambda \begin{bmatrix} -q \\ R(p) \end{bmatrix} = 0$*;*
3. *There exists* $\gamma \in \mathbb{R}^S_{++}$ *such that* $q = \gamma R(p)$.

Proof. $[1 \Rightarrow 2]$ The desired result is a consequence of the following proposition[7].

For any $m \times n$ matrix A, the systems

$$Ax \geq 0$$

and

$$yA = 0 \text{ and } y \geq 0$$

have solutions x and y such that

$$Ax + y \gg 0$$

From the above proposition applied to the matrix $\begin{bmatrix} -q \\ R(p) \end{bmatrix}$, there exist $a \in \mathbb{R}^I$ and $\lambda \in \mathbb{R}^{S+1}$ such that

$$\begin{bmatrix} -q \\ R(p) \end{bmatrix} a \geq 0 \tag{10.5}$$

$$\lambda \begin{bmatrix} -q \\ R(p) \end{bmatrix} = 0 \tag{10.6}$$

$$\lambda \geq 0$$

$$\begin{bmatrix} -q \\ R(p) \end{bmatrix} a + \lambda \gg 0 \tag{10.7}$$

[7] See [85], pp. 24 and 22.

By Assumption and from (10.5), $\begin{bmatrix} -q \\ R(p) \end{bmatrix} a = 0$ and therefore, from (10.7),

$$\lambda \gg 0 \tag{10.8}$$

(10.6) and (10.8) are the desired result.

[1 ⇐ 2] Suppose otherwise. Then there is $a' \neq 0$ such that $\begin{bmatrix} -q \\ R(p) \end{bmatrix} a' > 0$, and since $\lambda \gg 0$, it must be the case that

$$\lambda \begin{bmatrix} -q \\ R(p) \end{bmatrix} a' \neq 0$$

On the other hand,

$$\lambda \begin{bmatrix} -q \\ R(p) \end{bmatrix} = 0 \text{ implies } \lambda \begin{bmatrix} -q \\ R(p) \end{bmatrix} a = 0, \text{ for each } a \in \mathbb{R}^I$$

a contradiction.

[2 ⇒ 3] Take $\gamma = \left(\frac{\lambda^s}{\lambda^0}\right)_{s=1}^S$.

[2 ⇐ 3] Take $\lambda = \left(1, (\gamma^s)_{s=1}^S\right)$. ■

2.2 Complete financial markets

In the financial economy model described in the above subsection, completeness of markets is assumed to hold inside each spot. To guarantee the same possibility among different spots, and therefore for the overall economy, consistently with what said in Remark 5, each household has to be able to freely move wealth from one spot to the other. The definition below formalizes that observation.

Definition 15 In the financial economy model (presented in the previous subsection), **markets are complete** if each household can move her wealth from one spot to the other, without changing her wealth in any other spot, i.e., if and only if for any $s, s' \in \{0, 1, ...S\}$ and any $\Delta w^s \in \mathbb{R}$, there exists $\Delta w^{s'} \in \mathbb{R}$ and $b_h \in \mathbb{R}^I$ such that

$$\begin{bmatrix} -q \\ R(p) \end{bmatrix} b_h = \Delta w^s \mathbf{e}^s + \Delta w^{s'} \mathbf{e}^{s'}$$

where in general $\mathbf{e}^s \in \mathbb{R}^{S+1}$ and has all 0 components apart from the i-th one which is equal to 1.

We can characterize market completeness by means of a rank condition on the return matrix $R(p)$, as described in the following definition and proposition.

Definition 16 Financial markets are complete at prices $(p, q) \in N$ if rank $R(p) = S$. Financial markets are complete if they are complete for each $(p, q) \in N$.

Remark 17 Observe that we made part of the definition the fact that q is a no-arbitrage price. Moreover, if the function R is constant with respect to p and, say, $R(p) = \overline{R}$, - see Chapter 12 - or its *rank* is constant with respect to p - see Chapter 11 - then markets are complete if and only if rank $\overline{R} = S$.

Proposition 18 *In the financial economy model,*

markets are complete $\Leftrightarrow$ *financial markets are complete*

Proof. Observe that, from Definition 15, markets are complete if and only if

$$\text{any } S \times S \text{ submatrix of } \begin{bmatrix} -q \\ R(p) \end{bmatrix} \text{ has full rank} \tag{10.9}$$

We are then left with showing that, if $(p, q) \in N$, the above condition (10.9) is equivalent to

$$\operatorname{rank} R(p) = S \tag{10.10}$$

Clearly (10.9) implies (10.10). To show the opposite implication is enough to show that for each $s' \in S$

$$\operatorname{rank} \begin{bmatrix} -q \\ [r^s(p)]_{s \neq s'} \end{bmatrix}_{S \times S} = S$$

Without loss of generality take $s' = S$. Suppose the rank condition does not hold, then there exists $(\alpha^s)_{s=0}^{S-1} \in \mathbb{R}^S \setminus \{0\}$ such that

$$\begin{aligned} 0 = &\; -\alpha^0 q + \textstyle\sum_{s=1}^{S-1} \alpha^s r^s(p) = -\alpha^0 \sum_{s=1}^{S} \lambda^s r^s(p) + \sum_{s=1}^{S-1} \alpha^s r^s(p) \\ = &\; \textstyle\sum_{s=1}^{S-1} \left(-\alpha^0 \lambda^s + \lambda^s\right) r^s(p) + \alpha^0 \lambda^S r^S(p) \end{aligned} \tag{10.11}$$

where the second equality follows from the fact that $(p, q) \in N$. If $\alpha^0 = 0$, from (10.11), we have that there exists $(\alpha^s)_{s=1}^{S-1} \in \mathbb{R}^{S-1} \setminus \{0\}$ such that $0 = \sum_{s=1}^{S-1} \lambda^s r^s + 0 \cdot r^S$, contradicting the fact that rank $R = S$. If $\alpha^0 \neq 0$, (10.11) contradicts that rank condition. ■

Remark 19 Observe that if rank $R(p) = S$ and $(p, q) \notin N$, and therefore markets are not financially complete, it may be the case that markets are not complete. It is enough to consider the case

$$R(p) = \begin{bmatrix} 1 & 0 \\ 0 & 1 \end{bmatrix} \text{ and } q = \begin{bmatrix} 0 & 0 \end{bmatrix}$$

$(p, q) \notin N$ simply because there exists no $\gamma \in \mathbb{R}^2_{++}$ such that

$$\begin{bmatrix} 0 & 0 \end{bmatrix} = \begin{bmatrix} \gamma^1 & \gamma^2 \end{bmatrix} \begin{bmatrix} 1 & 0 \\ 0 & 1 \end{bmatrix}$$

Markets are not complete, because using the available financial assets there is no way to move wealth in or away from state 0 and therefore no good in state 0 can be exchanged with other goods in period 1.

When financial markets are complete, equilibrium allocations in the exchange economy model with uncertainty and complete markets coincide with financial equilibrium allocations - see Definition 3 and 10, respectively.

Proposition 20 *Assume that* $(p, q) \in N$.

1 *If* (x, p) *is an equilibrium allocation-price vector in the exchange economy model with uncertainty and complete markets at the economy* (e, u), *then there exists* $\gamma \in \mathbb{R}^S_{++}$ *such that*

$$\left(x, \left(b_h' \equiv \left(R^{-1} \Phi^1 (p) \left(x_h^1 - e_h^1 \right) \right) \right)_{h=1}^{H}, \, p' \equiv \left(p^0, \left(\frac{p^s}{\gamma^s} \right)_{s \neq 0} \right), \, q' \equiv \gamma R(p) \right)$$

is a financial equilibrium demand-price vector at the economy (e, R, u), *and*

2 *if* (x, b, p', q) *is a financial equilibrium demand-price vector at* (e, R, u), *then there exists* $\gamma \in \mathbb{R}^S_{++}$ *such that*

$$\left(x, \left(p^{0\prime}, \left(\gamma^s p^{s\prime} \right) \right)_{s \neq 0} \right)$$

is an equilibrium allocation-price vector in the exchange economy model with uncertainty and complete markets at the economy (e, u).

Proof. As observed in the proof of Proposition 18, there is no loss of generality in assuming that $R(p) \in \mathcal{M}_{S,S}$. For simplicity of notation, denote $R(p)$ by R.

Consider $\gamma \in \mathbb{R}^S_{++}$ defined in Lemma 14 and observe that by assumption q is a no-arbitrage asset price. Define

$$\Phi^1 (p) \equiv \begin{bmatrix} p^1 & & \\ & \ddots & \\ & & p^S \end{bmatrix}_{S \times G}, \quad x_h^1 \equiv \left(x_h^s \right)_{s \neq 0}, \; e_h^1 \equiv \left(e_h^s \right)_{s \neq 0}$$

Define the sets of affordable allocations of problems (10.2) and (10.4) as follows

$$B_h (p, e_h) \equiv \left\{ x_h \in \mathbb{R}^G_{++} : p (x_h - e_h) = 0 \right\}$$

$$\widehat{B}_h(p,q,e_h,R) \equiv$$
$$\left\{x_h \in \mathbb{R}^G_{++} : \exists\, b_h \in \mathbb{R}^I \ s.\,th. \ -\Phi(p)(x_h - e_h) + \begin{bmatrix} -q \\ R \end{bmatrix} b_h = 0\right\}$$

Since the objective function for each household in both problems is the same, to show both claims of the proposition, we have to verify that:

a. the sets of affordable allocations for each households at the "right" prices are the same for both equilibria, i.e.,

$$\text{for } h \in \mathcal{H},\ \ B_h(p,e_h) = \widehat{B}_h(p',q',e_h,R)$$

b. market clearing conditions in the two models are equivalent, i.e.,

$$\left\langle \sum_{h=1}^{H}(x_h - e_h) = 0 \right\rangle \Leftrightarrow \left\langle \sum_{h=1}^{H}(x_h - e_h) = 0 \text{ and } \sum_{h=1}^{H} b'_h = 0 \right\rangle$$

a.

$$x_h \in \widehat{B}_h(p',q',e_h,R)$$

$\Leftrightarrow$

$$\begin{cases} p^0(x_h^0 - e_h^0) + \gamma R b'_h = 0 \\ \Phi^1(p')(x_h^1 - e_h^1) - R b'_h = 0 \end{cases}$$

$\Leftrightarrow$

$$\begin{cases} b'_h = R^{-1}\Phi(p')(x_h^1 - e_h^1) \\ p^0(x_h^0 - e_h^0) + \gamma R\left(R^{-1}\Phi(p')(x_h^1 - e_h^1)\right) = 0 \end{cases}$$

$\Leftrightarrow$

$$\begin{cases} b'_h = R^{-1}\Phi^1(p')(x_h^1 - e_h^1) \\ p^0(x_h^0 - e_h^0) + \gamma \begin{bmatrix} \frac{p^1}{\gamma^1} & & \\ & \ddots & \\ & & \frac{p^S}{\gamma^S} \end{bmatrix} (x_h^1 - e_h^1) = p(x_h - e_h) = 0 \end{cases} \tag{10.12}$$

$\Rightarrow$

$$x_h \in B(p,e_h)$$

Assume that $x_h \in B(p,e_h)$. Take $b'_h = R^{-1}\Phi^1(p')(x_h^1 - e_h^1)$. Then, we have

$$\begin{cases} b'_h = R^{-1}\Phi^1(p')(x_h^1 - e_h^1) \\ p(x_h - e_h) = p^0(x_h^0 - e_h^0) + \gamma \begin{bmatrix} \frac{p^1}{\gamma^1} & & \\ & \ddots & \\ & & \frac{p^S}{\gamma^S} \end{bmatrix} (x_h^1 - e_h^1) = 0 \end{cases}$$

and we are back to (10.12), which is equivalent to

$$x_h \in \widehat{B}_h\left(p', q', e_h, R\right)$$

2. It is enough to observe that

$$\sum_{h=1}^{H} b_h' = R\Phi^1\left(p'\right)\left(\sum_{h=1}^{H}\left(x_h^1 - e_h^1\right)\right) = 0$$

■

Proposition 20 gives conditions on financial markets under which the exchange possibilities in the pure exchange model can be "replicated" using assets. Thanks to Proposition 20, all the results we have already seen for an exchange economy extend to our financial economy, just using the same arguments presented in Chapter 8 and, in many cases, using the same words. Therefore time and uncertainty do not enter in an essential way in the model when markets are financially complete: the equilibrium allocations for an economy with two periods and S states in period 1 coincides with the equilibrium allocations of an exchange economy where everything happens in "one shot". In next section we are going to see that things are different when households cannot freely move their wealth from any state to any another one, i.e., when markets are financially incomplete.

2.3 Incomplete financial markets

We are now ready to present the definition of incomplete markets in term of the rank of the return matrix $R(p)$.

Definition 21 Financial markets are incomplete if there exists $(p, q) \in N$ such that rank $R(p)_{S,I} = I < S$.

Remark 22 If the function R is constant with respect to p and, say, $R(p) = \overline{R}$, and $\overline{R} \in \mathcal{M}_{S,I}$ with $I < S$, then markets are incomplete.

Clearly, if financial markets are incomplete, markets are not complete. To see that, consider the example presented at the end of Section 1. If only asset α is available, rank $R(p) = \text{rank}\begin{bmatrix} 1 \\ 0 \end{bmatrix} = 1$ and state β is "isolated": no good in that state can be exchanged with any other good in any other state.

Different results about equilibria are obtained for different specification of the return matrix function $R(p)$. In the next three chapters we are going to analyze three of those specifications, which have been widely investigated in the literature. We distinguish among financial models with nominal, real and numeraire assets. Briefly, *nominal* assets promise to deliver units of account; *real* assets a *bundle* of goods, and *numeraire* assets just some amount of a *given good* , the so called numeraire good. In general, the letter y with appropriate subscript denotes the number of units of economics objects (or yields) delivered by the specific asset under consideration.

More formally, we have what follows.

- **Nominal Assets.** The return matrix function is constant, i.e.,

$$R : \mathbb{R}^G_{++} \to \mathcal{M}_{S,I}, \quad p \mapsto \begin{bmatrix} y^{11} & y^{1i} & y^{1I} \\ \dots & \dots & \dots \\ y^{s1} & y^{si} & y^{sI} \\ & & \\ y^{S1} & y^{Si} & y^{SI} \end{bmatrix}$$

  where $y^{si} \in \mathbb{R}$ is the number of **units of account** that asset i promises to pay in state s. Observe that with little abuse of notation, we can identify the set $\mathcal{R}_{S,I}$ of return matrices with the set $\mathcal{M}_{S,I}$ of matrices of real numbers, or simply with $\mathbb{R}^{SI}$.

- **Real Assets.** The return matrix is such that in each state $s \in \mathcal{S}$, asset i pays a vector $y^{si} \equiv \left(y^{sic}\right)_{c=1}^{C} \in \mathbb{R}^C$ of goods, i.e., $y^{sic} \in \mathbb{R}$ is the number of units of good c delivered in state s. Therefore the return in state s is $r^{si}(p) = p^s y^{si}$ and the return matrix is

$$\begin{bmatrix} p^1 y^{11} & p^1 y^{1i} & p^1 y^{1I} \\ \dots & \dots & \dots \\ p^s y^{s1} & p^s y^{si} & p^s y^{sI} \\ & & \\ p^S y^{S1} & p^S y^{Si} & p^S y^{SI} \end{bmatrix}$$

  We can then identify the set $\mathcal{R}_{S,I}$ with $\mathbb{R}^{SCI}$.

- **Numeraire Assets.** They are a specific case of real assets. They pay in units of the numeraire commodity only. y^{si} is the number of units of the numeraire commodity that asset i promises to pay in state s. Taking the numeraire commodity as commodity C, the return in state s is $r^{si}(p) = p^{sC} y^{si}$ and the return matrix is

$$\begin{bmatrix} p^{1C} y^{11} & p^{1C} y^{1i} & p^{1C} y^{1I} \\ \dots & \dots & \dots \\ p^{sC} y^{s1} & p^{sC} y^{si} & p^{sC} y^{sI} \\ & & \\ p^{SC} y^{S1} & p^{SC} y^{Si} & p^{SC} y^{SI} \end{bmatrix}$$

In this case, we can identify $\mathcal{R}_{S,I}$ with $\mathbb{R}^{SI}$.

The importance of the models with incomplete markets can also be greatly appreciated analyzing the validity of each of the main results of standard Arrow-Debreu model (i.e., existence, Pareto optimality and generic local uniqueness of equilibria) in that more general model.

Incomplete market models studied in the literature differ one from the other for the types of returns promised by available assets. In fact, *nominal* assets deliver units of account or "money"; *numeraire* assets units of a pre-specified good; *real* assets bundles of goods.

Each of the three main results of the Arrow-Debreu model fails to hold in one or more of the above defined type of incomplete market frameworks, as roughly[8] summarized in the following table

	existence	generic local uniqueness	Pareto optimality
numeraire assets	Yes	Yes*	No*
nominal assets	Yes	No*	No*
real assets	Yes*	Yes*	No*

(10.13)

where

"Yes" means "it is always insured",

"Yes*" means "it is generically[9], but not always insured", and

"No*" means "it is generically not insured".

In fact, if either all types of assets are available, or existing assets have mixed returns - i.e., they promised to deliver both units of accounts and goods - then all main results of the complete market model fail to hold - see for example, [126], [74], [93] and [95].

One of the main goals of this book is to carefully show each result in the table - see Chapters - 11-13.

Remark 23 When uncertainty enters the general equilibrium model, the very goal of firms is not easily defined. If profit is uncertain "the argument for a unanimous desire for profit maximization breaks down. Because profit, and therefore derived wealth, are now uncertain, the risk attitudes and expectations of owners will influence their preferences with regard to production plans. For example, strong risk averters will prefer relatively less risky production plans than moderate risk averters"[10]. We decided not to analyze a model with

[8]The results in the real asset model with respect to generic local uniqueness and Pareto optimality refere to a subset of the equilibrium set - see Chapter 13.

[9]"Generically" means in an open and full measure subset of the endowment space for the case of nominal and numeraire assets and in an open and full measure subset of the space of endowments and *asset structures* in the case of real assets.

[10][87], p. 153.

production and uncertainty. A good reference for that topic is Chapter 6 in [82] and references quoted there.

Chapter 11

NUMERAIRE ASSETS

In this chapter, we describe the main results about the model with incomplete financial markets and numeraire assets. That model has been analyzed by [61] and it is the closest one to the exchange economy one in terms of basic results: both existence and generic regularity hold true. Therefore, it is the easiest to be studied and the most appropriate to start the analysis on incomplete markets.

In both Chapter 8 and 9, to show existence of equilibria and properties of regular economies, we followed a general strategy we hope the reader has become familiar with. Since we are going to use the same strategy to prove analogous results in the incomplete market models presented in this chapter and in Chapter 14, we try to summarize that strategy in the steps described below.

- Step 1. Set up of the model. Notation and time structure.
- Step 2. Normalization of commodity prices, i.e., redundant (endogenous) variables. Definition of equilibrium.

  Since we study equilibria in terms of a system of equations, it is crucial to do a simple count of "relevant" equations and unknowns.
- Step 3. Properties of demand function[1].
- Step 4. Walras' laws, i.e., redundant equations.

  The above two steps allow to introduce a system with endogenous and exogenous variables which is the main tool to analyze equilibria.
- Step 5. The function defining the extended equilibrium system of equations.

[1] As we already explained in our discussion on reduced and extended approach in Subsection 8.2, Chapter 8, an independent analysis of the solution function to the household's maximization problem is not needed in the proof of existence of equilibria.

Then results about existence and regular economies can be studied.

- Step 6. Existence of equilibria: a. test economy; b. homotopy; c. uniqueness of equilibria associated with the test economy; d. regularity of the test economy; e. compactness of the homotopy.
- Step 7. Regular economies: a. 0 is a regular value for the equilibrium function; b. properness of the projection.

Remark 1 In the model with real assets - Chapter 13 - we follow the above steps 1-4 with some differences due to the specific structure of that model. To prove the existence of the relevant kind of equilibria, we use the reduced approach described in Section 8, Chapter 8, adapting steps 5-7 to that approach.

The organization of the chapter is consistent with the above procedure.

Section 1 presents the set up of the model. In Section 2, we analyze the properties of the solutions to the household's maximization problem. In Section 3, we show existence of equilibria. Finally, in Section 4, we study properties of the regular economies.

Section 1 presents the set up of the model (Steps 1-3). In Section 2, we analyze the properties of the solutions to the household's maximization problem (Steps 4). In Section 3, we show existence of equilibria (Steps 5 and 6). Finally, in Section 4, we study properties of the regular economies (Steps 7).

1. Set up of the model

The model we present is a particular case of that one presented in Subsection 2.1 in Chapter 10: assets are of the numeraire kind.

Definition 2 A **numeraire asset** i is an asset which promises to deliver y^{si} units of the numeraire good C in state s. Therefore,

$$\text{for each } s \text{ and } i, \text{ and each } p, \quad r^{si}(p) = p^{sC} y^{si}$$

Definition 3 The $S \times I$ matrix

$$Y \equiv \begin{bmatrix} y^{11} & \cdots & y^{1i} & \cdots & y^{1I} \\ \cdots & & & & \\ y^{s1} & \cdots & y^{si} & \cdots & y^{sI} \\ \cdots & & & & \\ y^{S1} & \cdots & y^{Si} & \cdots & y^{SI} \end{bmatrix}$$

is called **yield matrix** and $y^s \equiv \left(y^{si}\right)_{i=1}^{I}$ is the yield vector in state $s \in S$.

Defined

$$P^{1\cdot} \equiv diag\ \left(p^{sC}\right)_{s=1}^{S}$$

the return matrix function is

$$R:\mathbb{R}^G_{++}\rightarrow\mathcal{M}_{S,I},\;\;p\mapsto P^{1\cdot}Y=\begin{bmatrix} p^{1C}y^{11} & \dots & p^{1C}y^{1i} & \dots & p^{1C}y^{1I} \\ \dots & & & & \\ p^{sC}y^{s1} & \dots & p^{sC}y^{si} & \dots & p^{sC}y^{sI} \\ \dots & & & & \\ p^{SC}y^{S1} & \dots & p^{SC}y^{Si} & \dots & p^{SC}y^{SI} \end{bmatrix}$$

Household h's maximization problem is as follows. For given $e_h\in\mathbb{R}^G_{++}$, $p\in\mathbb{R}^G_{++}$, $q\in\mathbb{R}^I$, $R\in\mathcal{R}$,

$$\begin{array}{lll} \max_{(x_h,b_h)\in\mathbb{R}^G_{++}\times\mathbb{R}^I} & u_h(x_h) & \\ s.t. & & p^0x^0_h+qb_h=p^0e^0_h,\text{ and} \\ & & p^sx^s_h=p^se^s_h+p^{sC}y^sb_h\;\;\text{for } s\geq 1 \end{array}\tag{11.1}$$

or

$$\begin{array}{lll} \max_{(x_h,b_h)\in\widetilde{X}_h\times\mathbb{R}^I} & u_h(x_h) & \\ s.t. & & -\Phi(p)(x_h-e_h)+\begin{bmatrix} -q \\ P^{1\cdot}Y\end{bmatrix}b_h=0 \end{array}\tag{11.2}$$

Observe that for any $\alpha\equiv(\alpha^s)_{s=0}^S\in\mathbb{R}^{S+1}_{++}$,

$$\begin{array}{ll}\{(x_h,b_h)\in\mathbb{R}^G_{++}\times\mathbb{R}^I: & -p^0\left(x^0_h-e^0_h\right)-qb_h=0 \\ & -p^s\left(x^s_h-e^s_h\right)+y^sp^{sC}b_h=0,\text{ for } s\in\mathcal{S}\}=\end{array}$$

$$\begin{array}{ll}=\{(x_h,b_h)\in\mathbb{R}^G_{++}\times\mathbb{R}^I: & -\alpha^0p^0\left(x^0_h-e^0_h\right)-\alpha^0qb_h=0 \\ & -\alpha^sp^s\left(x^s_h-e^s_h\right)+\alpha^sp^{sC}y^sb_h=0,\text{ for } s\in\mathcal{S}\}\end{array}$$

Therefore, there is no loss of generality in "normalizing" prices spot by spot using commodity C price in that spot, i.e., choosing the above α equal to $\left(1/p^{sC}\right)_{s=0}^S$.

With innocuous abuse of notation, for each s, we denote $\frac{p^s}{p^{sC}}$ by p^s and $\frac{q}{p^{0C}}$ by q; therefore, we have that for each $s=0,1,...,S$, $p^{sC}=1$. Define also

$$\text{for each } s,\;\; p^{s\backslash}\equiv\left(p^{sc}\right)_{c\neq C}\in\mathbb{R}^{C-1}_{++},\;\; p^s\equiv\left(p^{s\backslash},1\right)$$

$$p^{\backslash}\equiv\left(p^{s\backslash}\right)_{s=0}^S\in\mathbb{R}^{(S+1)(C-1)}_{++}\;\;\text{and}\;\; p=\left(p^s\right)_{s=0}^S$$

We can then rewrite problems (11.1) (11.2) as follows.

For given $e_h \in \mathbb{R}^G_{++}$, $p^{\backslash} \in \mathbb{R}^{G-(S+1)}_{++}$, $q \in \mathbb{R}^I$, $Y \in \mathcal{M}_{S,I}$,

$$\begin{array}{ll} \max_{(x_h,b_h)\in\mathbb{R}^{C(S+1)}_{++}\times\mathbb{R}^I} & u_h(x_h) \\ s.t. & \begin{array}{ll} -p^0\left(x_h^0 - e_h^0\right) - qb_h = 0 & \text{and} \\ -p^s\left(x_h^s - e_h^s\right) + y^s b_h = 0 & \text{for } s \in \mathcal{S} \end{array} \end{array} \tag{11.3}$$

and

$$\begin{array}{ll} \max_{(x_h,b_h)\in\mathbb{R}^{C(S+1)}_{++}\times\mathbb{R}^I} & u_h(x_h) \\ s.t. & -\Phi\left(p\right)\left(x_h - e_h\right) + \begin{bmatrix} -q \\ Y \end{bmatrix} b_h = 0 \end{array} \tag{11.4}$$

Remark 4 The proposed normalization has two important consequences. Both asset and commodity prices are now expressed in terms of numeraire commodity spot by spot, instead of being expressed in units of account. The dependence of asset returns from commodity prices has been shown to be only "nominal", in the sense that it disappear after the normalization.

An economy is completely characterized by an element $(e, Y, u) \in \mathbb{R}^{GH}_{++} \times \mathcal{M}_{S,I} \times \mathcal{U}$. In this chapter, we take the utility function vector u and the return matrix as given. Therefore, we can give the following definition.

Definition 5 An economy in financial economy model with numeraire assets is an element $e \in \mathbb{R}^{GH}_{++}$.

Definition 6 The vector $\left(x, b, p^{\backslash}, q\right) \in \mathbb{R}^{GH}_{++} \times \mathbb{R}^{IH} \times \mathbb{R}^{G-(S+1)}_{++} \times \mathbb{R}^I$ is a **demand-price financial numeraire asset equilibrium** vector for an economy $e \in \mathbb{R}^{GH}_{++}$ if

1 for $h = 1, ..., H$, for given $\left(p^{\backslash}, q, e\right)$, (x_h, b_h) solves problem (11.4), i.e., households maximize, and

2
$$\begin{array}{ll} \sum_{h=1}^H x_h = \sum_{h=1}^H e_h & \text{and} \\ \sum_{h=1}^H b_h = 0 & \end{array}$$

i.e., both good and asset markets clear.

Since we have seen that the dependence of asset returns from commodity prices is only "nominal", in the sense explained in Remark 4, and consistently with Definitions 11, 13 and 21 discussed in Chapter 10, we make the following assumptions.

Assumption a1 (no-arbitrage) There is no $b \in \mathbb{R}^I$ such that $\begin{bmatrix} -q \\ Y \end{bmatrix} b > 0$.

We also define the no-arbitrage asset price set as

$$Q \equiv \left\{ q \in \mathbb{R}^I : \text{there is no } b \in \mathbb{R}^I \ s.\ th. \begin{bmatrix} -q \\ Y \end{bmatrix} b > 0 \right\}$$

Remark 7 From Lemma 14 in Chapter 10, Assumption $a1$ is equivalent to the following statement

$$\text{there exists } \gamma \in \mathbb{R}^S_{++} \quad \text{such that } q = \gamma \begin{bmatrix} -q \\ Y \end{bmatrix}$$

Assumption a2 (no-redundancy) rank $Y = I$.

Assumption a3 (incomplete financial markets) $0 \leq I < S$.

2. The demand function

Theorem 8 *For given $e_h \in \mathbb{R}^G_{++}$, $p^\backslash \in \mathbb{R}^{G-(S+1)}_{++}$, $q \in Q$, a solution to problem (11.4) exists and is unique.*

Proof. Existence.

For ease of notation drop the subscript h. Since u is C^2, to apply the extreme value theorem ([90] Theorem 6.4, p.175), we need to show that the budget set is compact. Following the same arguments used in Chapter 8, it easy to show that the set of solution to problem (11.4) is the same as the set of solutions to the problem

$$\begin{array}{lll} \max\limits_{(x,b) \in \mathbb{R}^{C(S+1)}_{++} \times \mathbb{R}^I} & u(x) & \\ s.t. & -\Phi(p)(x-e) + \begin{bmatrix} -q \\ Y \end{bmatrix} b = 0 & \text{and} \\ & u(x) \geq u(e) & \end{array} \tag{11.5}$$

We have therefore to show that the budget set of problem (11.5), which we denote by $B(p^\backslash, q, e)$, is closed in the topology of $\mathbb{R}^G$ and bounded. The first statement follows from Assumption u4 and Lemma 76 in Chapter 8. Boundedness is showed below.

Suppose $B(p^\backslash, q, e)$ is not bounded, i.e., there exists a sequence $(x^v, b^v)_{v \in \mathbb{N}}$ in $B(p^\backslash, q, e)$ such that $\|(x^v, b^v)\| \to +\infty$. Consider the sequence

$$\left(\frac{x^v, b^v}{\|(x^v, b^v)\|} \right)_{v \in \mathbb{N}}$$

in the sphere, a compact set. Then, up to a subsequence,

$$\left(\frac{x^v, b^v}{\|(x^v, b^v)\|} \right) \to (x, b) \neq 0$$

Consider

$$-\Phi(p)(x^v - e) - \begin{bmatrix} -q \\ Y \end{bmatrix} b^v = 0$$

and divide both sides by $\|(x^v, b^v)\|$ and take limits. You then get

$$\Phi(p)\, x = \begin{bmatrix} -q \\ Y \end{bmatrix} b \tag{11.6}$$

Since for $v \in \mathbb{N}$, $x^v \gg 0$, $x \geq 0$. We have then to study two cases keeping in mind that $(x, b) \neq 0$.

Case 1. $x = 0$ and therefore $b \neq 0$.

Then (11.6) becomes $Yb = 0$, which is impossible because of Assumption a2.

Case 2. $x \neq 0$ and therefore $x > 0$.

Since prices are strictly positive, (11.6) becomes

$$0 < \Phi(p)\, x = \begin{bmatrix} -q \\ Y \end{bmatrix} b$$

which contradicts Assumption $a1$.

Uniqueness.

Suppose otherwise, i.e., there exist two different vectors, (x, b) and (x', b'), both solutions to problem (11.4). If $x = x'$, from Assumption $a2$, and from the budget constraints, $b = b'$, a contradiction. Then assume $x \neq x'$, while $u(x) = u(x')$,

$$\Phi(p)(x - e) + \begin{bmatrix} -q \\ Y \end{bmatrix} b = 0, \text{ and } \quad -\Phi(p)(x' - e) + \begin{bmatrix} -q \\ Y \end{bmatrix} b' = 0$$

for some b and b'. Then consider $\widehat{x} = \frac{1}{2}(x + x')$. Multiplying the above budget constraints by $\frac{1}{2}$ and summing up, we get

$$\Phi(p)\left(\frac{x + x'}{2} - e\right) + \begin{bmatrix} -q \\ Y \end{bmatrix} \frac{b + b'}{2} = 0$$

i.e., $\left(\widehat{x}_h, \frac{b+b'}{2}\right)$ satisfies the budget constraint and from Assumption u3,

$$u(\widehat{x}) > u(x) = u(x')$$

a contradiction. ■

Definition 9 Household h's demand function for goods and assets is the solution function to (11.4), i.e.,

$$(x_h, b_h) : \mathbb{R}_{++}^{G-(S+1)} \times Q \times \mathbb{R}_{++}^{C} \to \mathbb{R}_{++}^{G}, \quad \left(p^{\backslash}, q, e_h\right) \mapsto \arg\max (11.4)$$

Theorem 10 *The following statements hold.*

1 $(x_h, b_h) = (x_h, b_h)\left(p^{\setminus}, q, e_h\right)$ *if and only if there exists* $\lambda_h \in \mathbb{R}^{S+1}_{++}$ *such that* (x_h, b_h, λ_h) *is the unique solution to system*

$$\begin{cases} Du_h(x_h) - \lambda_h \Phi(p) = 0 & (1) \\ -\Phi(p)(x_h - e_h) + \left[\begin{smallmatrix} -q \\ Y \end{smallmatrix}\right] b_h = 0 & (2) \\ \lambda_h \left[\begin{smallmatrix} -q \\ Y \end{smallmatrix}\right] = 0 & (3) \end{cases} \tag{11.7}$$

in the exogenous variables $\left(p^{\setminus}, q, e_h\right)$*;*

2 $(x_h, b_h)\left(p^{\setminus}, q, e_h\right)$ *is a* C^1 *function.*

Proof. 1. The result follows showing that (Ph) satisfies necessary and sufficient conditions for Lagrange theorem (see Subsection 3.2.1 in Chapter 1). The lagrangean function for problem (11.4) is

$$\mathcal{L}\left(x_h, b_h, \lambda_h, p^{\setminus}, q, e_h\right) = u_h(x_h) + \lambda_h \left(-\Phi(p)(x_h - e_h) + \begin{bmatrix} -q \\ Y \end{bmatrix} b_h\right)$$

and the associated first order conditions are just system (11.7).

[⇒] Necessary conditions are satisfied because

$$D_{x_h}\left[-\Phi(p)(x_h - e_h) + \begin{bmatrix} -q \\ Y \end{bmatrix} b_h\right] = -\Phi(p)$$

whose rank is $S+1$. Therefore, if $(x_h, b_h) = (x_h, b_h)\left(p^{\setminus}, q, e_h\right)$, then there exists $\lambda_h \in \mathbb{R}^{S+1}$ such that (x_h, b_h, λ_h) is the unique solution to system (11.7). The fact that $\lambda_h \gg 0$ follows from equation (1) in that system, Assumption u2 and thc fact that $p \in \mathbb{R}^G_{++}$.

[⇐] Sufficient conditions for Lagrange theorem are satisfied because, from Assumption u3, u_h is differentiably strictly quasiconcave and the constraint functions are linear.

2. Define

$$F_h : \mathbb{R}^G_{++} \times \mathbb{R}^{S+1}_{++} \times \mathbb{R}^I \times \mathbb{R}^{G-(S+1)}_{++} \times Q \times \mathbb{R}^G \to \mathbb{R}^G_{++} \times \mathbb{R}^{S-1} \times \mathbb{R}^I$$

$$F_h : \left(x_h, \lambda_h, b_h, p^{\setminus}, q, e_h\right) \mapsto \text{left hand side of (11.7)}$$

As an application of the implicit function theorem, it is enough to show that $D_{(x_h,\lambda_h,b_h)}F_h$ has full row rank $G + (S+1) + I$.

The computation of $D_{(x_h,\lambda_h,b_h)}F_h$ is described below.

	x_h	λ_h	b_h
$Du_h(x_h) - \lambda_h \Phi(p)$	$D^2 u_h(x_h)$	$-\Phi(p)^T$	0
$-\Phi(p)(x_h - e_h) + \left[\begin{smallmatrix} -q \\ Y \end{smallmatrix}\right] b_h$	$-\Phi(p)$	0	Y
$\lambda_h \left[\begin{smallmatrix} -q \\ Y \end{smallmatrix}\right]$	0	Y^T	0

Suppose that $D_{(x_h,\lambda_h,b_h)}F_h$ does not have full rank, that is, suppose there exists $\Delta = (\Delta x_h, \Delta\lambda_h, \Delta b_h) \neq 0$, where $\Delta x_h \in \mathbb{R}^G$, $\Delta\lambda_h \in \mathbb{R}^{S+1}$ and $\Delta b_h \in \mathbb{R}^I$, such that

$$\left[D_{(x_h,\lambda_h,b_h)}F_h\right]\Delta = 0$$

or

$$\left\{\begin{array}{lcll} D^2u_h(x_h)\,\Delta x_h - \Phi(p)\,\Delta\lambda_h & = & 0 & (1) \\ -\Phi(p)\,\Delta x_h + Y\Delta b_h & = & 0 & (2) \\ Y^T\Delta\lambda_h & = & 0 & (3) \end{array}\right. \tag{11.8}$$

Claim 1. $\Delta x_h \neq 0$.
Suppose otherwise. Then from (1) in (11.8), $-\Phi(p)\,\Delta\lambda_h = 0$, and since $\operatorname{rank}\Phi(p) = S+1$, $\Delta\lambda_h = 0$. Moreover, from (2) in (11.8), $Y\Delta b_h = 0$, and since $\operatorname{rank} Y = I$, $\Delta b_h = 0$. But then $\Delta = 0$, a contradiction.

Claim 2. $Du_h(x_h)\Delta x_h = 0$.
From (11.7.1), $Du_h(x_h)\Delta x_h = \lambda_h\Phi(p)\,\Delta x_h$;
from (2) in (11.8), $\lambda_h\Phi(p)\,\Delta x_h = \lambda_h Y\Delta b_h$;
from (3) in (11.7), $\lambda_h Y = 0$.

Claim 3. $\Delta x_h D^2u_h(x_h)\,\Delta x_h \not< 0$.
From (1) in (11.8), $\Delta x_h D^2u_h(x_h)\,\Delta x_h = \Delta x_h\Phi(p)\,\Delta\lambda_h$;
from (2) in (11.8), $\Delta x_h\Phi(p)\,\Delta\lambda_h = \Delta b_h Y\Delta\lambda_h$;
from (3) in (11.8), $Y\Delta\lambda_h = 0$.

Then Claims 1, 2 and 3 contradict Assumption u3. ■

3. Existence of equilibria

Usually, in equilibrium models with several budget constraints with the same structure for all households, an analogue of Walras' law [2] holds for each budget constraint. We formally show that result for the present model in Proposition 11 below. Define

$$x_h^{\setminus} \equiv \left(x_h^{sc}\right)_{c\neq C} \in \mathbb{R}_{++}^{G-(S+1)} \quad \text{and} \quad e_h^{\setminus} \equiv \left(e_h^{sc}\right)_{c\neq C} \in \mathbb{R}_{++}^{G-(S+1)}$$

Proposition 11 $\left(x, b, p^{\setminus}, q\right)$ *is a demand-price financial numeraire asset equilibrium for e if and only if* $\left(x, b, p^{\setminus}, q\right)$ *satisfies Condition 1 in Definition 6 and the following version of Condition 2*

$$\sum_{h=1}^{H} x_h^{\setminus} = \sum_{h=1}^{H} e_h^{\setminus} \quad \textit{and} \quad \sum_{h=1}^{H} b_h = 0 \tag{11.9}$$

Proof. If Condition 2 in Definition 6 holds then (11.9) is trivially satisfied. Suppose now that (11.9) holds. Summing up the budget constraints in problem

[2] See Proposition 29 in Chapter 8.

(11.2, we get

$$0 = -\Phi(p) \sum_h (x_h - e_h) + \begin{bmatrix} -q \\ P^{1\cdot}Y \end{bmatrix} \sum_h b_h$$

Using (11.9) and the fact that $p \gg 0$, we get that for $s \in S'$,

$$\sum_{h=1}^{H} (x_h^{sC} - e_h^{sC}) = 0$$

■

Following exactly the same procedure as in the previous chapters, we describe equilibria using the following "extended" equilibrium system of first order conditions and market clearing conditions .

# equations	matching variables		equations	
G	x_h	$(h1), h \in \mathcal{H}$	$Du_h(x_h) - \lambda_h \Phi(p)$	$= 0$
$S+1$	λ_h	$(h2), h \in \mathcal{H}$	$-\Phi(p)(x_h - e_h) + \left[\begin{smallmatrix} -q \\ Y \end{smallmatrix}\right] b_h$	$= 0$
I	b_h	$(h3), h \in \mathcal{H}$	$\lambda_h \left[\begin{smallmatrix} -q \\ Y \end{smallmatrix}\right]$	$= 0$
$G-(S+1)$	$p^{\setminus}$	(Mg)	$-\sum_{h=1}^{H} x_h^{\setminus} + \sum_{h=1}^{H} e_h^{\setminus}$	$= 0$
I	q	(Ma)	$\sum_{h=1}^{H} b_h$	$= 0$

(11.10)

Observe that the number of equations is equal to the number of endogenous variables.

Definition 12 Define

$$\xi \equiv \left((x_h, \lambda_h, b_h)_{h \in \mathcal{H}}, p^{\setminus}, q) \right) \in \Xi$$

where

$$\Xi \equiv \mathbb{R}_{++}^{GH} \times \mathbb{R}_{++}^{(S+1)H} \times \mathbb{R}^{IH} \times \mathbb{R}_{++}^{G-(S+1)} \times \mathbb{R}^{I}$$

and

$$F : \Xi \times \mathbb{R}_{++}^{GH} \to \mathbb{R}^{\dim \Xi}, \quad (\xi, e) \mapsto \text{left hand side of (11.10)}$$

Remark 13 Observe that we let q belong to $\mathbb{R}^I$ instead of the no-arbitrage set Q: in fact equations $(h3)$ in (11.10) insure that $q \in Q$.

Definition 14 $\xi \in \Xi$ is an **extended financial numeraire asset equilibrium vector** for the economy $e \in \mathbb{R}_{++}^{GH}$ if $F(\xi, e) = 0$.

With innocuous abuse of terminology, we will call ξ simply an equilibrium.

According to Theorem 57 in Chapter 7, we identify

$$\begin{array}{lll} f & \text{with} & F_e \\ y & \text{with} & 0 \\ M & \text{with} & \Xi \\ N & \text{with} & \mathbb{R}^{\dim \Xi} \end{array}$$

To verify the assumptions of the above theorem, we proceed as follows. In the present section, we first construct the needed homotopy H - see (11.11), using a Pareto optimal allocation as the test economy . Then, we show the other needed results: there exists $\xi^* \in \Xi$ such that $\xi^* \in G^{-1}(0)$ - Lemma 15 -, $\{\xi^*\} = G^{-1}(0)$ - Lemma 17 -, $DG(\xi^*)$ has full rank - Lemma 18 -, $H^{-1}(0)$ is compact - Lemma 19.

Consider an arbitrary economy $e \in \mathbb{R}^{GH}_{++}$, a Pareto optimal allocation x^* feasible at e, i.e., such that $\sum_{h=1}^{H} x_h^* = \sum_{h=1}^{H} e_h$. *For given e and* x^*, define

$$\left\{ \begin{array}{lll} Du_h(x_h) - \lambda_h \Phi(p) & = 0 & (h.1) \\ \lambda_h \left[\begin{smallmatrix} -q \\ Y \end{smallmatrix}\right] & & (h.2) \\ -\Phi(p)\left(x_h - ((1-\tau)e_h + \tau x_h^*)\right) + \left[\begin{smallmatrix} -q \\ Y \end{smallmatrix}\right] b_h & = 0 & (h.3) \\ \cdots & & \\ -\sum_{h=1}^{H} x_h^{\backslash} + \sum_{h=1}^{H} \left((1-\tau)e_h^{\backslash} + \tau x_h^{\backslash *}\right) & = 0 & (Mg) \\ \sum_{h=1}^{H} b_h & = 0 & (Mb) \end{array} \right. \tag{11.11}$$

$$H_e : \Xi \times [0,1] \to \mathbb{R}^n, \quad (\xi, \tau) \mapsto \text{ left hand side of system (11.11)}$$

and

$$G : \Xi \to \mathbb{R}^{\dim \Xi}, \quad \xi \mapsto H_e(\xi, 1)$$

Observe that

$$H_e(\xi, 0) = F(\xi, e)$$

and

$$H_e(\xi, \tau) = F(\xi, (1-\tau)e + \tau x^*)$$

We have now to construct *the* equilibrium associated with the economy x^*. Reformulated Lemma 40 in Chapter 8 replacing C with G, we know that there exists $\left(\theta^*_{\backslash 1}, \gamma^*\right) \in \mathbb{R}^{H-1}_{++} \times \mathbb{R}^{G}_{++}$ such that $\left(x^*, \theta^*_{\backslash 1}, \gamma^*\right)$ is the unique solution to system (8.13) at $\left(\sum_{h=1}^{H} x_h^*, (u_h(x_h^*))_{h \neq 1}\right)$, i.e.,

$$\left\{ \begin{array}{lll} Du_1(x_1) - \gamma & = 0 & (1) \\ \theta_h Du_h(x_h) - \gamma & = 0 & (2) \\ (u_h(x_h) - u_h(x_h^*))_{h \neq 1} & = 0 & (3) \\ -\sum_{h=1}^{H} x_h + \sum_{h=1}^{H} x_h^* & = 0 & (4) \end{array} \right. \tag{11.12}$$

As done in Chapter 8, let $\theta_1^* \equiv 1$.

Observe that $p^{sC*} = 1$ for any $s \in S'$.

Lemma 15 $\xi^* \in G^{-1}(0)$, *where*

$$\xi^* = \left(\left(x_h^*, \lambda_h^* = \left(\frac{\gamma^{sC*}}{\theta_h^*} \right)_{s=0}^{S}, b_h^* = 0 \right)_{h \in H}, p^* = \left(\frac{\gamma^{s*}}{\gamma^{sC*}} \right)_{s=0}^{S}, q^* = \sum_{s=1}^{S} \left(\frac{\gamma^{sC*}}{\gamma^{0C*}} \right) y^s \right)$$

Proof. We first give an idea of how to construct ξ^* and then we check that the statement of the lemma is correct.

Since form system (11.12) we know that $Du_h(x_h^*) - \frac{\gamma^*}{\theta_h^*} = 0$, we also have

$$D_{x_h^s} u_h(x_h^*) - \frac{\gamma^{s*}}{\theta_h^*} = 0, \quad \text{for } s \geq 0$$

"To get as close as we can to the first order conditions of the household's problem", we can rewrite it as

$$D_{x_h^s} u_h(x_h^*) - \left(\frac{\gamma^{sC*}}{\theta_h^*} \right) \left(\frac{\gamma^{s*}}{\gamma_h^{sC*}} \right) = 0$$

which clearly suggests the natural choice of λ_h^* and p^*.

We should also have

$$\lambda_h^* \begin{bmatrix} -q^* \\ Y \end{bmatrix} = 0$$

and therefore

$$q^* = \sum_{s=1}^{S} \frac{\lambda_h^{s*}}{\lambda_h^{0*}} y^s = \sum_{s=1}^{S} \frac{\frac{\gamma^{sC*}}{\theta_h^*}}{\frac{\gamma^{0C*}}{\theta_h^*}} y^s = \sum_{s=1}^{S} \frac{\gamma^{sC*}}{\gamma^{0C*}} y^s$$

which gives the right choice of q^*.

Let's now check that, using the proposed choice of $\xi^*$and (11.12), we get the desired result.

For each h and s,

$$D_{x_h^s} u_h(x_h^*) - \lambda_h^{s*} p^{s*} = D_{x_h^s} u_h(x_h^*) - \frac{\gamma^{sC*}}{\theta_h^*} \frac{\gamma^{s*}}{\gamma^{sC*}} = D_{x_h^s} u_h(x_h^*) - \frac{\gamma^{s*}}{\theta_h^*} = 0$$

$$\lambda_h^* \begin{bmatrix} -q^* \\ Y \end{bmatrix} = \left(\frac{\gamma^{sC*}}{\theta_h^*} \right)_{s=0}^{S} \begin{bmatrix} -\sum_{s=1}^{S} \left(\frac{\gamma^{sC*}}{\gamma^{0C*}} \right) y^s \\ Y \end{bmatrix} = 0$$

Finally, $b_h^* = 0$ allows to get both budget equations and market clearing conditions satisfied. ∎

Remark 16 From the fact that for $s \in \mathcal{S}$, we have that

$$D_{x_h^s} u_h (x_h^*) - \lambda^{s*} p^{s*} = 0$$

we also have that

$$D_{x_h^s C} u_h (x_h^*) = \lambda^{s*}$$

Therefore, for $s \in \mathcal{S}'$

$$p^{s*} = \frac{D_{x_h^s} u_h (x_h^*)}{\lambda^{s*}} = \frac{D_{x_h^s} u_h (x_h^*)}{D_{x_h^s C} u_h (x_h^*)}$$

Lemma 17 $\{\xi^*\} = G^{-1}(0)$.

Proof. From the previous lemma, it is enough to show that if $F\left(\widehat{\xi}, x^*\right) = 0$, then $\widehat{\xi} = \xi^*$. Define $\widehat{\xi} \equiv \left(\widehat{x}, \widehat{\lambda}, \widehat{b}, \widehat{p}, \widehat{q}\right)$.

1. $\widehat{x} = x^*$.

Suppose otherwise, i.e., $\widehat{x} \neq x^*$. Consider $\widetilde{x} \equiv \frac{1}{2}(\widehat{x} + x^*)$. Since $F\left(\widehat{\xi}, x^*\right) = 0$, $\sum_h \widehat{x}_h = \sum_h x_h^*$ and

$$\sum_h \widetilde{x}_h = \frac{1}{2}\left(\sum_h \widehat{x}_h + \sum_h x_h^*\right) = \sum_h x_h^* \tag{11.13}$$

From Assumption u3, we then have

$$u(\widetilde{x}) \gg u(x^*) \tag{11.14}$$

But (11.13) and (11.14) contradict the Pareto optimality of x^*. Therefore, we have that for $h = 1, ..., H$

$$\begin{cases} Du_h (x_h^*) - \widehat{\lambda}_h \Phi(\widehat{p}) = 0 & (1) \\ Y(\widehat{q})\, \widehat{b}_h = 0 & (2) \\ \widehat{\lambda}_h Y(\widehat{q}) = 0 & (3) \end{cases} \tag{11.15}$$

2. $\widehat{\lambda} = \lambda^*$.

For any h and s, from (1) in (11.15), $D_{x^s C} u_h (x_h^*) - \widehat{\lambda}_h^s = 0$. Therefore, from Remark 16, $\widehat{\lambda}_h^s = \lambda_h^{s*}$.

3. $\widehat{p} = p^*$.

From (1) in (11.15), point 2 above and Remark 16,

$$\widehat{p^s} = \frac{D_{x^s} u_h (x_h^*)}{\widehat{\lambda}_h^s} = \frac{D_{x_h^s} u_h (x_h^*)}{D_{x_h^s C} u_h (x_h^*)} = p^{s*}$$

4. $\widehat{q} = q^*$.

From (3) in (11.15), point 2 above and from Lemma 15,

$$\widehat{q} = \sum_{s=1}^{S} \frac{\widehat{\lambda}_h^s}{\widehat{\lambda}_h^0} y^s = \sum_{s=1}^{S} \frac{\lambda_h^{s*}}{\lambda_h^{0*}} y^s = q^*$$

5. $\widehat{b} = b^*$.

From (2) in (11.15) and from Assumption $a2$, we get that for any h, $\widehat{b}_h = 0 = b_h^*$.
■

Lemma 18 *$D_\xi F(\xi^*, x^*)$ has rank* $\dim \Xi$.

Proof. The computation of $D_\xi F(\xi^*, x^*)$ is described below

$$\begin{array}{l} \\ \ldots \\ Du_h - \lambda_h \Phi(p) \\ -\Phi(p)(x_h - x_h^*) + \left[\begin{smallmatrix} -q \\ Y \end{smallmatrix}\right] b_h \\ \lambda_h \left[\begin{smallmatrix} -q \\ Y \end{smallmatrix}\right] \\ \ldots \\ \sum_h \left(x_h^{\backslash} - x_h^{*\backslash} \right) \\ \sum_h b_h \end{array}
\begin{array}{c} \begin{array}{cccccc} x_h & \lambda_h & b_h & \ldots & p^{\backslash} & q \end{array} \\ \left(\begin{array}{cccccc}
 & & & & & \\
D^2 u_h(x_h^*) & -\Phi(p^*)^T & & & -\Lambda_h^* & \\
-\Phi(p^*) & & \left[\begin{smallmatrix} -q^* \\ Y \end{smallmatrix}\right] & & & \\
 & \left[\begin{smallmatrix} -q^* \\ Y \end{smallmatrix}\right]^T & & & & -\lambda_h^{0*} I_I \\
 & & & \ddots & & \\
\widehat{I} & & & & & \\
 & & I & & &
\end{array} \right) \end{array}$$

where

$$\widehat{I} = \begin{bmatrix} I_{C-1} \;\; 0 & & \\ & \ddots & \\ & & I_{C-1} \;\; 0 \end{bmatrix}_{G-(S+1) \times G}$$

$$\Lambda_h^* \equiv \begin{bmatrix} \lambda_h^{0*} I_{C-1} & & \\ 0 & & \\ & \ddots & \\ & & \lambda_h^{S*} I_{C-1} \\ & & 0 \end{bmatrix}_{G \times G-(S+1)}$$

Since $\lambda_h^{s*} \equiv \frac{\gamma^{sC*}}{\theta_h^*}$, we also have

$$\Lambda_h^* = \frac{1}{\theta_h^*} \begin{bmatrix} \gamma^{0C*} I_{C-1} & & \\ 0 & & \\ & \ddots & \\ & & \gamma^{SC*} I_{C-1} \\ & & 0 \end{bmatrix} \equiv \frac{1}{\theta_h^*} \Gamma^* \qquad (11.16)$$

Observe that Γ^* does not depend upon h.

Take $\Delta \equiv \left((\Delta x_h, \Delta\lambda_h, \Delta b_h)_{h=1}^{H}, \Delta p^{\setminus}, \Delta q \right)$, where for each h, $\Delta x_h \in \mathbb{R}^G, \Delta\lambda_h \in \mathbb{R}^{S+1}, \Delta b_h \in \mathbb{R}^I$, and $\Delta p^{\setminus} \in \mathbb{R}^{G-(S+1)}, \Delta q \in \mathbb{R}^I$. We now show that $D_\xi F(\xi^*, x^*)\,\Delta = 0$ implies that $\Delta = 0$. $D_\xi F(\xi^*, x^*)\,\Delta = 0$ is rewritten below.

$$\begin{cases} D^2 u_h(x_h^*)\,\Delta x_h - \Phi(p^*)\,\Delta\lambda_h - \Lambda_h^* \Delta p^{\setminus} = 0 & (h.1) \\ -\Phi(p^*)\,\Delta x_h + \left[\begin{smallmatrix} -q^* \\ Y \end{smallmatrix}\right] \Delta b_h = 0 & (h.2) \\ \theta_h^* \left[\begin{smallmatrix} -q^* \\ Y \end{smallmatrix}\right] \Delta\lambda_h - \gamma^{0C*}\Delta q = 0 & (h.3) \\ \sum_h \Delta x_h^{\setminus} = 0 & (Mx) \\ \sum_h \Delta b_h = 0 & (Mb) \end{cases} \qquad (11.17)$$

where in equation $(h.3)$ we used the fact that $\lambda_h^{0*} \equiv \frac{\gamma^{0C*}}{\theta_h^*}$.

We are also going to use the first order conditions for household h's problem, which are rewritten below.

$$\begin{cases} Du_h(x_h^*) - \lambda_h^* \Phi(p^*) = 0 & = 0 & (h1) \\ -\Phi(p)\,(x_h^* - x_h^*) + \left[\begin{smallmatrix} -q^* \\ Y \end{smallmatrix}\right] b_h^* = 0 & = 0 & (h2) \\ \lambda_h^* \left[\begin{smallmatrix} -q^* \\ Y \end{smallmatrix}\right] = 0 & = 0 & (h3) \end{cases} \qquad (11.18)$$

Claim 1. There exists h' such that $\Delta x_{h'} \neq 0$.

Suppose otherwise, i.e., for any h, $\Delta x_h = 0$. Then (h.2) in (11.17) becomes $\left[\begin{smallmatrix} -q^* \\ Y \end{smallmatrix}\right] \Delta b_h = 0$, and since Y has full rank, we get $\Delta b_h = 0$. From (h.1) in (11.17)

$$\begin{bmatrix} p^{0C*} & & \\ & \ddots & \\ & & p^{SC*} \end{bmatrix} \begin{bmatrix} \Delta\lambda_h^0 \\ \cdots \\ \Delta\lambda_h^S \end{bmatrix} = 0$$

and therefore $\Delta\lambda_h = 0$. But then from (h.1) in (11.17), we get that for any s, $\lambda_h^s \Delta p^{s\setminus} = 0$ and therefore $\Delta p^{\setminus} = 0$. Finally, from (h.3) in (11.17), we get $\Delta q = 0$. But then $\Delta = 0$, a contradiction.

Claim 2. $Du_h(x_h^*)\,\Delta x_h = 0$, for $h = 1, \dots, H$.

Postmultiplying (h.1) in (11.18) by Δx_h, we get

$$Du_h(x_h^*)\,\Delta x_h = \lambda_h^* \Phi(p^*)\,\Delta x_h$$

Premultiplying (h.2) in (11.17) by λ_h^*, we get

$$\lambda_h^* \Phi(p^*)\,\Delta x_h = \lambda_h^* \begin{bmatrix} -q^* \\ Y \end{bmatrix} \Delta b_h = 0$$

where the last equality comes from (h.3) in (11.18).

Claim 3. $\sum_{h=1}^{H}\left(\theta_h^*\Delta x_h D^2 u_h\left(x_h^*\right)\Delta x_h\right)=0.$
Premultiplying (h.1) in (11.17) by $\theta_h^*\Delta x_h$, we get

$$0=\theta_h^*\Delta x_h D^2 u_h\left(x_h^*\right)\Delta x_h-\theta_h^*\Delta x_h\Phi\left(p^*\right)\Delta\lambda_h-\theta_h^*\Delta x_h\Lambda_h^*\Delta p^{\backslash}$$

From (h.2) in (11.17),

$$\theta_h^*\Delta x_h\Phi\left(p^*\right)=\theta_h^*\Delta b_h\begin{bmatrix}-q^*\\ Y\end{bmatrix}$$

and therefore

$$0=\theta_h^*\Delta x_h D^2 u_h\left(x_h^*\right)\Delta x_h-\theta_h^*\Delta b_h\begin{bmatrix}-q^*\\ Y\end{bmatrix}\Delta\lambda_h-\theta_h^*\Delta x_h\Lambda_h^*\Delta p^{\backslash}$$

Summing up with respect to h, we get

$$\begin{aligned}0=&\sum_{h=1}^{H}\left(\theta_h^*\Delta x_h D^2 u_h\left(x_h^*\right)\Delta x_h\right)-\sum_{h=1}^{H}\left(\theta_h^*\Delta b_h\left[\begin{smallmatrix}-q^*\\ Y\end{smallmatrix}\right]\Delta\lambda_h\right)\\&-\left(\sum_{h=1}^{H}\theta_h^*\Delta x_h\Lambda_h^*\right)\Delta p^{\backslash}\end{aligned}\tag{11.19}$$

From (h.3) in (11.17),

$$\theta_h^*\begin{bmatrix}-q^*\\ Y\end{bmatrix}\Delta\lambda_h=\gamma^{0C*}\Delta q$$

From (11.16), we also have that

$$\theta_h^*\Delta x_h\Lambda_h^*=\Delta x_h\Gamma^*$$

Substituting in (11.19), we then get

$$\begin{aligned}0=\;&\sum_{h=1}^{H}\left(\Delta x_h D^2 u_h\left(x_h^*\right)\Delta x_h\right)-\left(\sum_{h=1}^{H}\Delta b_h\right)\gamma^{0C*}\Delta q\\&-\left(\sum_{h=1}^{H}\Delta x_h^{\backslash}\Gamma^*\right)\Delta p^{\backslash}\\=\;&\sum_{h=1}^{H}\left(\Delta x_h D^2 u_h\left(x_h^*\right)\Delta x_h\right)\end{aligned}$$

where the last equality comes from (Mx) and (Mb) in (11.17).
Claim 4. $\sum_{h=1}^{H}\left(\theta_h^*\Delta x_h D^2 u_h\left(x_h^*\right)\Delta x_h\right)<0$, (contradicting Claim 3).
Partition $\mathcal{H}$ in the two sets $\mathcal{H}_0$ and $\mathcal{H}_1$ such that
If $h\in\mathcal{H}_0$, then $\Delta x_h=0$ and if $h\in\mathcal{H}_1$, then $\Delta x_h\neq 0$. From Claim 1, $\mathcal{H}_1\neq\emptyset$.
If $h\in\mathcal{H}_0$, then $\theta_h^*\Delta x_h D^2 u_h\left(x_h^*\right)\Delta x_h=0$.
If $h\in\mathcal{H}_1$, from Claim 2, $\theta_h^*>0$ and Assumption u3, we get $\theta_h^*\Delta x_h D^2 u_h\left(x_h^*\right)\Delta x_h<0$ and finally

$$\begin{aligned}&\sum_{h=1}^{H}\left(\theta_h^*\Delta x_h D^2 u_h\left(x_h^*\right)\Delta x_h\right)=\\&=\sum_{h\in H_0}\left(\theta_h^*\Delta x_h D^2 u_h\left(x_h^*\right)\Delta x_h\right)+\sum_{h\in H_1}\left(\theta_h^*\Delta x_h D^2 u_h\left(x_h^*\right)\Delta x_h\right)<0\end{aligned}$$

■

Lemma 19 *For each* $e \in \mathbb{R}^{CH}_{++}$, $H_e^{-1}(0)$ *is compact.*

Proof. We want to show that any sequence $(\xi^v, t^v)_{v \in \mathbb{N}}$ in $H_e^{-1}(0)$ admits a convergent subsequence inside that set. Using the same argument presented for the case of exchange economies, we can show that, up to a subsequence, $(x^v)_{v \in \mathbb{N}}$ converges say to $\overline{x} \in \mathbb{R}^G_{++}$.
From $(h.1)$ in (11.11),

$$\lambda_h^{sv} = D_{x_h^{sC}} u_h(x_h^v) \to D_{x_h^{sC}} u_h(\overline{x}_h) \equiv \overline{\lambda}_h^s > 0$$

where the strict inequality follows from Assumption u2.
Then, again from $(h.1)$ in (11.11),

$$p^{sv} = \frac{D_{x_h^s} u_h(x_h^v)}{\lambda_h^{sv}} \to \frac{D_{x_h^s} u_h(\overline{x}_h)}{\overline{\lambda}_h^s} \equiv \overline{p}^s \in \mathbb{R}^C_{++}$$

From $(h.2)$ in (11.11),

$$q^v = \frac{1}{\lambda_h^{0v}} \sum_{s=1}^{S} \lambda_h^{sv} y^s \to \frac{1}{\overline{\lambda}_h^0} \sum_{s=1}^{S} \overline{\lambda}_h^s y^s \equiv \overline{q} \in \mathbb{R}^I$$

From $(h.3)$ in (11.11), and using the fact that Y has full rank I, we also get that $(b^v)_{v \in \mathbb{N}}$ converges. Finally, since $\{(\xi^v, \tau^v) : v \in \mathbb{N}\} \subseteq H_e^{-1}(0)$ means that for every v

$$F(\xi^v, (1 - \tau^v) e + \tau^v e^*) = 0$$

taking limits of both sides and using the continuity of F, we get

$$F\left(\overline{\xi}, (1 - \overline{\tau}) e + \overline{\tau} e^*\right) = 0$$

i.e., $\left(\overline{\xi}, \overline{\tau}\right) \in H_e^{-1}(0)$, as desired. ■

From the previous lemmas, we finally get the following existence theorem.

Theorem 20 *For every economy* $e \in \mathbb{R}^{GH}_{++}$, *there exists* $\xi \in \Xi$ *such that* $F(\xi, e) = 0$.

4. Regular economies

In this Section, we introduce the concept of regular economies in the case of incomplete markets with numeraire assets and, as in the case of exchange economies, we show that they have nice properties (see Theorems 25 and 26) and constitute a "very large" subset of the set of all economies (see Theorem 27).

As we have already seen in the analysis of regular economies for the models presented in Chapters 8 and 9, the proofs of the above results go through the following steps.

1 Show that

(a) equilibria exists for each economy,

(b) $zero$ is a regular value for F, and therefore $F^{-1}(0)$ is a manifold,

(c) $pr : F^{-1}(0) \to \Pi$, $(\xi, \pi) \mapsto \pi$ is proper - where Π is the set of economies.

2 Use the above results to apply mathematical results proved in Chapter 4 and 6 in order to analyze regular economies as follows.

(a) Corollaries 16 and 44 in Chapter 4 to show the "nice" properties of regular economies - see Theorems 25 and 26.

(b) Theorem 26 and Corollary 27 in Chapter 6 to show that the set of regular economies is open and of full measure - see Theorem 27

In the remaining part of the section, we first introduce the concept of regular economies in the case of incomplete markets with numeraire assets. We then state and show results 1b and 1c above - result 1a being Theorem 20 of the previous section. Finally, we simply state the results referred to in 2a and 2b above; observe that, given our choice of notation,their statements and proofs are almost verbatim those in the analogous theorem in Chapter 8.

Definition 21 $e \in \mathbb{R}^{CH}_{++}$ is a regular economy for an incomplete market model with numeraire asset if 0 is a regular value for F. Let $\mathcal{R}$ be the set of regular economies.

Lemma 22 *0 is a regular value for F.*

Proof. It suffices to prove that for every $(\xi, e) \in F^{-1}(0)$, $DF(\xi, e)$ has full row rank. The computation of a relevant partial jacobian matrix of $DF(\xi, e)$ is described below

$$
\begin{array}{ll}
 & \\
(1.1) & Du_1 - \lambda_1 \Phi(p) \\
(1.2) & -\Phi(p)(x_1 - e_1) + \left[\begin{smallmatrix}-q\\ Y\end{smallmatrix}\right] b_1 \\
(1.3) & \lambda_1 \left[\begin{smallmatrix}-q\\ Y\end{smallmatrix}\right] \\
\ldots & \\
(H.1) & Du_H - \lambda_H \Phi(p) \\
(H.2) & -\Phi(p)(x_H - e_H) + \left[\begin{smallmatrix}-q\\ Y\end{smallmatrix}\right] b_H \\
(H.3) & \lambda_H \left[\begin{smallmatrix}-q\\ Y\end{smallmatrix}\right] \\
(Mx) & \sum_h \left(x_h^{\backslash} - e_h^{\backslash}\right) \\
(Mb) & \sum_h b_h
\end{array}
\qquad
\begin{array}{ccccccccc}
x_1 & \lambda_1 & b_1 & \ldots & x_H & \lambda_H & b_H & e_H^{\backslash} & e_H
\end{array}
$$

$$
\left(
\begin{array}{ccccccccc}
D^2 u_1 & -\Phi(p)^T & & & & & & & \\
-\Phi(p) & & \left[\begin{smallmatrix}-q\\ Y\end{smallmatrix}\right] & & & & & & \\
 & \left[\begin{smallmatrix}-q\\ Y\end{smallmatrix}\right]^T & & & & & & & \\
 & & & \ddots & & & & & \\
 & & & & D^2 u_H & -\Phi(p)^T & & & \\
 & & & & -\Phi(p) & & \left[\begin{smallmatrix}-q\\ Y\end{smallmatrix}\right] & \Phi(p)^{\backslash} & I \\
 & & & & & \left[\begin{smallmatrix}-q\\ Y\end{smallmatrix}\right]^T & & & \\
\widehat{I} & & & & \widehat{I} & & & -I & \\
 & & I & & & & I & &
\end{array}
\right)
$$

From Theorem 10, each submatrix corresponding to the derivatives of the functions defining the first order conditions for household h with respect to "her" variables (x_h, λ_h, b_h) has full rank. Therefore, using the results described in Section 4 in Chapter 2, we are left with showing that we can perturb the last two superrows. Following the notation of that section, that result is accomplished as follows.

$$\begin{array}{ll} (Mx) \leftarrow e_H^{\backslash} & \rightsquigarrow (H.2) \\ (H.2) \leftarrow e_H^{\cdot} & \end{array}$$

$$\begin{array}{ll} (Mb) \leftarrow b_H & \rightsquigarrow (H.2) \\ (H.2) \leftarrow e_H^{\cdot} & \end{array}$$

■

Proposition 23 *$F^{-1}(0)$ is a smooth submanifold of $\Xi \times \mathbb{R}^{CH}$ of dimension CH.*

Proof. From Theorem 20, $F^{-1}(0) \neq \emptyset$. From the previous lemma and from Theorem 9 in Chapter 4, the desired result follows. ■

Lemma 24

$$pr : F^{-1}(0) \to \mathbb{R}^{CH}_{++}, \ (\xi, e) \mapsto e$$

is proper.

Proof. Take a sequence $(e^v)_{v \in \mathbb{N}}$ in $\mathbb{R}^{CH}_{++}$. By assumption, it admits a convergent subsequence, say to $\overline{e} \in \mathbb{R}^{CH}_{++}$. We want to show that considered a sequence $(\xi^v)_{v \in \mathbb{N}}$ such that $F(\xi^v, e^v) = 0$, it does converge to $(\overline{\xi}, \overline{e}) \in F^{-1}(0)$. The proof follows exactly the same steps as the proof of Lemma 19. ■

Theorem 25 *If $e \in \mathcal{R}$, then*

1 *there exists $r \in \mathbb{N}$ such that*

$$(F)^{-1}(0) = \{\xi^i\}_{i=1}^{r}$$

2 *for each i, there exists an open set $U \subseteq \mathbb{R}^{CH}_{++}$, an open set $V_i \subseteq \mathbb{R}^{\dim \Xi}$ and a unique function $g_i : U \to V_i$such that*

   (a) $(\xi^i, e) \in V_i \times U$,

   (b) *g_i is C^1,*

   (c) $g_i(e) = \xi^i$,

   (d) *for every $e' \in U$, $F(g_i(e'), e') = 0$.*

Theorem 26 *Consider $e \in \mathcal{R}$. Then*

1 *there exists* $r \in \mathbb{N}$ *such that* $pr^{-1}(e) = \{\xi_1, ..., \xi_i, ..., \xi_r\}$;

2 *there exist an open neighborhood* Y *of* e *in* $\mathbb{R}^{CH}_{++}$, *and for each* i *an open neighborhood* U_i *of* ξ_i *in* $F^{-1}(0)$, *such that:*

   (a) $U_j \cap U_k = \emptyset$ *if* $j \neq k$;

   (b) $pr^{-1}(Y) = \cup_{i=1}^{r} U_i$;

   (c) $pr_{|U_i} : U_i \to Y$ *is a diffeomorphism.*

Lemma 27 $\mathcal{R}$ *is*

1 *of full Lebesque measure in* $\mathbb{R}^{CH}_{++}$, *and*

2 *open in* $\mathbb{R}^{CH}_{++}$.

Chapter 12

NOMINAL ASSETS

1. Introduction

In this chapter we describe the main results about the model with nominal assets.

In Section 2, existence of equilibria is obtained simply observing that each equilibrium in a model with numeraire assets is basically an equilibrium in the present model. The main distinctive and interesting feature of the model with nominal assets is *indeterminacy of equilibria.*

In Section 3, we introduce what Cass calls "the most striking - as well the most troublesome - property of these models: Incomplete markets typically lead to significant price or *nominal* indeterminacy (...) which also naturally translates into substantial allocation or *real* indeterminacy"[1].

After having presented a definition of nominal and real indeterminacy, we describe a simple procedure to conjecture the degrees of those indeterminacies. That procedure, consisting in a careful count of the "relevant" unknowns and equations defining equilibria, is applied to basically all the models contained in the book.

In Section 4, following [29], we discuss both fundamental implications of the indeterminacy result and its "robustness"to variations and generalizations of the incomplete market model.

Finally, in Section 5 we prove the indeterminacy result, following Geneakoplos and Mas-Colell's approach[2] reformulated in terms of the extended approach we have been using throughout the book.

[1] [29], pp.263-264.
[2] See [60].

2. Existence of equilibria

The model we present is a particular case of that one presented in Subsection 2.1 in Chapter 10: assets are of the nominal kind.

Definition 1 A **nominal asset** i is an asset which promises to deliver y^{si} units of the account or money in state s. Therefore,

$$\text{for each } s \text{ and } i, \text{ and each } p, \quad r^{si}(p) = y^{si}$$

Definition 2 The $S \times I$ matrix

$$Y \equiv \begin{bmatrix} y^{11} & \cdots & y^{1i} & \cdots & y^{1I} \\ \cdots & & & & \\ y^{s1} & \cdots & y^{si} & \cdots & y^{sI} \\ \cdots & & & & \\ y^{S1} & \cdots & y^{Si} & \cdots & y^{SI} \end{bmatrix}$$

is called **yield matrix** and $y^s \equiv \left(y^{si}\right)_{i=1}^{I}$ is the yield vector in state $s \in \mathcal{S}$.

In the case of nominal assets, the return matrix function coincides with the yield matrix:

$$R : \mathbb{R}^G_{++} \to \mathcal{M}_{S,I}, \quad p \mapsto Y$$

While in Chapter 11 we have taken Y as given and we have defined an economy just as $e \in \mathbb{R}^{CH}_{++}$, in this chapter, to compare the nominal asset model with the numeraire one, we define an economy as a pair (e, Y) for both models. Consistently with that choice, definitions of equilibria in the two models are given with respect to (e, Y). We can then give the following definition.

Definition 3 An economy in a financial economy model with nominal assets is an element $(e, Y) \in \mathbb{R}^{GH}_{++} \times \mathcal{M}_{S,I}$.

Household h's maximization problem is as follows. For given $e_h \in \mathbb{R}^G_{++}$, $p \in \mathbb{R}^G_{++}$, $q \in \mathbb{R}^I$, $Y \in \mathcal{M}_{S,I}$,

$$\begin{array}{ll} \max_{(x_h, b_h) \in \mathbb{R}^G_{++} \times \mathbb{R}^I} & u_h(x_h) \\ s.t. & p^0 x^0_h + q b_h = p^0 e^0_h, \\ & p^s x^s_h = p^s e^s_h + y^s b_h \text{ for } s \geq 1 \end{array} \tag{12.1}$$

or

$$\begin{array}{ll} \max_{(x_h, b_h) \in \widetilde{X}_h \times \mathbb{R}^I} & u_h(x_h) \\ s.t. & -\Phi(p)(x_h - e_h) + \begin{bmatrix} -q \\ Y \end{bmatrix} b_h = 0 \end{array} \tag{12.2}$$

Definition 4 The vector $(x, b, p, q) \in \mathbb{R}^{GH}_{++} \times \mathbb{R}^{IH} \times \mathbb{R}^{G}_{++} \times \mathbb{R}^{I}$ is a **demand-price financial nominal asset equilibrium vector** for the economy $(e, Y) \in \mathbb{R}^{GH}_{++} \times \mathcal{M}_{S,I}$ if

1 for $h = 1, ..., H$, for given (p, q, e, Y), (x_h, b_h) solves problem (12.2), i.e., households maximize, and

2

$$\sum_{h=1}^{H} x_h = \sum_{h=1}^{H} e_h \quad \text{and}$$
$$\sum_{h=1}^{H} b_h = 0$$

i.e., both good and asset markets clear .

Since asset returns are constant with respect to commodity prices is only "nominal", consistently with Definitions 11, 13 and 21 discussed in Chapter 10, we make the following assumptions.

Assumption a1 (no arbitrage) There is no $b \in \mathbb{R}^I$ such that $\left[\begin{smallmatrix}-q\\ Y\end{smallmatrix}\right] b > 0$.

We also define the no arbitrage asset price set as

$$Q \equiv \left\{ q \in \mathbb{R}^I : \text{there is no } b \in \mathbb{R}^I \text{ such that } \begin{bmatrix} -q \\ Y \end{bmatrix} b > 0 \right\}$$

Remark 5 From Lemma 14 in Chapter 10, Assumption a1 is equivalent to the following statement

$$\text{there exists } \gamma \in \mathbb{R}^S_{++} \text{ such that } q = \gamma Y$$

Assumption a2 (no redundancy) $\operatorname{rank} Y = I$.

Assumption a3 (incomplete financial markets) $0 \leq I < S$.

We now show that basically each numeraire equilibrium - see Definition 6 - is a nominal equilibrium. To see that it is enough to compare the maximization problem (11.3) for the numeraire model in Chapter 11 with above problem (12.1).

Proposition 6 *If* $\left((x_h, b_h)_{h=1}^{H}, p^{\backslash}, q\right)$ *is a demand-price equilibrium in a numeraire asset model for the economy* (e, Y)*, then* $\left((x_h, b_h)_{h=1}^{H}, (p^{\backslash s}, 1)_{s=1}^{S}, q\right)$ *is a demand-price equilibrium in a nominal asset model for the economy* (e, Y).

Proof. By assumption, for each h, (x_h, b_h) solves the maximization problem (11.3) at $(p^{\backslash}, q, e, Y)$. Comparing the constraint budget equations of both problem (11.3) and problem (12.1), we have that (x_h, b_h) solves problem (12.1) at $\left((x_h, b_h)_{h=1}^{H}, (p^{\setminus s}, 1)_{s=1}^{S}, q, e, Y\right)$. Market clearing conditions in the two definitions of equilibria are the same. ■

Remark 7 Observe that if (x_h, b_h) solves problem (12.1), it does not solve problem (11.1) unless $p^{sC} = 1$ for $s \in \mathcal{S}$.

Theorem 8 *For each economy (e, Y) an equilibrium in a nominal asset economy exists.*

Proof. It is a consequence of the above Proposition 6 and of Theorem 20 in Chapter 11. ■

3. Real indeterminacy

In this section we describe the main distinctive characteristics of the model with nominal assets: real indeterminacy of equilibria. We start from a definition of degree of nominal, and allocative or real indeterminacy. Then we present an intuitive step-by-step procedure to determine those degrees. That procedure turns out to be effective in each model we present in this book.

3.1 Definition of nominal and real indeterminacy.

Definition 9 Let prices and returns be expressed in units of account. A general equilibrium model exhibits nominal indeterminacy or price indeterminacy if the following condition holds.

For each economy (e, Y), there exists a subset $N \subseteq \mathcal{S}' \times \mathcal{C}$ such that for each $(\overline{p}^{sc})_{(s,c)\in N} \in \mathbb{R}^{\#N}_{++}$, there exists $(p^{sc})_{(s,c)\in(\mathcal{S}'\times\mathcal{C})\setminus N}$ such that

$$\left((\overline{p}^{sc})_{(s,c)\in N}, (p^{sc})_{(s,c)\in(\mathcal{S}'\times\mathcal{C})\setminus N}\right)$$

is an equilibrium price for the economy (e, Y).

The cardinality of N is called the degree of nominal indeterminacy.

Example 10 We have seen that in the case of pure exchange economies, the existence of equilibrium prices is insured no matter how the price of an arbitrary good is chosen. Therefore the degree of nominal indeterminacy of that model is 1.

A simple way to get a good conjecture about the degree of nominal indeterminacy of a model is to subtract the number of the non redundant equations from the number of unknowns, or endogenous variables, in the system of equations defining equilibrium. A set of equations is redundant if they hold true if all the other equations hold true; more formally, we have the following definition.

Definition 11 Consider two sets X and Y, a function

$$f : X \times Y \to \mathbb{R}^n, \quad (x, z) \mapsto (f_i(x, z))_{i=1}^n$$

and the system

$$f(x, z) = 0 \tag{12.3}$$

in the unknowns, or endogenous variables , x and parameters, or exogenous variables , z. A set $I \subseteq \{1, ..., n\}$ of equations of system (12.3) is redundant if

$$\forall z \in Z, \quad f(x, z) = 0 \text{ if and only if } f_i(x, z) = 0 \text{ for } i \in \{1, ..., n\} \setminus I$$

Example 12 In the case of pure exchange economies, the reduced equilibrium system in the unknowns $p \in \mathbb{R}^G$ is

$$\sum_{h=1}^{H} (x_h(p, e_h) - e_h) = 0$$

The number of equations is G. As we have seen in Proposition 29 in Chapter 8, one of these equations is redundant. Therefore, the degree of nominal indeterminacy is $G - (G - 1) = 1$. Observe that the number of redundant equations is equal to the number of budget constraints.

Nominal or price indeterminacy may or may not translate in real or allocative indeterminacy.

Definition 13 A general equilibrium model exhibits real indeterminacy or allocation indeterminacy if the following condition holds.

There exists an open and full measure subset O of the set of economies such that if $e \in O$, then the set of equilibrium allocations associated with e contains the image of an open subset of $\mathbb{R}^d$ via a C^1 one-to-one function, with $d > 0$. d is called the degree of real or allocation indeterminacy.

If $d = 0$, i.e., the set of equilibrium allocations is made up by isolated points, we say that the model does not exhibit real indeterminacy.

As we will see in next section, the model with nominal assets does exhibit a degree of real indeterminacy equal to $(S - 1)$.

Moreover, it is easy to check that if the economy e is a Pareto optimal allocation, then both in the model with numeraire and nominal assets, there is a unique equilibrium allocation x associated with it, and in fact $x = e$.

Therefore, we can have real indeterminacy at most for a "large set" of economies, not for all of them. That simple observation motivates the presence of the expression "open and full measure" in the above definition.

Example 14 We have seen that the model of pure exchange economies exhibits no real indeterminacy. As discussed in Remark 18 in Chapter 8, households' budget set does not change if the price vector is multiplied by a positive constant. That property gives the possibility of normalizing the price of an arbitrary good without affecting the demand of any household.

To get a good conjecture about the degree of real indeterminacy, it is enough to count the number of "normalizations" - defined below - and subtract that number from the degree of nominal indeterminacy.

Definition 15 Assume that there exists a partition $\mathcal{D} = \{D_1, ..., D_n\}$ of the set $\mathcal{S} = \{0, 1, ..., S\}$ containing nonempty sets and such that for every $\left(\alpha^j\right)_{j=1}^n \in \mathbb{R}^n_{++}$ the budget set of each household at p and at $\left(\left(\alpha^j p^s\right)_{s \in D_j}\right)_{j=1}^n$ coincide. The cardinality of the finest of such partitions is called the number of admissible normalizations .

We will try to clarify the above definition with several examples in the following subsection.

Example 16 We have seen that in the model of a pure exchange economy p is an equilibrium price if there exists $\alpha \in \mathbb{R}_{++}$ such that αp is an equilibrium price (see again Remark 18 in Chapter 8). Observe that in this case, $\mathcal{D} = \{\{0, 1, ..., S\}\}$, and therefore the number of admissible normalizations is 1.

3.2 A step-by-step procedure to conjecture the degree of nominal and real indeterminacy

Let's summarize the way we suggested to compute nominal and real indeterminacy and then apply it to the case of nominal, numeraire and real assets models. It turns out that the intuitive results so obtained hold true.

1 Write the equations and the variables defining the equilibrium.

2 Compute the number of

(a) endogenous variables,

(b) equations, and

(c) redundant equations.

The number under c. is equal to the number of equations defining the budget constraints which are common to each households, i.e., the number of the so called "Walras' laws" [3].

[3]In fact, summing up each budget equation for all households and using the fact that prices are strictly positive, we get that the market clearing condition for one good follows from the market clearing conditions of all other goods in the set of goods included in the analyzed budget equation.

3

$$\text{degree of nominal indeterminacy} =$$

$$= \text{\# endogenous variables - (\# equations - \# redundant equations).}$$

4 Compute the number of admissible normalizations as follows.

(a) Can you divide the budget constraint in period 0 by p^{0C} without changing the budget set? If so, you get 1 normalization; if not, 0 normalizations;

(b) Can you divide each budget constraint in each state $s > 0$ by p^{sC} without changing the budget set? If so, you get S normalizations; if not, 0 normalizations, and then consider the following question. Can you divide all budget constraints in each state $s > 0$ by p^{1C} without changing the budget set? If so, you get 1 normalizations; if not, 0 normalizations.

To compute the total number of normalizations, sum up all the normalizations you got in a. and b. above. Finally,

$$\text{degree of real indeterminacy} =$$

$$= \text{degree of nominal indeterminacy - (\# normalizations)}$$

Let's now apply the above procedure to the models with different types of assets in terms of aggregate excess demand functions. For the sake of simplicity, we omit the subscript h where there is no possibility of confusion.

3.2.1 The model with nominal assets

1 $\sum_{h=1}^{H} (x_h (p, q) - e_h) = 0;$

$\sum_{h=1}^{H} b_h (p, q) = 0,$

in the variables p, q.

2 Number of equations: $G + I$;

number of variables: $G + I$;

number of Walras' laws: $S + 1$.

3 *(degree of nominal indeterminacy)* =

$= G + I - [(G + I) - (S + 1)] = S + 1.$

4 The budget equations are:

$p^0 \left(x^0 - e^0\right) + qb = 0$

$p^s (x^s - e^s) + y^s b = 0$, for $s > 0$.

(a) divide the state 0 constraint by p^{0C} :
$\frac{p^0}{p^{0C}}\left(x^0 - e^0\right) + \frac{q}{p^{0C}}b = 0;$

(b) divide the *period* 1 *constraints* by p^{1C} :
$\frac{p^s}{p^{1C}}\left(x^s - e^s\right) + y^s\frac{b}{p^{1C}} = 0$, for $s > 0$;

divide the state 0 constraint by p^{1C} to take into account that the demand for assets has to be redefined as $\frac{b}{p^{1C}}$:

$\frac{p^0}{p^{0C}}\left(x^0 - e^0\right) + \frac{qp^{1C}}{p^{0C}}\frac{b}{p^{1C}} = 0.$

Therefore, the total number of normalization is 2.

5 *(degree of real indeterminacy)* = $(S+1) - 2 = S - 1$.

[10] analyze a model with nominal assets where asset prices are exogenous variables, i.e., an economy is an element $(q, e, Y) \in \mathbb{R}^I \times \mathbb{R}^{GH}_{++} \times \mathcal{M}_{S,I}$.

3.2.2 The model with nominal assets and exogenous asset prices

1 $\sum_{h=1}^{H}\left(x_h\left(p\right) - e_h\right) = 0;$
$\sum_{h=1}^{H} b_h\left(p\right) = 0,$
in the variables p.

2 a. Number of equations: $G + I$;
b. number of variables: G;
c. number of Walras' laws: $S + 1$.

3 *(degree of nominal indeterminacy)* =
$= G - [(G+I) - (S+1)] = S + 1 - I.$

4 The budget equations are:
$p^0\left(x^0 - e^0\right) + qb = 0$
$p^s\left(x^s - e^s\right) + y^s b = 0$, for $s > 0$.

(a) divide the state 0 constraint by p^{0C}:

$\frac{p^0}{p^{0C}}\left(x^0 - e^0\right) + q\frac{b}{p^{0C}} = 0;$

Observe that since q is fixed, b has to be redefined as $\frac{b}{p^{0C}}$. Then, all budget constraints for $s \geq 1$ have to be changed as follows:
$\frac{p^s}{p^{0C}}\left(x^s - e^s\right) + y^s\frac{b}{p^{0C}} = 0$, for $s > 0$.
Therefore, the total number of normalization is 1.

5 *(degree of real indeterminacy)* = $S + 1 - I - 1 = S - I$.

3.2.3 The model with numeraire assets

1 $\sum_{h=1}^{H} (x_h(p,q) - e_h) = 0;$

$\sum_{h=1}^{H} b_h(p,q) = 0,$

in the variables p, q.

2 a. Number of equations: $G + I$;

b. number of variables: $G + I$;

c. number of Walras' laws: $S + 1$.

3 *(degree of nominal indeterminacy)* =

$= (G + I) - [(G + I) - (S + 1)] = S + 1$

4 The budget equations are:

$p^0 (x^0 - e^0) + qb = 0$

$p^s (x^s - e^s) + p^{sC} y^s b = 0$, for $s > 0$.

(a) divide the state 0 constraint by p^{0C}:

$\frac{p^0}{p^{0C}} (x^0 - e^0) + \frac{q}{p^{0C}} b = 0;$

(b) divide the *each* state $s > 0$ *constraints* by p^{sC}:

$\frac{p^s}{p^{sC}} (x^s - e^s) + y^s b = 0$, for $s > 0$.

Therefore, the total number of normalization is $S + 1$.

5 *(degree of real indeterminacy)* = $(S + 1) - (S + 1) = 0$.

3.2.4 The model with real assets

1 $\sum_{h=1}^{H} (x_h(p,q) - e_h) = 0;$

$\sum_{h=1}^{H} b_h(p,q) = 0,$

in the variables p, q.

2 a. Number of equations: $G + I$;

b. number of variables: $G + I$;

c. number of Walras' laws: $S + 1$.

3 *(degree of nominal indeterminacy)* =

$= (G + I) - [(G + I) - (S + 1)] = S + 1.$

4 The budget equations are:

$p^0 (x^0 - e^0) + qb = 0$

$p^s(x^s - e^s) + \sum_{i=1}^{I}(p^s y^{si})\, b^i = 0$, for $s > 0$.

(a) divide the state 0 constraint by p^{0C}:
$\frac{p^0}{p^{0C}}(x^0 - e^0) + \frac{q}{p^{0C}} b = 0$;

(b) divide the *each* state $s > 0$ *constraints* by p^{sC}:
$\frac{p^s}{p^{sC}}(x^s - e^s) + \sum_{i=1}^{I}\left(\frac{p^s}{p^{sC}} y^{si}\right) b = 0$, for $s > 0$.

Therefore, the total number of normalization is $S + 1$.

5 *(degree of real indeterminacy)* = $(S + 1) - (S + 1) = 0$.

3.2.5 The model with a type of mixed assets

We conclude this discussion with the case in which the return matrix function - see Definition 7 in Subsection 2.1, Chapter 10 - is such that for each (s, i)

$$r^{si} = \overline{y}^{si} + y^{si} p^s$$

where $\overline{y}^{si} \in \mathbb{R}$ denotes the number of units of account and $y^{si} \in \mathbb{R}^C$ the vector of units of commodities delivered by one unit of asset i in state s.

1 $\sum_{h=1}^{H}(x_h(p, q) - e_h) = 0$;
$\sum_{h=1}^{H} b_h(p, q) = 0$,
in the variables p, q.

2 a. Number of equations: $G + I$;

b. number of variables: $G + I$;

c. number of Walras' laws: $S + 1$.

3 *(degree of nominal indeterminacy)* =
$= (G + I) - [(G + I) - (S + 1)] = S + 1$.

4 The budget equations are:
$p^0(x^0 - e^0) + qb = 0$
$p^s(x^s - e^s) + \sum_{i=1}^{I}(\overline{y}^{si} + y^{si} p^s)\, b^i = 0$, for $s > 0$.
divide the state 0 constraint by p^{0C}:
$\frac{p^0}{p^{0C}}(x^0 - e^0) + \frac{q}{p^{0C}} b = 0$.
In fact, that is the only possible normalization.
Therefore, the total number of normalization is 1.

5 *(degree of real indeterminacy)* = $(S + 1) - 1 = S$.

The procedure we have applied to equilibrium conditions in terms of aggregate excess demand functions (the reduced approach) can be applied to equilibria defined in terms of the extended approach and (of course) gives the same results. Furthermore, observe that normalizations of prices amount to "elimination" of (endogenous) variable by fixing them to 1. Normalization of variables different from prices are in general possible and sometimes useful. In what follow we apply the procedure to conjecture degree of indeterminacy to the real assets case and we propose a different normalization for the case of the model with real assets which will turn out to be useful in the proof of existence of equilibria for that model.

3.2.6 The model with real assets with the extended approach and a different normalization

1 Given our maintained assumption, it is possible to show that if (x, b, p, q) is an equilibrium in the real asset model, then there exists $\lambda \equiv (\lambda_h)_{h=1}^{H} \in \mathbb{R}_{++}^{(S+1)H}$ such that (x, b, λ, p, q) is a solution to the following system.

$$\begin{array}{lll} D_{x_h^s} u_h(x_h) - \lambda_h^s p^s = 0 & \text{for } s = 0, ..., S & (h.1) \\ -\lambda_h^0 q + \sum_{s=1}^{S} \lambda_h^s \left(p^s y^{si}\right)_{i=1}^{I} = 0 & & (h.2) \\ -p^0 \left(x_h^0 - e_h^0\right) - q b_h = 0 & & (h.3) \\ -p^s \left(x_h^s - e_h^s\right) + \left(p^s y^{si}\right)_{i=1}^{I} b_h = 0 & \text{for } s = 0, ..., S & (h.4) \\ \sum_{h=1}^{H} (x_h - e_h) = 0 & & (M.1) \\ \sum_{h=1}^{H} b_h = 0 & & (m.2) \end{array} \qquad (12.4)$$

where equation $(h.1) - (h4)$ hold for $h = 1, ..., H$.

2 a. Number of equations: $(G + I + S + 1) H + G + I$;

b. number of variables: $GH + IH + (S + 1) H + G + I$;

c. number of Walras' laws: $S + 1$.

3 *(degree of nominal indeterminacy)* =

$= ((G + I + S') H + G + I) - [(GH + IH + S'H + G + I) - S'] = S'$
where $S' = S + 1$.

(a) divide state 0 constraint by p^{0C}:
$\frac{p^0}{p^{0C}} \left(x_h^0 - e_h^0\right) + \frac{q}{p^{0C}} b_h = 0$;

(b) divide *each* state $s > 0$ *constraints* by $\frac{\lambda_1^0 p^{0C}}{\lambda_1^s}$:

$$\frac{p^s}{\left(\lambda_1^0 p^{0C}\right)/\lambda_1^s} \left(x_h^s - e_h^s\right) + \sum_{s=1}^{S} \left(\frac{p^s}{\left(\lambda_1^0 p^{0C}\right)/\lambda_1^s} y^{si} \right)_{i=1}^{I} b_h = 0, \text{ for } s > 0;$$

(c) rewrite $(h.1) - (h.2)$ $h \neq 1$ in system (12.4), consistently with b).

$$D_{x_h^0} u_h(x_h) - \left(\lambda_h^0 p^{0C}\right) \frac{p^0}{p^{0C}} = 0$$
$$D_{x_h^s} u_h(x_h) - \left(\lambda_h^s \frac{\lambda_1^0 p^{0C}}{\lambda_1^s}\right) \frac{p^s}{\left(\lambda_1^0 p^{0C}\right)/\lambda_1^s} = 0, \text{ for } s \geq 1$$
$$-\left(\lambda_h^0 p^{0C}\right) \frac{q}{p^{0C}} + \sum_{s=1}^S \left(\lambda_h^s \frac{\lambda_1^0 p^{0C}}{\lambda_1^s}\right) \left(\frac{p^s}{\left(\lambda_1^0 p^{0C}\right)/\lambda_1^s} y^{si}\right)_{i=1}^I = 0.$$

Specifying the above conditions for $h = 1$ we obtain

$$\frac{q}{p^{0C}} = \sum_{s=1}^S \left(\frac{p^s}{\left(\lambda_1^0 p^{0C}\right)/\lambda_1^s} y^{si}\right)_{i=1}^I \tag{12.5}$$

Therefore, the total number of normalization is $S + 1$.

4 *(degree of real indeterminacy)* = $(S + 1) - (S + 1) = 0$.

5 Observe that the above argument shows that if (x, b, λ, p, q) is a solution to system (12.4), then

$$\left(\widehat{x}, \widehat{b}, \widehat{\lambda}, \widehat{p}, \widehat{q}\right) \equiv \left(x, b, \left(\lambda_h^0 p^{0C}, \left(\lambda_h^s \frac{\lambda_1^0 p^{0C}}{\lambda_1^s}\right)_{s=1}^S\right)_{h=1}^H, \left(\frac{p^0}{p^{0C}}, \left(\frac{p^s}{\left(\lambda_1^0 p^{0C}\right)/\lambda_1^s}\right)_{s=1}^S\right), \frac{q}{p^{0C}}\right)$$

is a solution to that system *without equations*

$$\left(\sum_{h=1}^H \left(x_h^{sC} - e_h^{sC}\right) = 0\right)_{s=0,\ldots,S}$$

Observe also that $\widehat{p}^{0C} = 1$ and $\left(\frac{\widehat{\lambda}_1^s}{\widehat{\lambda}_1^0}\right)_{s=1}^S = \mathbf{1}$, i.e., we normalized $p^{0C} = 1$ and $\left(\frac{\lambda_1^s}{\lambda_1^0}\right)_{s=1}^S$. Using (12.5), this quite involuted normalization allows to write

$$\widehat{q} = \mathbf{1} R\left(\widehat{p}\right)$$

In Section 5 of this chapter, we present a proof of indeterminacy in the case of nominal assets and endogenous asset prices. In that model, as conjectured above, the degree of nominal indeterminacy is equal to $S - 1$. Therefore, as argued by [60] "the dimension of indeterminacy is independent of the number of bonds I, as long as $0 < I < S$. If $I = 0$, the model is obviously determinate. If $I \geq S$, one can apply Arrow's logic [4] to show that equilibrium commodity allocations are Arrow-Debreu equilibrium allocations, and again there is no real indeterminacy. Theorem 1[4] points out an intriguing discontinuity. If markets

[4]Theorem 1 by [60] shows the indeterminacy result.

are financially complete, then the model is determinate. Let just one financial asset be missing and the model becomes highly indeterminate. Thus, in this sense, the complete markets hypothesis ($I = S$) lacks robustness. (Probably what this means is that the hypothesis has to be interpreted as $I > S$, i.e., one better have some redundancy)"[5].

4. Relevance and robustness of the indeterminacy result

4.1 Why is indeterminacy important?

We believe there are two basic reasons which qualify the indeterminacy result as quite important from an economics viewpoint.

First of all, "Nominal indeterminacy per se presents a severe practical hurdle for the rational expectation hypothesis. In short, is it plausible to maintain that households are capable of concentrating their beliefs (correctly) on one among a surfeit of possible market outcomes?"[6].

In other words, "indeterminacy creates conceptual difficulties with the assumption that agents correctly anticipate future spot prices - and hence, in particular, future spot price levels. When price levels matter, all agent must anticipate the same price levels to arrive at an equilibrium. But since there is nothing in the concept of equilibrium (...) for determining the price levels, it is difficult to see how the agents' expectations regarding future price levels come to be co-ordinated. Even if agents have a good understanding of how the economy functions, the model does not make any data available to them on the basis of which they could calculate future price levels. Thus when the contracts are nominal, if the assumption of correct price anticipations is to retained, then some mechanism must be introduced to determining price levels[7].

In the second place, "a model with a continuum of outcomes is severely limited in its predictive and explanatory power: looked at another way, indeterminacy suggests that there is perhaps an important element which is missing in the model"[8].

Of course, indeterminacy could also be seen as a *virtue* of the model. When so many equilibria are possible, there is room for some institute to "guide" the economy towards the choice of a particular equilibrium with desirable properties. Very informally, we could think of a central authority which is able to control inflation through money supply and then announcing its levels for the future it can steer the economy from one "bad" equilibrium to a better one - as suggested in [38]. Or we could image workers' unions imposing a new contract - changing the price of labor - without leading to any disruptive change of the

[5] [60], p. 26.
[6] [29], p. 264.
[7] [82], pp. 438-439.
[8] [82], p. 438.

functioning of the economy, but just moving from one equilibrium to another one more favorable to a subset of the households.

4.2 Why is indeterminacy robust?

Let us discuss what are the effects on real indeterminacy of some possible different specifications and some generalizations of the model.

4.2.1 Sunspots

"To even the most casual reader of the daily business pages, it must be fairly obvious that financial markets react (or overreact) to much news which clearly (or, at least, arguably) has little or no direct bearing on fundamentals"[9], i.e., households' endowments and preferences and firms' technologies. Economies seem then to be heavily influenced by apparently completely irrelevant phenomena, like sunspots. "The inherent nature of sunspot beliefs (or, more generally, of uncertainty about the future) militates against ever having a complete array of Arrow-Debreu markets"[10], i.e., complete markets. "The logic buttressing this position is simple, but compelling: suppose that, as today, the range of future markets were complete. (...) Then what prevents some agency from successfully instilling an updated, expanded view of the world?"[11] Cass was the first to try to carefully formalize the above observations in a general equilibrium model and to analyze the consequences of sunspots on equilibria. The model can be easily described as follows. Consider a two period exchange economy with uncertainty. In the second period, economies - endowments and utility functions - are constant across different spots, i.e.,

$$\text{for each } h, \quad e_h^s = e_h^1 \qquad \text{for } s = 1, ..., S$$
$$u_h(x_h) = \sum_{s=1}^{S} \pi^s v_h^s\left(x_h^0, x_h^s\right) \text{ with } v_h^s = v_h^1 \quad \text{for } s = 1, ..., S$$

where π^s is the probability of occurrence of state s. Economies satisfying the above properties are called "sunspot economies".

Observe that the set of economies of the above model is non generic in the whole set of economies. The main result in the sunspot literature is that in a model with incomplete markets and nominal assets, generically in the set of sunspot economies equilibrium allocations exhibit the same degree of indeterminacy as in the standard nominal asset model[12].

[9] [23], p. 677.
[10] [23], p. 677.
[11] [23], p. 677.
[12] Besides the path-breaking paper [23], other relevant contributions on the topic are [29], [105], [94], [115].

4.2.2 Assets other than nominal

"A common misconception is that the phenomenon of extensive real indeterminacy requires having bonds, that is, financial instruments whose yields are specified in units of account. Nothing could be farther from the truth - a point already emphatically underlined in both [10], and [60]" [13].

As we have seen in the Subsection 2.1, asset returns r^{si} can be seen as a function of prices, i.e.,

$$r^{si} = r^{si}(p, ...) \tag{12.6}$$

We have also discussed the following specification of that function:

$$r^{si}(p) = \overline{y}^{si} \tag{12.7}$$

$$r^{si}(p) = y^{si}p^{sC} \quad \text{and} \quad r^{si}(p) = y^{si}p^{s} \tag{12.8}$$

$$r^{si}(p) = \overline{y}^{si} + y^{si}p^{s} \tag{12.9}$$

i.e., the cases of nominal, numeraire and real, mixed returns, respectively.

As we have seen in Section 3, "*only* the special case (12.8) entails that there is necessarily generic local uniqueness, and hence no extensive real indeterminacy; indeed, the general case (12.9) entails even *one more* degree of real indeterminacy than the special case (12.7) which has been the main focus of attention!" [14]. For a formal analysis of the above statement, see [74] and [94].

Even more than that, the above analysis suggests that, "Except in one very special situation - where all yields from financial instruments depend *linearly* homogeneously on future spot goods prices - competitive equilibrium with incomplete or, even more generally, otherwise imperfect financial markets exhibits pervasive real indeterminacy"[15].

4.2.3 Endogeneizing yields

"The degree of indeterminacy - nominal or real - depends on which financial parameters are treated as endogenous or *variable* (as opposed to exogenous or *fixed*). This means, taking a broad view, that the problem associated with indeterminacy will be only mitigated (and not eliminated) by elaborating the structure of the institution and the behavior of the organizations (public or private) which constitute the financial sector"[16].

More specifically, "even in the special case (12.8), when (some or all of) the parameters y^{sic} are treated as variables, there will be extensive real indeterminacy. More generally, suppose that the parameters in (12.6) reflect the

[13][29], p. 271.
[14][29], p. 272.
[15][29], p. 264.
[16][29], p. 264.

characteristics of some particular array of financial instruments. Then, if various of these parameters are also taken as variable, this will usually contribute to increasing the degree of real indeterminacy beyond S. And there is just no convincing argument for taking all such parameters as fixed, since they essentially correspond - in abstract - to the terms on which credit is transacted between households"[17].

4.2.4 Outside money

A second misconception or, perhaps better, oversimplification concerning this issue is the belief that, since "the reason" for indeterminacy is that the future "price level" is not tied down, introducing the institution of outside money[18] per se will eliminate the problem. While there is some basis for this conjecture, its validity depends on how one conceives the operation of a monetary system and, even more critically, on what one takes as variable in a monetary economy"[19].

[24] describes a model in which outside money directly enters the utility function. In [81], outside money is instead required in order to buy goods - a so called "cash-in-advance" requirement. [121] analyzes a model where outside money is required to pay terminal taxes which are proportional to wealth in each state of the world.

Moreover, [119] analyzes a model where outside money is required to pay terminal taxes which are of the *lump sum* type, and [47] consider a general form of the tax function.

Following the strategy explained in Section 2, it is easy to see that while in the first group of papers mentioned above there is no real indeterminacy, in the second group actually equilibria exhibit *one more* degree of significant nominal indeterminacy, and hence real indeterminacy.

Consider next what happens when asset yields are endogenous variable, therefore combining the topic of the previous subsection with the analysis of the presence of outside money. "It turns out (...) that now there is *absolutely no* reduction in the degree of significant nominal indeterminacy, and hence real indeterminacy; in this situation there are $(S - I)\,(S - 1)$ degrees of real indeterminacy" [20].

We can conclude with Cass' comment. "Since I find it quite reasonable (even compelling) to believe that in a monetary economy, bond yields (as a

[17] [29], p. 277.

[18] With the term *outside money* we refer to money which is a direct debt of the public sector, e.g. circulating currency, or is based on such debt, e.g. commercial bank deposits matched by bank holdings of public sector debt. Examples are fiat money, gold and foreign exchange reserves. On the other hand, *inside money* is a form of money which is based on private sector debt , the prime modern example being commercial bank deposits to the extent that they are matched by bank lending to private sector borrowers - see [92].

[19] [29] p. 273.

[20] [29], p. 274.

proxy for the yields on a variety of financial instruments) as well a spot goods and bond prices are endogenous, I conclude the following (from the analysis of these extraordinary rudimentary models): The institution of outside money may reduce the degree of real indeterminacy. After all, it is likely that there is some connection (no matter how loose) between "monetary policy" and bond yields, so that these yields are not free to vary arbitrarily. However, at this time there are simply no acceptable grounds for asserting that outside money completely eliminates real indeterminacy" [21].

4.2.5 Restricted participation

"While there might be some disagreement over whether, in a modern developed economy, financial markets are actually incomplete, there can hardly be any disagreement over whether at least some economic agents are variously constrained in transacting on those financial markets"[22]. In fact, it is still possible to extend the model with incomplete markets to incorporate the presence of restrictions to the access to financial markets which are different across households. It is enough to add a restriction of the form

$$b_h \in B_h \subseteq \mathbb{R}^I$$

to the household's maximization problem, where B_h represents the set of possible portfolios available to the households.

[11] show that the indeterminacy result extends to the case where B_h is an I_h-dimensional linear subspace, with $0 \leq I_h \leq I$. Chapter 14, following [32] analyzes the case where the portfolio set is defined by C^2 quasiconcave inequality constraints [23].

5. Real indeterminacy in the nominal asset model

In this section, we want to show that typically in the space of endowments, the set of equilibrium allocations for the model with nominal assets contains the image of an open subset of $\mathbb{R}^{S-1}$ via a C^1 one-to-one function.

The proof goes through the following informally described steps.

Step 1. For each economy (e, Y), the set of equilibrium allocations in the model with nominal assets contains the set of equilibrium allocations of the model with numeraire assets associated with the economy $(e, diag\ (1, \gamma) Y)$, where γ is an arbitrary element of $\mathbb{R}^{S-1}_{++}$.

[21] [29] p. 274.

[22] [29], p. 274.

[23] In fact, in Chapter 14, we analyze a version of the model with numerarie assets. The indeterminacy result could be obtained applying the same techniques used in the present chapter.

Step 2. Different γ and associated yield matrices $[diag\ (1, \gamma)\, Y]$ may in principle be associated with identical equilibrium allocation. There are two sufficient conditions which guarantee that those equilibrium allocations are in fact different one from the other. Those two conditions are a *rank condition* on the households asset demands and an image or *span condition* on the yield matrices.

Step 3. If Y is in general position - see Assumption nom2 below - the span condition holds.

Step 4. Generically, the rank condition holds, if $H > I$.

Combining the above steps, we get the desired real indeterminacy result.

Besides the maintained Assumptions u on utility functions presented in Chapter 10, and Assumptions a1, a2 and a3 on asset structures introduced in Section 2, we also make the following two assumptions.

Assumption nom1 $H > I$.

Assumption nom2 Y is in general position , i.e., every $I \times I$ submatrix of Y has full rank.

The meaning of Assumption nom1 is self evident and it is central in the strategy used in the proof of the main result of indeterminacy, and it is in fact used in both Lemmas and 20 - for which the condition $H \geq I$ is sufficient - and 24 - for which $H > I$ is needed.

Assumption nom2 implies that "subject to the limitation of facing incomplete markets - households are capable of providing "full" wealth insurance (i.e., over any given subset of I future states)[24]. That assumption is used in Lemma 21[25].

Following the section on the definition of real indeterminacy, we have that

there is no loss of generality in normalizing p^{0C} and p^{1C} in the nominal model and $\left(p^{sC}\right)_{s=0}^{S}$ in the numeraire model;

market clearing condition for good C in each spot $s = 0, 1, ..., S$ are redundant in both models ("$S + 1$ Walras' laws apply") .

The above observations allow to characterize equilibria in the nominal and numeraire case in terms of the functions F_1 and F_2 defined below.

For a generic vector $v \equiv (v^{sc})_{s=0,1,...,S,c=1,...,C} \in \mathbb{R}^G$, define

$$v^{\backslash 01} \equiv v \equiv (v^{sc})_{(sc)\neq(0C),(1C)} \in \mathbb{R}^{G-2} \quad \text{and} \quad v^{\backslash} \equiv v \equiv (v^{sc})_{c\neq C} \in \mathbb{R}^{G-(S+1)}$$

Define

$$\widehat{\Xi} \equiv \mathbb{R}^{GH}_{++} \times \mathbb{R}^{IH} \times \mathbb{R}^{(S+1)H}_{++} \times \mathbb{R}^{G-2}_{++} \times \mathbb{R}^{I}$$

[24] [29] p. 269.

[25] In the proof of that lemma, we explain why it is not enough to assume $rank\ Y = I$, i.e., Assumption a2. For a discussion about weakening Assumption nom2, see, for example, Section III in [60].

with generic element

$$\widehat{\xi} \equiv \left(x, b, \lambda, p^{\setminus 01}, q\right)$$

and

$$\Xi \equiv \mathbb{R}_{++}^{GH} \times \mathbb{R}^{IH} \times \mathbb{R}_{++}^{(S+1)H} \times \mathbb{R}_{++}^{G-(S+1)} \times \mathbb{R}^{I}$$

with generic element

$$\xi \equiv \left(x, b, \lambda, p^{\setminus}, q\right)$$

Observe that $\dim \Xi = \dim \widehat{\Xi} - (S-1)$.

Finally, define

$$\overline{p}^{\setminus 01} = \left(\left(\left(p^{0c}\right)_{c \neq C}, 1\right), \left(\left(p^{1c}\right)_{c \neq C}, 1\right), \left(\left(p^{sc}\right)_{c \neq C}, 1\right)_{s=2}^{S} \right), \text{ and}$$

$$\text{for each } s,\ p^{s\setminus} \equiv \left(p^{sc}\right)_{c \neq C} \in \mathbb{R}^{C-1},\ \ p^{s} \equiv \left(p^{s\setminus}, 1\right)$$

$$p^{\setminus} \equiv \left(p^{s\setminus}\right)_{s=0}^{S} \in \mathbb{R}^{(S+1)(C-1)} \ \text{ and } \ p = \left(p^{s}\right)_{s=0}^{S}$$

$$F_1 : \widehat{\Xi} \times \mathbb{R}_{++}^{GH} \times \mathcal{M}_{S,I} \to \mathbb{R}^{\dim \widehat{\Xi} - (S-1)},$$
$$F_1 : \left(x, b, \lambda, p^{\setminus 01}, q, e, Y\right) \mapsto \text{ left hand side of (12.10)}$$

and

$$F_2 : \Xi \times \mathbb{R}_{++}^{GH} \times \mathcal{M}_{S,I} \to \mathbb{R}^{\dim \Xi},\ \left(x, b, \lambda, p^{\setminus}, q, e, Y\right) \mapsto \text{ left hand side of (12.11)}$$

where

$$\begin{cases} Du_h(x_h) - \lambda_h \Phi\left(\overline{p}^{\setminus 01}\right) & = 0 \quad (h1) \\ -\Phi\left(\overline{p}^{\setminus 01}\right)(x_h - e_h) + \left[\begin{smallmatrix} -q \\ Y \end{smallmatrix}\right] b_h & = 0 \quad (h2) \\ \lambda_h \left[\begin{smallmatrix} -q \\ Y \end{smallmatrix}\right] & = 0 \quad (h3) \\ & \\ -\sum_{h=1}^{H} x_h^{\setminus} + \sum_{h=1}^{H} e_h^{\setminus} & = 0 \quad (Mg) \\ \sum_{h=1}^{H} b_h & = 0 \quad (Ma) \end{cases} \tag{12.10}$$

and

$$\begin{cases} Du_h(x_h) - \lambda_h \Phi(p) & = 0 \quad (h1) \\ -\Phi(p)(x_h - e_h) + \left[\begin{smallmatrix} -q \\ Y \end{smallmatrix}\right] b_h & = 0 \quad (h2) \\ \lambda_h \left[\begin{smallmatrix} -q \\ Y \end{smallmatrix}\right] & = 0 \quad (h3) \\ & \\ -\sum_{h=1}^{H} x_h^{\setminus} + \sum_{h=1}^{H} e_h^{\setminus} & = 0 \quad (Mg) \\ \sum_{h=1}^{H} b_h & = 0 \quad (Ma) \end{cases} \tag{12.11}$$

With innocuous abuse of notation, we can then say what follows.

Definition 17 If $F_1\left(\widehat{\xi}, e, Y\right) = 0$, then $\widehat{\xi}$ is a nominal equilibrium associated with the economy (e, Y).

If $F_2(\xi, e, Y) = 0$, then ξ is a numeraire equilibrium associated with the economy (e, Y).

5.1 Step 1. Relationship between numeraire and nominal equilibria

A crucial role in the proof is played by the following set.

Definition 18

$$\mathcal{D}^S_{++} \equiv \left\{\Gamma \in \mathcal{M}_{S,S} : \exists\, \gamma \equiv (\gamma^s)_{s=2}^S \in \mathbb{R}^{S-1}_{++} \text{ such that } \Gamma = diag\left((1,\gamma)\right)\right\}$$

Set also $\gamma^0 = \gamma^1 = 1$.

We can clearly identify $\mathcal{D}^S_{++}$ with $\mathbb{R}^{S-1}_{++}$.

Proposition 19 *Take an arbitrary economy* (e, Y). *For each* $\Gamma \in \mathcal{D}^S_{++}$, *if* $\left(x, b, \lambda, p^\backslash, q\right)$ *is a numeraire equilibrium for the economy* $(e, \Gamma Y)$, *then*

$$\left(x, b, \left(\left(\lambda_h^s \gamma^s\right)_{s=0}^S\right)_{h=1}^H, \left(\left(p^{0c}\right)_{c \neq C}, \left(p^{1c}\right)_{c \neq C}, \left(\frac{p^s}{\gamma^s}\right)_{s=2}^S\right), q\right)$$

is a nominal equilibrium for the economy (e, Y).

Proof. By assumption,

$$\begin{cases} D_{x_h^s} u_h(x_h) - \lambda_h^s p^s = 0 & \text{for } s \geq 0 & (h1) \\ p^0\left(x_h^0 - e_h^0\right) + q b_h = 0 & & \\ p^s\left(x_h^s - e_h^s\right) - \gamma^s y^s b_h = 0 \quad \text{for } s \geq 1 & & (h2) \\ \lambda_h^0 q + \sum_{s=1}^S \lambda_h^s \left(\gamma^s y^s\right) = 0 & & (h3) \\ -\sum_{h=1}^H x_h^\backslash + \sum_{h=1}^H e_h^\backslash = 0 & & (Mg) \\ \sum_{h=1}^H b_h = 0 & & (Ma) \end{cases}$$

But then

$$\begin{cases} D_{x_h^s} u_h(x_h) - \left(\lambda_h^s \gamma^s\right) \frac{p^s}{\gamma^s} = 0 & \text{for } s \geq 0 & (h1) \\ p^0\left(x_h^0 - e_h^0\right) + q b_h = 0 & & \\ \frac{p^s}{\gamma^s}\left(x_h^s - e_h^s\right) - y^s b_h = 0 \quad \text{for } s \geq 1 & & (h2) \\ \lambda_h^0 q + \sum_{s=1}^S \left(\lambda_h^s \gamma^s\right) y^s = 0 & & (h3) \\ -\sum_{h=1}^H x_h^\backslash + \sum_{h=1}^H e_h^\backslash = 0 & & (Mg) \\ \sum_{h=1}^H b_h = 0 & & (Ma) \end{cases}$$

i.e., the desired result. ■

5.2 Step 2. Two sufficient conditions for different equilibrium allocations

Lemma 20 *Take and economy* (e, Y) *and* $\Gamma, \widetilde{\Gamma} \in \mathcal{D}^S_{++}$. *Let* $(x, b, \lambda, p^\backslash, q)$ *and* $\left(\widetilde{x}, \widetilde{b}, \widetilde{\lambda}, \widetilde{p}^\backslash, \widetilde{q}\right)$ *be numeraire equilibria for the economies* $(e, \Gamma Y)$ *and* $\left(e, \widetilde{\Gamma} Y\right)$, *respectively. If*

1 $\text{rank}\,[b_1, ..., b_H] = \text{rank}\left[\widetilde{b}_1, ..., \widetilde{b}_H\right] = I$, *and*

2 $\text{Im}\,\Gamma Y \neq \text{Im}\,\widetilde{\Gamma} Y$,

then $x \neq \widetilde{x}$.

Proof. Without loss of generality, from Assumption 2, it is not the case that $\text{Im}\,\widetilde{\Gamma} Y \subseteq \text{Im}\,\Gamma Y$ and therefore

$$\exists z^* \in \mathbb{R}^I \text{ such that } \forall z \in \mathbb{R}^I,\ \Gamma Y z^* \neq \widetilde{\Gamma} Y z \tag{12.12}$$

We now show that

$$\text{there exists } h \text{ such that } \Gamma Y b_h \neq \widetilde{\Gamma} Y \widetilde{b}_h \tag{12.13}$$

Suppose otherwise, i.e., for each h,

$$\Gamma Y b_h = \widetilde{\Gamma} Y \widetilde{b}_h \tag{12.14}$$

Then, from Assumption 1., there exist $\beta_1, ..., \beta_H \in \mathbb{R}$ such that

$$\begin{aligned} \Gamma Y z^* = {} & \Gamma Y \left(\textstyle\sum_{h=1}^H \beta_h b_h\right) = \textstyle\sum_{h=1}^H \beta_h (\Gamma Y b_h) \\ = {} & \textstyle\sum_{h=1}^H \beta_h \left(\widetilde{\Gamma} Y \widetilde{b}_h\right) = \widetilde{\Gamma} Y \left(\textstyle\sum_{h=1}^H \beta_h \widetilde{b}_h\right) \end{aligned} \tag{12.15}$$

where the third equality follows from (12.14). Hence 12.15 contradicts (12.12).

Suppose now that the conclusion of the lemma does not hold true, i.e., $x = \widetilde{x}$. Then, from equations $(h.1)$ in system (12.11) computed at $(e, \Gamma Y)$ and $\left(e, \widetilde{\Gamma} Y\right)$, respectively, $\lambda_h = \widetilde{\lambda}_h$ and then $p = \widetilde{p}$. But then from equations $(h.2)$, we have

$$\text{for each } h,\ \Gamma Y b_h = \widetilde{\Gamma} Y \widetilde{b}_h$$

contradicting (12.13). ■

5.3 Step 3. The span condition holds

In the following lemma, we use Assumption nom2.

Lemma 21 *Given* $\Gamma_1, \Gamma_2 \in \mathcal{D}^S_{++}$, *if* $\text{Im } \Gamma_1 Y = \text{Im } \Gamma_2 Y$, *then* $\Gamma_1 = \Gamma_2$.

Proof. Observe preliminarily that

$$\text{Im } \Gamma_1 Y = \text{Im } \Gamma_2 Y \Rightarrow \text{Im } Y = \text{Im } [\Gamma_2]^{-1} \Gamma_1 Y \tag{12.16}$$

as shown below.

The following statements are equivalent.

$$\begin{array}{l} z \in \text{Im } Y \\ \exists b \in \mathbb{R}^I \text{ such that } z = Yb \\ \exists b \in \mathbb{R}^I \text{ such that } \Gamma_2 z = \Gamma_2 Y b \\ \exists b' \in \mathbb{R}^I \text{ such that } \Gamma_2 z = \Gamma_1 Y b' \\ \exists b' \in \mathbb{R}^I \text{ such that } z = [\Gamma_2]^{-1} \Gamma_1 Y b' \\ z \in \text{Im } [\Gamma_2]^{-1} \Gamma_1 Y \end{array}$$

In the remaining part of the proof, we are going to show that

$$\text{Im } Y = \text{Im } \Gamma Y \Rightarrow \quad \Gamma = I$$

which, combined with (12.16), implies that $[\Gamma_2]^{-1} \Gamma_1 = I$, i.e., the desired result $\Gamma_1 = \Gamma_2$.

Since $\text{Im } \Gamma Y = \text{Im } Y$, taken the canonical basis $\mathbf{e}^1, ..., \mathbf{e}^I$ of $\mathbb{R}^I$, there exist $b_1, ..., b_I \in \mathbb{R}^I$ such that

$$\text{for each } i, \ \ \Gamma Y \mathbf{e}^i = Y b_i$$

or

$$\Gamma Y = Y \left[b_1^T, ..., b_I^T \right]$$

Defined

$$B \equiv \left[b_1^T, ..., b_I^T \right], \ \ \Gamma_I \equiv diag\left(1, (\gamma_i)_{i=2}^I\right) \ \ \text{and} \ \ Y_I \equiv \begin{bmatrix} y^{1T} \\ \dots \\ y^{IT} \end{bmatrix}$$

we have

$$\text{for } s \in \{1, ..., S\}, \ \ \gamma^s y^s = y^s B \tag{12.17}$$

Since, Y is in general position, for each $s' \in \{I+1, ..., S\}$, there exists $\mu^1_{s'}, ..., \mu^I_{s'}$ such that

$$y^{s'} = \sum_{s=1}^{I} \mu^s_{s'} y^s \ \ \text{with} \ \ \mu^s_{s'} \neq 0 \ \ \text{for each} \ \ s \in \{1, ..., I\} \tag{12.18}$$

- if not, and say, $\mu^1_{s'} = 0$, you would have $-y^{s'} + \sum_{s=2}^{I} \mu^s_{s'} y^s = 0$, and $\{y^s\}_{s=s',2,...,I}$ would be linearly dependent, contradicting that Y is in general position. Observe that the fact that $\operatorname{rank} Y = I$ would not suffice to get the above statement.

Therefore,for each $s' \in \{I+1, ..., S\}$

$$y^{s'} B = \left(\sum_{s=1}^{I} \mu^s_{s'} y^s \right) B = \sum_{s=1}^{I} \mu^s_{s'} \left(y^s B \right) = \sum_{s=1}^{I} \mu^s_{s'} \gamma^s y^s \tag{12.19}$$

where the last equality follows from (12.17). From (12.17) and (12.19),

$$\text{for each } s' \in \{I+1, ..., S\}, \quad \gamma^{s'} y^{s'} = \sum_{s=1}^{I} \mu^s_{s'} \gamma^s y^s$$

Then, from (12.18), $\gamma^{s'} \sum_{s=1}^{I} \mu^s_{s'} y^s = \sum_{s=1}^{I} \mu^s_{s'} \gamma^s y^s$, or

$$\text{for each } s' \in \{I+1, ..., S\}, \quad \sum_{s=1}^{I} \mu^s_{s'} \left(\gamma^s - \gamma^{s'} \right) y^s = 0$$

Since $y^1, ..., y^I$ are linearly independent and $\mu^s_{s'} \neq 0$ for $s \in \{1, ..., I\}$, it must be that for each $s' \in \{I+1, ..., S\}$, $\gamma^s = \gamma^{s'}$ for $s \in \{1..., I\}$. Therefore

$$\text{for } s \in \{1, ..., S\}, \quad \gamma^s = \gamma^1 = 1$$

as desired. ■

5.4 Step 4. Generically, the rank condition holds

Fix a matrix $Y \in \mathcal{M}_{S,I}$ and define

$$F : \Xi \times \mathbb{R}^{S-1}_{++} \times \mathbb{R}^{GH}_{++} \to \mathbb{R}^{\dim \Xi}, \quad \left(x, b, \lambda, p^{\backslash}, q, \gamma, e \right) \mapsto \text{left hand side of (12.20)}$$

where

$$\begin{cases} Du_h(x_h) - \lambda_h \Phi(p) & = 0 \quad (h1) \\ -\Phi(p)(x_h - e_h) + \begin{bmatrix} -q \\ \Gamma Y \end{bmatrix} b_h & = 0 \quad (h2) \\ \lambda_h \begin{bmatrix} -q \\ \Gamma Y \end{bmatrix} & = 0 \quad (h3) \\ -\sum_{h=1}^{H} x_h^{\backslash} + \sum_{h=1}^{H} e_h^{\backslash} & = 0 \quad (Mg) \\ \sum_{h=1}^{H} b_h & = 0 \quad (Ma) \end{cases} \tag{12.20}$$

and $\Gamma \equiv diag\left((1, \gamma)\right)$.

Lemma 22 *0 is a regular value for F.*

Proof. Since Γ is a diagonal full rank matrix, $\operatorname{rank} \Gamma Y = I$ and hence the result is the content of Lemma 22 in Chapter 11. ■

Lemma 23 *$pr : F^{-1}(0) \to \mathbb{R}^{S-1}_{++} \times \mathbb{R}^{GH}_{++}, \quad (x, b, \lambda, p^{\backslash}, q, \gamma, e) \mapsto (\gamma, e)$ is proper.*

Proof. The proof is basically the same as that one of Lemma 24 in Chapter 11; only observe that the convergence of q follows from the convergence of γ. ■

Using Assumption nom1, define

$$B(b) \equiv \left[b_1^T, ..., b_I^T\right] \in \mathcal{M}_{I,I}$$

and

$$G : \Xi \times \mathbb{R}^I \times \mathbb{R}^{S-1}_{++} \times \mathbb{R}^{GH}_{++} \to \mathbb{R}^{\dim \Xi + I + 1}, \; (\xi, c, \gamma, e) \mapsto \begin{pmatrix} F(\xi, \gamma, e) \\ cB(b) \\ \frac{1}{2}cc - 1 \end{pmatrix}$$

In the following lemma, we use Assumption nom1.

Lemma 24 *0 is a regular value for G.*

Proof. From Lemma 22 it suffices to show that it is possible to perturb the last two superrows of the jacobian matrix of G. The computation of a well chosen partial Jacobian matrix of G is described in next page.

$$
\begin{array}{l}
 \\ (1.1) \\ (1.2) \\ \\ (1.3) \\ \\ \cdots \\ (H.1) \\ (H.2) \\ \\ (H.3) \\ \\ (Mx) \\ (MR1) \\ (MR2)
\end{array}
\begin{array}{cccccccccccc}
x_1 & \lambda_1 & b_1 & \cdots & x_H & \lambda_H & b_H & e_1^{\setminus} & e_1^{\bullet} & e_H^{\setminus} & e_H^{\bullet} & c \\
\left(\begin{array}{c} D_1^2 u_1 \\ -\Phi(p) \end{array}\right. & -\Phi(p)^T & & & & & & & & & & \\
 & & \left[\begin{array}{c} -q \\ \Gamma Y \end{array}\right] & & & & & \Phi(p) & I & & & \\
 & \left[\begin{array}{c} -q \\ \Gamma Y \end{array}\right]^T & & & & & & & & & & \\
 & & & \ddots & & & & & & & & \\
 & & & & D_H^2 u_H & -\Phi(p)^T & & & & & & \\
 & & & & -\Phi(p) & & \left[\begin{array}{c} -q \\ \Gamma Y \end{array}\right] & & & \Phi(p) & I & \\
 & & & & & \left[\begin{array}{c} -q \\ \Gamma Y \end{array}\right]^T & & & & & & \\
\widehat{I} & & & & \widehat{I} & & & -I & & -I & & \\
 & & diag\,(c_1 \mathbf{1}) & & & & & & & & & B(b) \\
 & & & & & & & & & & & \left. c^T \right)
\end{array}
$$

where

$$\widehat{I} \equiv \begin{bmatrix} I_{C-1} & 0 & & & \\ & & \ddots & & \\ & & & I_{C-1} & 0 \end{bmatrix}_{G-(S+1)\times G}$$

Observe that from $(MR2)$, there exists i such that $c_i \neq 0$. Without loss of generality, take $i = 1$ and therefore $diag\,(c_1\mathbf{1})$ has full rank. The needed perturbations are illustrated below.

$$\begin{array}{lll} (MR1) & \leftarrow b_1 & \rightsquigarrow (1.2), (Mb) \\ (1.2) & \leftarrow e_1^{\bullet} & \end{array}$$

Since $H > I$, $D_{b_H}[cB(b)] = 0$ and hence we can "adjust" the superrow (Mb) using b_H. Then we obtain:

$$\begin{array}{lll} (Mb) & \leftarrow b_H & \rightsquigarrow (H.2) \\ (H.2) & \leftarrow e_H^{\bullet} & \end{array}$$

$$(MR2) \quad \leftarrow c \quad \rightsquigarrow (MR1)$$

and then proceed as above. ■

Roughly speaking, the following lemma shows that for almost all endowments, the set of equilibria contains the image of open subset of $\mathbb{R}^{S+1}$ via a C^1 one-to-one function and for all equilibria in that image the rank condition 1 in Lemma 20 does hold.

Lemma 25 *There exists an open full measure subset $\Omega^* \subseteq \mathbb{R}^{GH}_{++}$ such that for each $e^* \in \Omega^*$, there exists an open subset A_{e^*} of $\mathbb{R}^{S-1}_{++}$ and a C^1 function*

$$\xi_{e^*} : A_{e^*} \to \Xi, \quad \gamma \mapsto \xi_{e^*}(\gamma) \equiv \left(x_{e^*}(\gamma), b_{e^*}(\gamma), \lambda_{e^*}(\gamma), p^{\backslash}_{e^*}(\gamma), q_{e^*}(\gamma)\right)$$

such that for each $\gamma \in A_{e^*}$

1 $(\xi_{e^*}(\gamma), \gamma) \in F_{e^*}^{-1}(0)$;

2 $\operatorname{rank} B\,(b_{e^*}(\gamma)) = I$.

Proof. From Lemmas 22, Lemma 23, Theorem 26 and Corollary 27, part 3 in Chapter 6, there exists an open and full measure subset $D_1^* \subseteq \mathbb{R}^{S-1}_{++} \times \mathbb{R}^{GH}_{++}$ such that for each $(\gamma, e) \in D_1^*$, 0 is a regular value for $F_{(\gamma,e)}$.

From Lemma 23, Lemma 24 and Corollary 27 in Chapter 6, there exists an open and full measure subset $D_2^* \subseteq \mathbb{R}^{S-1}_{++} \times \mathbb{R}^{GH}_{++}$ such that for each $(\gamma, e) \in D_2^*$, 0 is a regular value for $G_{(\gamma,e)}$.

Define $D^* \equiv D_1^* \cap D_2^*$, an open and full measure subset of $\mathbb{R}^{S-1}_{++} \times \mathbb{R}^{GH}_{++}$.

From Lemma 22 and Lemma 23, the assumption of Corollary 44 in Chapter 4 are verified and then each $(\gamma^*, e^*) \in D^*$ is a regular values for pr. From the Conclusion 3 of that Corollary, it follows that for every $(\gamma^*, e^*) \in D^*$ there exists an open neighborhood $N(\gamma^*, e^*)$ of (γ^*, e^*) contained in D^*, an open neighborhood U^* of some $(\xi^*, \gamma^*, e^*) \in F^{-1}(0)$ and a C^1 diffeomorphism

$$f : N(\gamma^*, e^*) \to U^*$$

In fact, $f = \left(pr_{|U^*}\right)^{-1}$. Define

$$f(\gamma, e) \equiv \left(x(\gamma, e), b(\gamma, e), \lambda(\gamma, e), p^{\backslash}(\gamma, e), q(\gamma, e), \gamma, e\right)$$

Moreover, from Lemma 23, Lemma 24 and the transversality theorem (Theorem 26 in Chapter 6) for each $(\gamma, e) \in D^*$ and therefore for a sufficiently small neighborhood of (γ^*, e^*), without minor abuse of notation $N(\gamma^*, e^*)$ itself, for each $(\gamma, e) \in N(\gamma^*, e^*)$, $B(b(\gamma, e))$ has full rank.

Now, take $\Omega^* \equiv \Pi_{\mathbb{R}^{GH}_{++}}(D^*)$; from Proposition 42 in Chapter 6 Ω^* is an open and full measure subset of $\mathbb{R}^{GH}_{++}$. Since $(\gamma^*, e^*) \in D^*$, then $e^* \in \Omega^*$ and there exists an open neighborhood A_{e^*} of γ^* such that $A_{e^*} \times \{e^*\} \subseteq N(\gamma^*, e^*)$. Then, for each $\gamma \in A_{e^*}$, it is enough to define

$$\xi_{e^*}(\gamma) \equiv f(\gamma, e^*) \equiv \left(x_{e^*}(\gamma), b_{e^*}(\gamma), \lambda_{e^*}(\gamma), p^{\backslash}_{e^*}(\gamma), q_{e^*}(\gamma)\right)$$

By construction, f_{e^*} is C_1 and rank $B(b_{e^*}(\gamma)) = I$. ■

5.5 The theorem

Theorem 26 *There exists an open and full measure subset Ω^* in $\mathbb{R}^{GH}_{++}$ such that for each $e^* \in \Omega^*$ the set of nominal equilibrium allocations associated with it contains the image of an open subset of $\mathbb{R}^{S-1}$ via a C^1 one-to-one function.*

Proof. From Lemma 25 in Step 4, for each $e^* \in \Omega^*$, there exists a C^1 one-to-one function

$$\xi_{e^*} : A_{e^*} \to \Xi, \quad \gamma \mapsto \xi_{e^*}(\gamma)$$

where $\xi_{e^*}(\gamma) \equiv \left(x_{e^*}(\gamma), b_{e^*}(\gamma), \lambda_{e^*}(\gamma), p^{\backslash}_{e^*}(\gamma), q_{e^*}(\gamma)\right)$ and A_{e^*} is an open subset of $\mathbb{R}^{S-1}_{++}$.

By construction, for all $\gamma \in A_{e^*}$, $x_{e^*}(\gamma)$ is a numeraire equilibrium allocation associated with the economy $(e^*, \Gamma Y)$. Therefore, from Step 1, $x_{e^*}(\gamma)$ is a nominal equilibrium allocation associated with (e^*, Y). Then, the set of nominal equilibrium associated with (e^*, Y) contains $x_{e^*}(A_{e^*})$, the image of the open subset A_{e^*} of $\mathbb{R}^{S-1}_{++}$ via the C^1 function x_{e^*}.

We are then left with showing that the function x_{e^*} is one-to-one. Take $\gamma_1, \gamma_2 \in A_{e^*}$ with $\gamma_1 \neq \gamma_2$ and therefore $diag((1, \gamma_1)) \equiv \Gamma_1 \neq \Gamma_2 \equiv diag((1, \gamma_2))$.

Then, from Step 3, Im $\Gamma_1 Y \neq$ Im $\Gamma_2 Y$. Moreover, from Lemma 25 in Step 4, $\text{rank}\, B(b_{e^*}(\gamma_1)) = \text{rank}\, B(b_{e^*}(\gamma_2))$.

Finally, from Step 2, $x_{e^*}(\gamma_1) \neq x_{e^*}(\gamma_2)$, as desired. ■

Chapter 13

REAL ASSETS

In this chapter we analyze the model with real assets which, as in Chapter 10, promise to deliver a bundle of goods. Therefore, their yields are a function of the price vector, an endogenous variable.

In our analysis, we follow the seminal paper [56][1]. In their words, "As [67] has shown, equilibria need not exist in this setting even under the 'smooth preference' assumptions of [41] that we adopt. Hart's counter-example is based on a collapse in the span of security markets that occurs on an exceptional set of 'bad' spot prices. (...) Attempts to resurrect the existence of equilibria have thus concentrated on showing that the exceptional set of 'bad' spot prices is only relevant for an exceptional set of economies, a program of *generic existence*."[2]

In this chapter we prove existence of equilibria using the reduced approach instead of the extended approach followed in all the other cases analyzed in the book - see Section 8 in Chapter 8 for a discussion of the two approaches. On the other hand, we do use an extended *system* in several steps of the proofs. A relatively technical explanation of our choice is given in Remark 21, after having introduced enough terminology and notation.

The chapter is organized as follows. In Section 2, we present the set up of the model which also contains a list of the main steps of the proof. In Section 3, we define the two concepts of effective and pseudo-equilibria which are instrumental in the proof of existence. Then, Section 4 shows existence of pseudo-equilibria and Section 5 the main result of the chapter: in an open and full measure subset of the endowment and yield space, an equilibrium exists. Section 6 briefly discusses the definition and properties of regular economies.

[1] Other contributions are [17], [48] and [100].

[2] [56], p. 285.

1. Set up of the model

The model we present is a particular case of that one presented in Subsection 2.1 in Chapter 10: assets are of the real kind. The general notation used there is used in the present chapter as well.

To be able to apply (the existence) Theorem 59 in Chapter 7 , we need to deal with (at least) C^2 manifolds. Since we will characterize the equilibrium manifold in terms of gradients of utility function and of demand functions - see Theorem 22 - we have to substitute Assumption u1 introduced in Subsection 2.1 in Chapter 10 with the following one.

Assumption u1' For $h \in \mathcal{H}$, $u_h : \mathbb{R}^G_{++} \to \mathbb{R}$ is at least C^3.

Definition 1 A **real asset** i is an asset which promises to deliver a vector $y^{si} \equiv \left(y^{sic}\right)_{c=1}^{C} \in \mathbb{R}^C$ of goods in state s, where y^{sci} denotes the units of good c delivered in state s.
Therefore,

$$\text{for each } s \text{ and } i, \text{ and each } p \in \mathbb{R}^G_{++}, \quad r^{si}(p) = p^s y^{si}$$

and the return matrix function is

$$R : \mathbb{R}^G_{++} \to \mathcal{M}_{S,I}, \quad p \mapsto \begin{bmatrix} p^1 y^{11} & \cdots & p^1 y^{1i} & \cdots & p^1 y^{1I} \\ \cdots & & & & \\ p^s y^{s1} & \cdots & p^s y^{si} & \cdots & p^s y^{sI} \\ \cdots & & & & \\ p^S y^{S1} & \cdots & p^S y^{Si} & \cdots & p^S y^{SI} \end{bmatrix}$$

We can identify the set $\mathcal{R}_{S,I}$ of all return matrix functions defined above, (see also Subsection 2.1 in Chapter 10) with the euclidean space R^{SCI}.

On the other hand, to stress the non-trivial dependence of the return matrix from both p and y, with innocuous abuse of notation, we denote the return matrix by $R(p, y)$. As usually done we denote the s-th row of $R(p, y)$ by

$$r^s(p, y)$$

for $s = 1, .., S$, and its i-th column by

$${}^i r(p, y)$$

for $i = 1, ..., I$.

In this chapter, we fix the utility function vector u. Therefore, we can give the following definition of economy.

Definition 2 An **economy** in a financial economy model with real assets is an element $(e, y) \in \mathbb{R}^{GH}_{++} \times \mathbb{R}^{SCI}$.

Household h's maximization problem is as follows. For given $p \in \mathbb{R}^{G}_{++}$, $q \in \mathbb{R}^{I}$, $e_h \in \mathbb{R}^{G}_{++}$, $y \in \mathbb{R}^{SCI}$,

$$\begin{array}{lll} \max_{(x_h, b_h) \in \mathbb{R}^{G}_{++} \times \mathbb{R}^{I}} & u_h(x_h) & \\ s.t. & & p^0 x^0_h + q b_h = p^0 e^0_h, \text{ and} \\ & & p^s x^s_h = p^s e^s_h + \left(p^s y^{si}\right)^I_{i=1} b_h \ \text{ for } s \geq 1 \end{array}$$

or

$$\begin{array}{lll} \max_{(x_h, b_h) \in \mathbb{R}^{G}_{++} \times \mathbb{R}^{I}} & u_h(x_h) & \\ s.t. & & -\Phi(p)(x_h - e_h) + \begin{bmatrix} -q \\ R(p, y) \end{bmatrix} b_h = 0 \end{array} \tag{13.1}$$

Definition 3 The vector $(x, b, p, q) \in \mathbb{R}^{GH}_{++} \times \mathbb{R}^{IH} \times \mathbb{R}^{G}_{++} \times \mathbb{R}^{I}$ is a **demand-price financial real asset equilibrium** vector for the economy $(e, y) \in \mathbb{R}^{GH}_{++} \times \mathbb{R}^{SCI}$ if

1 for $h = 1, ..., H$, for given (p, q, e, y), (x_h, b_h) solves problem (13.1), i.e., households maximize, and

2
$$\begin{array}{l} \sum_{h=1}^{H} x_h = \sum_{h=1}^{H} e_h \quad \text{and} \\ \sum_{h=1}^{H} b_h = 0 \end{array}$$

i.e., both good and asset markets clear.

The main goal of this chapter is to show that in an open and full measure subset of the endowment and yield space, an equilibrium exists - see Theorem 29.

The basic steps in the proof of that result are described below.

1 We define the concept of effective equilibrium. In that equilibrium, household 1 is treated as a "Walrasian consumer", i.e., she has only one budget equation to satisfy:

$$px_1 = pe_1$$

The other households face the above "Walrasian constraint" *and* the constraints of the "true" maximization problem (13.1) with respect to states $s \neq 0$. Those constraints are written in terms of the following condition: the values of the excess demands for goods in period 1 has to be consistent with the available financial structure, i.e.,

$$\left(p^s\left(x^s_h - e^s_h\right)\right)^S_{s=1} \in \operatorname{Im} R(p, y)$$

Observe that $\text{Im}\, R(p, y)$ is a linear subspace of $\mathbb{R}^S$ of dimension equal to $\text{rank}\, R(p, y)$. When the endogenous variable p changes, that dimension may change. This the possibility of drop of rank mentioned in the introduction of the present chapter.

The definition of effective equilibrium is close to that one used by David Cass in his seminal paper on existence of equilibria with nominal assets - see [22]. That strategy of proof is known as Mr. 1's trick or, in fact, Cass' trick. It easily allows to show properness of a suitably defined projection, and more specifically the convergence of prices - see Proposition 23.

2 We show that each effective equilibrium is a (true) equilibrium - see Proposition 8.

3 We define the concept of pseudo-equilibrium. That concept is different from the effective one for two crucial reasons.

(a) For each household $h \neq 1$ the value of the excess demands in period 1 has to be contained in a linear subspace of $\mathbb{R}^S$ *which is independent of prices* and of fixed dimension I

$$\left(p^s\left(x_h^s - e_h^s\right)\right)_{s=1}^S \in L$$

where L is an endogenous variables belonging to $G_{I,S}$, a grassmannian manifold[3].

(b) The image of the return matrix $R(p, y)$ has to be contained in the linear subspace L. In fact, if $\text{rank}\, R(p, y) = I$ and therefore $\text{Im}\, R(p, y) = L$, then a pseudo-equilibrium is an effective equilibrium - see Proposition 9.

4 Exploiting the properties of the grassmannian manifold, we show that pseudo-equilibria exist for each economy - see Section 3.

5 Finally, using a standard Transversality argument, we show that typically in the space of economies, $\text{rank}\, R(p, y) = I$ and therefore (effective and true) equilibria do exist - see Section 4.

The use of the concept of pseudo-equilibrium rests upon the fact that for that concept of equilibrium the span of the asset structure does not change when prices change. Moreover, the use of the compact grassmannian manifold is crucial in the needed proof of compactness, in fact properness of a projection function - see Proposition 23.

[3]The structure of grassmannian manifolds is briefly described in the appendix to the present chapter.

2. Effective and pseudo-equilibria

Define $x_h^1 \equiv (x_h^s)_{s \neq 0} \in \mathbb{R}_{++}^{SC}$, $e_h^1 \equiv (e_h^s)_{s \neq 0} \in \mathbb{R}_{++}^{SC}$,

$$\Phi^1(p) \equiv \begin{bmatrix} p^1 & & \\ & \ddots & \\ & & p^S \end{bmatrix} \in \mathcal{M}_{S,SC}$$

Definition 4 A vector $(x, p) \in \mathbb{R}_{++}^{GH} \times \mathbb{R}_{++}^{G}$ is an **effective equilibrium** for the economy $(e, y) \in \mathbb{R}_{++}^{GH} \times \mathbb{R}^{SCI}$ if

1. for given (p, e, y), x_1 solves

$$\max_{x_1 \in \mathbb{R}_{++}^G} u_1(x_1) \quad s.t. \quad p(x_h - e_h) = 0 \tag{13.2}$$

2. for $h = 2, ..., H$, for given (p, e, y), x_h solves

$$\begin{array}{lll} \max\limits_{x_h \in \mathbb{R}_{++}^G} u_h(x_h) & & \\ s.t. & p(x_h - e_h) = 0 & (1) \\ & \Phi^1(p)\left(x_h^1 - e_h^1\right) \in \operatorname{Im} R(p, y) & (2) \end{array} \tag{13.3}$$

3. x satisfies the market clearing conditions

$$\sum_{h=1}^{H} (x_h - e_h) = 0 \tag{13.4}$$

Definition 5 A vector $(x, p, L) \in \mathbb{R}_{++}^{GH} \times \mathbb{R}_{++}^{G} \times G_{I,S}$ is a **pseudo-equilibrium** for the economy (e, y) if

1. for given (p, e, y), x_1 solves

$$\max_{x_1 \in \mathbb{R}_{++}^G} u_1(x_1) \quad s.t. \quad p(x_1 - e_1) = 0 \tag{13.5}$$

2. for $h = 1, ..., H$, for given (p, L, e), x_h solves

$$\begin{array}{lll} \max\limits_{x_h \in \mathbb{R}_{++}^G} u_h(x_h) & & \\ s.t. & p(x_h - e_h) = 0 & (1) \\ & \Phi^1(p)\left(x_h^1 - e_h^1\right) \in L & (2) \end{array} \tag{13.6}$$

3. x satisfies the market clearing conditions (13.4).
4. $\operatorname{Im} R(p, y) \subseteq L$.

Remark 6 Since households' budget constraint are not affected by multiplication by a positive constant of all prices, we can set one, and in fact only one, price equal to 1. Moreover, summing up the budget constraints of all households, we get Walras' law , i.e., we can characterize equilibria using all but one market clearing condition.

Consistently with the above remark, and for future use, we can give the following definitions.

Definition 7 $p^{\backslash} \equiv (p^{sc})_{(sc)\neq(0C)} \in \mathbb{R}^{G-1}_{++}$; for each h, $x^{\backslash}_h \equiv (x^{sc}_h)_{(sc)\neq(0C)}$ and $e^{\backslash}_h \equiv (e^{sc}_h)_{(sc)\neq(0C)}$.

Proposition 8 *If (x,p) is an effective equilibrium for an economy (e,y), then there exists (b,q) such that (x,p,b,q) is an equilibrium for (e,y).*

Proof. As we have seen in Subsection 3.2.6, there is no loss of generality in taking $q = \mathbf{1} \cdot R(p,y)$ (in fact that choice amounts to nothing but a particular normalization).

Define the set of affordable consumption vectors for household h associated with the economy (e,y) in the case of "true" and effective equilibria by $B_h(p,q)$ and $B^e_h(p)$, respectively, i.e.,

$$\begin{aligned} B_h(p,q) \equiv \ & \{x_h \in \mathbb{R}^G_{++} : \exists b_h \in \mathbb{R}^I \text{ such that } \ -p^0\left(x^0_h - e^0_h\right) + qb_h = 0 \\ & \text{and} - \Phi^1(p)\left(x^1_h - e^1_h\right) + R(p,y)\, b_h = 0\} \end{aligned}$$

$$B^e_1(p) \equiv \{x_1 \in \mathbb{R}^G_{++} : p(x_1 - e_1) = 0\}$$

and for $h \neq 1$,

$$\begin{aligned} B^e_h(p) \equiv \ & \{x_h \in \mathbb{R}^G_{++} : \ -p(x_h - e_h) = 0 \ \text{ and} \\ & \Phi^1(p)\left(x^1_h - e^1_h\right) \in \operatorname{Im} R(p,y)\} \end{aligned}$$

We will prove the desired result showing the following three steps.

1. For $h \neq 1$, a. $B^e_h(p) \subseteq B_h(p,q)$; b. $B^e_h(p) \supseteq B_h(p,q)$,
and therefore for $h \neq 1$, x_h solves the maximization problem in the definition effective equilibrium if and only if it solve the maximization problem in the definition of true equilibria.

2. a. if (x,p) is an effective equilibrium, then x_1 solves problem (13.1) for $h = 1$; b. if x_1 solves problem (13.1) for $h = 1$, then $x_1 \in B^e_1(p)$; therefore, from Lemma 72 in Chapter 8, and since we know that a solution to problem (13.2) exists and is unique, we can conclude that x_1 solves the maximization problem for household 1 in the definition of effective equilibria.

3. Market clearing conditions for the case of "true" equilibria do hold.

Observe that the above statements prove also that

if (x,p,b,q) is an equilibrium for (e,y), then (x,p) is an effective equilibrium for (e,y).

1a.

For $h > 1$, from equation (2) in the definition of $B_h^e(p)$, there exists $b_h \in \mathbb{R}^I$ such that

$$\Phi^1(p)\left(x_h^1 - e_h^1\right) = R(p,y)\,b_h. \tag{13.7}$$

For future use, define

$$b_1 \equiv \sum_{h\neq 1} b_h \tag{13.8}$$

For $h > 1$, from equation (1) in the definition of $B_h^e(p)$,

$$\begin{aligned} 0 &= p^0\left(x_h^0 - e_h^0\right) + \mathbf{1}\cdot\Phi^1(p)\left(x_h^1 - e_h^1\right) = \\ &= p^0\left(x_h^0 - e_h^0\right) + \mathbf{1}\cdot R(p,y)\,b_h = p^0\left(x_h^0 - e_h^0\right) + qb_h, \end{aligned} \tag{13.9}$$

Therefore, (13.9) and (13.7) are the two equations in the definition of $B_h(p,q)$ for $h > 1$.

1b.

If x_h satisfies the two equations in the definition of $B_h(p,q)$, then it clearly satisfies equation (2) in the definition of $B_h^e(p)$. Moreover from equation (2) in the definition of $B_h(p,q)$, we get

$$\mathbf{1}\cdot\Phi^1(p)\left(x_h^1 - e_h^1\right) = \mathbf{1}\cdot R(p,y)\,b_h$$

and using the fact that $q = \mathbf{1}\cdot R(p,y)$,

$$-\sum_{s=1}^{S} p^s\left(x_h^s - e_h^s\right) + qb_h = 0 \tag{13.10}$$

Adding up equation (1) in the definition of $B_h(p,q)$ and (13.10), we get equation (1) in the definition of $B_h^e(p)$.

2a.

From market clearing for goods, for $s = 0, 1, ..., S$

$$\sum_{h=1}^{H} p^s\left(x_h^s - e_h^s\right) = 0$$

Therefore, for $s \geq 1$,

$$\begin{aligned} 0 = &\ p^s\left(x_1^s - e_1^s\right) + \textstyle\sum_{h=2}^{H} p^s\left(x_h^s - e_h^s\right) \\ = &\ p^s\left(x_1^s - e_1^s\right) + \textstyle\sum_{h=2}^{H} r^s(p,y)\,b_h = p^s\left(x_1^s - e_1^s\right) + r^s(p,y)\textstyle\sum_{h=2}^{H} b_h \end{aligned}$$

where the second equality comes from (13.7). From (13.8), we then get

$$\Phi^1(p)\left(x_1^1 - e_1^1\right) = R(p,y)\,b_1$$

Finally, for $s = 0$, again using market clearing for goods,

$$\begin{aligned} 0 = & \; p^0 \left(x_1^0 - e_1^0\right) + \sum_{h=2}^H p^0 \left(x_h^0 - e_h^0\right) = p^0 \left(x_1^0 - e_1^0\right) + q \sum_{h=2}^H b_h \\ = & \; p^0 \left(x_1^0 - e_1^0\right) + q b_1 \end{aligned}$$

where the second equality comes from (13.9) and the last one again from (13.8).

2b.

It follows from the same argument given in 1b. above.

3.

Market clearing for goods is satisfied by assumption. Market clearing for assets holds from (13.8). ■

Proposition 9 *If* $\left(x, p^\backslash, L\right)$ *is a pseudo-equilibrium for an economy* (e, y) *and*

$$\operatorname{rank} R(p, y) = I, \tag{13.11}$$

then (x, p) *is an effective equilibrium for* (e, y).

Proof. Since $\operatorname{rank} R(p, y) = I$ and $L \in G_{I,S}$, then $\dim \operatorname{Im} R(p, y) = I = \dim L$. Since $\operatorname{Im} R(p, y) \subseteq L$, then $\operatorname{Im} R(p, y) = L$. ■

Because of the above Proposition, to show generic existence of equilibria we are first going to show that for each economy a pseudo-equilibrium exists and then that condition (13.11) holds in an open and full measure subset of the space of economies.

3. Existence of pseudo-equilibria

In this section, we use definitions, notations and results presented in the Appendix on grassmannian manifolds.

Since $L \in G_{I,S}$, there exists $\sigma \in \Sigma$ and a function $\psi_\sigma : V_\sigma \in G_{I,S} \to \mathcal{M}_{S-I,I}$ such that

$$\ker \left[I \mid \psi_\sigma(L)\right] P_\sigma = L$$

Define[4]

$$E \equiv \begin{bmatrix} e_{11} & \dots & e_{1I} \\ & & \\ e_{S-I,1} & \dots & e_{S-I,I} \end{bmatrix} \equiv \psi_\sigma(L)$$

Therefore, we can rewrite (13.6) as follows.

$$\begin{array}{lll} \max_{x_h \in \mathbb{R}_{++}^G} & u_h(x_h) & \\ s.t. & p(x_h - e_h) = 0 & (1) \\ & \left[I \mid \psi_\sigma(L)\right] P_\sigma \Phi^1(p)\left(x_h^1 - e_h^1\right) = 0 & (2) \end{array} \tag{13.12}$$

[4]No confusion should arise between the elements of the matrix E and endowments, which are denoted by the letter e.

Lemma 10 *A solution to problem (13.12) exists and it is unique.*

Proof. Existence.

Following the same arguments presented in Chapter 8, it easy to show that the set of solution to problem (13.12) is the same as the set of solutions to the problem

$$\begin{array}{lll} \max_{x_h \in \mathbb{R}^G_{++}} & u_h(x_h) & \\ s.t. & p(x_h - e_h) = 0 & (1) \\ & [I \mid \psi_\sigma(L)] P_\sigma \Phi^1(p)\left(x_h^{\mathbf{1}} - e_h^{\mathbf{1}}\right) = 0 & (2) \\ & u_h(x_h) \geq u_h(e_h) & (3) \end{array}$$

Then the desired result follows from Assumption u4.

Uniqueness. It follows from strict quasiconcavity of the utility function, which is implied by Assumption u3, and convexity of the constraint set. ■

Definition 11 The pseudo-demand function of household $h \neq 1$ is the solution function to (13.12), i.e., for $h \neq 1$,

$$x_h : \mathbb{R}^{G-1}_{++} \times G_{I,S} \times \mathbb{R}^G_{++} \to \mathbb{R}^G_{++}, \quad : \left(p^{\backslash}, L, e_h\right) \mapsto \arg\max (13.12)$$

Denote the component function (sc) of x_h by x_h^{sc}, and define

$$x_h^{\backslash} \equiv (x_h^{sc})_{(sc)\neq(0C)}$$

Define

$$p^{\mathbf{1}} \equiv (p^s)_{s=1}^S \in \mathbb{R}^{SC}_{++} \quad \text{and} \quad x_h^{\mathbf{1}} \equiv (x_h^s)_{s=1}^S \in \mathbb{R}^{SC}_{++}$$

where the superscript **1** is bold face.

Theorem 12 *For $h \neq 1$, the following statements hold.*

1 $x_h = x_h\left(p^{\backslash}, L, e_h\right) \Leftrightarrow$ *there exists* $(\mu_h, \eta_h) \in \mathbb{R}_{++} \times \mathbb{R}^{S-I}$ *such that* (x_h, μ_h, η_h) *is the unique solution to system*

$$\begin{cases} D_{x_h^0} u_h(x_h) - \mu_h p^0 & = & 0 & (h.1) \\ D_{x_h^{\mathbf{1}}} u_h(x_h) - \mu_h p^{\mathbf{1}} + \eta_h [I \mid \psi_\sigma(L)] P_\sigma \Phi^1(p) & = & 0 & (h.2) \\ p(x_h - e_h) & = & 0 & (h.3) \\ [I \mid \psi_\sigma(L)] P_\sigma \Phi^1(p)\left(x_h^{\mathbf{1}} - e_h^{\mathbf{1}}\right) & = & 0 & (h.4) \end{cases} \qquad (13.13)$$

2 x_h *is a* C^2 *function.*

Proof. Following Subsection 3.2.1 in Chapter 1, we want to show that problem (13.12) satisfies necessary and sufficient conditions illustrated there and then apply the implicit function theorem ([3], Theorem 13.7, p. 374) to the first order conditions.

The lagrangean function for problem (13.12) is

$$\mathcal{L}(x_h, \mu_h, \eta_h, p, L, e, y) =$$
$$u_h(x_h) + \mu_h(-p(x_h - e_h)) + \eta_h [I \mid \psi_\sigma(L)] P_\sigma \Phi^1(p)\left(x_h^1 - e_h^1\right)$$

and the associated first order conditions are just system (13.13).

[⇒] Necessary conditions described in Theorem 41 in Chapter 1 are satisfied because of the following argument. The computation of the partial jacobian matrix of the constraint functions with respect to the choice variables is described below.

$$\begin{array}{l} \\ p(x_h - e_h) \\ [I \mid \psi_\sigma(L)] P_\sigma \Phi^1(p)\left(x_h^1 - e_h^1\right) \end{array} \begin{array}{c} \begin{array}{cc} x_h^0 & \qquad x_h^1 \end{array} \\ \begin{pmatrix} p^0 & p^1 \\ 0 & [I \mid \psi_\sigma(L)] P_\sigma \Phi^1(p) \end{pmatrix} \end{array}$$

Observe that

$$[I \mid \psi_\sigma(L)] P_\sigma \Phi^1(p) =$$

$$= \begin{bmatrix} 1 & & e_{11} & \dots & e_{1I} \\ \dots & & \dots & & \dots \\ & 1 & e_{S-I,1} & \dots & e_{S-I,I} \end{bmatrix} \begin{bmatrix} p^{\sigma(1)} & & & & \\ & \ddots & & & \\ & & p^{\sigma(S-I)} & & \\ & & & p^{\sigma(S-I+1)} & \\ & & & & \ddots & \\ & & & & & p^{\sigma(S)} \end{bmatrix}$$

$$= \left[\begin{array}{cc|ccc} p^{\sigma(1)} & & e_{11}p^{\sigma(S-I+1)} & & e_{1I}p^{\sigma(S)} \\ & & & \ddots & \\ & p^{\sigma(S-I)} & e_{S-I,1}p^{\sigma(S-I+1)} & & e_{S-I,1}p^{\sigma(S)} \end{array}\right]$$

Therefore, the above mentioned partial jacobian matrix does have full rank.

We can then conclude that if $x_h = x_h\left(p^\backslash, L, e_h\right)$, then there exists $(\mu_h, \eta_h) \in \mathbb{R}_{++} \times \mathbb{R}^{S-I}$ such that (x_h, μ_h, η_h) is a solution to system (13.13).

The fact that $\mu_h > 0$ follows from equation $(h.1)$ in that system, Assumption u2 and the fact that $p \in \mathbb{R}^G_{++}$.

[⇐] Sufficient conditions described in Theorem 36 in Chapter 1 follow from the pseudoconcavity of the utility function and the linear affinity of the constraints.

The above two statements and the previous lemma allow to conclude that for each $\left(p^\backslash, L, e_h\right)$ the solution to system (13.13) is in fact unique. We denote it by $(x_h, \mu_h, \eta_h) = (x_h, \mu_h, \eta_h)\left(p^\backslash, L, e_h\right)$.

2. For each $\sigma \in \Sigma$, define

$$F_{h\sigma} : \mathbb{R}^G_{++} \times \mathbb{R}_{++} \times \mathbb{R}^{S-I} \times \mathbb{R}^{G-1}_{++} \times \mathcal{M}_{S-I,I} \times \mathbb{R}^G_{++} \to \mathbb{R}^G \times \mathbb{R} \times \mathbb{R}^{S-I}$$
$$F_{h\sigma} : \left(x_h, \mu_h, \eta_h, p^\backslash, E, e_h\right) \mapsto \text{ left hand side of system (13.14) below}$$

$$\begin{cases} D_{x_h^0} u_h(x_h) - \mu_h p^0 & = 0 \quad (h.1) \\ D_{x_h^1} u_h(x_h) - \mu_h p^1 + \eta_h [I \mid E] P_\sigma \Phi^1(p) & = 0 \quad (h.2) \\ p(x_h - e_h) & = 0 \quad (h.3) \\ [I \mid E] P_\sigma \Phi^1(p)(x_h^1 - e_h^1) & = 0 \quad (h.4) \end{cases} \tag{13.14}$$

and let

$$(\widetilde{x}_{h\sigma}, \widetilde{\mu}_{h\sigma}, \widetilde{\eta}_{h\sigma})\left(p^\backslash, E, e_h\right) \equiv (x_h, \mu_h, \eta_h)\left(p^\backslash, \phi_\sigma(E), e_h\right) \tag{13.15}$$

Observe that $(\widetilde{x}_{h\sigma}, \widetilde{\mu}_{h\sigma}, \widetilde{\eta}_{h\sigma})\left(p^\backslash, E, e_h\right)$ is the unique solution to the system (13.14) for $\left(p^\backslash, E, e_h\right) \in \mathbb{R}_{++}^{G-1} \times \mathcal{M}_{S-I,I} \times \mathbb{R}_{++}^G$.

As an application of the implicit function theorem, it is enough to show that the partial jacobian matrix $D_{\left(x_h,\mu_h,p^\backslash\right)} F_{h\sigma}$ has full row rank $G + 1 + (S - I)$. Then, by the implicit function theorem applied to $F_{h\sigma}$, $\widetilde{x}_{h\sigma}$ is C^2 on $\mathbb{R}_{++}^{G-1} \times \mathcal{M}_{S-I,I} \times \mathbb{R}_{++}^G$. Morevoer, by Proposition 69 in Chapter 3, x_h is C^2 on $\mathbb{R}^{G-1} \times V_\sigma \times \mathbb{R}_{++}^G$ for each $\sigma \in \Sigma$, and hence it is C^2 on $\mathbb{R}^{G-1} \times G_{I,S} \times \mathbb{R}_{++}^G$. Let's now show that $D_{\left(x_h,\mu_h,p^\backslash\right)} F_{h\sigma}$ has full row rank. The computation of that matrix is described below.

$$\begin{array}{l} \\ D_{x_h^0} u_h(x_h) - \mu_h p^0 \\ D_{x_h^1} u_h(x_h) - \mu_h p^1 + \eta_h [I \mid E] P_\sigma \Phi^1(p) \\ -p(x_h - e_h) \\ [I \mid E] P_\sigma \Phi^1(p)(x_h^1 - e_h^1) \end{array} \begin{array}{c} \begin{array}{cccc} \quad x_h^0 \quad & \quad x_h^1 \quad & \quad \mu_h \quad & \quad \eta_h \quad \end{array} \\ \begin{pmatrix} D^2_{x_h^0 x_h^0} u_h & D^2_{x_h^0 x_h^1} u_h & -p^{0T} & \\ D^2_{x_h^1 x_h^0} u_h & D^2_{x_h^1 x_h^1} u_h & -p^{1T} & [N_\sigma(p,E)]^T \\ -p^0 & -p^1 & & \\ & [N_\sigma(p,E)] & & \end{pmatrix} \end{array}$$

where

$$N_\sigma(p, E) \equiv [I \mid E] P_\sigma \Phi^1(p^1)$$

Suppose that $D_{\left(x_h,\mu_h,p^\backslash\right)} F_{h\sigma}$ does not have full row rank, i.e., there exists $\Delta \equiv \left(\Delta x_h^0, \Delta x_h^1, \Delta\mu_h, \Delta\eta_h\right) \neq 0$, with $\Delta x_h^0 \in \mathbb{R}^C, \Delta x_h^1 \in \mathbb{R}^{SC}, \Delta\mu_h \in \mathbb{R}, \Delta\eta_h \in \mathbb{R}^{S-I}$, such that

$$\left[D_{\left(x_h,\mu_h,p^\backslash\right)} F_{h\sigma}\right] \Delta = 0$$

or

$$\begin{bmatrix} D^2_{x_h^0 x_h^0} u_h & D^2_{x_h^0 x_h^1} u_h & -p^{0T} & \\ D^2_{x_h^1 x_h^0} u_h & D^2_{x_h^1 x_h^1} u_h & -p^{1T} & [N_\sigma(p,E)]^T \\ -p^0 & -p^1 & & \\ & N_\sigma(p,E) & & \end{bmatrix} \cdot \begin{bmatrix} \Delta x_h^{0T} \\ \Delta x_h^{1T} \\ \Delta\mu_h^T \\ \Delta\eta_h^T \end{bmatrix} = 0 \tag{13.16}$$

In what follows, we present the needed computations just for the case $\sigma = id$. The general case is left as an exercise. Then

$$\begin{aligned} [N_\sigma(p,E)] &= [I \mid E] \cdot \Phi^1(p) = \\ &= [I \mid E] \begin{bmatrix} \Phi^*(p) & \\ & \widehat{\Phi}(p) \end{bmatrix} = \left[\ \Phi^*(p) \quad E \cdot \widehat{\Phi}(p)\ \right] \end{aligned} \tag{13.17}$$

where

$$\begin{bmatrix} \Phi^*(p)_{(S-I)\times(S-I)C} & \\ & \widehat{\Phi}(p)_{I\times(IC)} \end{bmatrix} \equiv \left[\begin{array}{ccc|ccc} p^1 & & & & & \\ & \ddots & & & & \\ & & p^{S-I} & & & \\ - & - & - & - & - & - \\ & & & p^{S-I+1} & & \\ & & & & \ddots & \\ & & & & & p^S \end{array}\right]$$

Rewrite (13.16) as follows.

$$\begin{array}{lll} D^2_{x_h^0 x_h^0} u_h \Delta x_h^0 + D^2_{x_h^0 x_h^1} u_h \Delta x_h^1 - p^{0T} \Delta\mu_h & = 0 & (1) \\ D^2_{x_h^1 x_h^0} u_h \Delta x_h^0 + D^2_{x_h^1 x_h^1} u_h \Delta x_h^1 - p^1 \Delta\mu_h + \left[N_\sigma(p,E)\right]^T \Delta\eta_h & = 0 & (2) \\ -p^0 \Delta x_h^0 - p^1 \Delta x_h^1 & = 0 & (3) \\ N_\sigma(p,E)\, \Delta x_h^1 & = 0 & (4) \end{array} \tag{13.18}$$

Claim 1. $\Delta x_h \neq 0$. Suppose otherwise. Then, from (1) and (2) in (13.18), we have

$$\begin{bmatrix} -p^{01} & 0 \\ -\left(p^{s1}\right)_{s=1}^{S-I} & diag\left(p^{s1}\right)_{s=1}^{S-I} \end{bmatrix} \begin{bmatrix} \Delta\mu_h \\ \Delta\eta_h \end{bmatrix} = 0$$

and therefore $(\Delta\mu_h, \Delta\eta_h) = 0$.

Claim 2. $Du_h(x_h)\,\Delta x_h = 0$. From the first order condition (13.13) and from (3) and (4) in (13.18) we have

$$\begin{aligned} Du_h(x_h)\,\Delta x_h = & \; D_{x_h^0} u_h(x_h)\,\Delta x_h^0 + D_{x_h^1} u_h(x_h)\,\Delta x_h^1 \\ = & \; -\mu_h p \Delta x_h^0 - \mu_h p^1 \Delta x_h^1 - \eta_h N_\sigma(p,E)\,\Delta x_h^1 = 0 \end{aligned}$$

Claim 3. $\Delta x_h D^2 u_h(x_h)\,\Delta x_h \not< 0$. Premultiplying (1) and (2) in (13.18) by Δx_h^0 and Δx_h^1, and adding up, we get

$$\begin{aligned} 0 = & \; \Delta x_h^0 D^2_{x_h^0 x_h^0} u_h \Delta x_h^0 + \Delta x_h^0 D^2_{x_h^0 x_h^1} u_h \Delta x_h^1 - \Delta x_h^0 p^0 \Delta\mu_h + \\ & \Delta x_h^1 D^2_{x_h^1 x_h^0} u_h \Delta x_h^0 + + \Delta x_h^1 D^2_{x_h^1 x_h^1} u_h \Delta x_h^1 + \\ & -\Delta x_h^1 p^1 \Delta\mu_h + \Delta x_h^1 \left[N_\sigma(p,E)\right]^T \Delta\eta_h \end{aligned}$$

Then, using (3) and (4) in (13.18), we get

$$\begin{aligned} 0 = & \; \Delta x_h^0 D^2_{x_h^0 x_h^0} u_h \Delta x_h^0 + \Delta x_h^0 D^2_{x_h^0 x_h^1} u_h \Delta x_h^1 + \Delta x_h^1 D^2_{x_h^1 x_h^0} u_h \Delta x_h^0 + \\ + & \; \Delta x_h^1 D^2_{x_h^1 x_h^1} u_h \Delta x_h^1 = \Delta x_h D^2 u_h(x_h)\,\Delta x_h \end{aligned}$$

Claims 1, 2 and 3 contradict Assumption u3. ■

From the analysis in Chapter 8, we know that the solution to problem (13.5) exists, is unique and therefore we can give the following definition.

Definition 13 The demand function for household 1 is

$$x_1 : \mathbb{R}^{G-1}_{++} \times \mathbb{R}^{G}_{++} \to \mathbb{R}^{G}_{++}, \quad \left(p^{\backslash}, e_h\right) \mapsto \arg\max (13.5)$$

Denote the component function (sc) of x_1 by x_1^{sc}, and define

$$x_1^{\backslash} \equiv (x_1^{sc})_{(sc)\neq(0C)}$$

From Theorem 27 in Chapter 8 x_1 is a C^2 function.

Definition 14 The aggregate excess pseudo-demand function is

$$z^{\backslash} : \mathbb{R}^{G-1} \times G_{I,S} \times \mathbb{R}^{GH}_{++} \to \mathbb{R}^{G-1}$$
$$(p^{\backslash}, L, e) \mapsto \left(x_1^{\backslash}(p^{\backslash}, e_1) - e_1^{\backslash}\right) + \sum_{h=2}^{H} \left(x_h^{\backslash}((p^{\backslash}, L, e_h)) - e_h^{\backslash}\right)$$

Proposition 15 *$z^{\backslash}$ is a C^2 function.*

Proof. The result follows from Theorem 27 in Chapter 8, Theorem 12 and Remark 70 in Chapter 3. ■

We are going to apply Theorem 59 in Chapter 7, to the projection from the equilibrium set to the space of economies as illustrated below.

Remark 16 Given an arbitrary matrix $A \in \mathcal{M}_{S,I}$, it is easy to see that

$$\text{Im}\left[{}^1a, ..., {}^Ia\right] \subseteq L \Leftrightarrow \text{ for } i = 1, ..., I, \ \ {}^ia \subset L$$

where we recall that we denote the i-th column vector of A by ${}^ia$.

Therefore, we can rewrite Condition 4 in the definition of pseudo-equilibrium - see Definition 5 - as follows

$$[I \mid E]\, P_\sigma \cdot {}^ir(p, y) = 0 \quad \text{for } i = 1, ..., I$$

or

$$[I \mid E]\, P_\sigma \cdot R(p, y) = 0$$

We can then introduce the following definition.

Definition 17 For each $\sigma \in \Sigma$, define

$$K_\sigma : \mathbb{R}^{G-1}_{++} \times V_\sigma \times \mathbb{R}^{SCI} \to \mathbb{R}^{(S-I)I}$$
$$(p^{\backslash}, L, y) \mapsto vec\left([I \mid \psi_\sigma(L)]\, P_\sigma \cdot R(p, y)\right)$$

and

$$\begin{array}{l} f_\sigma : \mathbb{R}^{G-1}_{++} \times V_\sigma \times \mathbb{R}^{GH}_{++} \times \mathbb{R}^{SCI} \to \mathbb{R}^{G-1} \times \mathbb{R}^{(S-I)I} \\ \left(p^\backslash, L, e, y\right) \mapsto \left(z^\backslash \left(p^\backslash, L, e\right), K_\sigma \left(p^\backslash, L, y\right)\right) \end{array}$$

Moreover, for each $\sigma \in \Sigma$, define

$$\begin{array}{l} \widetilde{z}^\backslash_\sigma : \mathbb{R}^{G-1} \times \mathcal{M}_{S-I,I} \times \mathbb{R}^{GH}_{++} \to \mathbb{R}^{G-1} \\ \left(p^\backslash, E, e\right) \mapsto z^\backslash \left(p^\backslash, \phi_\sigma (E), e\right) \end{array}$$

$$\begin{array}{l} \widetilde{K}_\sigma : \mathbb{R}^{G-1}_{++} \times \mathcal{M}_{S-I,I} \times \mathbb{R}^{SCI} \to \mathbb{R}^{(S-I)I} \\ \left(p^\backslash, E, y\right) \mapsto vec\left(\left[I \mid E\right] P_\sigma \cdot R(p, y)\right) \end{array}$$

and

$$\begin{array}{l} \widetilde{f}_\sigma : \mathbb{R}^{G-1}_{++} \times \mathcal{M}_{S-I,I} \times \mathbb{R}^{GH}_{++} \times \mathbb{R}^{SCI} \to \mathbb{R}^{G-1} \times \mathbb{R}^{(S-I)I} \\ \left(p^\backslash, E, e, y\right) \mapsto \left(\widetilde{z}^\backslash_\sigma \left(p^\backslash, E, e\right), \widetilde{K}_\sigma \left(p^\backslash, E, y\right)\right) \end{array}$$

As a direct consequence of the above definition, we have that $\left(x, p^\backslash, L\right)$ is a pseudo-equilibrium for the economy (e, y) if and only if there exists $\sigma \in \Sigma$ such that $L \in V_\sigma$, $f_\sigma \left(p^\backslash, L, e, y\right) = 0$ and

$$x = \left(x_1 \left(p^\backslash, e_1\right), \left(x_h \left(p^\backslash, L, e_h\right)\right)_{h \neq 1}\right)$$

that is, if and only if there exists $\sigma \in \Sigma$ such that $E \in \mathcal{M}_{S-I,I}$,

$$\begin{array}{l} \widetilde{f}_\sigma \left(p^\backslash, E, e, y\right) = 0 \text{ and} \\ x = \left(x_1 \left(p^\backslash, e_1\right) \left(\widetilde{x}_{h\sigma} \left(p^\backslash, E, e_h\right)\right)_{h \neq 1}\right) \end{array}$$

Definition 18 The equilibrium set is $\mathcal{E} \equiv \cup_{\sigma \in \Sigma} f_\sigma^{-1}(0)$ and

$$pr : \mathcal{E} \to \mathbb{R}^{GH}_{++} \times \mathbb{R}^{SCI}, \quad \left(p^\backslash, L, e, y\right) \mapsto (e, y)$$

To show existence of equilibria we use Theorem 59 in Chapter 7 identifying the notation of that theorem with the notation in this model as summarized in the following table.

Theorem 59	
M	$\mathcal{E}$
N	$\mathbb{R}^{GH}_{++} \times \mathbb{R}^{SCI}$
g	pr
y	well chosen Test Economy (e^*, y^*)

In the remaining part of this section we check that all the assumption of Theorem 59 in Chapter 7 are verified.

Even if in this chapter we do not use the extended approach the way we do in most of the other chapters of the economics part of this book, the extended system does still play a role in the analysis of present model, as explained below in Remarks 20 and 21.

Definition 19 For each $\sigma \in \Sigma$ and associated parametrization $(\phi_\sigma, U_\sigma, V_\sigma)$ the extended system associated with σ is the following one.

$$\left\{\begin{array}{lcll} Du_1(x_1) - \lambda_1 p & = & 0 & (1.1) \\ p(x_1 - e_1) & = & 0 & (1.2) \\ D_{x_h^0} u_h(x_h) - \mu_h p^0 & = & 0 & (h.1) \\ D_{x_h^1} u_h(x_h) - \mu_h p^1 + \eta_h [I \mid E] P_\sigma \Phi^1(p) & = & 0 & (h.2) \\ p(x_h - e_h) & = & 0 & (h.3) \\ [I \mid E] P_\sigma \Phi^1(p)(x_h^1 - e_h^1) & = & 0 & (h.4) \\ \sum_{h=1}^H \left(x_h^\setminus - e_h^\setminus\right) & = & 0 & (M.1) \\ vec([I \mid E] P_\sigma \cdot R(p,y)) & = & 0 & (L) \end{array}\right. \tag{13.19}$$

where $E \in U_\sigma$ and therefore there exists L such that $L = \phi_\sigma(E) \in V_\sigma$. Define also

$$\Xi_\sigma \equiv \mathbb{R}^{GH}_{++} \times \mathbb{R}^{G-1}_{++} \times U_\sigma \times \mathbb{R}_{++} \times \mathbb{R}^{H-1}_{++} \times \mathbb{R}^{(S-I)(H-1)}$$

with generic element

$$\xi \equiv \left(x, p^\setminus, E, \lambda_1, \mu, \eta\right),$$

and

$$F_\sigma : \Xi_\sigma \times \mathbb{R}^{GH}_{++} \times \mathbb{R}^{SCI} \to \mathbb{R}^{\dim \Xi}, \quad (\xi, e, y) \mapsto \text{left hand side of (13.19)}$$

Remark 20 As a consequence of Theorem 27 in Chapter 8 and Theorem 12, we have that for each $\sigma \in \Sigma$,

1 if $f_\sigma\left(p^\setminus, L, e, y\right) = 0$, then there exists $(x, \lambda_1, \mu, \eta) \in \mathbb{R}^{GH}_{++} \times \mathbb{R}_{++} \times \mathbb{R}^{H-1}_{++} \times \mathbb{R}^{(S-I)(H-1)}$ such that $F_\sigma\left(x, p^\setminus, \psi_\sigma(L), \lambda_1, \mu, \eta, e, y\right) = 0$, and

2 if $F_\sigma\left(x, p^\setminus, E, \lambda_1, \mu, \eta, e, y\right) = 0$, then $f_\sigma\left(p^\setminus, \phi(E), e, y\right) = 0$.

Remark 21 In other words, *for given* σ, the extended system does represent equilibria and we can use F_σ to get information about f_σ. Therefore, F_σ can, at least in principle, be used to study *local properties* of equilibria. To appreciate the relationship between those two functions, let's briefly describe the proofs of the results needed to show existence of pseudo-equilibria and typical existence of equilibria.

Step 1 in Lemma 22 shows that for each σ, $f_\sigma^{-1}(0)$ is a C^2 manifold using F_σ and Theorem 40 in Chapter 4; Step 2 shows that the union of those manifold is in fact a manifold, a *global* result.

Proposition 23 shows properness of the projection pr, again a global result; observe however that a crucial part in the proof uses first order conditions of household 1's maximization problem, i.e., equations (1.1) and (1.2) in the extended system (13.19).

Lemma 24 shows existence of equilibria for the test economy using F_{id}.

Lemma 25 shows uniqueness of that equilibrium. Here the extended system fails even though the argument is local. It is in fact possible to show that uniqueness of the multipliers η is not insured - see Remark 26.

Lemma 27 shows that the test economy is a regular value for pr applying Corollary 43 in Chapter 4 basically to f_σ and pr, and Theorem 40 to a system related to the extended one.

Finally, Theorem 29 is shown using f_σ, but a similar proof using $F_\sigma$would have been possible.

The use of the extended approach in any other proof of existence in the book and the important role played by the extended function in the proof described above may lead to conjecture the possibility of using that approach in the present model. Unfortunately, at the best of our knowledge, several ways of adapting the extended approach to the real asset model do fail. Below we present three of such attempts.

We could introduce an extended equilibrium function, say $\widetilde{F}$, *equal* to F_σ presented in Definition 13.19, *but defined on all* $\mathbb{R}^{(S-I)I}$ instead of $U_\sigma \subseteq \mathbb{R}^{(S-I)I}$, therefore eliminating the dependence on σ. We could then follow the usual line of reasoning and defining a test economy and a related homotopy H. Unfortunately, the proof of compactness of $H^{-1}(0)$ is prevented by the lack of compactness of $\mathbb{R}^{(S-I)I}$. Observe that the counterpart in the reduced approach of that proof is the proof of properness of the projection from the equilibrium manifold to the economy space; there, it is crucial to exploit the compactness of the grassmannian manifold $G_{I,S}$.

We could choose to define the extended function on $G_{I,S}$ itself, i.e.,

$$\widetilde{F} : \mathbb{R}^{GH}_{++} \times \mathbb{R}^{G-1}_{++} \times G_{I,S} \times \mathbb{R}_{++} \times \mathbb{R}^{H-1}_{++} \times \mathbb{R}^{(S-I)(H-1)} \times \mathbb{R}^{GH}_{++} \times \mathbb{R}^{SCI} \to \mathbb{R}^{\dim \Xi}$$
$$\widetilde{F} : \left(x, p^\setminus, L, \lambda_1, \mu, \eta\right) \mapsto \text{(left hand side of (13.19) with } \psi_\sigma(L) \text{ in the place of } E\text{)}$$

Unfortunately, the above function is not well defined: in general, given L there exist more than one chart $(\psi_\sigma, V_\sigma, U_\sigma)$ such that $L \in V_\sigma$. Observe also that defining $\widetilde{F}$ on V_σ, or U_σ, would encounter the same difficulties mentioned above in the proof of needed compactness.

Finally, it is relatively easy to find a well chosen compact set $\mathcal{A}$ of matrices such that $\forall A \in \mathcal{A}$, $\ker A \in G_{I,S}$ and then define an extended equilibrium function using that set. In fact, all steps of the proofs goes through apart from the uniqueness of equilibria associated with the test economy.

Lemma 22 *$\mathcal{E}$ is a C^2 manifold of dimension equal to* $\dim \mathbb{R}^{GH}_{++} \times \mathbb{R}^{SCI}$.

Proof. The proof proceeds in the following two steps.

- Step 1. For each $\sigma \in \Sigma$, we prove that $\mathcal{E}_\sigma \equiv f_\sigma^{-1}(0)$ is a C^2 manifold of dimension equal to $\dim \mathbb{R}^{GH}_{++} \times \mathbb{R}^{SCI}$.
- Step 2. $\mathcal{E} = \cup_{\sigma\in\Sigma}\mathcal{E}_\sigma$ is a C^2 manifold.

Step 1.

Fix $\sigma \in \Sigma$ and consider the function $\widetilde{f}_\sigma$ introduced in Definition 17. Let us prove that 0 is a regular value for $\widetilde{f}_\sigma$.

It is enough to show that the partial jacobian matrix of $\widetilde{f}_\sigma$ with respect to e_1 and the subvector $\widehat{y} \equiv \left(y^{si1}\right)_{s=1,\ldots,S-I,\, i=1,\ldots,I} \in \mathbb{R}^{(S-I)I}$of y, computed at each $\left(p^\backslash, L, e\right) \in f_\sigma^{-1}(0)$, has full row rank. The computation of that jacobian matrix is described below

$$\begin{array}{r} \\ \widetilde{z}_\sigma^\backslash\left(p^\backslash, E, e\right) \\ \widetilde{K}_\sigma\left(p^\backslash, E, y\right) \end{array} \begin{array}{c} \quad e_1 \qquad\qquad\qquad\qquad \widehat{y} \\ \begin{pmatrix} D_{e_1}\left(x_1\left(p^\backslash, e_1\right) - e_1\right) & 0 \\ 0 & D_{\widehat{y}}\widetilde{K}_\sigma\left(p^\backslash, E, y\right) \end{pmatrix} \end{array}$$

We are then left with showing that both $D_{e_1}\left(x_1\left(p^\backslash, e_1\right) - e_1\right)$ and $D_{\widehat{y}}K_\sigma\left(p^\backslash, L, y\right)$ have full rank, as proved in Step 1.1 and Step 1.2 below, respectively.

Step 1.1

We are going to use Theorem 40 in Chapter 4. Taken an arbitrary value of $p^\backslash$, define

$$F_{1,p^\backslash} : \mathbb{R}^G_{++} \times \mathbb{R}_{++} \times \mathbb{R}^G_{++} \to \mathbb{R}^G \times \mathbb{R}$$

$F_{1,p^\backslash} : (x_1, \lambda_1, e_1) \mapsto$ left hand side of equations (1.1) and (1.2) in (13.19)

$$Z_1^\backslash : \mathbb{R}^G_{++} \times \mathbb{R}_{++} \times \mathbb{R}^G_{++} \to \mathbb{R}^{G-1}$$

$$Z_1^\backslash : (x_1, \lambda_1, e_1) \mapsto \left(x_1^\backslash - e_1^\backslash\right)$$

We can identify the notation of Theorem 40 in Chapter 4 with the notation used here as described in the following table.

Theorem 40 in Chapter 4	
X	$\mathbb{R}^G_{++} \times \mathbb{R}_{++}$
Z	$\mathbb{R}^G_{++}$
x	x_1, λ_1
z	e_1
f	$F_{1,p^\backslash}$
g	$Z_1^\backslash$
n	$G+1$
m	G
l	$G-1$
y	0

Assumption 1 of Theorem 40 in Chapter 4 is clearly satisfied. Given an arbitrary $\overline{e}_1$, from Lemma 27 in Chapter 8, there exists a unique $\left(\overline{x}_1, \overline{\lambda}_1\right)$ such that $F_{1,p^\backslash}\left(\overline{x}_1, \overline{\lambda}_1, \overline{e}_1\right) = 0$ and Assumption 2 is satisfied. Assumption 3, i.e.,

$$\operatorname{rank} D_{x_1,\lambda_1} F_{1,p^\backslash}\left(\overline{x}_1, \overline{\lambda}_1, \overline{e}_1\right) = G+1 \tag{13.20}$$

follows from the proof of Theorem 27 in Chapter 8. Therefore, Conclusion 1 of Theorem 40 in Chapter 4 holds, and there exist an open neighborhood $U \subseteq \mathbb{R}^G_{++}$ of $\overline{e}_1$ and a unique C^2 function $\eta \equiv \left(\overline{x}_1, \overline{\lambda}_1\right) : U \to \mathbb{R}^{G+1}$ such that $\left(\overline{x}_1, \overline{\lambda}_1\right)\left(\overline{e}_1\right) = \left(\overline{x}_1, \overline{\lambda}_1\right)$, and

$$\text{for every } e_1 \in U,\ F_{1,p^\backslash}\left(\left(\overline{x}_1, \overline{\lambda}_1\right)(e_1), e_1\right) = 0, \text{ i.e.}$$
$$\left(\left(\overline{x}_1, \overline{\lambda}_1\right)(e_1), \overline{e}_1\right) \in F^{-1}_{1,p^\backslash}(0)$$

In other words, for every $e_1 \in U$, the function $\overline{x}_1$ is nothing but the demand function x_1 at given $p^\backslash$ - see Definition 13- i.e., $\overline{x}_1(e) = x_1\left(p^\backslash, e_1\right)$. The function $k \equiv g(\eta, id)$ of Theorem 40 in Chapter 4 corresponds here to

$$Z_1^\backslash\left(\left(\overline{x}_1, \overline{\lambda}_1\right)(e_1), e_1\right) = \overline{x}_1^\backslash(e_1) - e_1^\backslash = \overline{x}_1^\backslash\left(p^\backslash, e_1\right) - e_1^\backslash$$

Therefore, the statement in Conclusion 2b of that Theorem simply says that

$$\operatorname{rank} D_{e_1}\left(x_1^\backslash\left(p^\backslash, e_1\right) - e_1^\backslash\right) = G-1$$

i.e., the desired result. We are then left with checking Condition 2a of Theorem 40 in Chapter 4 rank $D\left(F_{1p^\backslash}, Z_1^\backslash\right)(x_1, \lambda_1, e_1) = G+1+G-1$.

The computation of the above jacobian matrix is described below.

$$\begin{array}{ll} & \\ (1) & Du_1(x_1) - \lambda_1 p \\ (2) & -p(x_1 - e_1) \\ (3) & x_1^\backslash - e_1^\backslash \end{array}
\begin{array}{c} \begin{array}{cccc} x_1 & \lambda_1 & e_1^\backslash & e_1^{0C} \end{array} \\ \left(\begin{array}{cccc} D^2u_1(x_1) & -p^T & & \\ -p & & -p & 1 \\ \widehat{I} & & -I & \end{array}\right) \end{array}$$

where $\widehat{I}$ is the $(C-1)\,S \times SC$ matrix

$$\widehat{I} \equiv \begin{bmatrix} I_{C-1} & 0 & \\ & & I_{SC} \end{bmatrix}$$

Since rank $D_{x_1,\lambda_1} F_{1,p^\backslash}\left(\overline{x}_1, \overline{\lambda}_1, \overline{e}_1\right) = G+1$ - see (13.20) above - we are left with perturbing the last superrow, which is done below.

$$\begin{array}{lll} (3) & \leftarrow e_1^\backslash & \rightsquigarrow (2) \\ (2) & \leftarrow e_1^{0C} & \end{array}$$

Step 1.2

In what follows, we present the needed computations just for the case $\sigma = id$. The general case is left as an exercise and requires rearranging the elements of $\widehat{y}$. Then

$$[I \mid E] \cdot \begin{bmatrix} R^*(p,y) \\ \widehat{R}(p,y) \end{bmatrix} = R^*(p,y) + E\widehat{R}(p,y) \tag{13.21}$$

where

$$[R^*(p,y)] \equiv \begin{bmatrix} p^1 y^{11} & p^1 y^{1i} & p^1 y^{1I} \\ \dots & \dots & \dots \\ p^{S-I} y^{S-I,1} & p^{S-I} y^{S-I,i} & p^{S-I} y^{S-I,I} \end{bmatrix} \in \mathcal{M}_{S-I,I}$$

and

$$\widehat{R}(p,y) \equiv \begin{bmatrix} p^{S-I+1} y^{S-I+1,1} & p^{S-I} y^{S-I+1,i} & p^{S-I+1} y^{S-I,I} \\ & & \\ p^S y^{S1} & p^S y^{S1} & p^S y^{SI} \end{bmatrix} \in \mathcal{M}_{I,I}$$

Observed that $D_{\widehat{y}} vec\left(R^*(p,y) + E\widehat{R}(p,y)\right) = D_{\widehat{y}} vec\left(R^*(p,y)\right)$, the computation of $A = D_{\widehat{y}} \widetilde{K}_{id}\left(p^\backslash, E, e\right)$ is described below

$$\begin{array}{c} \\ p^1 y^{11} \\ \dots \\ p^1 y^{1i} \\ \dots \\ p^{S-I} y^{S-I,i} \\ \dots \\ p^{S-I} y^{S-I,I} \end{array} \begin{array}{c} \begin{array}{ccccccc} y^{111} & \dots & y^{1i1} & \dots & y^{S-I,i1} & \dots & y^{S-I,I1} \end{array} \\ \begin{pmatrix} p^{11} & & & & & & \\ & \ddots & & & & & \\ & & p^{11} & & & & \\ & & & \ddots & & & \\ & & & & p^{S-I,1} & & \\ & & & & & \ddots & \\ & & & & & & p^{S-I,1} \end{pmatrix} \end{array}$$

which clearly has full rank.

Finally applying Theorem 71 in Chapter 3, $\mathcal{E}_\sigma$ is a manifold contained in $\mathbb{R}^{G-1}_{++} \times V_\sigma \times \mathbb{R}^{GH}_{++} \times \mathbb{R}^{SCI}$.

Step 2.

Since it is well known that the union of manifolds is not necessarily a manifold, we still have to show that $\mathcal{E} = \cup_{\sigma\in\Sigma}\mathcal{E}_\sigma$ is in fact a manifold. First of all, let us show that, given $\sigma, \sigma' \in \Sigma$, $\mathcal{E}_\sigma \cap (V_\sigma \cap V_{\sigma'})$ coincides with $\mathcal{E}_{\sigma'} \cap (V_\sigma \cap V_{\sigma'})$. In other words, let us prove that, if $L \in V_\sigma \cap V_{\sigma'}$, then $f_\sigma \left(p^\backslash, L, e, y\right) = 0$ is equivalent to $f_{\sigma'} \left(p^\backslash, L, e, y\right) = 0$. Observe that the function z does not depend on the chart $(\psi_\sigma, V_\sigma, U_\sigma)$. Therefore, we have simply to check that $K_\sigma \left(p^\backslash, L, y\right) = 0$ if and only if $K_{\sigma'} \left(p^\backslash, L, y\right) = 0$. But from Remark 16, we have that $K_\sigma \left(p^\backslash, L, y\right) = 0$ if and only if $\operatorname{Im} R\,(p, y) = L$, and this latter equality is independent of the choice of the chart for L.

Now, recall that, for each σ, $\mathcal{E}_\sigma$ is a C^2 manifold of dimension equal to $\dim \mathbb{R}^{GH}_{++} \times \mathbb{R}^{SCI}$, contained in $G_{I,S}$. For each σ, there exists an atlas

$$\mathcal{A}_\sigma \equiv \left\{\left(\widehat{\psi}^\lambda_\sigma, \widehat{V}^\lambda_\sigma, \widehat{U}^\lambda_\sigma\right)\right\}_{\lambda\in\Lambda_\sigma}$$

If we show that the family of all charts of all atlases, i.e., $\mathcal{A} \equiv \cup_{\sigma\in\Sigma}\mathcal{A}_\sigma$, is an atlas of $\mathcal{E}$, we have proved that $\mathcal{E}$ is a C^2 manifolds, by Definition 42. Let $\sigma, \sigma' \in \Sigma$, and consider two charts of $\mathcal{A}$, $\left(\widehat{\psi}_\sigma, \widehat{V}_\sigma, \widehat{U}_\sigma\right)$ and $\left(\widehat{\psi}_{\sigma'}, \widehat{V}_{\sigma'}, \widehat{U}_{\sigma'}\right)$ such that $V_\sigma \cap V_{\sigma'} \neq \emptyset$. Without loss of generality, we can suppose $V_\sigma = V_{\sigma'}$. By Proposition 69 in Chapter 3, it follows that $\psi_{\sigma'} \circ (\psi_\sigma)^{-1} : U_\sigma \to U_{\sigma'}$ is C^2 and this concludes the proof. ■

Proposition 23 *pr is proper.*

Proof. We want to show that given a compact set $C \subseteq \mathbb{R}^{GH}_{++} \times \mathbb{R}^{SCI}$, an arbitrary sequence $\left(p^{\backslash v}, L^v, e^v, y^v\right)_{v\in\mathbb{N}}$ in $pr^{-1}(C)$ does admit a convergent subsequence converging to an element of $pr^{-1}(C)$.

Since $\{L^v : v \in \mathbb{N}\} \subseteq G_{I,S}$, is sequentially compact - see Proposition 51, and $\{(e^v, y^v) : v \in \mathbb{N}\} \subseteq C$,

$$(L^v, e^v, y^v) \to \left(\overline{L}, \overline{e}, \overline{y}\right) \in G_{I,S} \times \mathbb{R}^{GH}_{++} \times \mathbb{R}^{SCI}$$

Therefore, there exists σ such that $\overline{L} \in V_\sigma$ and for sufficiently large v, $\{L^v : v \in \mathbb{N}\} \subseteq V_\sigma$. Therefore, considering such subsequence, we have that $\left\{\left(p^{\backslash v}, L^v, e^v, y^v\right) : v \in \mathbb{N}\right\}_v \subseteq f_\sigma^{-1}(0)$, and, by definition of $x_1\left(p^\backslash, e_1\right)$, for each v

$$x_1^v = x_1\left(p^{\backslash v}, e_1^v\right)$$

is the unique solution to household 1's maximization problem at $(p^{\backslash v}, e_1^v)$, and there exists a unique λ_1^v such that

$$\begin{cases} Du_1(x_1^v) - \lambda_1^v p^v & = 0 \\ -p^v(x_1^v - e_1^v) & = 0 \end{cases} \tag{13.22}$$

Using a standard argument - see for example Lemma 48 in Chapter 8 - $(x_1^v)_{v \in \mathbb{N}}$ converges to $\overline{x}_1 \in \mathbb{R}^G_{++}$. Since $p^{0C} = 1$, from[5] (13.22), $(\lambda_1^v)_{v \in \mathbb{N}}$ converges to $D_{x_1^{01}} u_1(\overline{x}_1) > 0$ and, finally, $(p^{\backslash v})_{v \in \mathbb{N}}$ converges to $\overline{p}^{\backslash} \in \mathbb{R}^{G-1}_{++}$. ■

Now we want to choose (e^*, y^*) which is a regular value for *pr* and for which there exists a unique pseudo-equilibrium. Observe that we have to use the definition of regular value for a function whose domain is an abstract manifold - see Definition 58, Subsection 5 in Chapter 7.

We first construct the "test economy" (e^*, y^*). Then we prove that there exists a pseudo equilibrium ξ^* associated with it, and that equilibrium is unique. Finally we show that (e^*, y^*) is a regular value for *pr*.

From Lemma 40, part 2, in Chapter 8, there exists $\left(x^*, \theta^*_{\backslash 1}, \gamma^*\right)$ which is the unique solution to the following system in the unknown $(x, \theta_{\backslash 1}, \gamma)$,

$$\begin{cases} Du_1(x_1) - \gamma & = 0 \quad (1) \\ \theta_h Du_h(x_h) - \gamma & = 0 \quad (2) \\ (u_h(x_h) - u_h(x_h^*))_{h \neq 1} & = 0 \quad (3) \\ -\sum_{h=1}^H x_h + \sum_{h=1}^H x_h^* & = 0 \quad (4) \end{cases} \tag{13.23}$$

Choose $e^* = x^*$, a Pareto Optimal allocation. For $s \geq 1$, define $p^* = \frac{\gamma^*}{\gamma^{0C*}}$. We want then to choose y^* to have

$$R(p^*, y^*) \equiv \begin{bmatrix} R^*(p^*, y^*) \\ \widehat{R}(p^*, y^*) \end{bmatrix} = \begin{bmatrix} 0 \\ I \end{bmatrix} \tag{13.24}$$

It is then enough to set

$$y^{sci*} \equiv \begin{cases} \frac{1}{p^{sc*}} & \text{if} \quad (sc) = (S - I + i, 1) \text{ with } i = 1, ..., I \\ 0 & \qquad \text{otherwise.} \end{cases}$$

Then a natural candidate for the equilibrium associated with (e^*, y^*) is

$$\left(p^{\backslash *}, L^*\right) \equiv \left(\left(\frac{\gamma^{sc*}}{\gamma^{0C*}}\right)_{sc \neq 0C}, \operatorname{Im} R(p^*, y^*)\right) \tag{13.25}$$

Observe that since $\operatorname{rank} R(p^*, y^*) = I$, then $L^* \in G_{I,S}$.

[5] It is then clear that Mr. 1's trick is used to get convergence of prices.

Lemma 24 $\left(p^{\setminus *}, L^*, e^*, y^*\right) \in pr^{-1}\left(e^*, y^*\right)$.

Proof. Observe that

$$L^* \equiv \operatorname{Im} R\left(p^*, y^*\right) \equiv \operatorname{Im}\begin{bmatrix} 0 \\ I \end{bmatrix} = \ker\begin{bmatrix} I & 0 \end{bmatrix}$$

as shown below:

$$y \in \operatorname{Im}\begin{bmatrix} 0 \\ I \end{bmatrix} \Leftrightarrow \exists x \in \mathbb{R}^S \text{ such that } y = \begin{bmatrix} 0 \\ I \end{bmatrix} x = \begin{bmatrix} 0 \\ x \end{bmatrix} \Leftrightarrow$$

$$\Leftrightarrow \begin{bmatrix} I & 0 \end{bmatrix} y = 0 \Leftrightarrow y \in \ker\begin{bmatrix} I & 0 \end{bmatrix}$$

Therefore,

$$L^* = \ker\begin{bmatrix} I & 0 \end{bmatrix}$$

$$L^* = \phi_{id}(0) \in W_{id} \quad \text{and} \quad E^* \equiv \psi_{id}\left(L^*\right) = 0 \tag{13.26}$$

To prove the desired result, we have to show that $f_{id}\left(p^{\setminus *}, L^*, e^*, y^*\right) = 0$. Using Remark 20, it is in fact enough to check that

$$F_{id}\left(x_1^*,\ \lambda_1^* = \gamma^{0C*}, p^{\setminus *}, L^*, \mu_h^* = \frac{\gamma^{0C*}}{\theta_h^*}, \eta_h^* = 0, e^*, y^*\right) = 0 \tag{13.27}$$

which is done below, using the definition of F_{id} - see system (13.19)- and the definition of $\left(x^*, \theta_{\setminus 1}^*, \gamma^*\right)$ as a solution to system (13.23).

$$D_{x_1} u_1\left(x_1^*\right) - \lambda_1^* p^* = D_{x_1} u_1\left(x_1^*\right) - \gamma^{0C*}\frac{\gamma^*}{\gamma^{0C}} = 0$$

$$p^*\left(x_1^* - x_1^*\right) = 0$$

$$D_{x_h^0} u_h\left(x_h^*\right) - \mu_h^* p^* = D_{x_h^0} u_h\left(x_h^*\right) - \frac{\gamma^{0C*}}{\theta_h^*}\frac{\gamma^*}{\gamma^{0C}} = 0$$

$$D_{x_h^1} u_h\left(x_h^*\right) - \mu_h^* p^{1*} + \eta_h^*\left[I \mid \psi_{id}\left(L^*\right)\right] P_{id}\Phi^1\left(p^*\right) =$$
$$= D_{x_h^1} u_h\left(x_h^*\right) - \frac{\gamma^{0C*}}{\theta_h^*}\frac{\gamma^{1*}}{\gamma^{0C}} + 0\cdot\left[I \mid 0\right]\Phi^1\left(p^*\right) = 0$$

$$p^*\left(x_h^* - x_h^*\right) = 0$$

$$\left[I \mid \psi_{id}\left(L^*\right)\right] P_{id}\Phi^1\left(p^*\right)\left(x_h^{1*} - e_h^{1*}\right) = \left[I \mid 0\right]\Phi^1\left(p^*\right)\left(x_h^{1*} - x_h^{1*}\right) = 0$$

$$\textstyle\sum_{h=1}^H \left(x_h^{*\setminus} - x_h^{*\setminus}\right) = 0$$

$$\left[I \mid \psi_{id}\left(L^*\right)\right] P_{id}\cdot R\left(p^*, y^*\right) = \left[I \mid 0\right]\cdot\begin{bmatrix} 0 \\ I \end{bmatrix} = 0$$

■

Lemma 25 $\left\{\left(p^{\backslash *}, L^*, e^*, y^*\right)\right\} = pr^{-1}\left(e^*, y^*\right).$

Proof. Take $\left(\widehat{p}^{\backslash}, \widehat{L}, e^*, y^*\right) \in pr^{-1}\left(e^*, y^*\right)$. Then there exists $\widehat{\sigma} \in \Sigma$ such that $f_{\widehat{\sigma}}\left(\widehat{p}^{\backslash}, \widehat{L}, e^*, y^*\right) = 0$. Define $\widehat{x}_1 \equiv x_1\left(\widehat{p}^{\backslash}, e_1^*\right)$ and $\widehat{x}_h \equiv x_h\left(\widehat{p}^{\backslash}, \widehat{L}, e^*\right)$. Then $\widehat{x} = x^*$. Suppose otherwise, i.e., $\widehat{x} \neq x^*$. Consider $\widetilde{x} \equiv \frac{1}{2}\left(\widehat{x} + x^*\right)$. By definition, $\sum_h \widehat{x}_h = \sum_h x_h^*$ and

$$\sum_h \widetilde{x}_h = \frac{1}{2}\left(\sum_h \widehat{x}_h + \sum_h x_h^*\right) = \sum_h x_h^* \tag{13.28}$$

From the strict quasiconcavity of the utility functions, we then have

$$u\left(\widetilde{x}_h\right)_{h=1}^H > u\left(x_h^*\right)_{h=1}^H \tag{13.29}$$

But (13.28) and (13.29) contradict the Pareto Optimality of x^*.

By definition of $\widehat{x}_1$ there exists $\widehat{\lambda}_1$ such that

$$Du_1\left(x_1^*\right) - \widehat{\lambda}_1 \widehat{p}^{\backslash} = 0$$

From (1.1) in (13.23), we have that

$$D_{x_1^{0C}} u_1\left(e_1^*\right) = \gamma^{0C*} = \widehat{\lambda}_1$$

and

$$\widehat{p} = \frac{D_{x_1} u_h\left(e_h^*\right)}{\widehat{\lambda}_1} = \frac{\gamma^*}{\gamma^{0C*}} = p^*$$

We are left with showing that $\widehat{L} \in W_{\widehat{\sigma}}$ is equal to L^*.

From $K_{\widehat{\sigma}}\left(p^{\backslash *}, \widehat{L}, y^*\right) = 0$ and Remark 16, we have that

$$\operatorname{Im} R\left(p^*, y^*\right) \subseteq \ker\left[I | \psi_{\widehat{\sigma}}\left(\widehat{L}\right)\right] P_{\widehat{\sigma}} = \widehat{L}$$

Since $R\left(p^*, y^*\right) = \begin{bmatrix} 0 \\ I \end{bmatrix}$, $\dim \operatorname{Im} R\left(p^*, y^*\right) = I$ and $\operatorname{Im} R\left(p^*, y^*\right) = \widehat{L}$. Then the result follows from the definition of L^* - see Definition 13.25. ■

Remark 26 Observe that there could be $\widehat{\sigma} \neq id$ such that $L^* = \ker\left[I | \psi_{\widehat{\sigma}}\left(L^*\right)\right] P_{\widehat{\sigma}}$. Therefore, while from $(h.2)$ in system (13.19) computed at the test economy and the associated equilibrium variables, we would have

$$\begin{aligned} D_{x_h^1} u_h\left(x_h^*\right) - \mu_h^* p^{1*} = \;& \eta_h^*\left[I \mid 0\right] \Phi^1\left(p^*\right) = \eta_h^*\left[I \mid 0\right] \begin{bmatrix} \Phi^*\left(p^*\right) & \\ & \widehat{\Phi}\left(p^*\right) \end{bmatrix} \\ = \;& \widehat{\eta}_h\left[\Phi^*\left(p^*\right) \mid 0\right] \end{aligned}$$

and

$$\eta_h^* = \left[diag \, \left(p^{sC*} \right)_{s=1}^{S-I} \right]^{-1} \left(D_{x_h^{sC}} u_h \left(x_h^* \right) - \mu_h^* p^{sC*} \right)_{s=1}^{S-I}$$

we would also have

$$D_{x_h^1} u_h \left(x_h^* \right) - \mu_h^* p^{1*} = \widehat{\eta}_h \left[I \mid \psi_{\widehat{\sigma}} \left(L^* \right) \right] P_{\widehat{\sigma}} \Phi^1 \left(p \right)$$

Therefore, in general it would be $\eta_h^* \neq \widehat{\eta}_h$. In other words, uniqueness of the equilibrium multipliers η associated to the test economy is not insured.

Lemma 27 *(e^*, y^*) is a regular value for pr.*

Proof. Since $L^* \in W_{id}$,

$$\left(p^{\backslash *}, L^*, e^*, y^* \right) \in \mathcal{E} \cap \left(\mathbb{R}_{++}^{G-1} \times V_{id} \times \mathbb{R}_{++}^{GH} \times \mathbb{R}^{SCI} \right) = f_{id}^{-1} \left(0 \right)$$

Defined

$$\widetilde{f} : \mathbb{R}_{++}^{G-1} \times U_{id} \times \mathbb{R}_{++}^{GH} \times \mathbb{R}^{SCI} \to \mathbb{R}^{(G-1)+(S-I)I}$$

$$\widetilde{f} : \left(p^{\backslash}, E, e, y \right) \mapsto f_{id} \left(p^{\backslash}, \phi_{id} \left(E \right), e, y \right)$$

we have that

$$f_{id}^{-1} \left(0 \right) \text{ and } \widetilde{f}^{-1} \left(0 \right)$$

are diffeomorphic, the diffeomorphism from the former to the latter being

$$\delta \equiv \left(id_{\mathbb{R}_{++}^{G-1}}, \phi_{id}, id_{\mathbb{R}_{++}^{GH}}, id_{\mathbb{R}^{SCI}} \right)_{| f_{id}^{-1}(0)}$$

For lighter notation we identify E with $vecE$. Then, defined

$$\widetilde{pr} : \widetilde{f}^{-1} \left(0 \right) \to \mathbb{R}_{++}^{GH} \times \mathbb{R}^{SCI}, \quad \left(p, E, e, y \right) \mapsto \left(e, y \right)$$

(e^*, y^*) is a regular value for *pr* if and only if (e^*, y^*) is a regular value for $\widetilde{pr}$ - see Definition 58 Subsection 5 in Chapter 7.

Since the domain of $\widetilde{f}$ is an open subset of an euclidean space, we can then apply Corollary 43 in Chapter 4, as follows. Define

$$proj : \mathbb{R}_{++}^{G-1} \times \mathcal{M}_{S-I,I} \times \mathbb{R}_{++}^{GH} \times \mathbb{R}^{SCI} \to \mathbb{R}_{++}^{GH} \times \mathbb{R}^{SCI}, \quad \left(p, E, e, y \right) \mapsto \left(e, y \right)$$

We can identify the notation of that theorem, with the notation used here as described in the following table.

Corollary 43 in Chapter 4	
X	$\mathbb{R}_{++}^{G-1} \times U_{id}$
Z	$\mathbb{R}_{++}^{GH} \times \mathbb{R}^{SCI}$
x	$\left(p^{\backslash}, E \right)$
z	$\left(e^*, y^* \right)$
f	$\widetilde{f}$
pr	$\widetilde{pr}$
y	0

Assumption 1 of Corollary 43 requires 0 to be a regular value for $\widetilde{f}$, which is done following closely the proof of Step 1 in Lemma 22- observe that there we computed the partial jacobian with respect to a subvector of (e, y).

Assumption 2 of the Corollary requires 0 to be a regular value for $\widetilde{f}_{(e^*,y^*)}$. If that is the case, we can conclude that (e^*, y^*) is a regular value for $\widetilde{pr}$, as desired.

We are then left with showing $D_{(p^\setminus,E)}\widetilde{f}\left(p^{\setminus *}, E^*, e^*, y^*\right)$ has full rank, which is done below applying Theorem 40 in Chapter 4.

Define

$$F_{\mathcal{H}} : \Xi \longrightarrow \mathbb{R}^{(G+1)+(G+1+S-I)(H-1)}$$

$F_{\mathcal{H}} : \xi \mapsto$ left hand side of all equations but $(M.1)$ and (L) in system (13.19) with $\sigma = id$ and $(e, y) = (e^*, y^*)$

i.e., $F_{\mathcal{H}}(\xi) =$

$$\begin{array}{lcll}
D_{x_1} u_1(x_1) - \lambda_1 p & = & 0 & (1.1) \\
p(x_1 - x_1^*) & = & 0 & (1.2) \\
D_{x_h^0} u_h(x_h) - \mu_h p^0 & = & 0 & (h.1) \\
D_{x_h^1} u_h(x_h) - \mu_h p^1 + \eta_h [I \mid E] P_\sigma \Phi^1(p) & = & 0 & (h.2) \\
p(x_h - e_h) & = & 0 & (h.3) \\
[I \mid E] P_\sigma \Phi^1(p)\left(x_h^1 - x_h^{*1}\right) & = & 0 & (h.4)
\end{array}$$

$$\widehat{G} : \Xi \longrightarrow \mathbb{R}^{(G-1)+(S-I)I}$$

$\widehat{G} : \xi \mapsto$ left hand side of equations $(M.1)$ and (L) in system (13.19) with $\sigma = id$ and $(e, y) = (e^*, y^*)$

i.e., $\widehat{G}(\xi) =$

$$\left\{\begin{array}{lcll}
\sum_{h=1}^{H}\left(x_h^\setminus - e_h^{*\setminus}\right) & = & 0 & (M.1) \\
vec\left([I \mid E] P_\sigma \cdot R(p, y^*)\right) & = & 0 & (L)
\end{array}\right.$$

We can identify the notation of Theorem 40 in Chapter 4 with the notation used here as described in the following table.

Theorem 40 in Chapter 4	
X	$\mathbb{R}_{++}^{GH} \times \mathbb{R}_{++} \times \mathbb{R}_{++}^{H-1} \times \mathbb{R}^{(S-I)(H-1)}$
Z	$\mathbb{R}_{++}^{G-1} \times \mathcal{M}_{S-I,I}$
x	$(x, \lambda_1, \mu, \eta)$
z	$p^\setminus, E$
f	$F_{\mathcal{H}}$
g	$\widehat{G}$
n	$(G+1) + (G+1+S-I)(H-1)$
m	$G - 1 + (S-I) I$
l	$G - 1 + (S-I) I$
y	0

Assumptions of Theorem 40 in Chapter 4 is clearly satisfied; Assumption 2 is satisfied at $\left(x^*, \lambda_1^*, \mu^*, \eta^*, p^{\backslash *}, E^*\right)$ [6]; Assumption 3, i.e.,

$$\operatorname{rank} D_{(x,\lambda_1,\mu,\eta)} F_{\mathcal{H}} \left(x^*, \lambda_1^*, \mu^*, \eta^*, p^{\backslash *}, E^*\right) = G + 1$$

follows from the proof of Theorem 27 in Chapter 8 and the argument in the proof of Theorem 12.

Therefore, Conclusion 1 of Theorem 40 in Chapter 4 holds, and there exist an open neighborhood of $U \subseteq \mathbb{R}^{G-1}_{++} \times \mathbb{R}^{(S-I)I}$ containing $\left(p^{\backslash *}, E^*\right)$ and a unique C^2 function $\left(\widehat{x}, \widehat{\lambda}_1, \widehat{\mu}, \widehat{\eta}\right) : U \to \mathbb{R}^{GH}_{++} \times \mathbb{R}_{++} \times \mathbb{R}^{H-1}_{++} \times \mathbb{R}^{(S-I)(H-1)}$ such that $\left(\widehat{x}, \widehat{\lambda}_1, \widehat{\mu}, \widehat{\eta}\right)\left(p^{\backslash *}, E^*\right) = \left(x^*, \lambda_1^*, \mu^*, \eta^*\right)$, and

$$\text{for every } \left(p^{\backslash}, e\right) \in U,\ F_{\mathcal{H}}\left(\left(\widehat{x}, \widehat{\lambda}_1, \widehat{\mu}, \widehat{\eta}\right)\left(p^{\backslash}, E\right), p^{\backslash}, E\right) = 0 \qquad (13.30)$$

The function $k \equiv g\left(\eta, id\right)$ of Theorem 40 in Chapter 4 corresponds here to

$$\widehat{G} \circ \left(\left(\widehat{x}, \widehat{\lambda}_1, \widehat{\mu}, \widehat{\eta}\right), id_U\right) : U \to \mathbb{R}^{G-1+(S-I)I}$$

$$\left(p^{\backslash}, E\right) \mapsto \left(\sum_{h=1}^{H}\left(\widehat{x}_h^{\backslash}\left(p, E\right) - e_h^{*\backslash}\right), vec\left(\left[I \mid E\right] P_\sigma \cdot R\left(p, y^*\right)\right)\right)$$

and, in fact, from (13.30) and (13.15)

$$\sum_{h=1}^{H}\left(\widehat{x}_h^{\backslash}\left(p, E\right) - e_h^{*\backslash}\right) = \left(x_1^{\backslash}\left(p^{\backslash}, e_1^*\right) + \sum_{h \neq 1} \widetilde{x}_{h,id}^{\backslash}\left(p^{\backslash}, E, e_h^*\right)\right) - \sum_{h=1}^{H} e_h^*$$

Then Conclusion 2b of Theorem 40 in Chapter 4 simply says that

$$\operatorname{rank} D_{\left(p^{\backslash}, E\right)} \widetilde{f}\left(p^{\backslash *}, E^*, e^*, y^*\right) = G - 1 + \left(S - I\right) I$$

i.e., the desired result. We are then left with showing the following equivalent Condition 2a of Theorem 40 in Chapter 4:

$$D\left(F_{\mathcal{H}}, \widehat{G}\right)\left(x^*, \lambda_1^*, \mu^*, \eta^*, p^{\backslash *}, E^*\right) \quad \text{has full rank}$$

Recall that from (13.26), $\psi_{id}\left(L^*\right) = 0$. The computation of the above jacobian matrix is described below.

[6] For ease of notation, we changed the way we ordered the subvectors of ξ.

$$
\begin{array}{ll}
(1.1) & Du_1\left(x_1\right)-\lambda_1 p \\
(1.2) & -p\left(x_1-x_1^*\right) \\
(h.1) & D_{x_h^0} u_h\left(x_h\right)-\mu_h p^0 \\
(h.2) & D_{x_h^1} u_h\left(x_h\right)-\mu_h p^{\mathbf{1}}+\eta_h\,[I \mid E]\,\mathbf{\Phi}^{\mathbf{1}}(p) \\
(h.3) & p\left(x_h-x_h^*\right) \\
(h.4) & [I \mid E]\,\mathbf{\Phi}^{\mathbf{1}}(p)\left(x_h^1-e_h^1\right) \\
(M) & \sum_{h=1}^{H}\left(x_h^{\backslash}-e_h^{*\backslash}\right) \\
(L) & vec\left([I \mid E] \cdot R\left(p, y^*\right)\right)
\end{array}
\qquad
\begin{array}{c}
\begin{array}{ccccccc}
x_1 & \lambda_1 & x_h & \mu_h & \eta_h & p^{\backslash} & E
\end{array} \\
\left(\begin{array}{ccccccc}
D^2 u_1 & -p^{*T} & & & & -\lambda_1^* \widetilde{I}^T & \\
-p^* & & & & & & \\
 & & & & 0 & & \\
 & & D^2 u_h & -p^{*T} & N^T & -\mu_h^* \widetilde{I}^T & \\
 & & -p^* & & & & \\
 & & 0 \quad N & & & & \\
\widetilde{I} & & \widetilde{I} & & & & \\
 & & & & & * & K
\end{array}\right)
\end{array}
$$

where

$$\widetilde{I} \equiv \begin{bmatrix} I_{C-1} & 0_{(C-1)\times 1} & 0_{(C-1)\times(SC)} \\ 0_{(SC)\times(C-1)} & 0_{(SC)\times 1} & I_{SC} \end{bmatrix}$$

K is defined in Step 1 below, and

$$N \equiv [I|E^*]\,\Phi^{\mathbf{1}}(p^*)$$

We want now to show the following steps: Step1. K has full row rank; Step 2. The remaining top-left block in the matrix has full rank.

Step 1.

Recalling, see Definition (13.21),

$$[I \mid E] \cdot R(p,y) = R^*(p,y) + E\widehat{R}(p,y)$$

and

$$\widehat{R}(p,y) \equiv \begin{bmatrix} p^{S-I+1}y^{S-I+1,1} & p^{S-I}y^{S-I+1,i} & p^{S-I+1}y^{S-I,I} \\ \\ p^S y^{S1} & p^S y^{S1} & p^S y^{SI} \end{bmatrix} \in \mathcal{M}_{I,I}$$

Moreover, for $i = 1, ..., I$

$$E \cdot {}^i\widehat{r}(p,y) = \begin{bmatrix} e^1 \cdot {}^i\widehat{r}(p,y) \\ ... \\ e^{S-I} \cdot {}^i\widehat{r}(p,y) \end{bmatrix} \tag{13.31}$$

where for $s = 1, ..., S-I$, e^s is the s-th row of the matrix E, and for $i = 1, ..., I$, ${}^i\widehat{r}(p,y)$ is the i-th column of $\widehat{R}(p,y)$.

Rearranging terms in (13.31), the computation of the jacobian matrix

$$vec([I \mid E] \cdot R(p,y))$$

with respect to $(e^s)_{s=1}^{S-I}$ is described below.

$$\begin{array}{c} \\ e^1 \cdot {}^1\widehat{r}(p,y) \\ ... \\ e^1 \cdot {}^I\widehat{r}(p,y) \\ ... \\ e^{S-I} \cdot {}^1\widehat{r}(p,y) \\ ... \\ e^{S-I} \cdot {}^I\widehat{r}(p,y) \end{array} \begin{array}{c} \begin{array}{ccc} e^1 & \quad\cdots\quad & e^{S-I} \end{array} \\ \begin{pmatrix} \left[{}^1\widehat{r}(p,y)\right]^T & & \\ ... & & \\ \left[{}^I\widehat{r}(p,y)\right]^T & & \\ & & \\ & & \left[{}^1\widehat{r}(p,y)\right]^T \\ & & ... \\ & & \left[{}^I\widehat{r}(p,y)\right]^T \end{pmatrix} \end{array}$$

Therefore, using (13.24),

$$K = \begin{bmatrix} \left[\widehat{R}(p^*,y^*)\right]^T & & \\ & \ddots & \\ & & \left[\widehat{R}(p^*,y^*)\right]^T \end{bmatrix} = \begin{bmatrix} I & & \\ & \ddots & \\ & & I \end{bmatrix}$$

Step 2.

We want to show that

$$D_{(x,\lambda_1,\mu,\eta,p^\backslash)}\left(F_{\mathcal{H}},\widehat{G}\right)\left(x^*,\lambda_1^*,\mu^*,\eta^*,p^{\backslash *},E^*\right) \text{ has full rank}$$

Suppose otherwise, i.e., there exists $\Delta \equiv (\Delta x_1, \Delta\lambda_1, \Delta x_h, \Delta\mu_h, \Delta\eta_h, \Delta p^\backslash)$, with $\Delta x_1 \in \mathbb{R}^G, \Delta\lambda_1 \in \mathbb{R}, \Delta x_h \in \mathbb{R}^{GH}, \Delta\mu_h \in \mathbb{R}, \Delta\eta_h \in \mathbb{R}^{S-I}, \Delta p^\backslash \in \mathbb{R}^{G-1}$, such that

$$D_{(x,\lambda_1,\mu,\eta,p^\backslash)}\left(F_{\mathcal{H}},\widehat{G}\right)\left(x^*,\lambda_1^*,\mu^*,\eta^*,p^{\backslash *},E^*\right)\cdot\Delta = 0 \tag{13.32}$$

(13.32) is rewritten below.

$$\begin{array}{lll}
(1.1) & D^2u_1\Delta x_1 - p^*\Delta\lambda_1 - \lambda_1^*\widetilde{I}^T\Delta p^\backslash & = 0\\
(1.2) & -p^*\Delta x_1 & = 0\\
(h.1), (h.2) & D^2u_h\Delta x_h - p^*\Delta\mu_h + [0 \mid N]^T\Delta\eta_h - \mu_h^*\widetilde{I}^T & = 0\\
(h.3) & -p^*\Delta x_h & = 0\\
(h.4) & [0 \mid N]\,\Delta x_h & = 0\\
(M) & \sum_{h=1}^H \widetilde{I}\Delta x_h & = 0
\end{array} \tag{13.33}$$

Step 1. $\Delta x_1 \neq 0$.

Otherwise, from (1.1) in (13.33), $\Delta\lambda_1 = 0$ and $\Delta p^\backslash = 0$. From $(h.1)-(h.4)$ in (13.33), and using the rank computation in the proof of Theorem 12, $(\Delta x_h, \Delta\mu_h, \Delta\eta_h) = 0$ and therefore $\Delta = 0$.

Step 2. $Du_h(x_h^*)\,\Delta x_h = 0$ for $h = 1, ..., H$.

From system (13.23) and from (1.2) in (13.33)

$$Du_1(x_1^*)\,\Delta x_1 = \lambda_1^* p^*\Delta x_1 = 0$$

For $h \neq 1$, from $(h.1)$, $(h.2)$ in system (13.19) and from $(h.3)$, $(h.4)$ in (13.33),

$$Du_h(e_h^*)\,\Delta x_h = \mu_h^* p^*\Delta x_h - \eta_h\,[0 \mid N]\,\Delta x_h = 0$$

Step 3. $\sum_{h=1}^H \left(\theta_h^*\Delta x_h D^2u_h(e_h^*)\,\Delta x_h\right) = 0$.

Premultiplying (1.1) in (13.33) by Δx_1, we get

$$0 = \Delta x_1 D^2u_1\Delta x_1 - \Delta x_1 p^*\Delta\lambda_1 - \Delta x_1\lambda_1^*\widetilde{I}^T\Delta p^\backslash =$$

$$= \Delta x_1 D^2u_1\Delta x_1 - \gamma^{0C*}\Delta x_1\widetilde{I}^T\Delta p^\backslash \tag{13.34}$$

where the second equality follows from (1.2) in (13.33), and the definition of λ_1^*.

Premultiplying $(h.1)$ and $(h.2)$ in (13.33) by Δx_h, we get

$$\begin{aligned} 0 = & \Delta x_h D^2 u_h \Delta x_h - \Delta x_h p^* \Delta \mu_h + \Delta x_h \begin{bmatrix} 0 \\ N^T \end{bmatrix} \Delta \eta_h - \Delta x_h \mu_h^* \widetilde{I}^T \\ = & \Delta x_h D^2 u_h \Delta x_h - \frac{\gamma^{0C*}}{\theta_h^*} \Delta x_h \widetilde{I}^T \Delta p^\backslash \end{aligned} \tag{13.35}$$

where the second equality follows from $(h.3)$, $(h.4)$ in (13.33), and the definition of μ_h^*. Summing up (13.34) and (θ_h^* times(13.35))with respect to h, we get

$$\begin{aligned} 0 = & \sum_{h=1}^H \left(\theta_h^* \Delta x_h D^2 u_h \Delta x_h\right) - \gamma^{0C*} \sum_{h=1}^H \Delta x_h \widetilde{I}^T \Delta p^\backslash \\ = & \sum_{h=1}^H \left(\theta_h^* \Delta x_h D^2 u_h \Delta x_h\right) \end{aligned}$$

where the last equality comes from (M) in (13.33). Steps 1, 2 and 3 contradict Assumption u3. ■

From the previous lemmas, we finally get the following existence theorem.

Theorem 28 *For any economy* (e, y), *pr*$^{-1}(e, y) \neq \emptyset$, *i.e., a pseudo-equilibrium exists.*

4. Generic existence of equilibria

Theorem 29 *In an open and full measure subset of* $\mathbb{R}^{GH}_{++} \times \mathbb{R}^{SCI}$ *an equilibrium exists.*

Proof. From Propositions 8 and 9, it is enough to show that in an open and full measure subset of $\mathbb{R}^{GH}_{++} \times \mathbb{R}^{SCI}$, rank $R(p, y) = I$. In fact, we are going to show that $rank\ \widehat{R}(p, y) = I$. As usual, openness follows from the properness of the projection *pr* and the measure result from the application of the Transversality Theorem described below.

For each $\sigma \in \Sigma$, define

$$G_\sigma : \mathbb{R}^{G-1} \times \mathbb{R}^I \times \mathbb{R}^{SCI} \to \mathbb{R}^{I+1}, \quad \left(p^\backslash, d, y\right) \mapsto \left(d \cdot \widehat{R}(p, y), dd - 1\right)$$

We show that 0 is a regular value for $\left(\widetilde{f}_\sigma, G_\sigma\right)$ and then, by definition of regular value for a function whose domain is an abstract manifold - see Definition 58 in Chapter 7 - the desired result follows.

The computation of the partial jacobian matrix of $\left(\widetilde{f}_\sigma, G_\sigma\right)$ with respect to $\left(e_1, \widehat{y}, \widehat{\widehat{y}}, d\right)$ is described below for the case $\sigma = id$, the general case being left to the reader.

$$\begin{array}{l} \\ \widetilde{z}^{\backslash}_{id}\left(p^{\backslash},E,e\right) \\ \widetilde{K}_{id}\left(p^{\backslash},E,y\right) \\ d^{T}\cdot\widehat{R}\left(p,y\right) \\ d^{T}d-1 \end{array}\begin{array}{c} \begin{array}{cccc} e_1 & \widehat{y} & \widehat{\widehat{y}} & d \end{array} \\ \left(\begin{array}{cccc} D_{e_1}\widetilde{z}^{\backslash} & 0 & 0 & 0 \\ 0 & A & * & 0 \\ 0 & 0 & B & \widehat{R}\left(p,y\right) \\ 0 & 0 & 0 & 2d^{T} \end{array}\right) \end{array}$$

where the definition of $\widehat{\widehat{y}}$ and B are given below. From the proof of Lemma 22, both $D_{e_1}z$ and A have full row rank; observe that $d \neq 0$. Therefore, it is enough to show that B has full rank.

$$d\cdot\widehat{R}\left(p,y\right)=d\cdot\left[\begin{array}{ccc} p^{S-I+1}y^{S-I+1,1} & p^{S-I}y^{S-I+1,i} & p^{S-I+1}y^{S-I,I} \\ \\ p^{S}y^{S1} & p^{S}y^{S1} & p^{S}y^{SI} \end{array}\right]=$$

$$=\left[\begin{array}{c} \sum_{s=S-I+1}^{S} d^{s}p^{s}y^{s1} \\ \cdots \\ \sum_{s=S-I+1}^{S} d^{s}p^{s}y^{si} \\ \cdots \\ \sum_{s=S-I+1}^{S} d^{s}p^{s}y^{sI} \end{array}\right]$$

For consistency of notation with p and y, define $d \equiv (d^{s})_{s=S-I+1}^{S}$. Since $dd-1=0$, there exists $\overline{s} \in \{S-I+1,...,S\}$ such $d^{\overline{s}} \neq 0$. Then define $\widehat{\widehat{y}} \equiv \left(y^{\overline{s}i1}\right)_{i=1,...,I}$. The matrix B is computed below.

$$\begin{array}{l} \\ \sum_{s=S-I+1}^{S} d^{s}\sum_{c=1}^{C}p^{sc}y^{s1c} \\ \cdots \\ \sum_{s=S-I+1}^{S} d^{s}\sum_{c=1}^{C}p^{sc}y^{sic} \\ \cdots \\ \sum_{s=S-I+1}^{S} d^{s}\sum_{c=1}^{C}p^{sc}y^{sIc} \end{array}\begin{array}{c} \begin{array}{ccc} y^{\overline{s}11} & y^{\overline{s}i1} & y^{\overline{s}I1} \end{array} \\ \left(\begin{array}{ccccc} d^{\overline{s}}p^{\overline{s}1} & & & & \\ & \ddots & & & \\ & & d^{\overline{s}}p^{\overline{s}1} & & \\ & & & \ddots & \\ & & & & d^{\overline{s}}p^{\overline{s}1} \end{array}\right) \end{array}$$

i.e., $B = d^{\overline{s}}p^{\overline{s}1}I$, a full rank matrix.

To analyze the case of $\sigma \neq id$, just reorder the states according to σ and proceed as above. ∎

5. Regular economies

Definition 30 (x,b,p,q) is a **full rank equilibrium** for the economy (e,y) if it is an equilibrium and $rank\ R(p,y)=I$.

Theorem 29 says that there exists an open and full measure subset Π^* of the set of economies $\mathbb{R}^{GH}_{++} \times \mathbb{R}^{SCI}$ such that for each $(\overline{e}, \overline{y}) \in \Pi^*$ all associated pseudo-equilibria $(\overline{x}, \overline{p}^{\setminus}, \overline{L})$ are full rank equilibria. On the basis of the above analysis, nothing can be said about equilibria which are not full rank equilibria. In this section, we present definition and result on regular economies in terms of full rank equilibria.

Choose the normalizations presented in Subsection 3.2.4 in Chapter 12 and define

$$\text{for each } s,\ \ p^{s\setminus\prime} \equiv (p^{sc})_{c \neq C} \in \mathbb{R}^{C-1},\ \ p^{s\prime} \equiv \left(p^{s\setminus}, 1\right)$$

$$p^{\setminus\prime} \equiv \left(p^{s\setminus}\right)_{s=0}^{S} \in \mathbb{R}^{(S+1)(C-1)} \ \text{ and } \ p' = (p^{s})_{s=0}^{S}$$

$$x_h^{\setminus\prime} \equiv (x_h^{sc})_{c \neq C} \in \mathbb{R}^{G-(S+1)}_{++} \ \text{ and } \ e_h^{\setminus\prime} \equiv (e_h^{sc})_{c \neq C} \in \mathbb{R}^{G-(S+1)}_{++}$$

Following the same arguments presented in Chapter 11, we can then prove that a full rank equilibrium is characterized by the following system.

$$\begin{cases} Du_h(x_h') - \mu_h' p' & = & 0 & (h.1) \\ \mu_h' \begin{bmatrix} -q' \\ R(p', y) \end{bmatrix} & = & 0 & (h.2) \\ -\Phi(p')(x_h' - e_h) + \begin{bmatrix} -q' \\ R(p', y) \end{bmatrix} b_h' & = & 0 & (h.3) \\ \sum_{h=1}^{H} \left(x_h^{\setminus\prime} - e_h^{\setminus}\right) & = & 0 & (M.1) \\ \sum_{h=1}^{H} b_h' & = & 0 & (M.2) \end{cases} \tag{13.36}$$

Define

$$\xi' \equiv \left((x_h', \lambda_h', b_h')_{h \in \mathcal{H}}, p^{\setminus\prime}, q'\right) \in \Xi'$$

where

$$\Xi' \equiv \mathbb{R}^{GH}_{++} \times \mathbb{R}^{(S+1)H}_{++} \times \mathbb{R}^{IH} \times \mathbb{R}^{G-(S+1)}_{++} \times \mathbb{R}^{I}$$

and

$$F : \Xi' \times \mathbb{R}^{GH}_{++} \to \mathbb{R}^{\dim \Xi},\quad (\xi', e) \mapsto \text{left hand side of (13.36)}$$

Then, we can give the following definition.

Definition 31 $(e, y) \in \mathbb{R}^{CH}_{++} \times \mathbb{R}^{SCI}$ is a pseudo-regular economy for an incomplete market model with real assets if 0 is a regular value for F. Let $\mathcal{R}_p$ be the set of pseudo-regular economies.

Then all the results presented in Section 4 in Chapter 11 apply to the above case with respect to pseudo-regular economies and full rank equilibria.

Appendix. Grassmannian manifolds

In this appendix we introduce the notion of grassmannian manifold and we present some important properties of this class of abstract manifolds.

Given an euclidean space $\mathbb{R}^S$ and an integer number $I \leq S$, a grassmannian manifold is the set $G_{I,S}$ of the I-dimensional linear subspaces of $\mathbb{R}^S$ with a suitable topological structure and a suitable atlas.

Before assigning a topology and a manifold structure on $G_{I,S}$, let us recall some notions on spaces of matrices.

Proposition 32 *Given two positive integer numbers m and n, the space $\mathcal{M}_{m,n}$ of $m \times n$ matrices is a vector space and a topological space linearly isomorphic and homeomorphic to $\mathbb{R}^{mn}$.*

Proof. First of all, it is immediate to verify that $\mathcal{M}_{m,n}$ is a vector space of dimension mn with the sum of matrices and the scalar multiplication as basic operations (see [78], Section 1 in Chapter 3). There is a natural bijection f between $\mathcal{M}_{m,n}$ and $\mathbb{R}^{mn}$. Given $A = (a_{ij}) \in \mathcal{M}_{m,n}$, define

$$f(A) \equiv (a_{11}, ..., a_{1n}, a_{21}, ..., a_{2n}, ..., a_{m1}, ..., a_{mn})$$

f is an isomorphism (clearly not the unique isomorphism between $\mathcal{M}_{m,n}$ and $\mathbb{R}^{mn}$).

To define a topology on $\mathcal{M}_{m,n}$ it is sufficient define the family of open sets ([90] p. 76). This can be done by taking as open sets all the sets of the form $f^{-1}(O)$, for each open subset O of $\mathbb{R}^{mn}$. ■

Definition 33 Given two positive integers m, n, let $\mathcal{M}^*_{m,n}$ be the set of $m \times n$ matrices of full rank.

Proposition 34 *$\mathcal{M}^*_{m,n}$ is open in $\mathcal{M}_{m,n}$.*

Proof. Suppose $m \leq n$; the case $m \geq n$ is analogous. If $A \in \mathcal{M}^*_{m,n}$, there exists a $m \times m$ submatrix A' of A having nonzero determinant. Since the determinant is continuous as a real function defined in any space of square matrices, there exists a neighborhood U of A in $\mathcal{M}^*_{m,n}$ such that, for each $B \in U$, the submatrix B' of B with entries in the same position as A' has nonzero determinant. It follows that $B \in \mathcal{M}^*_{m,n}$. ■

Definition 35 The **kernel** of $A \in \mathcal{M}_{m,n}$, $\ker A$, is the kernel of the linear function $L : \mathbb{R}^n \to \mathbb{R}^m$ represented by the matrix A.

Definition 36 Let $G_{I,S}$ be the set of all I-dimensional linear subspaces of $\mathbb{R}^S$.

Remark 37 If $A \in \mathcal{M}^*_{S-I,S}$, then $\ker A \in G_{I,S}$ ([78], Theorem 3, p. 99).

Remark 38 If $L \in G_{I,S}$, then there exists $A \in \mathcal{M}^*_{S-I,S}$ such that $L = \ker A$. In fact, consider any basis $\{l^1, ..., l^I\}$ of L. The $I \times S$ matrix C having $l^1, ..., l^I$ as rows has full rank and $\ker C$ has dimension $S - I$. Let $\{a^1, ..., a^{S-I}\}$ be a basis of $\ker C$ and let A be the matrix having $a^1, ..., a^{S-I}$ as rows. A has full rank and it is straightforward to check that $L = \ker A$.

In other words, A is any matrix with rows composed by the vectors of a basis of an orthogonal space of L in $\mathbb{R}^S$.

Proposition 39 *Take A and $A' \in \mathcal{M}^*_{S-I,S}$. $\ker A = \ker A'$ if and only if there exists $B \in \mathcal{M}^*_{S-I,S-I}$ such that $A' = BA$.*

Proof. [⇒] Let $L \equiv \ker A = \ker A'$; then the orthogonal space $L^\perp$ coincides with $\text{Im } A^T = \text{Im } A'^T$, that is, the rows of A and those of A' are basis of $L^\perp$. Therefore B is defined as follows: the i-th row is the $(S-I)$-tuple of coefficients which represent i-th row A' as linear combination of rows of A. B turns out to be invertible.

[⇐] $\ker A \subseteq \ker A'$ since $Ax = 0$ implies $A'x = BAx = 0$. In addition, $\ker A' \subseteq \ker A$, since $A'x = BA' = 0$ implies $B^{-1}BAx = Ax = 0$. ∎

By the above result we can define the following equivalence relation on $\mathcal{M}^*_{S-I,S}$.

Definition 40 Given $A, A' \in \mathcal{M}^*_{S-I,S}$, $A \sim A'$ if $\ker A = \ker A'$.

Definition 41 The quotient space $\mathcal{M}^*_{S-I,S}/\sim$ is the space of equivalence class of matrices having the same I-dimensional linear subspace of $\mathbb{R}^S$ as kernel. Denote by

$$p : \mathcal{M}^*_{S-I,S} \to \mathcal{M}^*_{S-I,S}/\sim \tag{13.37}$$

the canonical quotient projection.

By Proposition 39, there is a natural bijection between $\mathcal{M}^*_{S-I,S}/\sim$ and $G_{I,S}$, precisely the function

$$f : \mathcal{M}^*_{S-I,S}/\sim \to G_{I,S}, \quad [A] \mapsto \ker A \tag{13.38}$$

Therefore, from now on, $G_{I,S}$ is a topological space, where the open sets are the images by f of the open sets in $\mathcal{M}^*_{S-I,S}/\sim$; consequently f becomes a homeomorphism.

Let us now see two important properties of $G_{I,S}$.

Proposition 42 *$G_{I,S}$ is a Hausdorff topological space*[7].

[7] [90] p. 195.

Proof. The proof follows from the fact that $G_{I,S}$ is a quotient space of $\mathcal{M}^*_{S-I,S}$, which is a Hausdorff space, because it is open in an euclidean space. The reader can verify the details, observing that the projection p is an open function (see also [55] p. 140). ■

Proposition 43 *$G_{I,S}$ is a compact topological space.*

Proof. Define

$$\mathcal{B} \equiv \{A \in \mathcal{M}_{S-I,S} : \text{the row vectors of } A \text{ are an orthonormal set}\} =$$

$$\left\{ A \equiv \begin{bmatrix} a_1 \\ \vdots \\ a_{S-I} \end{bmatrix} \in \mathcal{M}_{S-I,S} : a_i a_j = \begin{cases} 1 & \text{if} \quad i = j \\ 0 & \text{if} \quad i \neq j \end{cases} \right\}$$

Observe that $\mathcal{B} \subseteq \mathcal{M}^*_{S-I,S}$. We are going to show that $\mathcal{B}$ is compact and $p(\mathcal{B}) = G_{I,S}$. The continuity of p and the identification between $G_{I,S}$ and $\mathcal{M}^*_{S-I,S}/\sim$ implies the desired result.

By the identification of $\mathcal{M}_{S-I,I}$ with $\mathbb{R}^{(S-I)I}$, it follows that $\mathcal{B}$ is bounded. Then, consider a sequence (A_n) in $\mathcal{B}$ which converges to $\widetilde{A} \in \mathcal{M}_{S-I,I}$. Since the scalar product is continuous, it follows that $\widetilde{A} \in \mathcal{B}$ (the reader can check the details). Therefore $\mathcal{B}$ is closed and, being bounded, it is compact.

Now let $L \in G_{I,S}$ and take an orthonormal basis $(v_1, ..., v_I)$ for L. Consider the orthonormal base of $\mathbb{R}^S$obtained adding $(S - I)$ vectors $(v_{I+1}, ..., v_S)$ making $(v_1, ..., v_S)$ an orthonormal basis of L. Define the following full rank matrix

$$A \equiv \begin{bmatrix} v_{I+1} \\ \dots \\ v_S \end{bmatrix}$$

Of course, for each $s = 1, ..., S - I$, $Av_s = 0$, that is, $(v_1, ..., v_I)$ is a basis for $\ker A$. ■

Let us now define an atlas which makes $G_{I,S}$ a C^∞ $I(S - I)$-manifold.

Remark 44 Consider the set $\mathcal{S} \equiv \{1, ..., s, ..., S\}$ and the set of all bijective functions $\sigma : \mathcal{S} \to \mathcal{S}$. Observe that any such σ can be seen as a permutation of the elements of $\mathcal{S}$. Recall that the set Σ of all permutation functions on $\mathcal{S}$ has cardinality $S!$ ([78], Section 5 in Chapter 7).

Definition 45 Let σ a permutation of $\mathcal{S}$. A **permutation matrix** P_σ is a $S \times S$ matrix such that, for $s = 1, ..., S$, each entry of s-th row is 0 apart from that in the $\sigma(s)$-th position which is 1.

Remark 46 For each $m \in \mathbb{N}$ and each $m \times S$ matrix A, $AP_\sigma = \left({}^{\sigma(1)}a, \ldots, {}^{\sigma(S)}a\right)$. For each $n \in \mathbb{N}$, and for each $S \times n$ matrix C,

$$P_\sigma C = \begin{pmatrix} c^{\sigma(1)} \\ \vdots \\ c^{\sigma(S)} \end{pmatrix}$$

In other words, AP_σ is equal to the matrix A with columns reordered according to the permutation σ, and similarly for $P_\sigma C$ and the rows of C.

Remark 47 $P_{\sigma^{-1}} = P_\sigma^{-1} = P_\sigma^T$. In particular, notice that any permutation matrix has full rank.

Proposition 48 *Given $L \in G_{I,S}$, there exists $\overline{\sigma} \in \Sigma$ and $E \in \mathcal{M}_{S-I,I}$ such that*

$$\ker \begin{bmatrix} I & E \end{bmatrix} P_{\overline{\sigma}} = L$$

Proof. Let $A \in \mathcal{M}^*_{S-I,S}$ be such that $L = \ker A$. Since A has rank $S - I$, it has linearly independent $S - I$ columns. Let $\sigma \in \Sigma$ be a permutation which carries those columns in the first $S - I$ positions of a new matrix, that is, let

$$AP_\sigma \equiv \begin{bmatrix} (AP_\sigma)^{(1)} & (AP_\sigma)^{(2)} \end{bmatrix}$$

be such that $(AP_\sigma)^{(1)}$ has full rank equal to $S - I$. Then,

$$\left[(AP_\sigma)^{(1)}\right]^{-1} A = \begin{bmatrix} I & \left[(AP_\sigma)^{(1)}\right]^{-1} (AP_\sigma)^{(2)} \end{bmatrix} P_{\sigma^{-1}}.$$

From Proposition 39, it follows

$$\ker A = \ker \begin{bmatrix} I & \left[(AP_\sigma)^{(1)}\right]^{-1} (AP_\sigma)^{(2)} \end{bmatrix} P_{\sigma^{-1}}$$

and therefore the claim follows with $E \equiv \left[(AP_\sigma)^{(1)}\right]^{-1} (AP_\sigma)^{(2)}$ and $\overline{\sigma} = \sigma^{-1}$. ■

Proposition 49 *Assume that* $\ker \begin{bmatrix} I & E \end{bmatrix} P_\sigma = \ker \begin{bmatrix} I & E' \end{bmatrix} P_{\sigma'}$. *Define*

$$P \equiv P_{\sigma'} P_{\sigma^{-1}} \equiv \begin{matrix} \\ S-I \\ I \end{matrix} \begin{matrix} \begin{matrix} S-I & \quad I \end{matrix} \\ \begin{pmatrix} P^{(1)} & P^{(2)} \\ P^{(3)} & P^{(4)} \end{pmatrix} \end{matrix}$$

Then

1 $E = \left(P^1 + E'P^3\right)^{-1} \left(P^2 + E'P^4\right)$;

2 *if* $\sigma = \sigma'$, *then* $E = E'$.

Proof. 1. By Proposition 39, there exists $B \in \mathcal{M}^*_{S-I,S-I}$ such that

$$\left[\begin{array}{cc} I & E \end{array}\right] P_\sigma = B \left[\begin{array}{cc} I & E' \end{array}\right] P_{\sigma'}$$

and, using the definition of P,

$$\begin{aligned} \left[\begin{array}{cc} I & E \end{array}\right] &= B \left[\begin{array}{cc} I & E' \end{array}\right] P = B \left[\begin{array}{cc} I & E' \end{array}\right] \left[\begin{array}{cc} P^{(1)} & P^{(2)} \\ P^{(3)} & P^{(4)} \end{array}\right] \\ &= \left[\begin{array}{cc} B\left(P^{(1)} + E'P^{(3)}\right) & B\left(P^{(2)} + E'P^{(4)}\right) \end{array}\right] \end{aligned}$$

Therefore B and $\left(P^{(1)} + E'P^{(3)}\right)$ are one the inverse of the other and also

$$E = \left(P^1 + E'P^{(3)}\right)^{-1} \left(P^{(2)} + E'P^{(4)}\right)$$

2. If $\sigma = \sigma'$, then

$$\left[\begin{array}{cc} P^{(1)} & P^{(2)} \\ P^{(3)} & P^{(4)} \end{array}\right] = P = P_\sigma \cdot P_{\sigma^{-1}} = \left[\begin{array}{cc} I_{S-I} & \\ 0 & I_I \end{array}\right]$$

and

$$E = \left(I + E'0\right)^{-1} \left(0 + E'I\right) = E'$$

■

Let us fix $\sigma \in \Sigma$. By the two above propositions, given $L \in G_{I,S}$, it is uniquely determined the matrix $E \in \mathcal{M}_{S-I,I}$ such that $\ker \left[\begin{array}{cc} I & E \end{array}\right] P_\sigma = L$. Therefore, define

$$V_\sigma \equiv \left\{L \in G_{I,S} : \exists\, E \in \mathcal{M}_{S-I,I} \text{ such that } \ker \left[\begin{array}{cc} I & E \end{array}\right] P_\sigma = L\right\}$$

and the following function turns out to be well defined:

$$\psi_\sigma : V_\sigma \to \mathcal{M}_{S-I,I}, \quad L \mapsto E \text{ such that } L = \ker \left[\begin{array}{cc} I & E \end{array}\right] P_\sigma$$

Proposition 50 $\left\{\left(\psi_\sigma, V_\sigma, \mathcal{M}_{S-I,I}\right)\right\}_{\sigma \in \Sigma}$ *is a atlas which makes* $G_{I,S}$ *a* C^∞ *abstract manifold.*

Proof. First of all we have to show that V_σ is open, for any σ. Let $L \in V_\sigma$ and let $A \in \mathcal{M}^*_{S-I,S}$ be such that $\ker A = L$. By the identification of $G_{I,S}$ with $\mathcal{M}^*_{S-I,S}/\sim$, let us show that V_σ is an open subset of $\mathcal{M}^*_{S-I,S}/\sim$. To do this, by the definition of quotient topology, it is sufficient to prove that

$$p^{-1}\left(V_\sigma\right) = \left\{A \in \mathcal{M}^*_{S-I,S} : \ker A = \ker \left[\begin{array}{cc} I & E \end{array}\right] P_\sigma\right\}$$

is open, where p is the projection defined by the (13.37). Let $A \in p^{-1}(W_\sigma)$. By Proposition 48, in particular, by the proof of that proposition

$$A = \left[\begin{array}{cc} A^{(1)} & A^{(2)} \end{array}\right] P_\sigma$$

where $A^{(1)}$ has full rank. Let now U be a neighborhood of A in $\mathcal{M}^*_{S-I,S}$ such that, for any $\widetilde{A} \in U$,

$$\widetilde{A} = \left[\begin{array}{cc} \widetilde{A}^{(1)} & \widetilde{A}^{(2)} \end{array}\right] P_\sigma$$

where $\widetilde{A}^{(1)}$ has full rank. Recalling again the proof of Proposition 48, there exist a full rank $S - I \times S - I$ matrix B such that

$$B\widetilde{A} = \left[\begin{array}{cc} I & E \end{array}\right] P_\sigma$$

Finally, by Proposition 39, $\widetilde{A} \in p^{-1}V_\sigma$ and the the claim follows.

Let us prove that ψ_σ is a C^∞ diffeomorphism of W_σ onto $\mathcal{M}_{S-I,I}$.

By point 2 in Proposition 49, ψ_σ is one-to-one. To see that ψ_σ is onto, take $E \in \mathcal{M}_{S-I,I}$ and Consider $L \equiv \ker \left[\begin{array}{cc} I & E \end{array}\right] P_\sigma$. Clearly, $\psi_\sigma(L) = E$.

Let us now prove that ψ_σ and ψ_σ^{-1} are continuous (this makes $G_{I,S}$ a topological manifold - see Definition 41).

Fix $L \in V_\sigma$ and $E \equiv \psi_\sigma(L)$. Consider $A_L \in p^{-1}(L)$. By the proof of Proposition 48, $A_L = \left[\begin{array}{cc} B_L & C_L \end{array}\right] P_\sigma$, where B_L is nonsingular. Given any $A \in \mathcal{M}^*_{S-I,S}$, $A \equiv \left[\begin{array}{cc} B_L & C_L \end{array}\right] P_\sigma$, such that B is nonsingular, the function

$$f : A \mapsto B^{-1}C$$

is well defined. In particular, f is well defined in a suitable neighborhood of A_L, where it is continuous and it is invertible with continuous inverse. Now, the continuity of ψ_σ follows from the following argument. Let U be an open neighborhood of E and consider an open neighborhood W of A_L such that $f_{|W} : W \to U$ is well defined. Clearly $p(W)$ is an open neighborhood of L in $G_{I,S}$, contained in V_σ, and $\psi_\sigma(p(W))) \subseteq U$, since $\psi_\sigma \circ p_{|W} = f_{|W}$. Similarly one can verify that ψ_σ^{-1}.

Finally, we prove that for any $\sigma, \sigma' \in \Sigma$ such that $V_\sigma \cap V_{\sigma'} \neq \emptyset$

$$\phi_\sigma \circ \phi_{\sigma'}^{-1} : \psi_{\sigma'}(V_\sigma \cap V_{\sigma'}) \to \psi_\sigma(V_\sigma \cap V_{\sigma'})$$

is C^∞. By definition, $\psi_{\sigma'}^{(-1)}(E') = \ker \left[\begin{array}{cc} I & E \end{array}\right] P_{\sigma'}$ and $\psi_\sigma\left(\psi_{\sigma'}^{(-1)}(E')\right) = E$, where

$$\ker \left[\begin{array}{cc} I & E \end{array}\right] P_\sigma = \psi_{\sigma'}^{-1}(E')$$

By Proposition (49), and using the same notation,

$$E = \left(P^{(1)} + E'P^{(3)}\right)^{-1}\left(P^{(2)} + E'P^{(4)}\right)$$

, which is a C^∞ function of E'. ■

We conclude with two other useful properties of grassmannian manifolds.

Proposition 51 *$G_{I,S}$ is sequentially compact.*

Proof. Since we have proved that $G_{I,S}$ is compact and hausdorff, the claim follows if we show that $G_{I,S}$ satisfies the second countability axiom, that is, if it has a countable basis for its topology[8].

It is straightforward to prove that $G_{I,S}$ countable basis, since it is the finite union of the open sets V_σ each of them homeomorphic to $\mathcal{M}_{S-I,I}$ which satisfies the second countability axiom, being identified with an euclidean space. ■

[8] The reader can see [90], Section 4.1 in Chapter 4 for the countability axioms. The equivalence between compactness and sequential compactness for hausdorff compact spaces is presented in Exercise 6 in [90], p. 194.

Chapter 14

RESTRICTED PARTICIPATION

Following [29], we can say that "While there might be some disagreement over whether, in a modern developed economy, financial markets are actually incomplete, there can hardly be any disagreement over whether at least some economic agents are variously constrained in transacting on those financial markets. Without attempting a detailed explanation of how particular constraints come about (for example, in order to resolve problems arising from moral hazard), it is still possible to extend the model of Section 2[1] to incorporate, in a very general way, their implication for financial equilibrium by adding a restriction of the form

$$b_h \in B_h \subseteq \mathbb{R}^I$$

to problem (1)[2]. Here B_h represents the *portfolio set*,the possible credit transactions available to the household"[3]. The above model of "*restricted participation* constitutes a bona fide generalization of the mode with incomplete markets, but the latter potentially embodies far more inte..sting institutional features (and not just the flavor of restricted participation) since it permits, for instance, modeling short sales bounds or market margin requirements. Of course, in principle such constraints should themselves be determined endogenously"[4].

The restricted participation model is the object of this chapter. Markets may in principle be complete in the sense that the return matrix has full rank equal to the number of states in the first period, but each household has her own specific restricted way to access to it. There are several ways of modelling those restrictions. [11] and [104] deal with linear homogenous constraints. We follow

[1] The author refers to the model with incomplete markets.

[2] The author refers to the household's maximization problem.

[3] [29] p. 274.

[4] [29], p. 275.

[32] who describe restrictions in terms of a differentiable restriction function which has to be non negative. In their words, "The obvious leading example, nonnegativity constraints, represents the widespread practice of barring short sales[5]. But other, commonly encountered market restrictions are also modeled (albeit in somewhat simplified form), for example, obligations to carry life insurance, or hold compensating balances, or meet margin requirements. While it seems likely that our present analysis can be sharpened to incorporate explicit dependence of the constraints on endogenous variables other than just portfolio holdings, this question remains to be more thoroughly investigated.

A secondary purpose of the note is to illustrate a fairly general method for encompassing various sorts of constraints on individual behavior while still employing differential techniques. In particular, it is easily shown that the same sort of argument elaborated here can be utilized to establish finite, local uniqueness in the standard Walrasian model when the consumption set permits zero consumption, or involves joint consumption, or requires minimum consumption[6]"[7].

In conclusion, in the model of this chapter, households' maximizing choices have to satisfy not only budget constraints, but also financial constraints describing their different access to the assets market. Restricted participation models are a generalization of the incomplete market models. The latter coincide with a particular case of the former in which all households are restricted in the same way.

The chapter is basically organized as all the other chapters on complete and incomplete markets. We present the model, we show existence of equilibria and properties of regular economies.

1. Set up of the model

We consider a model of restricted participation with numeraire assets[8]. Therefore we adopt the same set-up described in Chapter 11 and, more in detail, in the financial economy model - see Subsection 2.1. We also make same assumptions endowments and utility functions presented there apart from Assumption u3, which is substituted by the following stronger one.

Assumption u3' For $h \in \mathcal{H}$, u_h is differentiably strictly concave, i.e., $D^2 u_h(x_h)$ is definite negative for every $x_h \in \mathbb{R}^G_{++}$.

[5]An alternative approach for dealing with this particular form of restriction is developed by [59] in the course of their analysis of constrained suboptimality in a model with an incomplete stock market.
[6]For works in this area, see [16], [102] and [121].
[7][32], p. 62.
[8]Along the lines of Chapter 12, conclusions reached in the present chapter can be easily applied to the case of nominal assets. For a restricted participation model in a real asset framework see [96].

In fact Assumption u3 suffices to show existence, while the above one is used to show properties of regular economies, see the proof of Lemma 19.

About the financial side of the economy, and consistently with the restricted participation framework, we assume that

1. there exists a set of financial instruments which can even complete the markets, and
2. each household h can access only partially and in a personalized manner to the available set of assets.

The above assumptions are formalized below and substitute Assumption a1, a2 and a3 presented in Chapter 11.

Assumption re1. $\operatorname{rank} Y = I \leq S.$

Remark 1 The previous assumption implies that there are no redundant assets in the economy. Observe that while in the model of incomplete markets, the analogous Assumption a2 is made without loss of generality, in the present framework it is not at all innocuous. When households' portfolio holdings are constrained, they may very well benefit from the opportunity afforded by the availability of additional bonds whose yields are not linearly independent.

Definition 2 Let J_h be a set of non negative integers between 1 and an arbitrary number smaller than I. Let $\mathcal{J}_h \equiv \{1, ..., J_h\}$, where J_h is a nonnegative integer smaller or equal to I. The restriction participation function for household h is

$$a_h : \mathbb{R}^I \to \mathbb{R}^{J_h}, \quad b_h \mapsto a_h(b_h) \equiv \left(a_h^j(b_h)\right)_{j \in \mathcal{J}_h}$$

For each subset $\mathcal{J}_h' \subseteq \mathcal{J}_h$, denote its cardinality by J_h' and define

$$a_h^{\mathcal{J}_h'} : \mathbb{R}^I \to \mathbb{R}^{J_h'}, \quad b_h \mapsto \left(a_h^j(b_h)\right)_{j \in \mathcal{J}_h'}$$

The portfolio set is

$$B_h \equiv \left\{b_h \in \mathbb{R}^I : a_h(b_h) \geq 0\right\}$$

Households are allowed to choose only $b_h \in B_h$.

We make the following assumption on a_h.

Assumption re2 For each h,

1 a_h is C^2 ;

2 a_h is componentwise concave , i.e., for each $j \in \mathcal{J}_h$, and for each $v \in \mathbb{R}^I$

$$vD^2 a_h^j(b_h)\, v \leq 0$$

3 a_h permits no participation on the asset market, i.e.,

$$a_h(0) \geq 0$$

4 For each $b_h \in \mathbb{R}^I$, and each $\mathcal{J}'_h \subseteq \mathcal{J}_h$,

$$\text{if } a_h^{\mathcal{J}'_h}(b_h) = 0, \text{ then } \operatorname{rank} Da^{\mathcal{J}'_h}(b_h) = J'_h$$

5 For every i there exists some h such that for every b_h,

$$D_{b_h^i} a_h(b_h) = 0$$

Assumption re2.1 allows to use differential techniques. Assumption re2.2 gives a relatively general characterization of the constraint on asset market participation; observe that non negativity constraints are a particular case of those studied in our framework. Assumption re2.3 allows the obvious possibility of avoiding to use the asset market. Assumption re2.4 is a technical one and it is needed both for characterizing household's maximization solutions in terms of Kuhn-Tucker conditions and to get the desired properties of the homotopy. Assumption re2.5 says that for each asset i there exists a household h for whom the i-th column of Da_h is 0. It then basically says that for each asset, there exists at least one household who is free to exchange it. That assumption is used in Lemma 12.

Remark 3 From Assumptions re2.3, for each j and h, we have that there exists a partition $\mathcal{J}_h^+, \mathcal{J}_h^0$ of $\mathcal{J}_h$, with cardinality J_h^+ and J_h^0, respectively, such that

$$\begin{array}{ll} \forall j \in \mathcal{J}_h^+, & a_h^j(0) > 0 \\ \forall j \in \mathcal{J}_h^0, & a_h^j(0) = 0 \end{array}$$

Therefore, there exists a neighborhood N_h^+ of zero such that

$$\forall b_h \in N_h^+, \quad \left(a_h^j(b_h)\right)_{j \in \mathcal{J}_h^+} > 0 \tag{14.1}$$

From Assumption re2.4,

$$\operatorname{rank} D\left(a_h^j(0)\right)_{j \in \mathcal{J}_h^0} = J_h^0$$

Therefore, from Definition 62 and Theorem 66 in Chapter 3,the image of each open neighborhood of 0 in $\mathbb{R}^I$ contains an open neighborhood of 0 in $\mathbb{R}^{J_h^0}$. Then,

$$\text{there exists } b_h \in N_h^+ \text{ such that } \left(a_h^j(b_h)\right)_{j \in \mathcal{J}_h^0} > 0 \tag{14.2}$$

From (14.1) and (14.2), we have that

$$\forall h,\ \exists\, \widetilde{b}_h \in \mathbb{R}^I \text{ such that } a_h\left(\widetilde{b}_h\right) > 0 \tag{14.3}$$

Observe that Assumptions re2.4 and re2.5 imply that for household h described in the latter one it is the case that $J_h < I$.

In our analysis we take for given utility functions, restricted participation functions and yield matrices. Therefore we can give the following definitions of economy and equilibrium.

Definition 4 An economy is an endowment vector $e \in \mathbb{R}^{GH}_{++}$.

As we have done in the numeraire asset model, we normalize the price of good C in each state s. Recall that

$$\text{for each } s,\ \ p^{s\backslash} \equiv (p^{sc})_{c \neq C} \in \mathbb{R}^{C-1}_{++},\ \ p^s \equiv \left(p^{s\backslash}, 1\right)$$

$$p^{\backslash} \equiv \left(p^{s\backslash}\right)_{s=0}^{S} \in \mathbb{R}^{(S+1)(C-1)}_{++} \quad \text{and} \quad p = (p^s)_{s=0}^{S}$$

Definition 5 The vector

$$\left((x_h, b_h)_{h=1}^{H}, p^{\backslash}, q\right) \in \left(\mathbb{R}^{G}_{++} \times \mathbb{R}^{I}\right)^{H} \times \mathbb{R}^{G-(S+1)}_{++} \times \mathbb{R}^{I}$$

is a **demand-price financial equilibrium in a numeraire asset - restricted participation model** for the economy $e \in \mathbb{R}^{GH}_{++}$ if

1 households maximize, i.e., for every h, for given $\left(p^{\backslash}, q, e\right)$, (x_h, b_h) solves

$$\max_{(x_h, b_h)} u_h(x_h)$$
$$\begin{aligned} -\Phi(p)(x_h - e_h) + \left[\begin{smallmatrix} -q \\ Y \end{smallmatrix}\right] b_h &= 0 \\ a_h(b_h) &\geq 0 \end{aligned} \tag{14.4}$$

2 markets clear, i.e.,

$$\begin{aligned} \textstyle\sum_{h=1}^{H} (x_h - e_h) &= 0 \\ \textstyle\sum_{h=1}^{H} b_h &= 0 \end{aligned}$$

2. Existence of equilibria

As in most of the previous chapters we want to characterize equilibria in terms of an "extended system".

It can be easily shown that a solution to problem (14.4) exists and is unique, exploiting the fact problem (14.4) is the same as household h's problem in the numeraire asset case[9] apart from the participation constraint. Existence follows from the fact that the constraint set in problem (14.4) is a nonempty - from Assumption re2.3 - closed subset of the constrained set of problem (11.4) in Chapter 11, a compact set. Uniqueness follows from the usual assumptions and Assumption re2.2.

Proposition 6 *For every h, and every $(p^\backslash, q, e)$, (x_h, b_h) is a solution to problem (14.4) if and only if there exists $(\lambda_h, \mu_h) \in \mathbb{R}^{S+1}_{++} \times \mathbb{R}^{J_h}$ such that $(x_h, b_h, \lambda_h, \mu_h)$ is a solution of the following system (of Kuhn-Tucker conditions of problem (14.4))*

$$\begin{cases} D_{x_h} u_h(x_h) - \lambda_h \Phi(p) & = 0 & & (h.1) \\ -\Phi(p)(x_h - e_h) + \left[\begin{smallmatrix}-q\\Y\end{smallmatrix}\right] b_h & = 0 & & (h.2) \\ \lambda_h \left[\begin{smallmatrix}-q\\Y\end{smallmatrix}\right] + \mu_h D a_h(b_h) & = 0 & & (h.3) \\ \min\left\{\mu_h^j, a_h^j(b_h)\right\} & = 0 & \text{for } j \in \mathcal{J}_h & (h.4) \end{cases} \tag{14.5}$$

Proof. The result follows showing that problem (14.4) satisfies necessary and sufficient conditions for Kuhn Tucker theorems (see Subsection 3.2.3 in Chapter 1). The lagrangean function for problem (14.4) is

$$\mathcal{L}\left(x_h, b_h, \lambda_h, p^\backslash, q, e_h\right) = u_h(x_h) + \lambda_h\left(-\Phi(p)(x_h - e_h) + \begin{bmatrix}-q\\Y\end{bmatrix} b_h\right)$$

and the associated first order conditions are just system (14.5).

[⇒] Necessary conditions for Theorem 41 are satisfied because the partial jacobian matrix of the binding constraint functions with respect to (x_h, b_h) has full row rank. The computation of that matrix is described below, where $\mathcal{J}_h^*$ is the set of binding constraints

$$\begin{matrix} & x_h & b_h \\ -\Phi(p)(x_h - e_h) + \left[\begin{smallmatrix}-q\\Y\end{smallmatrix}\right] b_h & -\Phi(p) & \left[\begin{smallmatrix}-q\\Y\end{smallmatrix}\right] \\ a_h^{\mathcal{J}_h^*}(b_h) & 0 & D a_h^{\mathcal{J}_h^*}(b_h) \end{matrix}$$

The full rank result follows from Assumption re2.4. Therefore, if $(x_h, b_h) = (x_h, b_h)\left(p^\backslash, q, e_h\right)$ is a solution to problem (14.4), then there exists $(\lambda_h, \mu_h) \in \mathbb{R}^{S+1} \times \mathbb{R}^I$ such that $(x_h, b_h, \lambda_h, \mu_h)$ is the unique solution to system (14.5). The fact that $\lambda_h \gg 0$ follows from equation $(h.1)$ in that system, Assumption u2 and the fact that $p \in \mathbb{R}^G_{++}$.

[⇐] Sufficient conditions for Theorem 43 in Chapter 1 are satisfied because, from Assumption u3', u_h is differentiably strictly concave and the constraint

[9] See problem (11.4) in Chapter 11.

functions for the equality and inequality constraints are linear and concave, respectively. ■

Remark 7 Observe that $(h.4)$ in (14.5) is equivalent to the standard way of writing Kuhn-Tucker conditions involving inequality constraints and associated multipliers. We will discuss in detail the difficulty related to the non-differentiability of the min function.

We describe equilibria using the following extended equilibrium system of Kuhn-Tucker and market clearing conditions . As usual, due to "Walras' laws", one market clearing condition per spot is redundant. Define

$$x_h^{\backslash} \equiv \left(x_h^{sc}\right)_{c\neq C} \in \mathbb{R}_{++}^{G-(S+1)} \quad \text{and} \quad x_h^{\cdot} \equiv \left(x_h^{sC}\right)_{s=0}^{S} \in \mathbb{R}_{++}^{S+1}$$

and similarly

$$e_h^{\backslash} \equiv \left(e_h^{sc}\right)_{c\neq C} \in \mathbb{R}_{++}^{G-(S+1)} \quad \text{and} \quad e_h^{\cdot} \equiv \left(e_h^{sC}\right)_{s=0}^{S} \in \mathbb{R}_{++}^{S+1}$$

# eq.	matching var.		eq.	
G	x_h	$(h.1)\,, h \in \mathcal{H}$	$D_{x_h} u_h (x_h) - \lambda_h \Phi (p)$	$= 0$
$S+1$	λ_h	$(h.2)\,, h \in \mathcal{H}$	$-\Phi (p) (x_h - e_h) + \left[\begin{smallmatrix} -q \\ Y \end{smallmatrix}\right] b_h$	$= 0$
I	b_h	$(h.3)\,, h \in \mathcal{H}$	$\lambda_h \left[\begin{smallmatrix} -q \\ Y \end{smallmatrix}\right] + \mu_h D a_h (b_h)$	$= 0$
J_h	μ_h	$(h.4)\,, h \in \mathcal{H}$	$\left(\min\left\{\mu_h^j, a_h^j (b_h)\right\}\right)_{j\in\mathcal{J}_h}$	$= 0$
$G-(S+1)$	$p^{\backslash}$	$(M.x)$	$\sum_{h=1}^{H} \left(x_h^{\backslash} - e_h^{\backslash}\right)$	$= 0$
I	q	$(M.b)$	$\sum_{h=1}^{H} b_h$	$= 0$

(14.6)

Definition 8

$$\Xi \equiv \mathbb{R}_{++}^{GH} \times \mathbb{R}^{(S+1)H} \times \mathbb{R}^{IH} \times \mathbb{R}^{\sum_{h=1}^{H} J_h} \times \mathbb{R}_{++}^{G-(S+1)} \times \mathbb{R}^{I}$$

$$\xi \equiv \left(\left(x_h, \lambda_h, b_h, \mu_h\right)_{h=1}^{H}, p^{\backslash}, q\right) \in \Xi$$

$$F : \Xi \times \mathbb{R}_{++}^{GH} \to \mathbb{R}^{\dim \Xi}, \quad (\xi, e) \mapsto \text{left hand side of (14.6)}$$

$$E \equiv F^{-1}(0).$$

Definition 9 $\xi \equiv \left(\left(x_h, \lambda_h, b_h, \mu_h\right)_{h=1}^{H}, p^{\backslash}, q\right) \in \Xi$ is an **extended financial equilibrium in a numeraire asset - restricted participation model** for the economy $e \in \mathbb{R}_{++}^{GH}$ if $F(\xi, e) = 0$.

According to Theorem 57 in Chapter 7,we identify

f	with	F_e
y	with	0
M	with	Ξ
N	with	$\mathbb{R}^{\dim \Xi}$

To verify the assumptions of the above theorem, we proceed as follows. We first construct the needed homotopy H - see system (14.7), using a Pareto optimal allocation as the test economy . This construction turns out to be more difficult than that one presented in the other chapters where we used the extended approach; the difficulty arising from the nonnegativity constraint. Then, we show the other needed results: there exists $\xi^* \in \Xi$ such that $\{\xi^*\} = G^{-1}(0)$ - Lemma 11 -, $DG(\xi^*)$ has full rank - Lemma 12 -, $H^{-1}(0)$ is compact - Lemma 13.

Consider an arbitrary economy $e \in \mathbb{R}^{CH}_{++}$, a Pareto optimal allocation x^* feasible at e, i.e., such that $\sum_{h=1}^{H} x_h^* = \sum_{h=1}^{H} e_h$. *For given e and* x^*, define

$$\begin{cases} Du_h(x_h) - \lambda_h \Phi(p) & = 0 & (h.1) \\ -\Phi(p)(x_h - ((1-\tau)e_h + \tau x_h^*)) + \left[\begin{smallmatrix} -q \\ Y \end{smallmatrix}\right] b_h & = 0 & (h.2) \\ \lambda_h \left[\begin{smallmatrix} -q \\ Y \end{smallmatrix}\right] + \mu_h (1-\tau) Da_h\left((1-\tau) b_h + \tau \widetilde{b}_h\right) & = 0 & (h.3) \\ \min\left\{\mu_h^j, a_h^j\left((1-\tau) b_h + \tau \widetilde{b}_h\right)\right\}_{j \in \mathcal{J}_h} & = 0 & (h.4) \\ \sum_{h=1}^{H}\left(x_h^{\setminus} - \left((1-\tau) e_h^{\setminus} + \tau x_h^{*\setminus}\right)\right) & = 0 & (M.x) \\ \sum_{h=1}^{H} b_h & = 0 & (M.b) \end{cases} \tag{14.7}$$

where $h = 1, ..., H$, $\widetilde{b}_h$ is defined in (14.3). Define also

$$H_e : \Xi \times [0,1] \to \mathbb{R}^n, \quad (\xi, \tau) \mapsto \text{ left hand side of system (14.7)}$$

and

$$G : \Xi \to \mathbb{R}^{\dim \Xi}, \quad \xi \mapsto H_e(\xi, 1)$$

Observe that

$$H_e(\xi, 0) = F(\xi, e) = F_e(\xi)$$

Remark 10 Equations $(h.1) - (h.4)$ are Kuhn-Tucker condition of the following maximization problem. For given $(p^{\setminus}, q, e)$, (x_h, b_h) solves

$$\max_{(x_h, b_h)} u_h(x_h)$$

$$-\Phi(p)(x_h - ((1-\tau)e_h + \tau x_h^*)) + \left[\begin{smallmatrix} -q \\ Y \end{smallmatrix}\right] b_h = 0 \tag{14.8}$$

$$a_h\left((1-\tau) b_h + \tau \widetilde{b}_h\right) \geq 0$$

We have now to construct *the* equilibrium associated with the economy x^*. Reformulated Lemma 40 in Chapter 8 replacing C with $C(S+1)$, we know that there exists $\left(\theta^*_{\setminus 1}, \gamma^*\right) \in \mathbb{R}^{H-1}_{++} \times \mathbb{R}^{G}_{++}$ such that $\left(x^*, \theta^*_{\setminus 1}, \gamma^*\right)$ is the unique solution to system (8.13) at $\left(\sum_{h=1}^{H} x^*_h, (u_h(x^*_h))_{h\neq 1}\right)$, i.e.,

$$\begin{cases} Du_1(x_1) - \gamma & = 0 \quad (1) \\ \theta_h Du_h(x_h) - \gamma & = 0 \quad (2) \\ (u_h(x_h) - u_h(x^*_h))_{h\neq 1} & = 0 \quad (3) \\ -\sum_{h=1}^{H} x_h + \sum_{h=1}^{H} x^*_h & = 0 \quad (4) \end{cases} \tag{14.9}$$

As done in Chapter 8, let $\theta^*_1 \equiv 1$.

Since by definition of $\widetilde{b}_h$, $\forall h, j$ it is the case that $a^j_h\left(\widetilde{b}_h\right) > 0$, the system $G(\xi) = 0$ is

$$\begin{cases} Du_h(x_h) - \lambda_h \Phi(p) & = 0 \quad (h.1) \\ -\Phi(p)(x_h - x^*_h) + \left[\begin{smallmatrix} -q \\ Y \end{smallmatrix}\right] b_h & = 0 \quad (h.2) \\ \lambda_h \left[\begin{smallmatrix} -q \\ Y \end{smallmatrix}\right] & = 0 \quad (h.3) \\ \mu^j_h & = 0 \quad (h.4) \\ \sum_{h=1}^{H}\left(x^{\setminus}_h - x^{\setminus *}_h\right) & = 0 \quad (M.x) \\ \sum_{h=1}^{H} b_h & = 0 \quad (M.b) \end{cases} \tag{14.10}$$

We are now ready to show that $G^{-1}(0)$ is odd and 0 is a regular value for G. That result is obtained showing that $G^{-1}(0) = \{\xi^*\}$ in Lemma 11 , and that rank $D_\xi G(\xi^*) = \dim \Xi$ in Lemma 12.

Lemma 11 *$\{\xi^*\} = G^{-1}(0)$, where*

$$\xi^* = \left(\begin{array}{c} \left(x^*_h = x^*_h,\ \lambda^*_h = \left(\frac{\gamma^{*sC}}{\theta^*_h}\right)_{s=0}^{S},\ b^*_h = 0, \mu^* = 0\right)_{h\in H}, \\ p^* = \left(\frac{\gamma^{*s}}{\gamma^{*sC}}\right)_{s=0}^{S}, q^* = \sum_{s=1}^{S}\left(\frac{\gamma^{*sC}}{\gamma^{*0C}}\right) y^s \end{array} \right)$$

Proof. The fact $\xi^* \in G^{-1}(0)$ follows simply from the definition of ξ^*, the properties of system (14.9) and system (14.10) itself. Consider an arbitrary $\widehat{\xi} \equiv \left(\widehat{x}, \widehat{\lambda}, \widehat{b}, \widehat{\mu}, \widehat{p}, \widehat{q}\right)$ such that $F\left(\widehat{\xi}, x^*\right) = 0$. The fact that $\widehat{\xi} = \xi^*$ can be shown following closely the proof of the same result in the case of the numeraire asset case - see Lemma 15- after having realized that equation $(h.4)$ in (14.10) imposes that $\mu = 0$. ∎

Lemma 12 rank $D_\xi G(\xi^*) = \dim \Xi$.

Proof. The computation of $DG(\xi^*)$ is described below

$$
\begin{array}{ll|cccccccc}
 & & \cdots & x_h & \lambda_h & b_h & \mu_h & \cdots & p^{\setminus} & q \\
\hline
\cdots & & \ddots & & & & & & & \\
(h.1) & D_{x_h} u_h(x_h) - \lambda_h \Phi(p) & & D^2 u_h(x_h^*) & -\Phi(p^*)^T & & & & -\Lambda_h^* & \\
(h.2) & -\Phi(p)(x_h - x_h^*) + \begin{bmatrix} -q \\ Y \end{bmatrix} b_h & & -\Phi(p^*) & & \begin{bmatrix} -q^* \\ Y \end{bmatrix} & & & & \\
(h.3) & \lambda_h \begin{bmatrix} -q \\ Y \end{bmatrix} + \mu_h Da_h(0) & & & \begin{bmatrix} -q^* \\ Y \end{bmatrix}^T & & [Da_h(0)]^T & & & -\lambda_h^{0*} I \\
(h.4) & \mu_h & & & & & I & & & \\
\cdots & & & & & & & \ddots & & \\
(Mx) & \sum_{h=1}^H \left(x_h^{\setminus} - x_h^{\setminus *} \right) & & \widehat{I} & & & & & & \\
(Mb) & \sum_{h=1}^H b_h & & & & I & & & &
\end{array}
$$

where as in Chapter 11

$$\widehat{I} \equiv \begin{bmatrix} I_{C-1} & 0 & & & \\ & & \ddots & & \\ & & & I_{C-1} & 0 \end{bmatrix}_{[G-(S+1)]\times G}$$

We can use the matrix I in superrow of μ_h to clean up the column of μ_h. We can then erase those superrow and supercolumn. Then we get the same matrix as in the case of numeraire asset economy, which has full rank from Lemma 18 in Chapter 11. ∎

Lemma 13 $H_e^{-1}(0)$ *is compact.*

Proof. We want to show that any sequence in $H_e^{-1}(0)$ admits a convergent subsequence inside that set. Due to Remark 10, we can show that $(x^n, \lambda^n, b^n, p^{\backslash n})_{n\in\mathbb{N}}$ converges to $(\overline{x}, \overline{\lambda}, \overline{b}, \overline{p}^{\backslash})$ following the same strategy used in the case of the model with numeraire assets - see Lemma 19 in Chapter 11. The convergence of $(\mu^n, q^n)_{n\in\mathbb{N}}$ is proved below

1. $(q^n)_{n\in\mathbb{N}}$ converges.

From Assumption re2.5, for every asset i, there exists a consumer h' such that $Da_{h'}^j\left((1-\tau)\, b_{h'} + \tau\widetilde{b}_{h'}\right) = 0$. Then $(h.3)$ in (14.7) can be written as

$$0 = \lambda_{h'}\left[\begin{smallmatrix}-q\\ Y\end{smallmatrix}\right] + \mu_{h'} Da_{h'}^j\left((1-\tau)\, b_{h'} + \tau\widetilde{b}_{h'}\right) =$$

$$= \lambda_{h'}\begin{bmatrix} -q^1 & \dots & -q^i & \dots & q^I \\ {}^1y & \dots & {}^iy & \dots & {}^Iy \end{bmatrix} +$$

$$+\mu_{h'}\begin{bmatrix} \left[D_{b_{h'}^1} a_{h'}^j\left((1-\tau)\, b_{h'} + \tau\widetilde{b}_{h'}\right)\right]^T & \dots & \left[D_{b_{h'}^I} a_{h'}^j\left((1-\tau)\, b_{h'} + \tau\widetilde{b}_{h'}\right)\right]^T \end{bmatrix}$$

which implies that for each n,

$$\lambda_{h'}^n\begin{bmatrix} q^{in} \\ {}^iy \end{bmatrix} = 0$$

and then

$$q^{in} \to \frac{\sum_{s=1}^{S}\overline{\lambda}_{h'}^s y^{si}}{\overline{\lambda}_h^0} = \overline{q}^i$$

2. $(\mu^n)_{n\in\mathbb{N}}$ converges.

Case 1. $\overline{\tau} \in [0, 1)$

Let $\left\{\mathcal{J}_h^A, \mathcal{J}_h^B\right\}$ be a partition of the set of indices $\mathcal{J}_h$ such that

$$\mathcal{J}_h^A = \left\{j \in \mathcal{J}_h : a_h^j\left((1-\overline{\tau})\,\overline{b}_h + \overline{\tau}\widetilde{b}_h\right) = 0\right\} \quad \text{and}$$

$$\mathcal{J}_h^B = \left\{j \in \mathcal{J}_h : a_h^j\left((1-\overline{\tau})\,\overline{b}_h + \overline{\tau}\widetilde{b}_h\right) > 0\right\}$$

with cardinality J_h^A and J_h^B, respectively. If $j \in \mathcal{J}_h^B$, there exists a n^* such that, for every $n > n^*$, $a_h^j\left(\left(1-\tau^n\right)b_h^n + \tau^n \widetilde{b}_h\right) > 0$. Hence for every $n > n^*$ we have $\mu_h^{jn} = 0$, i.e., $\mu_h^{jn} \to 0$ for every $j \in \mathcal{J}_h^B$.
If $j \in \mathcal{J}_h^A$, from Assumption re2.4, rank $\left(Da_h^{\mathcal{J}^A} a_h^j\left(\left(1-\overline{\tau}\right)\overline{b}_h + \overline{\tau}\widetilde{b}_h\right)\right) = J_h^A$. Then, there exists $n' > \hat{n}$ such that $\left|det\ M\left(b_h^n\right)\right| > 0$ for every $n > n'$, where $M\left(b_h^n\right)$ is a well chosen square J_h^A-dimensional submatrix of

$$Da_h\left(\left(1-\tau^n\right)b_h^n + \tau^n \widetilde{b}_h\right)$$

Let us take $n^{**} = \max\{n^*, n'\}$. Making the needed permutations of rows and columns of $Da_h\left(\left(1-\tau^n\right)b_h^n + \tau^n \widetilde{b}_h\right)$ in order to have $M\left(b_h^n\right)$ in the top-left corner and performing analogous permutation of the elements of μ_h and $\eta \equiv \left(\lambda_h \left[\begin{smallmatrix}-q\\Y\end{smallmatrix}\right]\right)$ in order to have the equalities in equation $(h.3)$ of (14.7) still satisfied we get:

$$\left[\mu_h^{An} \mid \mu_h^{Bn}\right]\begin{bmatrix} M\left(b_h^n\right) & M_{12}\left(b_h^n\right) \\ M_{21}\left(b_h^n\right) & M_{22}\left(b_h^n\right)\end{bmatrix} = \begin{bmatrix}\eta^{In} \\ \eta^{IIn}\end{bmatrix}$$

where $\mu_h^{An} \in \mathbb{R}^{J_h^A}$, $\mu_h^{Bn} \in \mathbb{R}^{J_h - J_h^A}, \eta^{In} \in \mathbb{R}^{J_h^A}, \eta^{IIn} \in \mathbb{R}^{I-\left(J_h^A\right)}$ and the partition of $Da_h\left(\left(1-\tau^n\right)b_h^n + \tau^n \widetilde{b}_h\right)$ is displayed below

$$\begin{array}{cc} & \begin{array}{cc} J_h^A & I - J_h^A \end{array} \\ \begin{array}{c} J_h^A \\ J_h - J_h^A \end{array} & \begin{pmatrix} M\left(b_h^n\right) & M_{12}\left(b_h^n\right) \\ M_{21}\left(b_h^n\right) & M_{22}\left(b_h^n\right) \end{pmatrix} \end{array}$$

Since $\mu_h^{Bn} = 0$, we have

$$\mu_h^{An}\left[D_{b_h^I} a^{\mathcal{J}_A}\left(\left(1-\tau^n\right)b_h^n + \tau^n \widetilde{b}_h\right)\right] = \eta^{In}$$

and

$$\mu_h^{An}\left[M\left(b_h^n\right)\right] = \eta^{In}$$

and also

$$\mu_h^{An} = \eta^{In}\left[M\left(b_h^n\right)\right]^{-1}$$

from which $\mu_h^{An} \to \overline{\mu}_h^A$.

Case 2. $\overline{\tau} = 1$

In this case, for every $h = 1, ..., H$

$$a_h^j\left(\left(1-\tau^n\right)b_h^n + \tau^n \widetilde{b}_h\right) \to a_h^j\left(\widetilde{b}_h\right) > 0$$

and hence $\mu_h^n \to 0$. ■

On the basis of the above lemma, we finally get the following Theorem.

Theorem 14 *For every economy* $e \in \mathbb{R}^{GH}_{++}$ *, there exists* $\xi \in \Xi$ *such that* $F(\xi, e) = 0$.

3. Regular economies

In this section we first present the definition of regular economies and then discuss the main problem we encounter in trying to prove the theorems concerning their properties we have shown in the other models.

Definition 15 $e \in \mathbb{R}^{CH}_{++}$ is a regular economy for the restricted participation model if 0 is a regular value for F_e. Let $\mathcal{R}$ be the set of regular economies.

As explained in detail at the beginning of Section 4 in Chapter 11, besides existence of equilibria, the two crucial steps to get the desired results are

1 0 is a regular value for F, and therefore $F^{-1}(0)$ is a manifold , and there exists an open and full measure subset $\mathcal{R}$ of $\mathbb{R}^{GH}_{++}$ such that

$$\forall\, e^* \in \mathcal{R},\ \ \forall \xi^* \in F^{-1}_{e^*}(0),\ \ \text{rank}\, DF_{e^*}(\xi^*) \text{ is full} \tag{14.11}$$

2 $pr : F^{-1}(0) \to \mathbb{R}^{GH}_{++}, (\xi, e) \mapsto e$ is proper.

As we have seen in the other models we analyzed, a preliminary step is to show that the jacobian matrix of the left hand side of the first order conditions of the household maximization problem has full row rank. In the case of inequality constraints, one of the equations of Kuhn-Tucker conditions is expressed in terms of min between each constraint and the associated multiplier. The obvious problem is that the min function is not even differential where both constraint and multiplier are equal to zero, a sort of "border line" case. To be able to show result (14.11), we have therefore to show that border line cases occur outside an open and full measure subset Π^* of the economy space. As usual that result is obtained showing the properness of the projection from the equilibrium set to the economy space and then verifying a rank condition crucial as an application of the transversality theorem (see Theorem 26 and Corollary 27 , see Theorem 26in Chapter 6). The further step is to show that in an open and full measure subset $\mathcal{R}$ of Π^* condition (14.11) does hold. That step turns out to be an easy consequence of the first one.

Before showing the desired result, let's introduce a crucial definition.

Definition 16 Given $(\xi, e) \in F^{-1}(0)$, household h is at a border line case in $\overline{\mathcal{J}}_h \subseteq \mathcal{J}_h$ if

$$\text{for each } j \in \overline{\mathcal{J}}_h,\ \ a^j_h(b_h) = 0 \ \ \underline{\text{and}} \ \ \mu^j_h = 0$$

Household h is in a border line case if the above $\overline{\mathcal{J}}_h \neq \emptyset$.

In other words, given $(\xi, e) \in F^{-1}(0)$, for each household h, we consider the following partition of $\mathcal{J}_h$:

$$\begin{array}{l} \mathcal{J}_h^1(\xi, e) = \{j \in \mathcal{J}_h : a_h^j(b_h) > 0, \mu_h^j = 0\} \\ \mathcal{J}_h^2(\xi, e) = \{j \in \mathcal{J}_h : a_h^j(b_h) = 0, \mu_h^j > 0\} \\ \mathcal{J}_h^3(\xi, e) = \{j \in \mathcal{J}_h : a_h^j(b_h) = 0, \mu_h^j = 0\} \end{array} \tag{14.12}$$

We can then say that, given $(\xi, e) \in F^{-1}(0)$, household h is at a *border line case* in $\overline{\mathcal{J}}_h \subseteq \mathcal{J}_h$ if $\overline{\mathcal{J}}_h = \mathcal{J}_h^3(\xi, e)$.

Lemma 17 *There exists an open and full measure subset Π^* of $\mathbb{R}^{GH}_{++}$ such that F is C^1 on Π^*.*

Proof. Define

$$E \equiv F^{-1}(0)$$

$$B \equiv \left\{(\xi, e) \in E : \exists\, h \text{ such that } \mathcal{J}_h^3(\xi, e) \neq \emptyset\right\}$$

$$pr : E \to \mathbb{R}^{GH}_{++}, \qquad pr : (\xi, e) \mapsto e$$

We want to prove that $pr(B)$ is closed and of measure zero in $\mathbb{R}^{GH}_{++}$.

For each h, define

- $\mathcal{P}(\mathcal{J}_h)$ as the set of all subsets of $\mathcal{J}_h$ with generic element $\mathcal{J}_h'$;
- $\widetilde{\mathcal{J}} \equiv \times_{h=1}^H \mathcal{P}(\mathcal{J}_h)$ with generic element $\mathcal{J}' \equiv (\mathcal{J}_h')_{h=1}^H$.

Consider the model in which household $h's$ maximization problem contains the constraint

$$\left(a_h^j(b_h)\right)_{j \in \mathcal{J}_h'} = 0 \text{ instead of } a_h(b_h) \geq 0 \tag{14.13}$$

Consider the associated extended equilibrium system, in which we added the "missing" variables $\left(\mu_h^j\right)_{j \in \mathcal{J}_h \setminus \mathcal{J}_h'}$ as identically equal to zero, to be able to compare solutions of that system with equilibria.

$$\left\{\begin{array}{lll} Du_h(x_h) - \lambda_h \Phi(p) & = 0 & (h.1) \\ -\Phi(p)(x_h - e_h) + \left[\begin{smallmatrix} -q \\ Y \end{smallmatrix}\right] b_h & = 0 & (h.2) \\ \lambda_h \left[\begin{smallmatrix} -q \\ Y \end{smallmatrix}\right] + \left(\mu_h^j\right)_{j \in \mathcal{J}_h'} Da_h^{\mathcal{J}_h'}(b_h) & = 0 & (h.3) \\ \left(\mu_h^j\right)_{j \in \mathcal{J}_h \setminus \mathcal{J}_h'} & = 0 & (h.4) \\ a_h^{\mathcal{J}_h'}(b_h) & = 0 & (h.5) \\ \sum_{h=1}^H \left(x_h^\setminus - e_h^\setminus\right) & = 0 & (M.x) \\ \sum_{h=1}^H b_h & = 0 & (M.b) \end{array}\right. \tag{14.14}$$

Define

$$F_{\mathcal{J}'} : \Xi \times \mathbb{R}^{GH}_{++} \to \mathbb{R}^{\dim \Xi}, \quad (\xi, e) \mapsto \text{left hand side of (14.14)}$$

$E(\mathcal{J}') \equiv E\left((\mathcal{J}'_h)_{h=1}^H\right) \equiv (F_{\mathcal{J}'})^{-1}(0)$ and $\xi \in \left(F_{\mathcal{J}'|e}\right)^{-1}(0)$ a fictitious equilibrium associated with e.

Observe that the multipliers μ_h associated with constraints $(h.4)$ in (14.14) can be negative. Define $(\xi, e) \in F^{-1}(0)$ an equilibrium pair.

We now want to prove that

$$E \subseteq \bigcup_{\mathcal{J}' \in \widetilde{\mathcal{J}}} E(\mathcal{J}') \tag{14.15}$$

Just inspecting system (14.6), we see that (ξ, e) is also a solution to

$$\begin{cases} Du_h(x_h) - \lambda_h \Phi(p) & = 0 \quad (h.1) \\ -\Phi(p)(x_h - e_h) + \left[\begin{smallmatrix} -q \\ Y \end{smallmatrix}\right] b_h & = 0 \quad (h.2) \\ \lambda_h \left[\begin{smallmatrix} -q \\ Y \end{smallmatrix}\right] + \mu_h D_{b_h} a_h(b_h) & = 0 \quad (h.3) \\ \left(\mu_h^j\right)_{j \in \mathcal{J}_h^1(\xi,e) \cup \mathcal{J}_h^3(\xi,e)} & = 0 \quad (h.4.1) \\ \left(a_h^j(b_h)\right)_{j \in \mathcal{J}_h^2(\xi,e) \cup \mathcal{J}_h^3(\xi,e)} & = 0 \quad (h.4.2) \\ \sum_{h=1}^H \left(x_h^\backslash - e_h^\backslash\right) & = 0 \quad (M.x) \\ \sum_{h=1}^H b_h & = 0 \quad (M.b) \end{cases} \tag{14.16}$$

Therefore, if $(\xi, e) \in F^{-1}(0)$, then there exists $\left((\mathcal{J}'_h)_{h=1}^H\right)$ such that $(\xi, e) \in E(\mathcal{J}')$, i.e., it is solve a particular system (14.14). In fact it is enough to take, for each h

$$\mathcal{J}'_h = \mathcal{J}_h^2(\xi, e) \cup \mathcal{J}_h^3(\xi, e)$$

Observe that since in system (14.16) $\left(\mu_h^j\right)_{j \in \mathcal{J}_h^1(\xi,e)} = 0$, then

$$\lambda_h \begin{bmatrix} -q \\ Y \end{bmatrix} + \mu_h D a_h(b_h) = \lambda_h \begin{bmatrix} -q \\ Y \end{bmatrix} + \left(\mu_h^j\right)_{j \in \mathcal{J}'_h} D a_h^{\mathcal{J}'_h}(b_h)$$

and system (14.16) is the same as system (14.14).

Therefore, if $(\xi, e) \in E$, then there exists $\mathcal{J}' \in \widetilde{\mathcal{J}}$ such that $(\xi, e) \in E(\mathcal{J}')$, and (14.15) follows.

Consider $\mathcal{J}' \in \widetilde{\mathcal{J}}$. Then for each $\mathcal{J}'_h$ take an arbitrary subset $\mathcal{J}_h^3$ with at least one of those which is nonempty, i.e., define

$$\mathcal{J}^3(\mathcal{J}') \equiv \left\{ (\mathcal{J}_h^3)_{h=1}^H \in \widetilde{\mathcal{J}} : \ \forall h \ \mathcal{J}_h^3 \subseteq \mathcal{J}'_h \text{ and } \exists\, h' \text{ such that } \mathcal{J}_{h'}^3 \neq \emptyset \right\}$$

Then consider the set

$$E_B\left(\mathcal{J}',\mathcal{J}^3\right)\equiv$$
$$\equiv\left\{(\xi,e)\in E\left(\mathcal{J}'\right):\text{for each }h,\ \left(\mu_h^j\right)_{j\in\mathcal{J}_h^3}=0\text{ with }\left(\mathcal{J}_h^3\right)_{h=1}^H\in\mathcal{J}^3\left(\mathcal{J}'\right)\right\} \tag{14.17}$$

i.e., $E_B\left(\mathcal{J}',\mathcal{J}^3\right)$ is the set of fictitious equilibria (characterized by $(\mathcal{J}')$) in which at least one household h' is at a border line case in $\mathcal{J}_{h'}^3\neq\emptyset$.

By definition of $E_B\left(\mathcal{J}',\mathcal{J}^3\right)$, we have that for each h and j,

$$a_h^j\left(b_h\right)=0\text{ if and only if }j\in\mathcal{J}_h'$$

and

$$\mu_h'=0\text{ if }j\in\mathcal{J}_h^3$$

Observe that

1. Consider a true equilibrium pair $(\xi,e)\in E$ for which at least one household h' is at a border line case. For each h, take

$$\begin{aligned}J_h^3&=J_{3h}\left(\xi,e\right)\\J_h'&=J_{2h}\left(\xi,e\right)\cup J_{3h}\left(\xi,e\right)\end{aligned} \tag{14.18}$$

Then $(\xi,e)\in E_B\left(\left(J_h',J_h^3\right)_{h=1}^H\right)$.

2. For an arbitrary $(\xi',e')\in E_B\left(\left(J_h',J_h^3\right)_{h=1}^H\right)$, it could be that if $j\in J_h\backslash J_h'$, then $a_h^j\left(b_h'\right)<0$, i.e., it is not the case that $(\xi',e')\in E$.

3. Moreover, $\mathcal{J}_h^3$ is the number of added equations of the form $\mu_h^j=0$ to system (14.14) to get a system whose solutions coincide with the set (14.17). Therefore, keeping in mind that "the more equations, the smaller the solution set", if for each h, $\widehat{\mathcal{J}}_h^3\subseteq\mathcal{J}_h^3$, it is the case that $E_B\left(\mathcal{J}',\widehat{\mathcal{J}}^3\right)\supseteq E_B\left(\mathcal{J}',\mathcal{J}^3\right)$. That simple observation explains the following set inclusions.

From Observation 1. above,

$$B\subseteq\bigcup_{\mathcal{J}'\in\widetilde{\mathcal{J}}}\ \bigcup_{\mathcal{J}^3\in\mathcal{J}^3(\mathcal{J}')}E_B\left(\mathcal{J}',\mathcal{J}^3\right)$$

Now given $\mathcal{J}'\in\widetilde{\mathcal{J}}$ and $\left(\mathcal{J}_h^3\right)_{h=1}^H\in\mathcal{J}^3\left(\mathcal{J}'\right)$, take h' and j' such that $j'\in\mathcal{J}_{h'}^3\neq\emptyset$ and define $\mathcal{J}_{h'j'}^3\equiv\left(\mathcal{J}_{hh'j'}^3\right)_{h=1}^H$ as follows

$$\mathcal{J}_{hh'j'}^3=\begin{cases}\{j'\}\subseteq\mathcal{J}_{h'}^3\neq\emptyset & \text{if }h=h'\\ \emptyset & \text{otherwise}\end{cases}$$

Define also

$$E_B\left(\left(\mathcal{J}'\right),\left(h',j'\right)\right)\equiv E_B\left(\mathcal{J}',\mathcal{J}_{h'j'}^3\right)$$

i.e., the set of fictitious equilibrium pairs for which household h' is at a border line case in $\{j\} \subseteq \mathcal{J}^3_{h'}$. Then clearly, for each $\mathcal{J}' \in \widetilde{\mathcal{J}}$, $\mathcal{J}^3 \in \mathcal{J}^3(\mathcal{J}')$, h' and j' such that $j' \in \mathcal{J}^3_{h'} \neq \emptyset$,

$$E_B(\mathcal{J}', \mathcal{J}^3) \subseteq E_B(\mathcal{J}', (h', j'))$$

and

$$B \subseteq \bigcup_{\mathcal{J}' \in \widetilde{\mathcal{J}}} \bigcup_{h'=1}^{H} \bigcup_{j' \in \mathcal{J}'_h} E_B(\mathcal{J}', (h', j'))$$

Now since

$$pr(B) \subseteq \bigcup_{\mathcal{J}' \in \widetilde{\mathcal{J}}} \bigcup_{h'=1}^{H} \bigcup_{j' \in \mathcal{J}'_h} pr(E_B(\mathcal{J}', (h', j')))$$

and the number of sets involved in the above unions is finite, it is enough to show that, taken an arbitrary h', say equal to H, and an arbitrary $j \in \mathcal{J}_H$, say equal to J_H, the set

$$pr(E_B(\mathcal{J}', (H, J_H)))$$

is closed and of measure zero.

$e \in pr(E_B(\mathcal{J}', J_H))$ if there exists ξ such that (ξ, e) is a solution to system

$$\begin{cases} Du_h(x_h) - \lambda_h \Phi(p) & = 0 \\ -\Phi(p)(x_h - e_h) + \left[\begin{smallmatrix} -q \\ Y \end{smallmatrix}\right] b_h & = 0 \\ \lambda_h \left[\begin{smallmatrix} -q \\ Y \end{smallmatrix}\right] + \left(\mu_h^j\right)_{j \in \mathcal{J}'_h} Da_h^{\mathcal{J}'_h}(b_h) & = 0 \\ \left(\mu_h^j\right)_{j \in \mathcal{J}_h \setminus \mathcal{J}'_h} & = 0 \\ a_h^{\mathcal{J}'_h}(b_h) & = 0 \\ \ldots & \\ Du_H(x_H) - \lambda_H \Phi(p) & = 0 \\ -\Phi(p)(x_H - e_H) + \left[\begin{smallmatrix} -q \\ Y \end{smallmatrix}\right] b_H & = 0 \\ \lambda_H \left[\begin{smallmatrix} -q \\ Y \end{smallmatrix}\right] + \left(\mu_H^j\right)_{j \in \mathcal{J}'_H} Da_H^{\mathcal{J}'_H}(b_H) & = 0 \\ \left(\mu_H^j\right)_{j \in \mathcal{J}_H \setminus \mathcal{J}'_H} & = 0 \\ a_H^{\mathcal{J}'_H}(b_H) & = 0 \\ \sum_h \left(x_h^{\backslash} - e_h^{\backslash}\right) & = 0 \\ \sum_h b_h & = 0 \\ \mu_H^{J_H} & = 0 \end{cases} \tag{14.19}$$

Defined

$$F_{BH} : \Xi \times \mathbb{R}^{GH}_{++} \to \mathbb{R}^{\dim \Xi + 1}, \quad F_{BH} : (\xi, e) \mapsto \text{left hand side of system (14.19)}$$

we want to show that $F_{BH}(\xi, e) = 0$ has solutions for each e in a closed and measure zero subset of $\mathbb{R}^{GH}_{++}$.

The measure zero result follows from Lemma 19 below and the transversality theorem, see Theorem 26 in Chapter 6 , see Theorem 26. Closedness follows from the fact that $E_B(\mathcal{J}', J_H)$ is closed in the closed set E and from Lemma 18. Then $\Pi^* = \mathbb{R}^{GH}_{++} \backslash pr(B)$ and we have proved the desired result. ■

Lemma 18 *The function* $pr : F^{-1}(0) \to \mathbb{R}^{GH}_{++}$, $(\xi, e) \mapsto e$ *is proper.*

Proof. Using a similar argument of those we have already presented in Lemma 13 we get the desired result. ■

Lemma 19 *zero is a regular value for* F_{BH}.

Proof. The computation of $D_{\left(\xi,\left(e_h\right)_{h=1}^{H},e_H\right)} F_{BH}$ is displayed below

$$
\begin{array}{r}
\\ \\
(h.1)\\
(h.2)\\
(h.3)\\
(h.4)\\
(h.5)\\
(H.1)\\
(H.2)\\
(H.3)\\
(H.4)\\
(H.5)\\
(M.x)\\
(M.b)\\
(H.7)
\end{array}
\begin{array}{cccccccccccccc}
x_h & \lambda_h & b_h & \left(\mu_h^j\right)_{j\in\mathcal{J}_h\setminus\mathcal{J}'_h} & \left(\mu_h^j\right)_{j\in\mathcal{J}'_h} & x_H & \lambda_H & b_H & \left(\mu_H^j\right)_{j\in\mathcal{J}_H\setminus\mathcal{J}'_H} & \left(\mu_H^j\right)_{j\in\mathcal{J}'^{\setminus}_H} & \mu_H^{J_H} & e_H^{\setminus} & e_1^{\bullet} & e_H^{\bullet}
\end{array}
$$

$$
\left(
\begin{array}{cccccccccccccc}
D^2u_h & -\Phi(p)^T & & & & & & & & & & & & \\
-\Phi(p) & & \begin{bmatrix}-q\\ Y\end{bmatrix} & & & & & & & & & & I & \\
 & \begin{bmatrix}-q\\ Y\end{bmatrix}^T & L_h & & \left[Da_h^{\mathcal{J}'_h}\right]^T & & & & & & & & & \\
 & & & I & & & & & & & & & & \\
 & & Da_h^{\mathcal{J}'_h} & & & & & & & & & & & \\
 & & & & & D^2u_H & -\Phi(p)^T & & & & & & & \\
 & & & & & -\Phi(p) & & \begin{bmatrix}-q\\ Y\end{bmatrix} & & & & \Phi(p)^{\setminus} & & I \\
 & & & & & & \begin{bmatrix}-q\\ Y\end{bmatrix}^T & L_H & & \left[Da_H^{\mathcal{J}'^{\setminus}_H}\right]^T & \left[Da_H^{J_H}\right]^T & & & \\
 & & & & & & & & I & & & & & \\
 & & & & & & & Da_H^{\mathcal{J}'_H} & & & & & & \\
\widehat{I} & & & & & \widehat{I} & & & & & & -I & & \\
 & & I & & & & & I & & & & & & \\
 & & & & & & & & & & 1 & & &
\end{array}
\right)
$$

where $\mathcal{J}_H'^{\setminus} \equiv \mathcal{J}_H' \setminus \{J_H\}$, and
L_h is the following $I \times I$ symmetric matrix

$$\begin{bmatrix} \sum_{j\in\mathcal{J}_h'} \mu_h^j D^2_{b_h^1} a_h^j & \cdots & \sum_{j\in\mathcal{J}_h'} \mu_h^j D^2_{b_h^1, b_h^I} a_h^j \\ & \ddots & \\ \sum_{j\in\mathcal{J}_h'} \mu_h^j D^2_{b_h^I b_h^1} a_h^j & & \sum_{j\in\mathcal{J}_h'} \mu_h^j D^2_{b_h^I} a_h^j \end{bmatrix}$$

Observe that from Assumption re2.5, if $h = h(i) \equiv i$ we have that

$$\forall j \in \mathcal{J}_i,\ \forall b_i \in \mathbb{R}^I,\ D_{b_i^i} a_i^j (b_i) = 0$$

and therefore

$$\forall j \in \mathcal{J}_i,\ \ \forall i',\ \forall b_i \in \mathbb{R}^I,\ D_{b_i^i, b_i^{i'}} a_i^j (b_i) = 0 = D_{b_i^{i'}, b_i^i} a_i^j (b_i)$$

i.e., the i-th rows and columns of L_i are equal to zero.

The desired full rank result is proved below.

$$(h.1) \leftarrow x_h$$

$$(h.2) \leftarrow e_h$$

$$\begin{array}{lll} (h.3) \leftarrow \lambda_h & \rightsquigarrow (h.1) & \\ (h.1) & \leftarrow x_h & \rightsquigarrow (h.2) \\ (h.2) & \leftarrow e_h^{\bullet} & \end{array}$$

$$(h.4) \leftarrow \left(\mu_h^j\right)_{j\in\mathcal{J}_h \setminus \mathcal{J}_h'}$$

$$\begin{array}{ll} (h.5) \leftarrow b_h & \rightsquigarrow (h.2), (h.3) \\ (h.3) \leftarrow \lambda_h & \rightsquigarrow (h.1) \\ (h.1) \leftarrow x_h & \rightsquigarrow (h.2) \\ (h.2) \leftarrow e_h^{\bullet} & \end{array}$$

In the above perturbations, we used Assumption re2.4..

$$\begin{array}{ll} (M.x) \leftarrow e_H^{\setminus} & \rightsquigarrow (H.2) \\ (H.2) \leftarrow e_H^{\bullet} & \end{array}$$

$$\begin{array}{ll} (M.b) \leftarrow \left(b_{h(i)}\right)_{i=1}^{I} & \rightsquigarrow (h(i).2) \\ (h(i).2) \leftarrow e_{h(i)}^{\bullet} & \end{array}$$

Observe that because of Assumption re2.5, equations $(h(i).5)$ and $(h(i).3)$ are not disturbed.

$$\begin{array}{ll} (H7) \leftarrow \mu_h^{J_H} & \rightsquigarrow (H.3) \\ (H.3) \leftarrow \lambda_H & \rightsquigarrow (H.1) \\ (H.1) \leftarrow x_H & \rightsquigarrow (H.2), (M.x) \\ (M.x) \leftarrow e_H^{\setminus} & \rightsquigarrow (H.2) \\ (H.2) \leftarrow e_h^{\bullet} & \end{array}$$

Observe that we can perturb $(h.1)$ with x_h since $D^2 u_h$ is negative definite, from Assumption u3'. ∎

We can now present the main result of the section.

Proposition 20 *There exists an open and full measure subset $\mathcal{R}$ of $\mathbb{R}^{GH}_{++}$ such that*

$$\forall\, e^* \in \mathcal{R},\ \ \forall \xi^* \in F_{e^*}^{-1}(0),\ \ \text{rank}\, DF_{e^*}(\xi^*) \text{ is full} \qquad (14.20)$$

Proof. Restrict F to the open and full measure subset Π^* of $\mathbb{R}^{GH}_{++}$. Then the result follow from Lemma 18 and the same argument presented in Lemma 19. ∎

Chapter 15

PLANNER INTERVENTION ON THE MARKET OUTCOME

1. Introduction

In Chapter 8 we have shown that equilibrium allocations in economies with complete (financial) markets are Pareto optimal . It is also well-known that with incomplete markets, equilibrium allocations typically are not Pareto optimal, as [67] first suggested and as we show for the cases of numeraire, nominal and real assets, and restricted participation in Section 2 [1]. The above result is relatively intuitive: while in the incomplete market framework households face multiple budget constraints, Pareto optimality presumes a planner who has access to the missing markets and therefore has a higher freedom in reallocating goods. It is then quite natural to investigate if equilibria are at least some sort of second best or, as many authors say, if they are *constrained* optimal. In other words, we want to compare equilibria with allocations that can be implemented by a planner who is *constrained* to use only existing assets. More precisely, we want to define what the planner may or may not do consistently with presence of incomplete markets and then compare equilibria without and with planner intervention. If equilibria of the latter type are Pareto superior to equilibria of the former type, then we say that equilibria are constrained *suboptimal* with respect to the chosen type of planner intervention. In that case, we also say that the planner can Pareto improve upon the competitive equilibria. As we are going to show in Section 6, typically in the space of economies the planner *can* Pareto improve using "few" taxes and subsidies. The intuition for that result is as follows. Even that limited good reallocations have

two effects on an individual's utility - a direct effect from the income transfer and an indirect effect due to the relative price change in the commodity spot markets. When

[1] See caveat about the result on real assets at the very beginning of Subsection 2.3.

markets are complete, the income reallocation caused by the price change can be decomposed into a combination of assets that have already been priced by the market. When the asset market is incomplete, it is generically the case that the price changes will cause an income redistribution that the market itself could not directly implement. In essence, the central planner has access to a wider class of assets than those directly traded.[2]

The first paper dealing with constrained suboptimality in a full fledged general equilibrium model is that one by Geneakoplos and Polemarchakis [61]. In this chapter we follow the approach presented in the paper by [36] (from now on CKV) - see also [30] and [31]- which has also been used by several other papers[3].

The methodology we describe can be applied to the following general.

- Consider a phenomenon which is absent in the standard general equilibrium model described in Chapters 8 and 9. Incomplete markets, asymmetric information, public goods, some form of strategic interaction and any kind of problem which goes under the name of "imperfection" are examples of the above mentioned phenomena.
- Construct a general equilibrium model which incorporates one of those imperfections. Chapters 11, 12, 13 and 14 do that exercise for the case of incomplete markets.
- Show existence of equilibria.
- Choose some economically relevant function of the equilibrium variables - the goal function. A very natural choice would be households' utility levels computed at the equilibrium allocations. A plethora of other choices is possible.
- Define some sort of planner intervention who uses a set of policy tools and satisfies constraints related to the assumption that the imperfection cannot be overthrown by the planner, and to the specific nature of the intervention.
- Show that in a large set of economies, the planner is able to change the value taken by the goal function.

It is crucial to warn the reader that all our analysis is done in local terms: the above mentioned changes are shown to be possible in a neighborhood of the goal function values "without planner intervention".

The chapter is organized as follows. In Section 2 we show that when markets are incomplete and irrespectively of the asset nature, typically equilibria are not

[2]The quotation is from [61]. In fact, they give intuition for a different kind of government intervention, but their observations apply to our case as well.
[3]See [34], [35], [38], [39], [123].

Pareto optimal. In Section 3, we first carefully define the problem we want to analyze. Then, we present a methodology to study that problem, and which consists of a well defined set of steps to be followed. As it will be clear, the main thrust in the proof is to show that a well chosen system of equations has no solutions. For a better understanding of the proposed strategy of proof, the relatively simple and already mentioned example of a Pareto improving intervention via few taxes and transfers in an incomplete market model is carefully spelled out.

2. Generic Pareto non-optimality

In this section we show that in each of the four models with incomplete markets we analyzed in previous chapters , typically equilibria are not Pareto optimal. The general strategy is to present a necessary condition for Pareto optimality and then, using the properness of the "natural projection" and a transversality argument, to show that necessary condition does not hold for all equilibria associated to each element in an open and full measure subset of the endowment space. In this section, we make Assumptions u1, u2 and u4 about utility functions - see Subsection 2.1 in Chapter 10 - and we substitute u3 (differentiably strictly quasiconcavity) with u3' (differentiably strictly concavity, i.e., Du_h is negative definite).

2.1 Numeraire assets

Lemma 1 *Consider the numeraire asset model presented in Chapter 11 and let be* $F : \Xi \times \mathbb{R}^{GH}_{++} \to \mathbb{R}^{\dim \Xi}$*, the function describing equilibria (see Definition 12 in Chapter 11). If* $\xi = (x, \lambda, b, p^{\backslash}, q)$ *is such that* $F(\xi, e) = 0$ *and* x *is a Pareto optimal allocation, then for any* s, s' *and* h, h'*,*

$$\frac{\lambda_h^s}{\lambda_h^{s'}} = \frac{\lambda_{h'}^s}{\lambda_{h'}^{s'}} \tag{15.1}$$

Proof. Since $F\left(x, \lambda, b, p^{\backslash}, q, e\right) = 0$, then $Du_h(x_h) - \lambda_h \Phi\left(p\right) = 0$ and therefore for $s = 0, 1, ..., S$,

$$D_{x^{sC}} u_h(x_h) - \lambda_h^s = 0$$

Since x is a Pareto optimal allocation, from Lemma 40 in Chapter 8 adapted to the case of $C\left(S+1\right)$ goods[4], for each h and for $s = 0, 1, ..., S$, there exists $\gamma^{sC*} \in \mathbb{R}_{++}$ such that

$$D_{x^{sC}} u_h(x_h) - \gamma^{sC*} = 0$$

[4]See also system (11.12) in Chapter 11.

Therefore, $\lambda_h^s = \gamma^{sC*}$, and

$$\frac{\lambda_h^s}{\lambda_h^{s'}} = \frac{\gamma^{sC*}}{\gamma^{s'C*}} = \frac{\lambda_{h'}^s}{\lambda_{h'}^{s'}}$$

■

Theorem 2 *There exists an open and full measure subset D of the endowment space $\mathbb{R}_{++}^{GH}$ such that for every $e \in D$, all equilibria associated with e are not Pareto optimal.*

Proof. As usual the full measure result comes from the transversality result (see Theorem 26 in Chapter 6) and the openness from the properness of the natural projection. Since the latter result was shown in Chapter 11 - see Lemma 24 - here we present just the former result.

Define

$$F_{nu} : \Xi \times \mathbb{R}_{++}^{GH} \to \mathbb{R}^{\dim \Xi + 1}, \quad (\xi, e) \mapsto \left(F(\xi, e), \lambda_H^1 - \frac{\lambda_1^1}{\lambda_1^0} \lambda_H^0 \right)$$

It is enough to show that zero is a regular value for F_{nu} to get the desired result. If that is the case it follows that in an open and full measure subset D of $\mathbb{R}_{++}^{GH}$, the system $F_{nu}(\xi, e) = 0$ has no solutions. Since, from the existence theorem, we know that for each $e \in \mathbb{R}_{++}^{GH}$, $F(\xi, e) = 0$ does have solutions, we get that

$$\forall e \in D \text{ and } \forall \xi \in F_e^{-1}(0), \lambda_H^1 - \frac{\lambda_1^1}{\lambda_1^0} \lambda_H^0 \neq 0$$

and therefore equilibrium allocations associated with $e \in D$ are not Pareto optimal. In fact, to simplify computations, observe that if $\overline{e}$ belongs to the open and full measure subset $\mathcal{R}$ of $\mathbb{R}_{++}^{GH}$ and $(\overline{\xi}, \overline{e}) \in F^{-1}(0)$, then $D_\xi F(\overline{\xi}, \overline{e})$ has full row rank - see Section 4 in Chapter 11. Therefore, it is enough to show that zero is a regular value for $F_{nu|\mathcal{R}}$ to get the desired result.

The computation of $D_{(\xi, e_H)} F_{nu|\mathcal{R}}(\overline{\xi}, \overline{e})$ is described below.

$$
\begin{array}{l}
(1.1)\ \ Du_1 - \lambda_1 \Phi(p) \\
(1.2)\ \ -\Phi(p)(x_1 - e_1) + \left[\begin{smallmatrix} -q \\ Y \end{smallmatrix}\right] b_1 \\
(1.3)\ \ \lambda_1 \left[\begin{smallmatrix} -q \\ Y \end{smallmatrix}\right] \\
\ldots \\
(H.1)\ \ Du_H - \lambda_H \Phi(p) \\
(H.2)\ \ -\Phi(p)(x_H - e_H) + \left[\begin{smallmatrix} -q \\ Y \end{smallmatrix}\right] b_H \\
(H.3)\ \ \lambda_H \left[\begin{smallmatrix} -q \\ Y \end{smallmatrix}\right] \\
(Mx)\ \ \sum_h \left(x_h^{\backslash} - e_h^{\backslash}\right) \\
(Mb)\ \ \sum_h b_h \\
(P)\ \ \lambda_H^1 - \frac{\lambda_1^1}{\lambda_1^0} \lambda_H^0
\end{array}
\quad
\begin{array}{cccccccccc}
x_1 & \lambda_1 & b_1 & \ldots & x_H & \lambda_H & b_H & \ldots & e_H^{\backslash} & e_H^{\bullet} \\
\end{array}
$$

$$
\left(
\begin{array}{cccccccccc}
D^2 u_1 & -\Phi(p)^T & 0 & & & & & & & \\
-\Phi(p) & & \left[\begin{smallmatrix} -q \\ Y \end{smallmatrix}\right] & & & & & & & \\
 & \left[\begin{smallmatrix} -q \\ Y \end{smallmatrix}\right]^T & & & & & & & & \\
 & & & \ddots & & & & & & \\
 & & & & D^2 u_H & -\Phi(p)^T & & & & \\
 & & & & -\Phi(p) & & \left[\begin{smallmatrix} -q \\ Y \end{smallmatrix}\right] & & \Phi(p)^{\backslash} & I \\
 & & & & & \left[\begin{smallmatrix} -q \\ Y \end{smallmatrix}\right]^T & & & & \\
\widehat{I} & & & & \widehat{I} & & & & -I & \\
 & & I & & & & I & & & \\
 & * & & & & \left(-\frac{\lambda_1^1}{\lambda_1^0}, 1, 0\right) & & & &
\end{array}
\right)
$$

where as in Chapter 11

$$\widehat{I} \equiv \begin{bmatrix} I_{C-1} & 0 & & & \\ & & \ddots & & \\ & & & I_{C-1} & 0 \end{bmatrix}$$

and some nonrelevant matrices are denoted by $*$.

Since we restricted the domain of F_{nu} to $\mathcal{R}$, to show that the above matrix has full rank, it is enough to perturb the last row without perturbing the other rows, which is done below [5]

$$\begin{array}{lll} (P) & \leftarrow \lambda_H^0 & \leadsto (H.1), (H.3) \\ (H.3) & \leftarrow (\lambda_H^s)_{s \neq 0,1} & \leadsto (H.1) \\ (H.1) & \leftarrow x_H & \leadsto (H.2), (M.x) \\ (M.x) & \leftarrow e_H^{\backslash} & \leadsto (H.2) \\ (H.2) & \leftarrow e_H^{\bullet} & \end{array}$$

■

2.2 Nominal assets

As we have seen in Chapter 12, the model with nominal assets (and variable asset prices) exhibits a degree of nominal indeterminacy equal to $S - 1$. Therefore, to get a result of generic Pareto non-optimality as a consequence of a transversality argument, we need to add at least S equations of the type described in (15.1) to those characterizing extended equilibria. That is done in the following theorem. The strategy of proof there can be used to get a similar result in the case of fixed asset prices.

Theorem 3 *In the model with nominal assets and variable asset prices, if $H > \frac{S}{(S-I)}$* [6]*, then there exists a full measure and open subset D of the endowment space such that the equilibrium allocations associated with every $e \in D$ are not Pareto optimal.*

Proof. As we have already explained in Chapter 12, in the model with nominal assets and variable asset prices, while, as usual, $S + 1$ Walras' laws hold, only two normalizations are possible. Therefore, the extended equilibrium function is described as follows. Let

$$\Xi \equiv \mathbb{R}_{++}^{GH} \times \mathbb{R}^{IH} \times \mathbb{R}_{++}^{S'H} \times \mathbb{R}_{++}^{G-2} \times \mathbb{R}_{++}^{I}$$

[5] In the perturbation of equation $(H.3)$, recall that rank $Y = I = (\#$ equations $(H.3))$.

[6] The reason of that requirement is explained in the proof - see (15.4) and below.

with generic element

$$\xi \equiv \left(x, b, \lambda, p^{\setminus 01}, q\right)$$

and as in Chapter 12,

$$p^{\setminus 01} \equiv \left(p^{sc}\right)_{(sc) \neq (01),(11)}$$

Define

$$\begin{array}{l} F : \Xi \times \mathbb{R}^{GH}_{++} \to \mathbb{R}^{\dim \Xi - (S-1)} \\ F : (\xi, e) \mapsto \text{left hand side of system (12.10) in Chapter 12} \end{array}$$

For $h \neq 1$, consider the following subset of necessary conditions 15.1 for Pareto optimality - with $d = (S - I)$:

$$\begin{cases} \frac{\lambda_1^0}{\lambda_1^1} = \frac{\lambda_h^0}{\lambda_h^1} \\ \frac{\lambda_1^0}{\lambda_1^2} = \frac{\lambda_h^0}{\lambda_h^2} \\ \cdots \\ \frac{\lambda_1^0}{\lambda_1^d} = \frac{\lambda_h^0}{\lambda_h^d} \end{cases}$$

or

$$\begin{cases} \lambda_h^0 - \lambda_h^1 \frac{\lambda_1^0}{\lambda_1^1} = 0 \\ \lambda_h^0 - \lambda_h^2 \frac{\lambda_1^0}{\lambda_1^2} = 0 \\ \cdots \\ \lambda_h^0 - \lambda_h^d \frac{\lambda_1^0}{\lambda_1^d} = 0 \end{cases} \tag{15.2}$$

For $h \neq 1$, define

$$\begin{array}{l} f_h : \Xi \times \mathbb{R}^{GH}_{++} \to \mathbb{R}^d \\ f_h : (\xi, e) \mapsto \text{left hand side of system (15.2)} \end{array}$$

Consider the following function

$$F_{no} : \Xi \times \mathbb{R}^{GH}_{++} \to \mathbb{R}^{\dim \Xi - (S-1)} \times \mathbb{R}^{(H-1)(S-I)}$$

$$F_{no} : (\xi, e) \mapsto \left(F(\xi, e), (f_h(\xi, e))_{h \neq 1}\right)$$

Applying the same general strategy used in Theorem 2, it is enough to show 0 is a regular value for F_{no}.

The computation of the jacobian matrix of function $(f_h)_{h\neq 1}$ with respect to $(\lambda_h)_{h\neq 1}$ is described below

$$
\begin{array}{c|ccccccccccc}
 & \lambda_2^0 & \lambda_2^1 & \cdots & \lambda_2^d & (\lambda_2^s)_{s=d+1}^S & \cdots & \lambda_H^0 & \lambda_H^1 & \cdots & \lambda_H^d & (\lambda_H^s)_{s=d+1}^S \\
\hline
\lambda_2^0 - \lambda_2^1 \frac{\lambda_1^0}{\lambda_1^1} & 1 & \frac{\lambda_1^0}{\lambda_1^1} & & & & & & & & & \\
\lambda_2^0 - \lambda_2^2 \frac{\lambda_1^0}{\lambda_1^2} & 1 & & & & & & & & & & \\
 & & & \ddots & & & & & & & & \\
\cdots & & & & & & & & & & & \\
\lambda_2^0 - \lambda_2^d \frac{\lambda_1^0}{\lambda_1^d} & 1 & & & \frac{\lambda_1^0}{\lambda_1^d} & & & & & & & \\
\vdots & & & & & & & & & & & \\
\lambda_H^0 - \lambda_H^1 \frac{\lambda_1^0}{\lambda_1^1} & & & & & & & 1 & \frac{\lambda_1^0}{\lambda_1^1} & & & \\
\lambda_H^0 - \lambda_H^2 \frac{\lambda_1^0}{\lambda_1^2} & & & & & & & 1 & & & & \\
 & & & & & & & & & \ddots & & \\
\cdots & & & & & & & & & & & \\
\lambda_H^0 - \lambda_H^d \frac{\lambda_1^0}{\lambda_1^d} & & & & & & & 1 & & & \frac{\lambda_1^0}{\lambda_1^d} & \\
\end{array}
\tag{15.3}
$$

Therefore, the partial jacobian matrix $D_h \equiv D_{(\lambda_h^s)_{s=1}^d} f_h(\xi, e)$ is a full rank matrix, and $(\lambda_h^s)_{s=1}^d$ can be used to perturb the associated equations.

The computation of the relevant partial jacobian of F_{no} is described below

$$
\begin{array}{c|ccccccccccccccc}
 & x_1 & \lambda_1 & b_1 & x_2 & \lambda_2 & b_2 & \dots & x_H & \lambda_H & b_H & \dots & e_1^{\backslash} & e_1^{\bullet} & e_2^{\backslash} & e_2^{\bullet} \\
\hline
(1.1) & D^2u_1 & \widetilde{\Phi}^T & & & & & & & & & & & & & \\
(1.2) & -\widetilde{\Phi} & & \begin{bmatrix} -q \\ Y \end{bmatrix} & & & & & & & & & \widetilde{\Phi}^{\backslash} & \widetilde{\Phi}^{\bullet} & & \\
(1.3) & & \begin{bmatrix} -q \\ Y \end{bmatrix}^T & & & & & & & & & & & & & \\
(2.1) & & & & D^2u_2 & -\widetilde{\Phi}^T & & & & & & & & & & \\
(2.2) & & & & -\widetilde{\Phi} & & \begin{bmatrix} -q \\ Y \end{bmatrix} & & & & & & & & \widetilde{\Phi}^{\backslash} & \widetilde{\Phi}^{\bullet} \\
(2.3) & & & & & \begin{bmatrix} -q \\ Y \end{bmatrix}^T & & & & & & & & & & \\
\dots & & & & & & & \ddots & & & & & & & & \\
(H.1) & & & & & & & & D^2u_H & -\widetilde{\Phi}^T & & & & & & \\
(H.2) & & & & & & & & -\widetilde{\Phi} & & \begin{bmatrix} -q \\ Y \end{bmatrix} & & & & & \\
(H.3) & & & & & & & & & \begin{bmatrix} -q \\ Y \end{bmatrix}^T & & & & & & \\
(Mx) & \widehat{I} & & & \widehat{I} & & & & \widehat{I} & & & & -I & & -I & \\
(Mb) & & & I & & & & & & & I & & & & & \\
(O.2) & & * & & & D_2 & & & & & & & & & & \\
\dots & & & & & & & & & & & & & & & \\
(O.H) & & * & & & & & & & D_H & & & & & &
\end{array}
$$

where $\widetilde{\Phi}^{\bullet} \equiv diag\ \left(p^{sC}\right)_{s=0}^{S}$, $\widetilde{\Phi} \equiv \Phi\left(\overline{p}^{\setminus 01}\right)$.

To get the desired result, the dimension of the value set of F_{no} has to be larger than the dimension of its domain, i.e.,

$$\dim \Xi - (S-1) + (H-1)(S-I) > \dim \Xi \tag{15.4}$$

or

$$(H-1)(S-I) \geq S$$

which is the condition contained in the statement of the Theorem. The reason for which we add only $(S-I)(H-1)$ equations is explained at the very end of the proof.

We can now show that the above matrix has full row rank.

$$(1.1) \leftarrow x_1$$

$$(1.2) \leftarrow e_1^{\bullet}$$

$$\begin{array}{lll} (1.3) & \leftarrow (\lambda_1^s)_{s=d+1}^{S} & \rightsquigarrow (1.1) \\ (1.1) & \leftarrow x_1 & \rightsquigarrow (1.2) \\ (1.2) & \leftarrow e_1^{\bullet} & \end{array}$$

Similar arguments apply to the perturbation of $(h.1)$,$(h.2)$,$(h.3)$ for $h \neq 1$.

$$\begin{array}{lll} (M.x) & \leftarrow e_1^{\setminus} & \rightsquigarrow (1.2) \\ (1.2) & \leftarrow e_1^{\bullet} & \end{array}$$

$$\begin{array}{lll} (M.b) & \leftarrow b_1 & \rightsquigarrow (1.2) \\ (1.2) & \leftarrow e_1^{\bullet} & \end{array}$$

$$\begin{array}{lll} (O.2) & \leftarrow (\lambda_2^s)_{s=1}^{d} & \rightsquigarrow (2.1),(2.3) \\ (2.3) & \leftarrow (\lambda_2^s)_{s=d+1}^{S} & \rightsquigarrow (2.1) \\ (2.1) & \leftarrow x_2 & \rightsquigarrow (2.2),(M.x) \\ (M.x) & \leftarrow e_1^{\setminus} & \rightsquigarrow (1.2) \\ (1.2) & \leftarrow e_1^{\bullet} & \\ (2.2) & \leftarrow e_2^{\bullet} & \end{array}$$

Similar arguments apply to the perturbation of $(O.h)$ for $h > 2$. Observe that in the perturbation of $(O.2)$, we need I elements of the vector λ_h to adjust the disturb in equations (2.3). We are left with $(S+1-I)$ elements of λ_h to perturb added equations of the form (15.1); on the other hand, we cannot use λ_h^0 because of the specific structure of the matrix (15.3). ■

Theorem 4 *In a model with nominal assets and fixed asset prices, if $H \geq 2$, then there exists a full measure and open subset D of the endowment space such* *that the equilibrium allocations associated with every $e \in D$ are not Pareto optimal.*

Proof. The only difference in the extended equilibrium systems between the case of variable and fixed asset prices is in fact that feature. Therefore, the equilibrium function in the latter case is of the type

$$F : \mathbb{R}^{\dim \Xi - I} \times \mathbb{R}^{GH}_{++} \rightarrow \mathbb{R}^{\dim \Xi - (S-1)}$$

Then condition (15.4) becomes

$$\dim \Xi - (S-1) + (H-1)(S-I) > \dim \Xi - I$$

or

$$H \geq 2$$

■

2.3 Real assets

Our nonoptimality result in the model with real assets deals with full rank equilibria - see Definition 30 in Chapter 13. Combining the analysis presented there and that in the Subsection 2.1 above, we can prove the following result.

Theorem 5 *In a model with real assets, there exists a full measure and open subset D of the endowment space such that the full rank equilibrium allocations associated with every $e \in D$ are not Pareto optimal.*

2.4 Restricted participation

As we said in Chapter 14, there are several ways of defining restricted participation. In the case presented in the book, the proof of generic Pareto non-optimality is almost the same as in the case of numeraire asset and is therefore omitted.

3. The planner intervention: tools and goals

In this section and in the following ones, we describe a general methodology to analyze the intervention of a planner in an economy formally described by a general equilibrium model. We will see that to show the possibility of successful intervention, it is sufficient to show that a well chosen system of equations has no solution. In fact, the goal of the remaining part of the chapter is to present a clearly stated, and relatively easy to check, set of conditions which insure that result - see Conditions from 1 to 8 below. Starting point of the analysis is a function whose zeros describe equilibria:

$$F : \Xi \times \Theta \times \mathcal{U} \rightarrow \mathbb{R}^{n_1}, \ \ (\xi, \theta, u) \mapsto F(\xi, \theta, u)$$

where

1 $\Xi \subseteq \mathbb{R}^{n_1}$ is the set of endogenous variables ,

2 $\Theta \subseteq \mathbb{R}^{l}$ is the set of exogenous variables ,

3 $\mathcal{U}$ is the space of functions, part of the definition of an economy.

Remark 6 In many models $\mathcal{U}$ is just the space of utility functions endowed with the topology of C^3 uniform convergence on compact sets[7]. For future use, we state some facts on that topology. The set $C^3\left(\mathbb{R}^G_{++}, \mathbb{R}\right)$ of C^3 functions with domain $\mathbb{R}^G_{++}$ and value set $\mathbb{R}$ can be endowed with a metric constructed as described below. Consider a family $\{K_n\}_{n\in\mathbb{N}}$ of compact subsets of $\mathbb{R}^G_{++}$ such that $\cup_{n\in\mathbb{N}} K_n = \mathbb{R}^G_{++}$. For each $n \in \mathbb{N}$, define the following norm on $C^3\left(K_n, \mathbb{R}\right)$

$$\|f\|_{3,K_n} \equiv \max_{x\in K_n} |f(x)| + \max_{x\in K_n} \|Df(x)\| + \max_{x\in K_n} \left\|D^2 f(x)\right\| + \max_{x\in K_n} \left\|D^3 f(x)\right\|$$

Define the following metric on $C^3\left(\mathbb{R}^G_{++}, \mathbb{R}\right)$:

$$d(f,g) \equiv \sum_{n=1}^{\infty} \frac{1}{2^n} \min\left\{\|f-g\|_{3,K_n}, 1\right\}$$

It can be shown that $\left(C^3\left(\mathbb{R}^G_{++}, \mathbb{R}\right), d\right)$ is a metric space (not a normed space). The topology induced by d on $C^3\left(\mathbb{R}^G_{++}, \mathbb{R}\right)$ is the topology of the C^3 uniform convergence on compact sets, or the C^3 compact open topology. Call $\mathcal{C}^3$ that space. It can be shown that it is a Fréchet space (i.e., metric, complete and locally convex) and separable - see [117], Example 1, pp. 85-89.

We can now define

$$\mathcal{U}_h \equiv \left\{u_h \in \mathcal{C}^3 : \ u \text{ satisfies Assumptions } u2, u3', u4\right\}$$

Since $\mathcal{U}_h$ is a subset of the metric space $\mathcal{C}^3$, it is a metric space itself - [90], p. 126 - and we can endow it with the relative. Observe then that in this space compactness and sequential compactness are equivalent - [90], Theorem 7.4, p. 181.

Referring to the all models we presented, we can identify the exogenous variable defining the economy with $\theta \in \Theta$. Furthermore, while in the previous chapters $u \in \mathcal{U}$ is given, from now we take u as an explicit argument of the function characterizing equilibria. We then give the following definitions.

[7]We need the assumption that u is C^3 to apply the Transversality Theorem to a function whose components contain the Hessian of the households' utility functions - see subsection 5.2.2.

Definition 7 An economy is an element $(\theta, u) \in \Theta \times \mathcal{U}$. For each (θ, u), $F^{-1}_{(\theta,u)}(0)$ is the set of equilibria associated with that economy.

We then add some new variables $t \in T \equiv \mathbb{R}^m$ to the model. T is the set of the planner's tools. The function F is modified by the planner intervention. Those changes are described by new specifications of market clearing conditions and agents' maximization problems - and associated first order conditions -, and formally summarized in the function

$$F_1 : \Xi \times T \times \Theta \times \mathcal{U} \to \mathbb{R}^{n_1}, \quad (\xi, t, \theta, u) \mapsto F_1(\xi, t, \theta, u) \tag{15.5}$$

In general, the values of the planner's tools have to satisfy some constraints which are describe by the function

$$F_2 : \Xi \times T \times \Theta \times \mathcal{U} \to \mathbb{R}^p, \quad (\xi, t, \theta, u) \mapsto F_2(\xi, t, \theta, u) \tag{15.6}$$

It is then possible to describe the (equilibria of the) economy with planner intervention by the function

$$\widetilde{F} : \Xi \times T \times \Theta \times \mathcal{U} \to \mathbb{R}^n, \quad (\xi, t, \theta, u) \mapsto (F_1(\xi, t, \theta, u), F_2(\xi, t, \theta, u)) \tag{15.7}$$

where $n = n_1 + p$. We will call $\widetilde{F}$ the function of extended equilibrium with planner intervention.

The set T of tools can be written as $T = T_1 \times T_2 = \mathbb{R}^{m-p} \times \mathbb{R}^p$, with $(t_1, t_2) \in T$, and where t_1 can be interpreted as the vector of independent tools and t_2 as the vector of dependent tools. The reason of this terminology lays in the fact that, in many applications, for given (ξ, θ, u), when t_1 is chosen by the planner, then the value of t_2 is, at least locally, uniquely determined by the equation

$$F_2(\xi, t, \theta, u) = 0$$

Observe that the dimension of the set of dependent tools T_2 is equal to the dimension of the range of F_2, i.e., the "number of constraints" on planner intervention.

We summarize the above discussion requiring that the following condition is satisfied.

Condition 1 $\Xi \subseteq \mathbb{R}^{n_1}, \Theta \subseteq \mathbb{R}^l, U_1 \subseteq \mathbb{R}^{m_1}$ are open subsets of the Euclidean space they are contained in. $u_h : U_1 \to \mathbb{R}$ is a C^3 function and $u_h \in \mathcal{U}_h$, where $\mathcal{U}_h$ is a space of functions endowed with the topology of the C^3 uniform converge on compact sets. Let $\mathcal{U} \equiv \times_{h=1}^{m_2} \mathcal{U}_h$, endowed with the product topology (see [90] p. 86). For each $u \in \mathcal{U}$, $\widetilde{F}_u$ is a C^1 function.

Definition 8 Given $(\theta, u) \in \Theta \times \mathcal{U}$, ξ is an equilibrium with respect to "an economy with planner intervention and values t of the planner tools" if $\widetilde{F}(\xi, t, \theta, u) = 0$.

Following [30] terminology we call "altered economy" an economy with planner intervention. It is natural to expect that the set of equilibria of an "altered economy" (θ, u, t) where the planner does not intervene reduces to the set of equilibria associated with the economy (θ, u) without planner's intervention. This observation can be formalized in the following way.

Definition 9 $\widetilde{F}$ is a variation of F if for each (θ, u) there exists a functions $\psi : \Xi \to T$, $\xi \mapsto \psi(\xi)$ such that $\widetilde{F}(\xi, \psi(\xi), \theta, u) = 0 \Leftrightarrow F(\xi, \theta, u) = 0$. Define also $\psi_1 \equiv (\psi_j)_{j=1}^{m-p}$ and $\psi_2 \equiv (\psi_j)_{j=m-p+1}^{m}$.

ψ can be interpreted as the function which lets t to assume the "non-intervention" value $\bar{t} \equiv \psi(\xi)$.

Condition 2 $\widetilde{F}$ is a variation of F.

The above condition simply says that "an equilibrium without planner intervention is an equilibrium with planner intervention when the planner decides not to intervene".

In order to better understand the model we consider the following example, presented in CKV (i.e., [36]).

Example 10 (CKV model) We consider the general equilibrium model with incomplete market and numeraire assets presented in Chapter 11. The space Ξ^{n_1} is identified with the space Ξ, whose generic element is

$$\xi = \left((x_h, \lambda_h, b_h)_{h \in H}, p^{\backslash}, q \right)$$

Θ is simply $\mathbb{R}^{GH}_{++}$, the spaces of endowments, and $\theta = e$. The function F is given by the first order conditions of households' maximization problems and by the non-redundant market equilibrium conditions on goods and asset markets, i.e.,

$$F : \Xi^n \times \mathbb{R}^{GH}_{++} \times \mathcal{U} \to \mathbb{R}^n$$

$$F : (\xi, e, u) \mapsto \begin{pmatrix} \begin{pmatrix} Du_h(x_h) - \lambda_h \Phi(p) \\ \Phi(p)(x_h - e_h) + \left[\begin{smallmatrix} -q \\ Y \end{smallmatrix}\right] b_h \\ \lambda_h \left[\begin{smallmatrix} -q \\ Y \end{smallmatrix}\right] \end{pmatrix}_{h \in \mathcal{H}} \\ \sum_h \left(x_h^{\backslash} - e_h^{\backslash} \right) \\ \sum_h b_h \end{pmatrix}$$

See Chapter 11 for details on the meaning of the symbols used above. In particular, observe that $\mathcal{U} = \times_{h=1}^{H} \mathcal{U}_h$, where $\mathcal{U}_h$ is the set of utility functions. As in Section 2, we make Assumption u1, u2, u3' and u4 on $u_h \in \mathcal{U}_h$, $h = 1, ..., H$.

Consider a redistribution of initial endowments, i.e., assume that there exists a lump-sum fiscal policy that the planner can use in order to transfer wealth across consumers. More precisely, the planner can modify all consumers' wealth in period 0, while in period 1, she limits herself to change the endowments of only two consumers. The planner can neither create nor destroy wealth. Hence we have

$$T_1 = \mathbb{R}^H, \quad T_2 = \mathbb{R}^2, \quad T = \mathbb{R}^{H+2},$$
$$t_1 = \left(\left(\tau_h^{0C} \right)_{h \neq 1}, \tau_2^{1C} \right), \quad t_2 = \left(\tau_1^{0C}, \tau_1^{1C} \right)$$

where C is the index of the numeraire good. In general, τ_h^{sC} is the tax of the numeraire good C imposed on household h in state s. Therefore, the constraints on the planner tools are described by the following function:

$$F_2 : \Xi \times T_1 \times T_2 \times \Theta \times \mathcal{U} \to \mathbb{R}^2$$
$$F_2 (\xi, t_1, t_2, \theta, u) = \begin{pmatrix} \sum_{h=1}^{H} \tau_h^{0C} \\ \\ \tau_1^{1C} + \tau_2^{1C} \end{pmatrix}$$

Equilibria in the "altered economy" are described by the following function

$$F_1 : \Xi \times T_1 \times T_2 \times \Theta \times \mathcal{U} \to \mathbb{R}^{\dim \Xi}$$
$$F_1 (\xi, t_1, t_2, \theta, u) = \begin{pmatrix} \begin{pmatrix} Du_h (x_h) - \lambda_h \Phi(p) \\ \Phi(p) (x_h - e_h - \tau_h) + \left[\begin{smallmatrix} -q \\ Y \end{smallmatrix} \right] b_h \\ \lambda_h \left[\begin{smallmatrix} -q \\ Y \end{smallmatrix} \right] \end{pmatrix}_{h \in \mathcal{H}} \\ \sum_h \left(x_h^{\backslash} - e_h^{\backslash} \right) \\ \sum_h b_h \end{pmatrix}$$

where for each h, $\tau_h \equiv (\tau_h^{sc})_{\substack{s=0,\ldots,S \\ c=1,\ldots,C}} \in \mathbb{R}^G$ and

for $h = 1, 2$, $\tau_h^{sc} = 0$ if $(sc) \neq (0C)$,$(1C)$,

for $h \geq 3$, $\tau_h^{sc} = 0$ if $(sc) \neq (0C)$.

The equilibrium without planner intervention corresponds to $t = 0$. In this case, ξ is a solution to $\widetilde{F} (\xi, 0, 0, \theta, u) = 0$ if and only if it is a solution to $F (\xi, \theta, u) = 0$, i.e., the function ψ is such that $\psi (\xi) = 0$ for any ξ. (- to be continued -)

The objective of the analysis is the effects of changes in t_1 around the "no intervention" value $\overline{t}_1$ on some economically relevant goal. The function

$$G : \Xi \times T \times \Theta \to \mathbb{R}^k, \quad (\xi, t, \theta) \mapsto G (\xi, t, \theta, u)$$

describes the goals of the planner. In order to apply our analysis, the following condition has to be satisfied.

Condition 3 For each $u \in \mathcal{U}$, the function G_u is C^1.

The following relevant example of the function G comes immediately to mind. In a model like that one with incomplete markets studied in the previous chapters, a basic goal is to Pareto improve upon the typically inefficient equilibrium. Given an economy and an associated equilibrium allocation x, does there exist a value t such that the equilibrium allocation x' in the altered economy is Pareto superior to the starting allocation x? In other words, can a planner Pareto improve upon the market outcome? We formally describe the above observations as follows.

Definition 11 Given an economy $(\theta, u) \in \Theta \times \mathcal{U}$ and a corresponding equilibrium vector $\xi \equiv (x, ...)$, i.e., such that $F(\xi, \theta, u) = 0$, ξ or x are Pareto improvable (with respect to the intervention described by $\widetilde{F}$) if there exist a vector $\left(\xi' \equiv (x', ...), t'\right) \in \Xi \times T$ such that:

1 $\widetilde{F}(\xi', t', \theta, u) = 0$

2 $(u_h(x'_h))_{h=1}^H > (u_h(x_h))_{h=1}^H$, i.e., x' Pareto dominates x.

If every equilibrium vector associated with the economy (θ, u) is Pareto improvable, (θ, u) is a Pareto improvable economy.

Consistently with the above definition, we can specify the function G as

$$G : \Xi^{n_1} \times T \times \Theta \times \mathcal{U} \to \mathbb{R}^H, \quad (\xi, t, \theta, u) \mapsto u(x)$$

Remark 12 It is not necessarily the case that G and u coincide, as in the discussion above and the Example we present - see for Example [123]. Other kinds of planner goals and related tools can be studied and in fact have been studied in [34], , [57], [123], following the general approach described in this chapter.

To proceed in our analysis, observe that the following statement is *false*:
for every $(\theta, u) \in \Theta \times \mathcal{U}$, and every $\xi \in \Xi$ such that $F(\xi, \theta, u) = 0$, ξ is Pareto improvable.

A Pareto optimal endowment is an equilibrium allocation and obviously it is not Pareto improvable. Therefore, at most, we can expect that there exists an open and full measure set of economies whose equilibria are improvable. This is true even when we consider different planner's goals.

Moreover, using our differentiable approach, it is quite natural to expect that our analysis is a local one.

With those two caveat in mind, we introduce the following condition.

Condition 4 For every $u \in \mathcal{U}$, there exists an open and full measure subset Θ_u of Θ such that for every $\theta' \in \Theta_u$ and for every $\xi' \in \Xi$ such that $\widetilde{F}(\xi', \psi(\xi'), \theta, u) = 0$,

$$\text{rank}\, D_{(\xi,t_2)}\widetilde{F}\left(\xi', \psi(\xi'), \theta', u\right) \text{ is full}$$

To verify Condition 4, the usual strategy is to verify that 0 is a regular value for $\widetilde{F}$. That result implies also that the equilibrium set is a manifold - see Theorem 9, Chapter 4.

Remark 13 $D_{(\xi,t_2)}\widetilde{F}$ can be rewritten as

$$\begin{bmatrix} D_\xi F_1 & D_{t_2} F_1 \\ D_\xi F_2 & D_{t_2} F_2 \end{bmatrix}$$

In many models analyzed in the literature, and also in the example we present, Condition 4 is clearly verified because the following statements are true.

1 from the generic regularity of equilibria, $D_\xi F_1$ has full row rank in an open and full measure subset of Θ_u,

2 $D_\xi F_2$ is equal to 0, and

3 $D_{t_2} F_2$ has full row rank.

Under Condition 4, from the implicit function theorem ([3] 13.7 p. 374), there exists a neighborhood N_1 of $\overline{t}_1$ and a unique C^1 function

$$h(t_1) \equiv (\xi(t_1), t_2(t_1))$$

defined on N_1, such that for $t_1 \in N_1$

$$\widetilde{F}(\xi(t_1), t_2(t_1), t_1, \theta, u) = 0$$

The function h describes how equilibrium variables and dependent tools adjust to planner intervention via the independent tools t_1. Then the function

$$g_{(\theta,u)} : N_1 \to \mathbb{R}^k,\ t_1 \mapsto G(\xi(t_1), t_2(t_1), t_1, \theta, u)$$

describes how the goal function changes when the planner uses her policy tools t_1 and variables move in the equilibrium set defined by $\widetilde{F}$.

Then, the goal of the analysis is to show that there exists an open and dense subset $S^* \subseteq \Theta \times \mathcal{U}$ such that for each $(\theta, u) \in S^*$, the planner can "move" the equilibrium value of the goal function in any directions locally around $g_{(\theta,u)}(\overline{t}_1)$, the value of the goal function in the case of no intervention. In other words, for any direction of movement away from $g_{(\theta,u)}\left(\overline{t}_1\right)$ (i.e., for any $v \in \mathbb{R}^k$) and for any neighborhood N_1 of $\overline{t}_1$, there exists a point $t_1^* \in N_1$ such that $g_{(\theta,u)}\left(t_1^*\right)$ belong to that directions (i.e., $g_{(\theta,u)}\left(t_1^*\right) = g_{(\theta,u)}\left(\overline{t}_1\right) + \varepsilon v$, for some $\varepsilon \in \mathbb{R}_{++}$). But the above statements are nothing but intuitive ways of stating the fact that $g_{(\theta,u)}$ is *essentially surjective at* $\overline{t}_1$- see Definition 62 in Chapter 3. We can then give the following definition.

Definition 14 Planner intervention is generically locally effective with respect to the goal function G and the function $\widetilde{F}$ describing equilibria with planner intervention at the economy (θ, u), if for each $\xi \in \Xi$ such that $\widetilde{F}\left(\xi, \psi\left(\xi\right), \theta, u\right) = 0$, $g_{(\theta,u)}$ is essentially surjective at $\overline{t}_1 \equiv \psi_1\left(\xi\right)$.

Planner intervention is effective with respect to G and $\widetilde{F}$ if there exists an open and dense subset $S^* \subseteq \Theta \times \mathcal{U}$ at which the intervention is successful.

Remark 15 From now on, we will simply speak of "effectiveness of planner intervention", omitting the adverbs "generically" and "locally".

4. Planner intervention and solutions to a system of equations

In this section we present two different approaches which lead to the same system of equations whose analysis may show that planner intervention is effective.

4.1 Submersion approach

As it is shown in Theorem 66, Chapter 3, a sufficient condition for $g_{(\theta,u)}$ to be essentially surjective (62 Chapter 3) is that

$$\text{rank}\left[D_{t_1} g(\overline{t}_1)\right]_{k\times(m-p)} = k \tag{15.8}$$

The above condition implies that it must be

$$k = \#\,\text{goals} \quad \leq \quad \text{independent tools} = m - p \tag{15.9}$$

In order to find some more manageable sufficient condition for 15.8 we can apply Theorem 40 in Chapter 4. Identify the variables and functions in that theorem - see first column below - with variables and functions here - see second column.

X	$\Xi \times T_2$
$\overline{x}$	(ξ, t_2)
Z	T_1
$\overline{z}$	t_1
f	$\widetilde{F}_{(\theta,u)}$
g	G
$\eta\left(\overline{z}\right)$	$h(t_1) = \left(\xi\left(t_1\right), t_2\left(t_1\right)\right)$

If Condition 1, 3 and 4 of the present chapter holds, then Assumptions 1, 2 and 3 in Theorem 40 in Chapter 4 are satisfied. Observe that Condition 2.b in Theorem 40 in Chapter 4 here takes the form specified by (15.8). Therefore, to get effectiveness of planner intervention we have to prove the following statement:

$$\begin{array}{l}\text{there exists an open and dense subset } S^* \subseteq \Theta \times \mathcal{U} \text{ such that} \\ \text{for every } (\theta, u) \in S^* \text{ and for every } \xi \in \Xi^{n_1} \text{ such that} \\ \widetilde{F}(\xi, \psi(\xi), \theta, u) = 0 \text{ it is the case that} \\ \text{rank}\left[D_{(\xi,t)}\left(\widetilde{F}, G\right)(\xi, \psi(\xi), \theta, u)\right]_{(n_1+p+k)\times(n_1+m)} = n_1 + p + k\end{array} \tag{15.10}$$

The above condition implies that it must be

$$p + k \leq m \tag{15.11}$$

that is

$$(\#\text{ constraints on planner}) + (\#\text{ goals}) \quad \leq \quad \#\text{ tools}$$

which is equivalent to Condition (15.9).

Condition (15.10) is equivalent to show that in the above defined set $S^* \subseteq \Theta \times \mathcal{U}$, the following system has no solutions $(\xi, c) \in \Xi^{n_1} \times \mathbb{R}^{n_1+p+k}$

$$\begin{cases} \widetilde{F}(\xi, \psi(\xi), \theta, u) & = & 0 \\ c\left[D_{(\xi,t)}\left(\widetilde{F}, G\right)(\xi, \psi(\xi), \theta, u)\right] & = & 0 \\ cc - 1 & = & 0 \end{cases}$$

which, from Condition 2, is equivalent to

$$\begin{cases} F(\xi, \theta, u) & = & 0 \quad (1) \\ c\left[D_{(\xi,t)}\left(\widetilde{F}, G\right)(\xi, \psi(\xi), \theta, u)\right] & = & 0 \quad (2) \\ \frac{1}{2}cc - 1 & = & 0 \quad (3) \end{cases} \tag{15.12}$$

In conclusion, we have proved the following theorem,

Theorem 16 *Consider the functions F, $\widetilde{F}$ and G defined in the previous section and suppose that Conditions 1-4 are verified. If*

$$\begin{array}{l}\textit{there exists an open and dense subset } S^* \subseteq \Theta \times \mathcal{U} \textit{ such that} \\ \textit{for every } (\theta, u) \in S^* \textit{ there are no solutions } (\xi, c) \in \Xi^{n_1} \times \mathbb{R}^{n_1+p+k} \textit{ to} \\ \textit{system (15.12)}\end{array} \tag{15.13}$$

then planner intervention is effective.

Remark 17 A necessary condition for condition (15.13) is the validity of either (15.9) or (15.11). It is then useful to preliminary check one of those inequalities.

4.2 Vector optimization approach

This approach starts from the analysis of the following problem. As in the previous subsection, suppose that Conditions 1-4 are verified. Given an economy $(\theta, u) \in \Theta \times \mathcal{U}$, assume that the planner has to solve

$$\begin{array}{l} \mathit{vector} \max\limits_{(\xi,t)\in\Xi\times T} \; G(\xi, t, \theta, u) \\ s.t. \\ \widetilde{F}(\xi, t, \theta, u) = 0 \end{array} \tag{15.14}$$

From Fritz-John conditions (see for example [73]) if (ξ, t) is a solution to problem (15.14), there exists a vector $c \in \mathbb{R}^{n_1+p+k} \backslash \{0\}$ such that

$$\begin{cases} \widetilde{F}(\xi, t, \theta, u) & = 0 \\ c\left[D_{\xi,t}\left(\widetilde{F}(\xi, t, \theta, u), G(\xi, t, \theta, u)\right)\right] & = 0 \end{cases} \tag{15.15}$$

Thus, if a vector (ξ, t) does not verify system (15.15) then it is not a solution to problem (15.14).

Consider the equilibrium vector $(\xi, \psi(\xi))$ of the "altered economy" where the planner decides to not intervene. If $(\xi, \psi(\xi))$ is not a solution to system (15.15) with $c \neq 0$, $(\xi, \psi(\xi))$ does not verify a necessary optimality condition and there exists a vector (ξ', t') such that $G(\xi', t'_1, t'_2) \geq G(\xi, \psi(\xi))$. Therefore, in order to prove the effectiveness of the policy instruments at equilibrium point ξ, it is sufficient to show that :

there exists an open and dense subset $S^* \subseteq \Theta \times \mathcal{U}$ such that for every $(\theta, u) \in S^*$ there are no solutions $(\xi, c) \in \Xi^{n_1} \times \mathbb{R}^{n_1+p+k}$ to the system

$$\begin{cases} \widetilde{F}(\xi, \psi(\xi), \theta, u) & = 0 & (1) \\ cD_{\xi,t}\left(\widetilde{F}(\xi, \psi(\xi), \theta, u), G(\xi, \psi(\xi), \theta, u)\right) & = 0 & (2) \\ c & \neq 0 & (3) \end{cases} \tag{15.16}$$

i.e., condition (15.13) in Theorem 16

We now apply our analysis to Example 10.

Example 18 (CKV model - resumed) First of all observe that Condition (15.9) is verified. Tools are $\left(\left(\tau_h^{0C}\right)_{h=1}^{H}, \tau_1^{1C}, \tau_2^{1C}\right)$, whose number is $H+2$. Goals are the given by the utility levels of the H households. Constraints are just the two redistribution requirements.

To verify Condition 4, following Remark 13, we have that from Lemma 22 in Chapter 11, $D_\xi F_1$ has full row rank in an open and full measure subset of Θ_u. Defined

$$t_1 = \left(\left(\tau_h^{0C}\right)_{h\neq 1}, \tau_2^{1C}\right), \quad t_2 = \left(\tau_1^{0C}, \tau_1^{1C}\right)$$

the computation of $D_{(\xi,t_2)}F_2$ is described below

$$\begin{array}{c} \\ \sum_{h=1}^{H}\tau_h^{0C} \\ \tau_1^{1C}+\tau_2^{1C} \end{array} \begin{array}{c} \begin{array}{ccc} \xi & \tau_1^{0C} & \tau_1^{1C} \end{array} \\ \begin{pmatrix} 0 & 1 & 0 \\ 0 & 0 & 1 \end{pmatrix} \end{array}$$

Therefore, $D_\xi F_2$ is equal to 0, and $D_{t_2}F_2$ has full row rank as desired. Let's now specify system (15.12). The computation of $D_{\xi,t}\left(\widetilde{F},U\right)$ is described below.

$$
\begin{array}{lll}
 & & \begin{matrix} x_1 & \lambda_1 & b_1 & \ldots & x_H & \lambda_H & b_H & p^{\backslash} & q & t_1 & t_2 \end{matrix} \\
\begin{matrix}
G \\ S' \\ I \\ \ldots \\ G \\ S' \\ I \\ G' \\ I \\ 2 \\ \\ H
\end{matrix}
&
\begin{matrix}
Du_1(x_1) - \lambda_1 \Phi(p) \\
-\Phi(p)(x_1 - e_1 - \tau_1) + \left[\begin{smallmatrix} -q \\ Y \end{smallmatrix}\right] b_1 \\
\lambda_1 \left[\begin{smallmatrix} -q \\ Y \end{smallmatrix}\right] \\
\\
Du_H(x_H) - \lambda_H \Phi(p) \\
-\Phi(p)(x_H - e_H - \tau_H) + \left[\begin{smallmatrix} -q \\ Y \end{smallmatrix}\right] b_H \\
\lambda_H \left[\begin{smallmatrix} -q \\ Y \end{smallmatrix}\right] \\
\sum_h \left(x_h^{\backslash} - e_h^{\backslash} \right) \\
\sum_h b_h \\
\sum_{h=1}^{H} \tau_h^{0C} \\
\tau_1^{1C} + \tau_2^{1C} \\
u(x)
\end{matrix}
&
\left(
\begin{matrix}
D^2 u_1 & -\Phi(p)^T & & & & & & -\Lambda_1 & & & \\
-\Phi(p) & & \left[\begin{smallmatrix} -q \\ Y \end{smallmatrix}\right] & & & & & Z_1 & \begin{bmatrix} -b_1 \\ 0 \end{bmatrix} & \begin{bmatrix} 0 \\ 0 \end{bmatrix} & \begin{bmatrix} I_2 \\ 0 \end{bmatrix} \\
 & \left[\begin{smallmatrix} -q \\ Y \end{smallmatrix}\right]^T & & & & & & & -\lambda_1^0 I & & \\
 & & & \ddots & & & & & & & \\
 & & & & D^2 u_H & -\Phi(p)^T & & -\Lambda_H & & & \\
 & & & & -\Phi(p) & & \left[\begin{smallmatrix} -q \\ Y \end{smallmatrix}\right] & Z_H & \begin{bmatrix} b_H \\ 0 \end{bmatrix} & \begin{bmatrix} \varphi_{\tau_H}(1) \\ 0 \end{bmatrix} & \begin{bmatrix} 0 \\ 0 \end{bmatrix} \\
 & & & & & \left[\begin{smallmatrix} -q \\ Y \end{smallmatrix}\right]^T & & & -\lambda_H^0 I & & \\
\widetilde{I} & & & & \widetilde{I} & & & & & & \\
 & I & & & & I & & & & & \\
 & & & & & & & & & \varphi_t(1) & I_2 \\
 & & & & & & & & & & \\
Du_1 & & & & Du_H & & & & & &
\end{matrix}
\right)
\end{array}
\tag{15.17}
$$

where $S' = S + 1$, $G' = G - S'$,

$$\varphi_{T_2}(1) \equiv \begin{bmatrix} 1 & 0 & \dots & 0 & 0 \\ 0 & 0 & \dots & 0 & 1 \end{bmatrix}_{2\times H}$$

for $h > 2$

$$\varphi_{T_h}(1) \equiv \begin{bmatrix} 0 & \dots & 0 & 1 & 0 & \dots & 0 \\ 0 & \dots & 0 & 0 & 0 & \dots & 0 \end{bmatrix}_{2\times H}$$

$$\varphi_t(1) \equiv \begin{bmatrix} 1 & 1 & \dots & 1 & 0 \\ 0 & 0 & \dots & 0 & 1 \end{bmatrix}_{2\times H}$$

$$\Lambda_h \equiv \begin{bmatrix} \begin{matrix} \lambda_h^0 I_{C-1} \\ 0 \end{matrix} & & \dots \\ & \ddots & \dots \\ .. & \dots & \begin{matrix} \lambda_h^S I_{C-1} \\ 0 \end{matrix} \end{bmatrix}_{G\times[G-S']}$$

$$Z_h \equiv \begin{bmatrix} (x_h^{0c} - e_h^{0c})_{c\neq C} & 0 & \dots \\ 0 & \ddots & \dots \\ .. & \dots & (x_h^{Sc} - e_h^{Sc})_{c\neq C} \end{bmatrix}_{(S+1)\times[G-S']}$$

We now partition c in order to make it conformable with the super rows of $\left[D_{\xi,t}\left(\widetilde{F}, G\right)\right]$:

$$c \equiv \left(\left(c_{x_h}, c_{\lambda_h}, c_{b_h} \right)_{h=1}^{H}, c_{p\backslash}, c_q, c_{t_2}, c_g \right)$$
$$c \in \left(\mathbb{R}^G \times \mathbb{R}^{S'} \times \mathbb{R}^I \right)^H \times \mathbb{R}^{G-S'} \times \mathbb{R}^I \times \mathbb{R}^2 \times \mathbb{R}^H$$

Then we can write

$$\begin{cases} c\left[D_{\xi,t}\left(\widetilde{F}, G\right)\right] = 0 \\ cc - 1 = 0 \end{cases}$$

as

$$\begin{cases} c_{x_1} D^2 u_1 - c_{\lambda_1} \Phi(p) + c_{p\backslash} \widetilde{I} + c_{g_1} D u_1 & = 0 \quad (1.1) \\ -c_{x_1} \Phi(p)^T + c_{b_1} \begin{bmatrix} -q \\ Y \end{bmatrix}^T & = 0 \quad (1.2) \\ c_{\lambda_1} \begin{bmatrix} -q \\ Y \end{bmatrix} + c_q I & = 0 \quad (1.3) \\ \dots & \\ c_{x_H} D^2 u_H - c_{\lambda_H} \Phi(p) + c_{p\backslash} \widetilde{I} + c_{g_1} D u_H & = 0 \quad (H.1) \\ -c_{x_h} \Phi(p)^T + c_{b_h} \begin{bmatrix} -q \\ Y \end{bmatrix} & = 0 \quad (H.2) \\ c_{\lambda_h} \begin{bmatrix} -q \\ Y \end{bmatrix} + c_q I & = 0 \quad (H.3) \\ -c_{x_1} \Lambda_1 + c_{\lambda_1} Z_1^T - \sum_{h \neq 1} c_{x_h} \Lambda_h + \sum_{h \neq 1} c_{\lambda_h} Z_h^T & = 0 \quad (4) \\ c_{\lambda_1} \begin{bmatrix} b_1 \\ 0 \end{bmatrix} - \lambda_1^0 c_{b_1} I + \dots + \sum_{h \neq 1} c_{\lambda_h} \begin{bmatrix} b_h \\ 0 \end{bmatrix} - \sum_{h \neq 1} \lambda_h^0 c_{b_h} I & = 0 \quad (5) \\ c_{\widetilde{\lambda}} I + c_{t_2} \varphi_t (1) & = 0 \quad (6) \\ \left[c_{\lambda_1^0}, c_{\lambda_1^1} \right] I + c_{t_2} \varphi_t (2) & = 0 \quad (6') \\ c_{x_1} c_{x_1} + c_{\lambda_1} c_{\lambda_1} + c_{b_1} c_{b_1} + c_{x_h} c_{x_h} + c_{\lambda_h} c_{\lambda_h} + c_{b_h} c_{b_h} + & \\ + c_{p\backslash} c_{p\backslash} + c_q c_q + c_{t_2} c_{t_2} + c_g c_g & = 0 \quad (7) \end{cases} \tag{15.18}$$

where

$$c_{\widetilde{\lambda}} \equiv \left(c_{\lambda_2^0}, c_{\lambda_3^0}, \dots, c_{\lambda_H^0}, c_{\lambda_2^1} \right) \tag{15.19}$$

Observe that

$$\sum_{h=2}^{H} c_{\lambda_h} \begin{bmatrix} \varphi_{T_h}(1) \\ 0 \end{bmatrix} = \left(c_{\lambda_2^0}, c_{\lambda_3^0}, \dots, c_{\lambda_H^0}, c_{\lambda_2^1} \right)$$

(to be continued)

Remark 19 In the previous example, since $\left[D_{(\xi,t_2)} \widetilde{F} \right]$ has full row rank, then system (15.12) has no solution with $c_g = 0$. Suppose otherwise. Then, from equation (15.12.2) we have:

$$\left[c_{\backslash g}, c_g \right] \begin{bmatrix} D_{(\xi,t)} \widetilde{F} \\ D_{(\xi,t)} G \end{bmatrix} = 0,$$

where $c_{\backslash g} \equiv \left((c_{x_h}, c_{\lambda_h}, c_{b_h})_{h=1}^{H}, c_{p\backslash}, c_q, c_t \right)$. The rank condition on $\left[D_{(\xi,t)} \widetilde{F} \right]$ implies that $c_{\backslash g} = 0$ and therefore $c = 0$, which contradicts equation (3) in system (15.12).

We are going to use the above result in Step 1 of Section 6.

5. Conditions for planner effectiveness

The goal of this section is to present a clearly stated, and relatively easy to check, set of conditions which insure that Condition 15.13 in Theorem 16 does hold.

Definition 20 [8]$\mathcal{E}_E \equiv \{(\theta, u) \in \Theta \times \mathcal{U}$: system (15.12) has no solution$\}$.

As we will see in Subsection 5.2, the verification of the following condition is a crucial step in the analysis of the effectiveness of planner intervention.

Condition 5 There exists an open and dense set $\mathcal{E}_r \subseteq \Theta \times \mathcal{U}$ such that for every $(\theta, u) \in \mathcal{E}_r$ and for every ξ such that $F(\xi, \theta, u) = 0$, we have

$$\operatorname{rank} D_\xi F(\xi, \theta, u) = n$$

Remark 21 Generic regularity of equilibria does imply that the above condition is satisfied as the following lemma shows.

Lemma 22 *If*

$$\forall\, u \in \mathcal{U},\ \exists \text{ a full measure subset } D_u \text{ of } \Theta, \text{ such that } \forall \theta \in D_u,$$
$$\forall \xi^* \in F^{-1}_{(\theta,u)}(0), \text{ a property } \mathcal{P} \text{ holds,}$$

then

$$\exists \text{ a dense subset } S^* \subseteq \Theta \times \mathcal{U} \text{ such that } \forall (\theta, u) \in S^*,$$
$$\forall \xi^* \in F^{-1}_{(\theta,u)}(0), \text{ property } \mathcal{P} \text{ holds.}$$

Proof. Suppose otherwise. Then there exists an open set $S^0 \subseteq \Theta \times \mathcal{U}$ such that $\mathcal{P}$ does not hold. Consider $(\theta, u) \in S^0$. Then there exists an open neighborhood N_θ of θ such that $N_\theta \times \{u\} \subseteq S^0$, contradicting the assumption. ■

Therefore, instead of Condition 5 we can consider the following one.

Condition 5' For each $u \in \mathcal{U}$, there exists an open and full measure set $\mathcal{E}_r^\Theta \subseteq \Theta$ such that for every $\theta \in \mathcal{E}_r^\Theta$ and for every ξ such that $F(\xi, \theta, u) = 0$, we have

$$\operatorname{rank} D_\xi F(\xi, \theta, u) = n$$

We are now ready to present some sufficient conditions for the openness and density of $\mathcal{E}_E$.

5.1 Openness

Condition 6

$$pr : F^{-1}(0) \to \Theta \times \mathcal{U}, \quad (\xi, \theta, u) \mapsto (\theta, u)$$

is a proper function.

[8]The letter E stands for "Effective".

The properness of *pr* gives a sufficient condition in order to guarantee the openness of the set $\mathcal{E}_E$ as it is shown below.

Theorem 23 *If pr is proper, i.e., Condition 6 is verified, then $\mathcal{E}_E$ is an open set.*

Proof. We define

$$M \equiv \{(\xi, t, \theta, u) \in \Xi \times T \times \Theta \times \mathcal{U} : F(\xi, \theta, u) = 0 \text{ and } \operatorname{rank}\left[D_{\xi,t}\left(\widetilde{F}(.), G(,)\right)\right] < n + p + k\}$$

Observe that $\mathcal{E}_E = [pr(M)]^c$; since F is a continuous function, $F^{-1}(0)$ is closed. Consider all the square submatrices of $\left[D_{\xi,t}\left(\widetilde{F}(.), G(,)\right)\right]$ of dimension $n+p+k$. The rank condition on $\left[D_{\xi,t}\left(\widetilde{F}(.), G(,)\right)\right]$ requires that the determinants of the previous submatrices are all zero. Since determinant can be seen as a continuous function, it follows that M a closed. The properness of *pr* implies that $pr(M)$ is closed and therefore $\mathcal{E}_E$ is open. ■

We now want to show that in the example analyzed by CKV, the projection is in fact proper. But some preliminary comments are necessary.

Observe that both domain and the value set of the projection function are infinite dimensional spaces. Since $\mathcal{U}$ is a metric space, - see Remark 6 - and hence Definition 20 in Chapter 4 of a proper function applies. We can now prove that our projection is proper.

Lemma 24 *pr is proper.*

Proof. Taken a compact set K in $\mathbb{R}^{GH}_{++} \times \mathcal{U}$, we want to show that $(\xi^n, e^n, u^n)_{n \in \mathbb{N}}$ in $pr^{-1}(K)$ is in fact sequentially compact. Observe that (e^n, u^n) converges to $(\overline{e}, \overline{u})$.

The main difficulty in the proof is to show that $x^n \to \overline{x} \gg 0$.

Observe that for any n, h and c,

$$x_h^{cn} \leq \sum_{h=1}^{H} e_h^{cn} \leq \left(\max_{e^{cn'} \in \{e^{cn}\} \cup \{\overline{e}^c\}} \sum_{h=1}^{H} e_h^{cn'} \right)$$

where the set $\{e^{cn} : n \in \mathbb{N}\} \cup \{\overline{e}^c\}$ is compact. Moreover, for any n, h and c, $x_h^{cn} \geq 0$. Therefore, $\{x^n : n \in \mathbb{N}\}$ is contained in a compact set C_x and, up to a subsequence, $x^n \to \overline{x} \geq 0$. We have now to show that $\overline{x} \gg 0$. In fact, it is enough to show that, for every h, there exists an utility function $\widetilde{u}_h$ and an endowment $\widetilde{e}_h$ such that $\widetilde{u}_h(\overline{x}_h) \geq \widetilde{u}_h(\widetilde{e}_h)$. Define the compact set

$$C_e \equiv \{e^n : n \in \mathbb{N}\} \cup \{\overline{e}\}$$

and

$$\widetilde{e} \equiv \arg \min_{e' \in C_e} \overline{u}\left(e'\right)$$

We are going to show that $\overline{u}\left(\overline{x}\right) > a\overline{u}\left(\widetilde{e}\right) \equiv k$, with $a \in \mathbb{R}_{++}$. Then the result follows from our assumptions on the utility functions. Since u^n converges uniformly on compact sets, we have that

taken $\varepsilon_1 = \frac{1}{4}\left|\overline{u}\left(\widetilde{e}\right)\right|$, $\exists N$ such that $\forall n > N$ and $\forall x \in C_x \quad \overline{u}\left(x\right) > u^n\left(x\right) - \varepsilon_1$

Since $\{x^n : n \in \mathbb{N}\} \subseteq C_x$, we have that

$$\forall n > N, \quad \overline{u}\left(x^n\right) > u^n\left(x^n\right) - \varepsilon_1 \tag{15.20}$$

By definition of u^n, x^n and e^n, we have that

$$\forall n > N, \quad u^n\left(x^n\right) - \varepsilon_1 \geq u^n\left(e^n\right) - \varepsilon_1 \tag{15.21}$$

Similarly, since u^n converges uniformly on compact sets, we have

$$\exists M \text{ such that } \forall m > M \text{ and } \forall e \in C_e \qquad u^m\left(e\right) > \overline{u}\left(e\right) - \varepsilon_1$$

Since $\{e^n : n \in \mathbb{N}\} \subseteq C_e$, we have that

$$\forall m > M, \quad u^m\left(e^m\right) > \overline{u}\left(e^m\right) - \varepsilon_1 \tag{15.22}$$

Then for $n > N,M$, using (15.20), (15.21), (15.22) ,and the definition of $\widetilde{e}$, we get

$$\overline{u}\left(x^n\right) > u^n\left(x^n\right) - \varepsilon_1 \geq u^n\left(e^n\right) - \varepsilon_1 > \overline{u}\left(e^n\right) - 2\varepsilon_1 \geq \overline{u}\left(\widetilde{e}\right) - 2\varepsilon_1$$

Now, using the definition of ε_1, consider

$$k \equiv \overline{u}\left(\widetilde{e}\right) - 2\varepsilon_1 = \begin{cases} \frac{1}{2}\overline{u}\left(\widetilde{e}\right) & \text{if } \overline{u}\left(\widetilde{e}\right) \geq 0 \\ \frac{3}{2}\overline{u}\left(\widetilde{e}\right) & \text{if } \overline{u}\left(\widetilde{e}\right) < 0 \end{cases}$$

Therefore, for sufficiently large n,

$$\overline{u}\left(x^n\right) > k$$

Taking limits of both sides with respect to n and using the continuity of the function $\overline{u}$, we get the desired result

$$\overline{u}\left(\overline{x}\right) > k$$

$u^n \to \overline{u}$ implies that $Du^n \to D\overline{u}$ uniformly on compact sets, so $Du^n(x^n) \to D\overline{u}(\overline{x})$ holds. The remaining part of the proof follows the standard case in which an economy is just e. ■

5.2 Density

Define the function

$$\widehat{F} : \Xi \times \mathbb{R}^{n_1+p+k} \times \Theta \times \mathcal{U} \to \mathbb{R}^{n_1} \times \mathbb{R}^{n_1+m} \times \mathbb{R}$$

$$\widehat{F} : (\xi, c, \theta, u) \mapsto \text{left hand side of system (15.12)}$$

The density result is established if we can show that

1 0 is a regular value for $\widehat{F}$,

2 the number of variables (ξ, c) is less than the number of equations.

If 1 holds, then we could conclude that by the transversality theorem (Theorem 26 in Chapter 6), for (θ, u) in a dense subset of $\Theta \times \mathcal{U}$, the set of (ξ, c) for which (15.12) holds has a manifold structure; but 2 implies that the manifold must be of negative dimension, i.e., empty.

Showing 2 is a straightforward exercise, but there is a problem with respect to the proof of 1. (θ, u) is an element of the infinite dimensional set $\Theta \times \mathcal{U}$, while the transversality theorem proved in Chapter 6 applies to finite dimensional manifolds. The way out we choose is to look at a finite dimensional subset (submanifold) of that set, by taking advantage of the generic regularity of equilibria, see Condition 5 or 5', in a way described in the following subsections[9].

5.2.1 Perturbation technique

In this subsection[10], to make our argument easier to follows, we specialize $\mathcal{U}$ to be the set of utility functions. Therefore, $u \equiv (u_h)_{h=1}^{H}$ and

$$u_h : \mathbb{R}^{G}_{++} \to \mathbb{R}, \quad : x_h \mapsto u_h (x_h)$$

Suppose that Conditions 1-5 are verified and fix an arbitrary $(\bar{e}, \bar{u}) \in \mathcal{E}_r$, where $\mathcal{E}_r$ is defined in Condition 5.
By construction of $\mathcal{E}_r$, $(\bar{e}, \bar{u})$ has finitely many, say K, equilibria, each denoted by vectors $\bar{x}^k$, $k = 1, ..., K$. Moreover, there exist finitely many disjoint open balls $B^1, ..., B^K$, of $\mathbb{R}^{HG}_{++}$, such that $\bar{x}^k \in B^k$ for every k. So each B^k contains exactly one equilibrium allocation. Moreover, we can choose B^k such that its closure $\bar{B}^k$ is compact and there exist disjoint open balls $\hat{B}^k \subset \mathbb{R}^{GH}_{++}$ such that $B^k \subset \bar{B}^k \subset \hat{B}^k$, for every k.

[9] As a matter of fact, an infinite dimensional version of transversality theorem is known and applicable here (see [107]).
We prefer the finite dimensional argument since it takes advantage of the structure of the model well and thus gives us a deeper understanding.
[10] In this subsection we just adapt the section with the same title in CKV to our framework.

For each h, take a C^∞ function ρ_h from $\mathbb{R}^G_{++}$ to $[0,1]$ such that for every k, $\rho_h = 1$ on B^k and $\rho_h = 0$ on the complement of $\hat{B}^k$ [11].

Condition 7. Take $(\overline{u}_h)_{h=1}^H \in \mathcal{U}$. Let $\mathcal{A}_h$ be the set of $G \times G$ symmetric matrices. Take $A_h \in \mathcal{A}_h$ and consider the function

$$\begin{aligned}&\widetilde{u}_h\left(.,A_h\right) : \mathbb{R}^G_{++} \to \mathbb{R}\\ &\widetilde{u}_h(x_h, A_h) \equiv \bar{u}_h(x_h) + \tfrac{1}{2}\rho_h(x_h)\textstyle\sum_k[\left(x_h - \bar{x}_h^k\right) A_h \left(x_h - \bar{x}_h^k\right)]\end{aligned} \tag{15.23}$$

For each h, there exists an open neighborhood $\mathcal{A}_h^0$ of $0 \in \mathcal{A}_h$ such that for each $A_h \in \mathcal{A}_h^0$ we have that $\widetilde{u}_h\left(.,A_h\right) \in \mathcal{U}_h$.

Let's check that Condition 7 is satisfied in the case of utility functions with "nice" properties.

Lemma 25 *Assume $u_h \in \mathcal{U}_h$ if and only if*

1 *u_h is C^3;*

2 *u_h is differentiably strictly increasing, i.e., $\forall x_h \in \mathbb{R}^G_{++}$, $Du_h\left(x_h\right) \gg 0$;*

3 *u_h is differentiably strictly concave, i.e., $\forall x_h \in \mathbb{R}^G_{++}$, $D^2 u_h\left(x_h\right)$ is negative definite;*

4 *$\forall$or any $\underline{u} \in \mathbb{R}$, the closure in $\mathbb{R}^C$ of $Cl_{\mathbb{R}^C}\left\{x \in \mathbb{R}^C_{++} : u_h\left(x\right) \geq \underline{u}\right\}$ is contained in $\mathbb{R}^C_{++}$.*

Then Condition 7 holds.

Proof. 1. $\widetilde{u}_h\left(.,A_h\right)$ is C^3 because $\overline{u}_h$ is C^3 and ρ_h is smooth.
2. and 3.
To simplify the notation take $K = 1$.

$$\begin{aligned}D_{x_h}\widetilde{u}_h(x_h, A_h) = \;& D_{x_h}\bar{u}_h(x_h) + \tfrac{1}{2}\left[D_{x_h}\rho(x_h)\right]\left[\left(x_h - \bar{x}_h\right) A_h \left(x_h - \bar{x}_h\right)\right] +\\ &+\tfrac{1}{2}\rho_h(x_h) A_h \left(x_h - \bar{x}_h\right)\end{aligned}$$

$$D_{x_h}\left[\rho_h(x_h) A_h \left(x_h - \bar{x}_h^k\right)\right] = M\left(x_h, \overline{x}_h, A_h\right) + \rho_h(x_h) A_h$$

[11] The existence of such a function, called "bump function" is proved in Lemma 41, Chapter 7. See also Remark 42.

where the computation of $M(x_h, \overline{x}_h, A_h)$ is described below

$$\begin{array}{c} \rho_h(x_h)a_h^1(x_h-\bar{x}_h) \\ \rho_h(x_h)a_h^2(x_h-\bar{x}_h) \\ \dots \\ \rho_h(x_h)a_h^G(x_h-\bar{x}_h) \end{array} \begin{array}{c} \quad x_h^1 \qquad\qquad x_h^2 \qquad \dots \qquad x_h^G \\ \begin{pmatrix} \frac{\partial \rho_h}{\partial x_h^1}a_h^1(x_h-\bar{x}_h) & \frac{\partial \rho_h}{\partial x_h^2}a_h^1(x_h-\bar{x}_h) & & \frac{\partial \rho_h}{\partial x_h^G}a_h^1(x_h-\bar{x}_h) \\ \frac{\partial \rho_h}{\partial x_h^1}a_h^2(x_h-\bar{x}_h) & \frac{\partial \rho_h}{\partial x_h^2}a_h^2(x_h-\bar{x}_h) & & \frac{\partial \rho_h}{\partial x_h^G}a_h^2(x_h-\bar{x}_h) \\ & & & \\ \frac{\partial \rho_h}{\partial x_h^1}a_h^G(x_h-\bar{x}_h) & \frac{\partial \rho_h}{\partial x_h^2}a_h^G(x_h-\bar{x}_h) & & \frac{\partial \rho_h}{\partial x_h^G}a_h^G(x_h-\bar{x}_h) \end{pmatrix} \end{array}$$

where a_h^i is the i-th row of matrix A_h. Then,

$$\begin{aligned} D^2_{x_h}\widetilde{u}_h(x_h, A_h) = \;& D^2\bar{u}_h(x_h) + \tfrac{1}{2}\left[D^2\rho(x_h)\right]\left[(x_h-\bar{x}_h)A_h(x_h-\bar{x}_h)\right] + \\ & \tfrac{1}{2}\left[D\rho(x_h)\right]\textstyle\sum_k A_h\left(x_h-\bar{x}_h^k\right) + M(x_h, \overline{x}_h, A_h) + \rho_h(x_h)A_h \end{aligned}$$

We first prove the gradient of $\widetilde{u}_h$ is strictly positive. Observe that

$$D\bar{u}_h(x_h) \gg 0$$

Define

$$\begin{aligned} & f_h : \mathbb{R}^G_{++} \times \mathcal{A}_h \to \mathbb{R}^G, \\ & f_h : (x_h, A_h) \mapsto \tfrac{1}{2}\left[D_{x_h}\rho(x_h)\right]\left[(x_h-\bar{x}_h)A_h(x_h-\bar{x}_h)\right] + \tfrac{1}{2}\rho_h(x_h)A_h(x_h-\bar{x}_h) \end{aligned}$$

and therefore

$$D_{x_h}\widetilde{u}_h(x_h, A_h) = D\bar{u}_h(x_h) + f_h(x_h, A_h)$$

Observe that f_h is C^1 and that

$$\forall \alpha \in \mathbb{R},\ \forall (x_h, A_h) \in \mathbb{R}^G_{++} \times \mathcal{A}_h,\quad f_h(x_h, \alpha A_h) = \alpha f_h(x_h, A_h)$$

Restrict A_h to belong to a closed ball $\overline{N}_0$ around 0, and x_h to a compact set $C \subseteq \mathbb{R}^G_{++}$ containing B [12]. Since, by construction, $\forall A_h \in \mathcal{A}_h$ and $\forall x_h \in \mathbb{R}^G_{++} \backslash B$, $f_h(x_h, A_h) = 0$ it is as if $f_{|\mathbb{R}^G_{++} \times \overline{N}_0}$ were defined on a compact set. Therefore, there exists $L_1 \in \mathbb{R}$ such that

$$\forall (x_h, A_h) \in \mathbb{R}^G_{++} \times \overline{N}_0,\quad f_h(x_h, A_h) > L_1\mathbf{1}$$

If $L_1 \geq 0$, then we are done. If $L_1 < 0$, we proceed as follows.

Observe that since C is compact and $\overline{u}_h$ is C^2, there exists $L_2 > 0$ such that $\forall x_h \in C$

$$D_{x_h}\overline{u}_h(x_h) \gg L_2\mathbf{1} \gg 0$$

Now consider

$$\overline{N}_0' \equiv \left\{A_h' \in \mathcal{A}_h : \exists A_h \in \overline{N}_0 \text{ such that } A_h' = \frac{1}{2}\frac{L_2}{(-L_1)}A_h\right\}$$

[12]In the case of $K > 1$, we consider $\cup_{k=1}^K B_k$.

and $N_0' \equiv \text{Int}\, \overline{N}_0'$. Observe that N_0' is an open neighborhood of 0. Then $\forall A_h' \in N_0'$, we have

$$f_h\left(x_h, A_h'\right) = \frac{1}{2}\frac{L_2}{(-L_1)} f_h\left(x_h, A_h\right) > \frac{1}{2}\left(-L_2\right)$$

Then $\forall\left(x_h, A_h'\right) \in \mathbb{R}^G_{++} \times N_0'$,

$$D_{x_h}\widetilde{u}_h(x_h, A_h') = D_{x_h}\bar{u}_h(x_h) + f_h\left(x_h, A_h'\right) \gg \left[L_2 + \frac{1}{2}\left(-L_2\right)\right]\mathbf{1} = \frac{1}{2}L_2\mathbf{1} \gg 0$$

as desired. The proof of 3. is very similar to the above proof: it is here that we need u_h to be C^3.

4. It follows from the fact that u_h coincides with $\overline{u}_h$ outside the compact set C which is contained in $\mathbb{R}^G_{++}$. ■

Let $\mathcal{A}^0 \equiv \prod \mathcal{A}_h^0$ with generic element $A \equiv (A_h)_{h=1}^H$, and

$$\widetilde{u}(x, A) \equiv \left(\widetilde{u}_h(x_h, A_h)\right)_{h=1}^H$$

Observe that $\widetilde{u}(., 0) = \overline{u}$.

By construction, for any x_h sufficiently close to $\overline{x}_h^k$ - in fact for x_h such that $(x_h)_{h=1}^H \in B^k$- we have $\rho_h\left(x_h\right) = 1$ and all the derivatives of ρ_h are equal to 0, and thus

$$D_x\widetilde{u}_h(x_h, A_h) = D\overline{u}_h(x_h) + A_h(x_h - \overline{x}_h^k)$$

and

$$D_{xx}\widetilde{u}_h(x_h, A_h) = D^2\overline{u}_h(x_h) + A_h$$

Moreover given the matrix

$$A_h \equiv \begin{bmatrix} a_h^{11} & \cdots & a_h^{1g} & \cdots & a_h^{1G} \\ \cdots & & & & \\ a_h^{1g} & \cdots & a_h^{gg} & \cdots & a_h^{gG} \\ \cdots & & & & \\ a_h^{1G} & \cdots & a^{gG} & \cdots & a_h^{GG} \end{bmatrix}_{G\times G}$$

denote by $\widehat{A}_h$ the vector of distinct elements of the symmetric matrix A_h, i.e.,

$$\widehat{A}_h \equiv \left(a_h^{11}, ..., a_h^{1G}, a_h^{22}, ..., a_h^{2G}, ..., a_h^{GG}\right) \in \mathbb{R}^{\frac{1}{2}G(G+1)}$$

Define $\widehat{\mathcal{A}}_h^0$ as the open neighborhood of 0. in $\mathbb{R}^{\frac{1}{2}G(G+1)}$corresponding to $\mathcal{A}_h^0$ [13] and $\widehat{\mathcal{A}}^0 \equiv \times_{h=1}^H \widehat{\mathcal{A}}_h^0$ with generic element $\widehat{A}^0 \equiv \left(\widehat{A}_h^0\right)_{h=1}^H$. With innocuous abuse of notation identify $\widetilde{u}(x, A)$ with $\widetilde{u}(x, \widehat{A})$ [14].

Then we can define

$$F_A : \Xi \times \Theta \times \widehat{\mathcal{A}}^0 \to \mathbb{R}^{n_1}, \quad \left(\xi, \theta, \widehat{A}\right) \mapsto F\left(\xi, \theta, \widetilde{u}(., \widehat{A})\right)$$

$$G_A : \Xi \times T \times \Theta \times \widehat{\mathcal{A}}^0 \to \mathbb{R}^k, \quad \left(\xi, t, \theta, \widehat{A}\right) \mapsto G\left(\xi, t, \theta, \widetilde{u}(., \widehat{A})\right)$$

$$\widetilde{F}_A : \Xi \times T \times \Theta \times \widehat{\mathcal{A}}^0 \to \mathbb{R}^n,$$
$$\widetilde{F}_A : \left(\xi, t, \theta, \widehat{A}\right) \mapsto \left(F_1\left(\xi, t, \theta, \widetilde{u}(., \widehat{A})\right), F_2\left(\xi, t, \theta, \widetilde{u}(., \widehat{A})\right)\right)$$

$$\widehat{F}_A : \Xi \times \mathbb{R}^{n_1+p+k} \times \Theta \times \widehat{\mathcal{A}}^0 \to \mathbb{R}^{n_1} \times \mathbb{R}^{n_1+m} \times \mathbb{R},$$
$$\widehat{F}_A : \left(\xi, c, \theta, \widehat{A}\right) \mapsto \widehat{F}\left(\xi, c, \theta, \widetilde{u}(., \widehat{A})\right)$$

Observe that the only difference between the above functions and the functions $F, G, \widetilde{F}$ and $\widehat{F}$ is that in the former ones $\widehat{\mathcal{A}}^0$ is in the place of $\mathcal{U}$ in the corresponding domains.

For future use we compute some partial derivatives of $\widetilde{u}_h$. Define $\widehat{x}_h^g \equiv (x_h^g - \overline{x}_h^g)$.

Observe that

$$\begin{bmatrix} \widehat{x}^1 & \widehat{x}^2 & \widehat{x}^3 & \ldots & \widehat{x}^G \end{bmatrix} \begin{bmatrix} a_h^{11} & a_h^{12} & a_h^{13} & \ldots & a_h^{1G} \\ a_h^{21} & a_h^{22} & a_h^{23} & \ldots & a_h^{2G} \\ a_h^{31} & a_h^{32} & a_h^{33} & \ldots & a_h^{3G} \\ \ldots & & & \ddots & \\ a_h^{G1} & a_h^{G12} & a_h^{G3} & & a_h^{GG} \end{bmatrix} \begin{bmatrix} \widehat{x}^1 \\ \widehat{x}^2 \\ \widehat{x}^3 \\ \ldots \\ \widehat{x}^G \end{bmatrix} =$$

$$= \sum_{g=1}^G a_h^{gg} (\widehat{x}^g)^2 + \sum_{g,g'=1,\ldots,G, g<g'} 2a_h^{gg'} \widehat{x}^g \widehat{x}^{g'}$$

Therefore the derivative of the above expression with respect to $a_h^{gg'}$ is

$$\begin{cases} (\widehat{x}^g)^2 & \text{if} \quad g = g' \\ 2\widehat{x}^g \widehat{x}^{g'} & \text{if} \quad g < g' \end{cases}$$

[13]More formally, defined

$$\phi : A_h \to \mathbb{R}^{\frac{1}{2}G(G+1)}, \phi : A_h \mapsto \widehat{A}_h,$$

$\widehat{\mathcal{A}}_h^0 \equiv \phi\left(\mathcal{A}_h^0\right)$.

[14]More formally, we should write $\widetilde{u}_h\left(x_h; \widehat{A}_h\right) \equiv \widetilde{u}_h\left(x_h; \phi^{-1}\left(A_h\right)\right)$.

Taking $K = 1$ to simplify notation, we then have what follows.

$$D_{\widehat{A}}\widetilde{u}_h\left(x_h, A_h\right) = \left(\frac{\partial \widetilde{u}_h}{\partial a_h^{gg'}}\right)_{g,g',=1,\dots,G,g<g'}$$

is such that

$$\frac{\partial \widetilde{u}_h}{\partial a_h^{gg'}} = \begin{cases} \frac{1}{2}\rho_h(x_h)\left(\widehat{x}^g\right)^2 & \text{if} \quad g = g' \\ \rho_h(x_h)\widehat{x}^g\widehat{x}^{g'} & \text{if} \quad g < g' \end{cases}$$

Therefore

$$\frac{\partial^2 \widetilde{u}_h}{\partial a_h^{gg'}\partial x^j} = \begin{cases} \frac{1}{2}\frac{\partial \rho_h(x_h)}{\partial x^j}\left(\widehat{x}^g\right)^2 & \text{if} \quad g = g', \text{ and } j \neq g \\ \frac{1}{2}\frac{\partial \rho_h(x_h)}{\partial x^g}\left(\widehat{x}^g\right)^2 + \rho_h(x_h)\widehat{x}^g & \text{if} \quad g = g', \text{ and } j = g \\ \frac{1}{2}\frac{\partial \rho_h(x_h)}{\partial x^j}\widehat{x}^g\widehat{x}^{g'} & \text{if} \quad g < g' \text{and } j \neq g, g' \\ \frac{1}{2}\frac{\partial \rho_h(x_h)}{\partial x^j}\widehat{x}^g\widehat{x}^{g'} + \rho_h(x_h)\widehat{x}^{g'} & \text{if} \quad g < g' \text{and } j = g \\ \frac{1}{2}\frac{\partial \rho_h(x_h)}{\partial x^j}\widehat{x}^g\widehat{x}^{g'} + \rho_h(x_h)\widehat{x}^g & \text{if} \quad g < g' \text{and } j = g' \end{cases}$$

Then, computing the above expression at the equilibrium values $\overline{x}_h$, recalling the definition of ρ_h, we get

$$D_{\widehat{A}_h}\widetilde{u}_h(\overline{x}_h, A_h) = 0$$

$$D_{A_h,x_h}u_h(x_h, A_h) = D_{x_h,A_h}\widetilde{u}_h(x_h, A_h) = 0$$

5.2.2 A rank condition for density

Since that the domain of F_A,$\widetilde{F}_A$ and $\widehat{F}_A$ are finite dimensional space, we can apply the transversality theorem of Chapter 6. The computation of the jacobian matrix of $\widehat{F}_A$ is described below.

$$\begin{array}{l} \\ F_A\left(\xi,\theta,\widehat{A}\right) \\ c\left[D_{\xi,t}\left(\widetilde{F}_A, G_A\right)\right] \\ \frac{1}{2}cc-1 \end{array} \begin{array}{c} \begin{array}{cccc} \xi & c & \widehat{A} & \theta \end{array} \\ \begin{pmatrix} D_\xi F & 0 & 0 & * \\ * & \left[D_{\xi,t}\left(\widetilde{F}_A, G_A\right)\right]^T & N & * \\ 0 & c & 0 & 0 \end{pmatrix} \end{array}$$

where N be the jacobian matrix of $c\left[D_\xi, t\left(\widetilde{F}, G\right)\right]$ with respect to $\widehat{A}$.

Remark 26 Our version of the transversality theorem can be applied to C^r functions, with r at least equal to 1 - see Theorem 26 in Chapter 6. Since the expression $c\left[D_{\xi,t}\left(\widetilde{F}_A, G_A\right)\right]$ contains the Hessian of the utility function and since we are computing derivatives with respect to x, we need to have u of class at least C^3.

At the equilibrium points $D_{\widehat{A}_h}\widetilde{u}_h(\overline{x}_h, \widehat{A}_h) = 0$ for every h, and therefore

$$D_{\widehat{A}}F_A\left(\xi, \theta, \widehat{A}\right) = 0$$

Example 27 (CKV model - resumed) The computation of N is described below. Observe that the only difference between $D_{(\xi,t)}(F, G)$ and $D_{(\xi,t)}(F_A, G_A)$ is in the superrows corresponding to $D_{x_h}u_h(x_h)$ and $D_{x_h}\widetilde{u}_h(x_h)$, respectively. In fact, those superrows are the only ones where $\widehat{A}_h$ appears. Therefore, N contains zero entries apart from those corresponding to the jacobian matrix of the left hand side of equations $(1.1), ..., (H.1)$ in system (15.18). Denote the partial jacobian of the left hand side of those equations for household h with respect to $\widehat{A}_h$ by $N(c_{x_h})$. Since

$$c_{x_h}A_h = \begin{bmatrix} c_{x_h}^1 & c_{x_h}^2 & \dots & c_{x_h}^G \end{bmatrix} \begin{bmatrix} a_h^{11} & a_h^{12} & \dots & a_h^{1G} \\ a_h^{12} & a_h^{22} & \dots & a_h^{2G} \\ & & \ddots & \\ a_h^{1G} & a_h^{2G} & \dots & a_h^{GG} \end{bmatrix} =$$

$$= \begin{bmatrix} \sum_{g=1}^G c_{x_h}^g a_h^{1g} & c_{x_h}^1 a_h^{12} + \sum_{g=2}^G c_{x_h}^g a_h^{2g} & \dots & \sum_{g=1}^G c_{x_h}^g a_h^{gG} \end{bmatrix}$$

the computation $N(c_{x_h})$ can be described as follows:

$$\begin{array}{l} \\ \sum_{g=1}^G c_{x_h}^g a_h^{1g} \\ c_{x_h}^1 a_h^{12} + \sum_{g=2}^G c_{x_h}^g a_h^{2g} \\ \sum_{g=1}^2 c_{x_h}^g a_h^{g3} + \sum_{g=3}^G c_{x_h}^g a_h^{3g} \\ \\ \ldots\ldots \\ \sum_{g=1}^G c_{x_h}^g a_h^{gG} \end{array} \quad \begin{array}{c} \begin{array}{ccccccccc} a_h^{11} & a_h^{12} & \dots & a_h^{1G} & a_h^{22} & .. & a_h^{2G} & .. & a_h^{GG} \end{array} \\ \left(\begin{array}{ccccccccc} c_{x_h}^1 & c_{x_h}^2 & \dots & c_{x_h}^G & & & & & \\ & c_{x_h}^1 & \dots & 0 & c_{x_h}^2 & .. & c_{x_h}^G & & \\ & & \dots & 0 & 0 & .. & 0 & & \\ & & \ddots & & & \ddots & & \dots & \\ & & & c_{x_h}^1 & 0 & \dots & c_{x_h}^2 & & c_{x_h}^G \end{array} \right) \end{array}$$

The following lemma is then obvious

Lemma 28 *$N(c_{x_h})$ has full rank if and only if $c_{x_h} \neq 0$.*

(to be continued).

We are now ready to introduce a further condition which plays a crucial role in proving that $\mathcal{E}_E$ is dense.

Condition 8. One of the following conditions holds.

1 0 is a regular value for $\widehat{F}_A$.

2 The matrix

$$M\left(\xi,\theta,\widehat{A}\right) \equiv \left[\begin{array}{cc} \left[D_{(\xi,t)}\left(\widetilde{F}_A, G_A\right)\right]^T & N \\ c & 0 \end{array}\right]$$

computed at points in $\widehat{F}_A^{-1}(0)$ has full row rank.

Remark 29 Recall that if $0 \notin \operatorname{Im} \widehat{F}_A$, i.e., $\widehat{F}_A\left(\xi, c, \theta, \widehat{A}\right) = 0$ has no solution, Condition 8.1 does hold.

Remark 30 If Conditions 1-4 are verified then the above two statements are equivalent.

Theorem 31 *If Conditions 1-8 are verified, then the planner intervention is effective, i.e., $\mathcal{E}_E$ is an open and dense subset in $\Theta \times \mathcal{U}$.*

Proof. Openess follows from Theorem 23. Denseness follows from the argument below.

Since Condition 8 is verified, and since

$$\dim\ (\text{domain of } \widehat{F}_A) = 2n_1 + p + k < 2n_1 + m + 1 = \dim(\text{value set of } \widehat{F}_A)$$

from the transversality theorem, we have that in a full measure subset D_A^* of $\Theta \times \widehat{\mathcal{A}}_0$ the following system has no solution (ξ, c)

$$\begin{cases} F_A\left(\xi,\theta,\widehat{A}\right) & = 0 \quad (1) \\ c\left[D_{(\xi,t)}\left(\widetilde{F}_A\left(\xi,\psi(\xi),\theta,\widehat{A}\right), G_A\left(\xi,\overline{t},\theta,\widehat{A}\right)\right)\right] & = 0 \quad (2) \\ \frac{1}{2}cc - 1 & = 0 \quad (3) \end{cases}$$

Since full measure implies denseness, for any arbitrary $\left(\theta, \widehat{A}\right) \in \Theta \times \widehat{\mathcal{A}}_0$ and for any open neighborhood $\mathcal{O}_\theta \times \mathcal{A}_u$ of $\left(\theta, \widehat{A}\right)$, there exists $(\theta', A') \in \mathcal{O}_\theta \times \mathcal{A}_u$ such that (15.12) has no solution i.e. $(\theta', A') \in D_A^*$. Take an arbitrary $(\theta, u) \in \Theta \times \mathcal{E}_r$ (this set has been defined in Condition 5) and an arbitrary open neighborhood $\mathcal{O}^*$ of (θ, u). We want to show that $\exists(\theta', u') \in \mathcal{O}^*$ such that system (15.12) has no solution. By definition of product topology there exist two open neighborhood $\mathcal{O}_\theta^*$ of θ and $\mathcal{O}_u^*$ of u such that $\mathcal{O}^* \supseteq \mathcal{O}_\theta^* \times \mathcal{O}_u^*$.

We now have to show that there exists a sufficiently small enough open ball $\mathcal{A}^{0*} \subseteq \mathcal{A}^0$ around $0 \in \mathcal{A}^0$ such that

$$\{u(x, A)\}_{A \in \mathcal{A}^{0*}} \subseteq \mathcal{O}_u^* \tag{15.24}$$

That result is shown in Lemma 32 below.

From definition of D_A^*, there exists (θ', A') such that

$$\begin{array}{ll} (1) & (\theta', A') \in D_A^* \cap \mathcal{O}_\theta^* \times A_u \\ (2) & \left(\theta', u(x, A')\right) \in \mathcal{O}_\theta^* \times \mathcal{O}_u^* \end{array}$$

This implies that for every arbitrary economy $(\theta, u) \in \mathcal{E}_r$ and for every neighborhood $\mathcal{O}^*$, we can find an economy $(\theta', A') \in \mathcal{O}^*$ such that (15.12) has no solution. Hence we have proved the existence of a dense set $\mathcal{E}_E$ in $\mathcal{E}_r$ .This completes the proof since $\mathcal{E}_r$ is a dense set in $\Theta \times \mathcal{U}$. ∎

Lemma 32 *There exists a sufficiently small open ball* $\mathcal{A}^{0*} \subseteq \mathcal{A}^0$ *around* $0 \in \mathcal{A}^0$ *such that* $\{u(x, A)\}_{A \in \mathcal{A}^{0*}} \subseteq \mathcal{O}_u^*$.

Proof. We know that the definitory subbasis $\mathcal{S}$ for the compact open topology on $\mathcal{F}(X, Y)$ is

$$\mathcal{S} = \{\mathcal{F}(A, G) : A \in \mathcal{A}, G \in \mathcal{G} \text{ and } f(A) \subseteq G\}$$

where $\mathcal{A}$ is the class of compact subsets of X and $\mathcal{G}$ is the class of the open subsets of Y, and an element of the basis $\mathcal{B}$ is the union of elements of the subbasis (see Definition p. 285, [90]). By definition of basis, for every $\overline{u} \in \mathcal{U}$ and for every open neighborhood $\mathcal{O}_{\overline{u}}^*$ of $\overline{u}$, there exists an element $B \in \mathcal{B}$ such that $u \in B \subseteq \mathcal{O}_{\overline{u}}^*$. We want now to show that there exists $\mathcal{A}^{0*} \subseteq \mathcal{A}^0$ around $0 \in \mathcal{A}^0$ such that

$$\{u(x, A)\}_{A \in \mathcal{A}^{0*}} \subseteq \mathcal{B}$$

and therefore (15.24) holds.

Recall that for each h, we defined

$$\widetilde{u}_h(x_h, A_h) \equiv \bar{u}_h(x_h) + \frac{1}{2}\rho_h(x_h) \sum_k [\left(x_h - \bar{x}_h^k\right) A_h \left(x_h - \bar{x}_h^k\right)]$$

and that, by definition of $\widetilde{u}_h$,

1. $\widetilde{u}_h$ and $\overline{u}_h$ agree outside the compact set $\cup_{k=1}^K \overline{B}^k$,
2. $\widetilde{u}_h(x_h, A_h) - \bar{u}_h(x_h)$ is a continuous function of A_h, and
3. $\widetilde{u}_h(x_h, A_h) - \bar{u}_h(x_h) = 0$ if $A_h = 0$. Then, $\forall \varepsilon' > 0$ there exists $\delta > 0$ such that if $\|A_h\| < \delta$ then $|\widetilde{u}_h(x_h, A_h) - \bar{u}_h(x_h)| < \varepsilon'$. Define

$$\{u \in \mathcal{U} : u(C_i) \subseteq G_i\} \equiv S_i \qquad \text{and} \qquad B \equiv \cap_{i=1}^n S_i$$

Since for every i, $\overline{u} \in S_i$, it follows that there exists $\varepsilon_i > 0$ such that

$$S_i^* \equiv \{u \in \mathcal{U} : u(C_i) \subseteq (\overline{u} - \varepsilon_i, \overline{u} + \varepsilon_i)\} \subseteq S_i$$

Then $\cap_{i=1}^n S_i^* \subseteq B$.

For each i, define the following set of numbers

$$K_i \equiv \left\{k \in \mathbb{N} : \widehat{B}_k \cap C_i\right\}$$

Take a small enough neighborhood $\mathcal{A}^{0*}$ of $0 \in \mathcal{A}^0$ such that[15]

$$\forall\, A \in \mathcal{A}^{0*},\ \|u(.,A) - \overline{u}\| < \min_i \varepsilon_i$$

Then

$$\{u(x,A)\}_{A \in \mathcal{A}^{0*}} \subseteq \cap_{i=1}^n S_i^* \subseteq B \subseteq \mathcal{O}_{\overline{u}}^*$$

as desired. ■

6. Verification of conditions for planner effectiveness

In the remaining part of this chapter, we present a procedure to check that Condition 8 is verified, using again the model by CKV.

In Step 2 below, to show that Condition 8.1 holds in the sense that $0 \notin \operatorname{Im} \widehat{F}_A$, we need to exploit a condition which is true only in an open and dense subset of the space of the economies. Therefore we restrict the domain of $\widehat{F}_A$ to $\Xi \times \mathcal{D}_{rr}$, where

$$\mathcal{D}_{rr} \equiv \left\{\left(e, \widehat{A}\right) \in \mathbb{R}_{++}^{GH} \times \widehat{\mathcal{A}}^0 : \left(e, \widetilde{u}\left(., \widehat{A}\right)\right) \in \mathcal{E}_{rr}\right\}$$

and

$$\mathcal{E}_{rr} \equiv \left\{(e,u) \in \mathcal{E}_r : F(\xi, e, u) = 0 \Rightarrow \forall\, h, h' \text{ with } h \neq h',\ \frac{\lambda_h^1}{\lambda_h^0} \neq \frac{\lambda_{h'}^1}{\lambda_{h'}^0}\right\}$$

We divide our analysis in steps.

6.1 Step 1

We first of all, specify the matrix $M\left(\xi, c, \theta, \widehat{A}\right)$ in CKV model. To prove that it does have full rank, we use the perturbation method. With this goal in mind, we write M, specifying the number of equations and matching variables next to each group of equations. We put a * next to each full row rank submatrix of matrix; then, you can perturb the corresponding superrow using the associated subvector of c. For each superrow you should have a starred matrix, otherwise we cannot perturb it. Therefore, we can write matrix M as follows

[15] The norm is the usual one.

$$
\begin{array}{l}
 \\
G\ (1.1) \\
S'\ (1.2) \\
I\ (1.3) \\
\\
\dots \\
G\ (H.1) \\
S'\ (H.2) \\
I\ (H.3) \\
G-S'\ (4) \\
I\ (5) \\
H\ (6) \\
2\ (6') \\
1\ (7)
\end{array}
\begin{array}{ccccccccccccc}
c_{x_1} & c_{\lambda_1} & c_{b_1} & \cdots & c_{x_H} & c_{\lambda_H} & c_{b_H} & c_{p\backslash} & c_q & c_{t_2} & c_g & \widehat{A}_1 & \widehat{A}_H \\
D^2u_1{}^* & -\Phi(p)^T & & & & & & \widetilde{I} & & & Du_1 & N\left(c_{x_1}\right)^{?*} & \\
-\Phi(p)^* & & \left[\begin{smallmatrix}-q\\Y\end{smallmatrix}\right] & & & & & & & & & & \\
 & \left[\begin{smallmatrix}-q\\Y\end{smallmatrix}\right]^{T*} & & & & & & & I^* & & & & \\
 & & & \ddots & & & & & & & & & \\
 & & & & D^2u_H^* & -\Phi(p)^T & & \widetilde{I} & & & Du_H & & N\left(c_{x_H}\right)^{?*} \\
 & & & & -\Phi(p)^* & & \left[\begin{smallmatrix}-q\\Y\end{smallmatrix}\right] & & & & & & \\
 & & & & & \left[\begin{smallmatrix}-q\\Y\end{smallmatrix}\right]^{T*} & & & I^* & & & & \\
-\Lambda_1^* & Z_1^T & & \cdots & -\Lambda_H^* & Z_H^T & & & & & & & \\
 & \left[b_1^T 0\right] & -\lambda_1^0 I^* & \cdots & & \left[b_H^T 0\right] & -\lambda_H^0 I^* & & & & & & \\
 & & & & & \left[\begin{array}{c}\varphi_{\tau_H}(1)\\0\end{array}\right]^T & & & & \left[\varphi_t\left(1\right)\right]^T & & & \\
 & I_2^* 0 & & & & & & & & I_2 & & & \\
c_{x_1} & c_{\lambda_1} & c_{b_1} & & c_{x_H} & c_{\lambda_H} & c_{b_H} & c_{p\backslash} & c_q & c_{t_2} & c_g^* & &
\end{array}
\qquad (15.25)
$$

From Remark 19, if $c_g = 0$ then Condition 8 is immediately verified. Therefore we are left to consider$c_g \neq 0$ (this justify the $(*)$ next to c_g in the above matrix). Furthermore, we wrote

$$N\left(c_{x_h}\right)^{?*}$$

to indicate the fact that $N\left(c_{x_h}\right)$ has full row rank if and only if $c_{x_h} \neq 0$, as explained in Lemma 28. In fact, the following step analyzes the cases related to the possibility that some or all c_{x_h} are equal to 0.

6.2 Step 2

Case 1. For every h, $c_{x_h} \neq 0$, and therefore $N\left(c_{x_h}\right)$ has full row rank.

Case 2. For every h, $c_{x_h} = 0$. We prove Condition 8.1 showing that there is no solution to $\widehat{F}_{A|\mathcal{D}_{rr}}\left(\xi, c, \theta, \widehat{A}\right) = 0$.

Case 3. There exists an h such that $c_{x_h} = 0$. In this case we show that a subsystem of $\widehat{F}_{A|\mathcal{D}_{rr}}\left(\xi, c, \theta, \widehat{A}\right) = 0$ has no solution satisfying the condition of this case.

Even though Case 2 is a subcase of Case 3, it is useful to analyze the former one as well simply because it gives useful hints for the analysis of the latter one.

6.2.1 Case 1.$c_{x_h} \neq 0$ for every h.

For $h = 1, ..., H$, define

$$c^{\bullet}_{x_h} \equiv \left(c_{x^{sC}_h}\right)_{s=0,1,\ldots,S}, \quad c^{\backslash}_{x_h} \equiv \left(c^{\backslash}_{x^{sc}_h}\right)_{c\neq C}$$

By using the perturbation methods, we show that matrix (15.25) has full rank.

$$(h.1) \ \leftarrow \widehat{A}_h$$

$$\begin{array}{lll} (h.2) & \leftarrow c^{\bullet}_{x_h} & \rightsquigarrow (h.1) \\ (h.1) & \leftarrow \widehat{A}_h & \end{array}$$

$$\begin{array}{lll} (h.3) & \leftarrow c_{\lambda_h} & \rightsquigarrow (h.1) \\ (h.1) & \leftarrow \widehat{A}_h & \end{array}$$

$$\begin{array}{lll} (4) & \leftarrow c^{\backslash}_{x_1} & \rightsquigarrow (1.1), (1.2), \\ (1.2) & \leftarrow c^{\bullet}_{x_1} & \rightsquigarrow (1.1) \\ (1.1) & \leftarrow \widehat{A}_1 & \end{array}$$

$$\begin{array}{lll} (5) & \leftarrow c_{b_1} & \rightsquigarrow (1.2) \\ (1.2) & \leftarrow c^{\bullet}_{x_1} & \rightsquigarrow (1.1) \\ (1.1) & \leftarrow \widehat{A}_1 & \end{array}$$

$$\begin{array}{lll}
(6) & \leftarrow c_{\widetilde{\lambda}} & \rightsquigarrow (h.1) \text{ for } h \neq 1, (h.3) \text{ for } h \neq 1, (4), (5) \\
(h.3) & \leftarrow (c_{\lambda_h})_{h=2}^{H} \backslash c_{\widetilde{\lambda}} & \rightsquigarrow (h.1), (4), (5) \\
(5) & \leftarrow c_{b_h} & \rightsquigarrow (h.2) \\
(4) & \leftarrow c_{\overset{\backslash}{x}_h} & \rightsquigarrow (h.1), (h.2) \\
(h.2) & \leftarrow c_{\overset{\bullet}{x}_h} & \rightsquigarrow (h.1) \\
(h.1) & \leftarrow \widehat{A}_h &
\end{array}$$

where $(c_{\lambda_h})_{h=2}^{H} \setminus c_{\widetilde{\lambda}}$ is the vector made up of components of $(c_{\lambda_h})_{h=2}^{H}$ we have not used in the perturbation of row (6)[16].

$$\begin{array}{lll}
(6') & \leftarrow \left(c_{\lambda_1^0}, c_{\lambda_1^1}\right) & \rightsquigarrow (1.1), (1.3), (4), (5) \\
(1.3) & \leftarrow \left(c_{\lambda_1^s}\right)_{s \neq 0,1} & \rightsquigarrow (1.1), (4), (5) \\
(5) & \leftarrow c_{b_1} & \rightsquigarrow (1.2) \\
(4) & \leftarrow c_{\overset{\backslash}{x}_1} & \rightsquigarrow (1.1), (1.2) \\
(1.2) & \leftarrow c_{\overset{\bullet}{x}_1} & \rightsquigarrow (1.1) \\
(1.1) & \leftarrow \widehat{A}_1 &
\end{array}$$

$$\begin{array}{lll}
(7) & \leftarrow c_g & \rightsquigarrow (h.1)_{h=1}^{H} \\
(h.1)_{h=1}^{H} & \leftarrow \widehat{A}_h &
\end{array}$$

Remark 33 In our example we are in fact able to get the above full rank result in an easier way. Consider the submatrix M' of (15.25) obtained by erasing the rows (6),$(6')$ and (7). M' contains a submatrix which is equal to $D_\xi \widetilde{F}$,which has rank equal to the rows of M' - provided the Condition 5 or 5' is verified-. Therefore, M' itself has full row rank. From Proposition 45 in Chapter 1, it is then enough to perturb just the last rows (6) $(6')$ and (7).

[16]See equation (6) in system (15.18) and definition of $c_{\widetilde{\lambda}}$ (15.19)

6.2.2 Case 2. $c_{x_h} = 0$ for every h

In this case, $N_h(c_x)$ does not have full rank for every h. Moreover, system (15.18) becomes

$$\left\{\begin{array}{lll}
G & -c_{\lambda_1}\Phi(p) + c_{p\backslash}\widetilde{I} + c_g D u_1 & = 0 \quad (1.1) \\
S' & c_{b_1}\left[\begin{smallmatrix}-q\\Y\end{smallmatrix}\right] & = 0 \quad (1.2) \\
G - S' & c_{\lambda_1}\left[\begin{smallmatrix}-q\\Y\end{smallmatrix}\right] + c_q I & = 0 \quad (1.3) \\
 & \ldots & \\
G & -c_{\lambda_H}\Phi(p) + c_{p\backslash}\widetilde{I} + c_g D u_H & = 0 \quad (H.1) \\
S' & c_{b_H}\left[\begin{smallmatrix}-q\\Y\end{smallmatrix}\right]^T & = 0 \quad (H.2) \\
I & c_{\lambda_H}\left[\begin{smallmatrix}-q\\Y\end{smallmatrix}\right] + c_q I & = 0 \quad (H.3) \\
G - S' & c_{\lambda_1} Z_1^T + \sum_{h\neq 1} c_{\lambda_h} Z_h^T & = 0 \quad (4) \\
I & c_{\lambda_1}\begin{bmatrix} b_1 \\ 0 \end{bmatrix} - \lambda_1^0 c_{b_1} I + \sum_{h\neq 1} c_{\lambda_h}\begin{bmatrix} b_h \\ 0 \end{bmatrix} - \sum_{h\neq 1}\lambda_h^0 c_{b_h} I & = 0 \quad (5) \\
H & c_{\widetilde{\lambda}} I + c_{t_2}\varphi_t(1) & = 0 \quad (6) \\
2 & \left[c_{\lambda_1^0}, c_{\lambda_1^1}\right] I + c_{t_2}\varphi_t(2) & = 0 \quad (6') \\
1 & c_{\lambda_1}^T c_{\lambda_1} + c_{b_1} c_{b_1} + c_{\lambda_h} c_{\lambda_h} + c_{b_h} c_{b_h} + c_{p\backslash} c_{p\backslash} + & \\
 & + c_q c_q + c_{t_2} c_{t_2} + c_g c_g & = 0 \quad (7)
\end{array}\right.$$

Using the first order conditions[17], equation $(h.1)$ becomes

$$-c_{\lambda_h}\Phi(p) + c_{p\backslash}\widetilde{I} + c_g\lambda_h\Phi(p) = 0$$

From the definition of the matrix $\widetilde{I}$, the previous equation implies that

$$\begin{aligned} c_{\lambda_h} &= \lambda_h c_{g_h} \\ c_{p\backslash} &= 0 \end{aligned}$$

From $(h.2)$, we have $c_{b_h} = 0$.

From $(h.3)$, exploiting the fact that $c_{\lambda_h} = \lambda_h c_{g_h}$ and the first order condition $\left[\begin{smallmatrix}-q\\Y\end{smallmatrix}\right]^T \lambda_h = 0$, we get

$$0 = \lambda_h c_{g_h}\begin{bmatrix} -q \\ Y \end{bmatrix} + c_q I = c_q$$

[17]The use of first order condition is extremely useful for the study of both cases 2 and 3.

Similar argument applies to household 1's variables. Hence, equations (6) and (6′) become

$$\begin{cases} c_{g_2}\lambda_2^0 + c_{T_1^0} = 0 \\ \vdots \\ c_{g_H}\lambda_H^0 + c_{T_1^0} = 0 \\ c_{g_2}\lambda_2^1 + c_{T_1^1} = 0 \\ c_{g_1}\lambda_0^1 + c_{T_1^0} = 0 \\ c_{g_1}\lambda_1^1 + c_{T_1^1} = 0 \end{cases}$$

or

$$\begin{cases} c_{g_1}\lambda_1^0 + c_{T_1^0} = 0 \\ c_{g_2}\lambda_2^0 + c_{T_1^0} = 0 \\ \cdots \\ c_{g_H}\lambda_H^0 + c_{T_1^0} = 0 \\ c_{g_1}\lambda_1^1 + c_{T_1^1} = 0 \\ c_{g_2}\lambda_1^1 + c_{T_1^1} = 0 \end{cases} \tag{15.26}$$

Now equation (15.26) admits the solution $(c_g, c_{t_2}) = 0$. If we can prove that $(c_g, c_{t_2}) = 0$ is the only solution of system (15.26), then we can conclude in case 2, system (15.12) has no solution since from (7) we have $0 - 1 = 0$

Notice that $(c_g, c_{t_2}) = 0$ is the only solution of (15.26) if the following matrix has full rank:

$$\begin{bmatrix} \lambda_1^0 & & 0 & & \lambda_1^1 & 0 \\ & \lambda_2^0 & & & 0 & \lambda_2^1 \\ 0 & & \ddots & & & \\ & & & \lambda_H^0 & & \\ 1 & 1 & \cdots & 1 & 0 & 0 \\ 0 & 0 & 0 & & 1 & 1 \end{bmatrix}$$

This is true if and only if the matrix

$$\begin{bmatrix} \lambda_1^0 & & & & \\ & \ddots & & & \\ & & \lambda_H^0 & & \\ & & & -\frac{\lambda_1^1}{\lambda_1^0} & -\frac{\lambda_2^1}{\lambda_2^0} \\ & & & 1 & 1 \end{bmatrix}$$

has full rank. That condition is verified if and only if

$$\frac{\lambda_1^1}{\lambda_1^0} \neq \frac{\lambda_2^1}{\lambda_2^0}$$

which is true by definition of $\mathcal{D}_{rr}$.

6.2.3 Case 3. There exists an h such that $c_{x_h} = 0$

Without any loss of generality we can assume that $c_{x_1} = 0$[18]. Consider again system (15.18). Since some variables are zero, we can eliminate the equations which are verified in the form $0 = 0$. Observe also that from case 2 we have :

$$\begin{cases} c_{\lambda_1} & = \lambda_1 c_{g_1} \\ c_{b_1} & = 0 \\ c_{p\backslash} & = 0 \\ c_q & = 0 \end{cases}$$

On the base of the above remarks, system (15.18) becomes

$$\left\{ \begin{array}{lll} G & -\lambda_1 c_{g_1} \Phi(p) + c_{g_1} Du_1 & = 0 \quad (1.1) \\ G - S' & c_{\lambda_1} \left[\begin{smallmatrix} -q \\ Y \end{smallmatrix} \right] & = 0 \quad (1.3) \\ & \cdots & \\ G & c_{x_H} D^2 u_H + -c_{\lambda_H} \Phi(p) + c_{g_H} Du_H & = 0 \quad (H.1) \\ S+1 & -c_{x_H} \Phi(p)^T + c_{b_H}^T \left[\begin{smallmatrix} -q \\ Y \end{smallmatrix} \right] & = 0 \quad (H.2) \\ I & c_{\lambda_H} \left[\begin{smallmatrix} -q \\ Y \end{smallmatrix} \right] & = 0 \quad (H.3) \\ G - S' & c_{\lambda_1} Z_1^T - \sum_{h \neq 1} c_{x_h} \Lambda_h + \sum_{h \neq 1} c_{\lambda_h} Z_h^T & = 0 \quad (4) \\ I & c_{\lambda_1} \begin{bmatrix} b_1 \\ 0 \end{bmatrix} - \lambda_1^0 c_{b_1} I + \sum_{h \neq 1} c_{\lambda_h}^T \begin{bmatrix} b_h \\ 0 \end{bmatrix} - \sum_{h \neq 1} \lambda_h^0 c_{b_h} I & = 0 \quad (5) \\ H & c_{\tilde{\lambda}} I + c_{t_2} \varphi_t (1) & = 0 \quad (6) \\ 2 & \left[c_{\lambda_1^0}, c_{\lambda_1^1} \right] I + c_{t_2} \varphi_t (2) & = 0 \quad (6') \\ 1 & c_{\lambda_1} c_{\lambda_1} + c_{b_1} c_{b_1} + c_{x_h} c_{x_h} + c_{\lambda_h} c_{\lambda_h} + c_{b_h} c_{b_h} + c_{t_2} c_{t_2} + c_g c_g & = 0 \quad (7) \end{array} \right.$$

We want to show that $c_{x_1} = 0 \Rightarrow c_{x_h} = 0$ for every h, since this statement allows us to go back to Case 2. Postmultiplying $(h.2)$ by λ_h, we get

$$0 = -c_{x_h} \Phi(p) \lambda_h + c_{b_h} \begin{bmatrix} -q \\ Y \end{bmatrix} \lambda_h = -c_{x_h} \Phi(p) \lambda_h \qquad (15.27)$$

since from first order condition we have $\lambda_h \left[\begin{smallmatrix} -q \\ Y \end{smallmatrix} \right] = 0$. Postmultiplying equation $(h.1)$ by c_{x_h}, we get

$$0 = c_{x_h} D^2 u_h c_{x_h} - c_{\lambda_h} \Phi(p) c_{x_h} + c_{g_h} Du_h c_{x_h} \qquad (15.28)$$

Transposing first order condition $Du_h - \lambda_h \Phi(p) = 0$ and premultiplying it by c_{x_h}, we have

$$c_{x_h} Du_h = c_{x_h} \Phi(p) \lambda_h = 0 \qquad (15.29)$$

where the last equality comes form (15.27).
Plugging (15.29) in (15.28), we get

$$0 = c_{x_h} D^2 u_h c_{x_h} - c_{\lambda_h} \Phi(p) c_{x_h} \qquad (15.30)$$

[18] The following argument does not involve equations (6) or (6′) with respect to which household 1 is treated differently with respect to the other households.

Transposing $(h.2)$ and premultiplying it by c_{λ_h}, we get

$$0 = -c_{\lambda_h}\Phi(p)c_{x_h} + c_{\lambda_h}\begin{bmatrix}-q \\ Y\end{bmatrix} c_{b_h} = -c_{\lambda_h}\Phi(p)c_{x_h}, \tag{15.31}$$

where the last equality comes from $(h.2)$. Plugging (15.31) in (15.30), we get

$$0 = c_{x_h} D^2 u_h c_{x_h}$$

Since $D^2 u_h$ is a negative definite matrix, we finally get that $c_{x_h} = 0$, as desired.

It is important to observe that in this example, the fact $c_{x_1} = 0 \Rightarrow c_{x_h} = 0$ for every h, plays a fundamental rule in handling case 3. Thanks to this implication we are able to reduce case 3 to case 2. Unfortunately, this is not generally true: in order to get the desired result, some further steps are necessary (see for example [20], [38] and [39]). We now try to give a general recipe to analyzes those more difficult cases.

1 By the assumption of the case you are analyzing, some components of the vector c are null. Replace those components with zeros in system (15.18). Some equations "disappear", in the sense that they are trivially satisfied $(0 = 0)$.

2 Count the number of equations and unknowns which are left.

3 The number of equations you can erase from the system obtained in step 1 is equal to (# equations) - (# variables) -1. You still need to have more equations than unknowns. Of course, the idea is to drop equation you are not able to perturb.

4 You can finally apply the perturbation method to the matrix associated with the system obtained in step 3.

References

[1] Abraham R., Marsden J. E., and T. Ratiu (1988), *Manifolds, Tensor Analysis and Applications*, Springer-Verlag, New York.

[2] Allen B. (1981), Utility perturbations and the equilirbium price set, *Journal of Mathematical Economics*, **8**, 277-307.

[3] Apostol T. M. (1974), *Mathematical Analysis*, second edition, Addison-Wesley Publishing Company, Reading, MA, 1974.

[4] Arrow K. J. (1953), Le Role de Valeurs Boursières pour la Répartition la Meilleure des Risques, *Econométrie, Colloques Internationaux du Centre National de la Recherche Scientifique*, **40**, 41–47; English translation: (1964), The Role of Securities in the Optimal Allocation of Risk Bearing, *Review of Economic Studies*, **31**, 91–96.

[5] Arrow K. J., and Debreu G. (1954), Existence of an Equilibrium for a Competitive Economy, *Econometrica*, **22**, 265–290.

[6] Arrow K. J., and Hahn F. (1971), *General Competitive Analysis*, Holden Day, San Francisco.

[7] Avriel M., Diewert W.E., Schaible S., and Zang I. (1988), *Generalized Concavity, Mathematical Concepts and Methods in Science and Engineering*, vol.36, Plenum Press, New York.

[8] Balasko Y.(1976), L'équilibre économique du point de vue différentiel, Doctoral thesis, Université de Paris, IX.

[9] Balasko Y. (1988), *Foundations of the Theory of General Equilibrium*, Academic Press, Boston.

[10] Balasko Y., and Cass D. (1989), The structure of financial equilibrium with exogenous yields: I. Unrestricted participation, *Econometrica*, **57**, 135-62.

[11] Balasko Y., Cass D., and Siconolfi P. (1990), The structure of financial equilibrium with exogenous yields: the case of restricted participations, *Journal of Mathematical Economics*, **19**, 195-216.

[12] Barten A. P., and Böhm V. (1982), Consumer Theory, in Arrow K. J., and Intriligator M. D., (eds.) *Handbook of Mathematical Economics*, North-Holland, New York, vol. 2, pp. 381-429.

[13] Bazaraa M.S., Sheraly H. D., and Shetty C.M. (1993), *Non linear programming*, second edition, John Wiley & Sons, Inc., New York.

[14] Berger M., and Gostiaux B. (1988), *Differential Geometry: Manifolds, Curves, and Surfaces*, Springer Verlag, New York.

[15] Billingsley P. (1986), *Probability and Measure*, second edition, John Wiley & Sons, New York.

[16] Bonisseau J. M., and F. Cayupi (1995), The Equilibrium Manifold with Boundary Constraints on the Consumption Sets, Working Paper, Universite' de Paris 1.

[17] Bottazzi J. M., 1995, Existence of equilibria with incomplete markets: The case of smooth returns, *Journal of Mathematical Economics* **24**, 61-72.

[18] Bourbaki N. (1958), *Elements de mathématique, topologie générale*, Hermann, Paris.

[19] Bruce J. W., and Giblin P. J. (1987), *Curves and Singularities*, Cambridge University Press, Cambridge.

[20] Carosi L. (2001), Optimality in a financial economy with restricted participation, *Decisions in Economics and Finance*, Springer Verlag, **24**, 1-19.

[21] Carosi L., and Villanacci A., (in progress), Endogeneous retricted participation, mimeo.

[22] Cass D. (1984), Competitive equilibria with incomplete financial markets, CARESS working paper 84-09, University of Pennsylvania, Philadelphia, PA.

[23] Cass D. (1989), Sunspots and Incomplete Financial Markets: the Leading Example, in G. Feiwel (ed.), *The Economics of Imperfect Competition and Employment: Joan Robinson and Beyond*, MacMillan, London.

[24] Cass D. (1990), Real Indeterminacy from Imperfect Financial Markets: Two Addenda, CARESS Working Paper, University of Pennsylvania.

[25] Cass D. (1991), Nonlinear Programming for Economists, University of Pennsylvania, Class Notes.

[26] Cass D. (1992), Two Observations About the Implicit Function Theorem in General Equilibrium Theory, University of Pennsylvania, mimeo.

[27] Cass D. (1992), The Structure of Sunspot Equilibria in the Presence of Incomplete Financial Markets: Introduction, *Economic Theory*, **2**, 307-08.

[28] Cass D. (1992), Sunspots and Incomplete Financial Markets: The General Case, *Economic Theory*, **2**, 341-58.

[29] Cass D. (1992), Incomplete Financial Markets and Indeterminacy of Competitive Equilibrium, in J. J. Laffont ed., *Advances in Economic Theory; Sixth World Congress*, vol. 2, Cambridge University Press, Cambridge, 262-88.

[30] Cass D. (1995), Notes on Pareto Improvement in Incomplete Financial Markets, *Rivista di Matematica per le Scienze Economiche e Sociali,* **18**, 3-14.

[31] Cass D., and Citanna A., (1998), Pareto improving financial innovation in incomplete markets, *Economic Theory*, **11**, 467-94.

[32] Cass D., Siconolfi P., and Villanacci A., (2001), Generic Regularity of Competitive Equilibria with Restricted Participation, *Journal of Mathematical Economics*, **36**, 61-76.

[33] Checcucci V., Tognoli A., and Vesentini E. (1968), *Lezioni di topologia generale*, Giangiacomo Feltrinelli Editore, Milano, Italia.

[34] Citanna A. (1996), Financial Innovation and Price Volatility, GSIA Working Paper #1996-E30, revised version HEC C.R. 685-1999.

[35] Citanna A., and Schmedders K. (2002), Controlling price volatility through financial innovation, HEC C.R. 749-2002.

[36] Citanna A., Kajii A., and Villanacci A. (1998), Constrained suboptimality in incomplete markets: a general approach and two applications, Economic Theory, Mini-Symposium on Pareto Improvement in Incomplete Financial Markets, *Economic Theory* **11**, 495-521.

[37] Citanna A., and Villanacci A. (1999), Existence and regularity of partially revealing rational expectations equilibria in finite economies, *Journal of Mathematical Economics*, **34**, 1-26.

[38] Citanna A., and Villanacci A. (2000), Incomplete Markets. Allocative Efficiency and the Information Revealed by Prices, *Journal of Economic Theory*, **90**, 222-53

[39] Citanna A., and Villanacci A., (2001) Competitive Equilibrium with Moral Hazard in Economies With Multiple Commodities , forthcoming on the *Journal of Mathematical Economics*.

[40] Debreu G. (1959), *Theory of Value*, John Wiley & Sons Inc., New York.

[41] Debreu G. (1970), Economies with a finite set of equilibria, *Econometrica*, **38**, 387-92, (reprinted in Debreu G. (1983), *Mathematical Economics*, Cambridge University Press, New York, 179-85).

[42] Debreu G. (1972), Smooth preferences, *Econometrica* **40**, 603-15, (reprinted in Debreu G. (1983), *Mathematical Economics* Cambridge University Press, New York, 186-201).

[43] Debreu G. (1974), Four Aspects of the Mathematical Theory of Economic Equilibrium, *Proceedings of the International Congress of Mathematicians*, Vancouver, 1, 65-77, (reprinted in Debreu G. (1983), *Mathematical Economics*, Cambridge University Press, New York, 203-09).

[44] Debreu G. (1976), Regular Differentiable Economies, *American Economic Review*, **66**, 280-87, (reprinted in Debreu G. (1983), *Mathematical Economics*, Cambridge University Press, New York, 232-41).

[45] Debreu G. (1983), *Mathematical Economics* Cambridge University Press, New York.

[46] Deimling K. (1985), *Nonlinear Functional Analysis*, Springer Verlag, Berlin.

[47] Del Mercato E., and Villanacci A. (in progress), Real Indeterminacy, Taxes and Outside Money in Incomplete Financial Markets, mimeo.

[48] Demarzo P.M. and B.C. Eaves (1996),Computing equilibria of GEI by relocalization on a Grassmann manifold, *Journal of Mathematical Economics*, **26**, 479-97.

[49] Diamond P. (1967), The role of a stock market in a general equilibrium model with technological uncertainty, *American Economic Review*, **57**, 759-76.

[50] Dieudonné J. (1960), *Foundation of modern analisys*, Academic Press, London.

[51] Dierker E. (1974), *Topological methods in Walrasian economics*, Lecture Notes in Economics and Mathematical Systems, 92, Springer-Verlag, Berlin.

[52] Dierker E. and Dierker H., (1972), The local uniqueness of equilibria, *Econometrica*, **40**, 867-881.

[53] Donzelli F. (1986), *Il concetto di equilibrio nella teoria economica neoclassica*, La Nuova Italia Scientifica, Roma.

[54] Dubey P., Geanakoplos J., and Shubik M. (1990), Default in a General Equilibrium Model with Incomplete Markets, Cowles Foundation Discussion Paper n. 773.

[55] Dugundji J. (1966), *Topology*, Allyn and Bacon Inc., Boston.

[56] Duffie D., and Shafer W., (1985), Equilibrium with incomplete markets: I. A basic model of generic existence, *Journal of Mathematical Economics*, **14**, 3, 285-300.

[57] Elul R. (1995), Welfare Effects of Financial Innovation in Incomplete Markets Economies with Several Consumption Goods, *Journal of Economic Theory*, **65**, 43-78.

[58] Geanakoplos J. (1990), An introduction to General Equilibrium with Incomplete Asset Markets, *Journal of Mathematical Economic*, **19**, 1-38.

[59] Geanakoplos J., Magill M., Quinzii M. and Drèze J. (1990), Generic Inefficiency of Stock Market Equilibrium When Markets are Incomplete, *Journal of Mathematical Economic*, **19**, 113-51.

[60] Geneakoplos J., and Mas-Colell A. (1989), Real Indeterminacy with Financial Assets, *Journal of Economic Theory*, **47**, 22-38.

[61] Geanakoplos J., and Polemarchakis H. M. (1986), Existence, regularity, and constrained suboptimality of competitive allocations when assets structure is incomplete, in W. P. Heller, R. M. Starr, and D. A. Starret Eds, *Essays in honor of K. J. Arrow*, Cambridge University Press, Cambridge, *Vol 3*, 65-95.

[62] Geanakoplos J., and Shafer W. (1990), Solving systems of simultaneous equations in economics, *Journal of Mathematical Economic*, **19**, 69-93.

[63] Gherardelli F., Rosati L.A., and Tomassini G. (1985), *Lezioni di geometria*, vol. I (second edition), Cedam, Padova, Italia.

[64] Gherardelli F., Rosati L.A., and Tomassini G. (1980), *Lezioni di geometria*, vol. II, Cedam, Padova, Italia.

[65] Golubitsky M., and Guillemin V. (1973), *Stable Mappings and Their Singularities*, Springer–Verlag, New York.

[66] Guillemin V., and Pollack A. (1974), *Differential Topology*, Prentice-Hall, Englewoods Cliffs.

[67] Hart O. D 1975, On the Optimality of Equilibrium when the Market Structure is Incomplete, *Journal of Economic Theory* **11**, 418-443.

[68] Herings P. J. J. (1996), *Static and Dynamic Aspects of General Disequilibrium Theory*, Boston Kluwer Academic, MA.

[69] Herings P. J. J., van der Laan G., and R. Venniker (1998), The transition from Drèze equilibrium to a Walrasian equilibrium, *Journal of Mathematical Economics,* **33**, 303-30.

[70] Hildebrand W., and Kirman A. (1976), *Introduction to Equilibrium Analysis*, North-Holland, Amsterdam.

[71] Hirsch M. (1976), *Differential Topology*, Springer–Verlag, New York, NY, 1976.

[72] Hirsch M., and Smale S. (1974), *Differential Equations, Dynamical Systems, and Linear Algebra*, Academic Press, New York.

[73] Jahn J. (1986), *Mathematical Vector Optimization in Partially Ordered Linear Spaces*, Verlag Peter Lang, Frankfurt am Main.

[74] Krasa S., and J. Werner (1991), Equilibria with options: existence and indeterminacy, *Journal of Economic Theory, 54, 305-20.*

[75] Kreyszig E. (1989), *Introductory Functional Analysis with Applications*, John Wiley & Sons, New York.

[76] Lang S. (1985), *Differential Manifolds*, Springer–Verlag, New York.

[77] Lang S. (1986), *A First Course in Calculus*, Springer Verlag, New York.

[78] Lang S. (1971), *Linear Algebra*, second edition, Addison Wesley, Reading.

[79] Lipschutz S. (1965), *General Topology*, McGraw-Hill Company, New York.

[80] Lloyd N. G. (1978), *Degree Theory*, Cambridge University Press, Cambridge.

[81] Magill M., and Quinzii M. (1992), Real Effects of Money in General Equilibrium, *Journal of Mathematical Economics,* **21**, 301-42.

[82] Magill M., and Quinzii M. (1996), *Theory of Incomplete Markets,* vol. I, The MIT Press, Cambridge, Massachusetts.

[83] Magill M., and Shafer W. (1991), Incomplete markets, in W. Hildebrand and H. Sonnenschein eds, *Handbook of Mathematical Economics,* vol. IV, Elsevier Science Publishers B.V..

[84] Malinvaud E. (1972), *Lectures on Microeconomic Theory*, North-Holland Publishing Company, Amsterdam.

[85] Mangasarian O. L. (1994), *Nonlinear Programming*, SIAM, Philadelphia.

[86] Mas Colell A., (1985), *The theory of general economic equilibrium. A differentiable approach*, Cambridge University Press, Cambridge, UK.

[87] Mas Colell A., Whinston M. D., and Green J. R. (1995), *Microeconomic Theory*, Oxford University Press, New York.

[88] McKenzie L.W. (1954), On Equilibrium in Graham's Model of World Trade and Other Competitive Systems, *Econometrica*, 22, pp. 147–161.

[89] Milnor J. (1966), *Topology from the Differentiable Viewpoint*, University of Virginia Press, Charlottesville.

[90] Munkres J. M. (1975), *General Topology, A First Course,* Prentice-Hall, Englewood Cliffs.

[91] Munkres J. M. (1991), *Analysis on Manifolds*, Addison-Wesley Publishing Company, Redwood City.

[92] Pearce D. W. (1992), *The MIT dictionary of modern economics,* fourth edition, edited by D. W. Pearce, Aberdeen Economics Consultants, McMillan Press, Great Britain.

[93] Pietra T. (1992), The Structure of the Set of Sunspot Equilibria in Economies with Incomplete Financial Markets, *Economic Theory*, **2**, 321-40.

[94] Pietra T. (1992), Indeterminacy in General Equilibrium Economies with Incomplete Financial Markets: Mixed Asset Returns, *Journal of Mathematical Economics,* **21**, 155-72.

[95] Pietra T. (2001), The set of sunspot equilibria in economies with incomplete markets: variables assets prices *Economic Theory,* **18**, 649-59.

[96] Polemarchakis H. M., and P. Siconolfi (1997), Generic Existence of Competitive Equilibria with Restricted Participation, *Journal of Mathematical Economics,* **28**, 289-311.

[97] Radner R. (1967), Equilibre des Marchés à Terme et au Comptant en Cas d'Incertitude, Cahier d'Econométrie, C.N.R.S., Paris.

[98] Radner R. (1972), Existence of equilibrium of plan, prices and price expectations, *Econometrica* **40**, 289-303.

[99] Reiter S. (1986), *Studies in mathematical economics*, The Mathematical Association of America.

[100] Schmedders K., (1999), A homotopy algorithm and an index theorem for the general equilibrium model with incomplete asset markets, *Journal of Mathematical Economics,* **32**, 225-41.

[101] Shafer W. (1988), Lecture 1, Introduction, in Lecture Notes in Incomplete Markets, BOWO Discussion Paper, University of Bonn.

[102] Shannon C. (1994), Regular Nonsmooth Equations, *Journal of Mathematical Economics* **23**, 147-65.

[103] Simmons G. F. (1963), *Topology and Modern Analysis*, McGraw-Hill Book Company, Singapore.

[104] Siconolfi, P., 1988, Equilibrium with Asymmetric constraints on Portfolio, in Galeotti M., Geronazzo L. and F. Gori, eds., Non-linear Dynamics in Economics and Social Science, (Società Pitagora, Bologna).

[105] Siconolfi P. (1991), Sunspot Equilibria and Incomplete Financial Markets, *Journal of Mathematical Economics*, **20**, 327-39.

[106] Simon C. S. (1986), Scalar and vector maximization: calculus techniques with economic applications, in Reiter S, *Studies in mathematical economics*, The Mathematical Association of America, p. 62-159-.

[107] Smale, S., 1965, An Infinite-dimensional Version of Sard's Theorem, *American Journal of Mathematics*, **87**, 861-66.

[108] Smale S. (1974), Global analysis and economics. IIA: Extension of a theorem of Debreu, *Journal of Mathematical Economics,* **1**, 1-14.

[109] Smale S. (1974), Global analysis and economics. III: Pareto Optima and price equilibria, *Journal of Mathematical Economics,* **1**, 107-17.

[110] Smale S. (1974), Global analysis and economics. IV: Finiteness and stability of equilibria with general consumption sets and production, *Journal of Mathematical Economics,* **1**, 119-27.

[111] Smale S. (1974), Global analysis and economics. V: Pareto theory with constraints, *Journal of Mathematical Economics,* **1**, 213-21.

[112] Smale S. (1974), Global analysis and economics. VI: Geometric analysis of Pareto Optima and price equilibria under classical hypotheses, *Journal of Mathematical Economics,* **3**, 1-14.

[113] Spivak M. (1980), *Calculus*, second edition, Publish or Perish, Houston.

[114] Stoer J., and Witzgall C. (1970), *Convexity and Optimization in Finite Dimensions I*, Springer–Verlag, Berlin-Heidelberg.

[115] Suda S., Tallon J-M, and Villanacci A. (1992), Real indeterminacy of equilibria, in a sunspot economy with inside money, *Economic theory*, **2**, p. 309-19.

[116] Tversky A. (1969), Intransitivity of preferences, *Psychological Review*, **76**, 31-48.

[117] Treves F. (1967), Topological vector spaces, distributions and kernel, Academic Press, New York.

[118] Varian H. R. (1999), Intermediate Microeconomics, A Modern Approach, 5th edition, W. W. Norton & Company, New York NY.

[119] Villanacci A. (1991), Real Indeterminacy, Taxes and Outside Money in Incomplete Financial Market Economies: I. The case of Lump Sum Taxes, in M. Galeotti, L. Geronazzo, F. Gori, eds., *Proceedings of the Second Meeting on Non-linear Dynamics in Economics*, Springer Verlag.

[120] Villanacci A., (1993), A Proof of Existence and Regularity for an Exchange Economy with Non Negativity Constraints, Working Paper, University of Florence, Italy.

[121] Villanacci A., (1993), Real Indeterminacy, Taxes and Outside Money in Incomplete Financial Market Economies: II. The Case of Taxes as Function of Wealth, DIMADEFAS Working Paper, University of Florence, Italy.

[122] Villanacci A., and Zenginobuz U., (2001a), Existence and Regularity of Equilibria in a General Equilibrium Model with Private Provision of a Public Good, Bogazici University working paper, Instanbul, Turkey.

[123] Villanacci A., and Zenginobuz U., (2001b), Private Provision of a Public Good in a General Equilibrium Model, Bogazici University working paper, Instanbul, Turkey.

[124] Walker D.A. (1997), *Advances in General Equilibrium Theory*, Edward Elgar, Cheltenham UK.

[125] Werner J. (1985), Equilibrium in Economies with Incomplete Financial Markets, *Journal of Economic Theory* **36**, 110-19.

[126] Werner J., (1990), Structure of financial markets and real indeterminacy of equilibria, *Journal of Mathematical Economics, **19**, 217-32.*

[127] Whitney H. (1935), A function not constant on a connected set of critcal points, *Duke Math. J.* **1**, 514–17.

[128] Whitney H. (1944), The selfintersections of a smooth n-manifold in $2n$-space, *Annals of Math.* **45**, 220–46.

Index

GPSR Compliance
The European Union's (EU) General Product Safety Regulation (GPSR) is a set of rules that requires consumer products to be safe and our obligations to ensure this.

If you have any concerns about our products, you can contact us on

ProductSafety@springernature.com

In case Publisher is established outside the EU, the EU authorized representative is:

Springer Nature Customer Service Center GmbH
Europaplatz 3
69115 Heidelberg, Germany

www.ingramcontent.com/pod-product-compliance
Ingram Content Group UK Ltd.
Pitfield, Milton Keynes, MK11 3LW, UK
UKHW020134160926
475607UK00003B/27

* 9 7 8 1 4 0 2 0 7 2 0 1 7 *